Selective Organic Transformations

VOLUME 1

Selective Organic Transformations

EDITOR:

B. S. THYAGARAJAN

UNIVERSITY OF IDAHO
MOSCOW, IDAHO

VOLUME 1

WILEY–INTERSCIENCE
A DIVISION OF JOHN WILEY & SONS
NEW YORK · LONDON · SYDNEY · TORONTO

10 9 8 7 6 5 4 3 2 1

Library of Congress Catalogue Card Number: 72–79147

ISBN 0–471–86687–3

Printed in the United States of America

This volume is dedicated to the many able organic chemists who have contributed to the growth of mechanistic concepts relating to selectivity and specificity in organic reactions.

Editor

"I have found you an argument; but I am not obliged to find you an understanding."

Samuel Johnson

"Or if you choose, a machine might be imagined where the assumptions were put in at one end, while the theorems came out at the other, like the legendary Chicago machine where the pigs go in alive and come out transformed into hams and sausages."

Henri Poincaré

"If now we imagine an observer who discovers that the future course of a certain phenomenon can be predicted by Mathieu's equation but who is unable for some reason to perceive the system which generated the phenomenon, then evidently he would be unable to tell whether the system in question is an elliptic membrane or a variety artist."

Sir Edmund T. Whittaker

"Reasoning draws a conclusion and makes us grant the conclusion, but does not make the conclusion certain, nor does it remove doubt so that the mind may rest on the intuition of truth, unless the mind discovers it by the path of experience."

Roger Bacon

PREFACE

The domain of organic chemistry abounds in asymmetric functionality. Many different chemical operations contribute to the construction or destruction of such dissymmetric assemblages of functional units in a unique and specific manner.

Varied though the reactive sites may be, dependent on the orientational factors, the configurational stability, or the population of proximate reactive centers, the strategy and skill of the organic chemist has found a tool for each specific task in any molecular environment. The art of organic synthesis has unveiled many complex transformations that involve specificity and selectivity in the skillful assemblage of structural units of staggering proportions. A masterly exposition of the challenges of this field was given in a scintillating commentary entitled, "General Methods for the Construction of Complex Molecules" by Professor E. J. Corey in his plenary lecture before the Fourth International Symposium on the Chemistry of Natural Products at Stockholm in 1966.

Concomitant with the challenges of immediate synthetic goals, selective organic transformations pose the parallel objective of determining answers to intriguing questions like, "Specificity—why and how? Selectivity—why and how?" The latter quest is often the slower and more arduous effort. The mechanisms of many such selective organic transformations have been unraveled by numerous investigators. The result is a body of knowledge that has grown rapidly in recent years.

Perceptive analyses of the stereochemical and electronic driving forces responsible for these unique transformations hold continuing fascination for investigators in this area. An interplay of mechanistic ideas between varied molecular environments often leads to complimentary and overlapping growth in mechanistic postulates.

As in every other area of organic chemistry, mechanistic studies in the field of selective organic transformations have proliferated to such proportions that specialized surveys are required to assist the active investigator, the more advanced student, and the teacher.

The present series is therefore an attempt to offer periodic critical evaluations of selective organic transformations, largely from the mechanistic point of view. The original investigations surveyed in these volumes may well seed the growth of others that will challenge, sustain, or enlarge some of these postulates. It is hoped future volumes will bear ample testimony to this possibility.

I heartily acknowledge the generous cooperation of the contributing authors in making possible the appearance of this volume. Their ready willingness to provide these stimulating surveys is a service rendered by them to the many who are interested in investigations in the respective topics of survey. I hope that this will indicate the need for a continuing series in *Selective Organic Transformations* which many will profitably turn to as a valuable secondary information source.

I shall welcome suggestions for appropriate surveys for inclusion in volumes to follow.

B. S. THYAGARAJAN

Moscow, Idaho
January 1970

CONTENTS

Selective Organic Transformations

VOLUME 1

Asymmetric Induction through Hydride Reductions

SHUN-ICHI YAMADA AND KENJI KOGA

Faculty of Pharmaceutical Sciences, University of Tokyo

I. INTRODUCTION

Asymmetric induction is defined as meaning the production of a new asymmetric atom or an entire dissymmetric molecule under conditions where the resulting two stereoisomers (diastereoisomers or enantiomers) are formed in unequal amounts (1). Reactions in biological systems are, in general, quite effective in this sense; almost complete stereoselective reactions in these systems may be the consequences of their being catalyzed by enzymes, which are composed of proteins having many asymmetric centers and are, themselves, highly asymmetric. It is reasonable to assume, therefore, that virtually all naturally occurring and/or biologically active substances having asymmetric centers possess definite configurations devoid of other theoretically possible stereoisomers. Although asymmetric induction in most nonbiological reactions cannot be performed with as high a degree of stereoselectivity as in biological reactions, studies concerned with the elucidation of their reaction mechanisms and their applications to organic syntheses have been carried out by many workers using a variety of reaction types and conditions (2).

In this chapter stereochemical studies on asymmetric induction through hydride reductions are described. The present authors do not intend to give an exhaustive review of the subject but rather to discuss basic principles that are proposed to interpret or predict the stereochemical results, and also to illustrate some examples of their applications.

II. HYDRIDE REDUCTIONS OF ACYCLIC KETONES DIRECTLY BONDED TO ONE ASYMMETRIC CARBON ATOM

A. Cram's Rule

Stereochemical studies on the kinetically controlled reactions of acyclic ketones having one asymmetric carbon atom adjacent to the carbonyl group (**1**) with nucleophiles (R′—Z) to the corresponding alcohols (**2**) (Fig. 1) were

$$\underset{\mathbf{1}}{R{-}CO{-}C(R_1)(R_2)(R_3)} \xrightarrow{R'-Z} \underset{\mathbf{2}}{R'{-}C(R)(HO){-}C(R_1)(R_2)(R_3)}$$

Figure 1. 1,2-Asymmetric induction.

studied by Cram and his co-workers mainly using Grignard and hydride reagents. By determining the configurations and the ratios of diastereoisomers produced, they proposed an empirical rule to correlate the substituents (R_1, R_2, and R_3) attached to the asymmetric carbon atom with the diastereoisomer produced in excess. This rule is applicable to qualitative interpretation or prediction of the stereochemical course in 1,2-asymmetric induction by considering, depending on the nature of the substituents (R_1, R_2, and R_3), one of the following three empirical models as the transition state of the lowest energy.

1. Open Chain Model

Table 1 represents some results of reductions of aldehydes and ketones (**1**) whose substituents (R_1, R_2, and R_3) attached to the asymmetric carbon

TABLE 1

$$\underset{\mathbf{1}}{RCO{-}CR_1R_2R_3} \xrightarrow{R'{-}Z} \underset{\mathbf{2}}{RR'C(OH){-}CR_1R_2R_3}$$

	1					Product			ΔΔG‡ (cal/mole)[b]	
No.	R_1	R_2	R_3	R	R′—Z	Threo:erythro	Prediction[a]	Ref.	Found	Calcd.
1	Ph	Me	H	H	MeMgI	1:2	Erythro	3	−410	−600
2	Ph	Me	H	Me	$LiAlH_4$	2.5:1	Threo	3	−550	−600
3	Ph	Me	H	H	PhMgBr	>4:1	Threo	3	−830	−600
4	Ph	Me	H	Ph	$LiAlH_4$	1:>4	Erythro	3	−830	−600
5	Ph	Et	Me	Ph	$LiAlH_4$	1:2.2	Erythro	4	−470	−500
6	Et	Me	H	Me	$LiAlH_4$	1:1	Threo	5	~0	−100
7	Ph	Et	H	Et	$LiAlH_4$	3:1	Threo	6	−650	−500
8	C_6H_{11}	Me	H	Me	$NaBH_4$	63:37	Threo	7	−320	−300
9	Ph	*i*-Pr	H	*i*-Pr	$LiAlH_4$	10:1	Threo	8	−1380	−200

[a] Prediction by open chain model (**3**).
[b] Taken from reference 17. See text (Section II-C).

atom are all aryl, alkyl, or hydrogen and give mixtures of diastereoisomerically related products (**2**). From Table 1, it is obvious that in reactions of the carbonyl compounds having the same asymmetric carbon atom, the configuration of the newly created asymmetric carbon atom of the predominant diastereoisomer is inverted by inverting R′ of the reagent and the substituent R (No. 1 and No. 2, No. 3 and No. 4). It may be concluded, therefore, that the direction of 1,2-asymmetric induction in this type of reaction is dependent on the direction of the reagent attack on the carbonyl group, and is not dependent on the relative stabilities of the diastereoisomers produced. The problem is then to find the correlation between substituents attached to the asymmetric carbon atom of the carbonyl compound (**1**) and the direction of reagent attack on the carbonyl group. These three substituents, R_1, R_2, and R_3, are represented as L (large), M (medium), and S (small) in order to indicate their effective bulkiness in the reaction.

Cram proposed the so-called open chain model (**3**) for the prediction of the stereochemical course of the reactions from the following considerations (3). The oxygen of the carbonyl group is expected to become effectively the bulkiest group in the molecule because of the coordination by reagent and its solvation in the reaction. For steric interactions, therefore, the oxygen of the carbonyl group will tend to orient itself between the two least bulky groups (M and S) attached to the adjacent asymmetric carbon atom as shown in **3**, which is considered to be the transition state of the lowest energy. The configuration of the newly created asymmetric carbon atom of the predominant diastereoisomer, therefore, may be predicted by the following rule (3):

In reactions of this type, that diastereoisomer (**4**) will predominate which would be formed by the approach of the entering group from the least hindered side of the double bond when the rotational conformation of the C—C bond is such that the double bond is flanked by the two least bulky groups (M and S) attached to the asymmetric carbon atom (**3**).

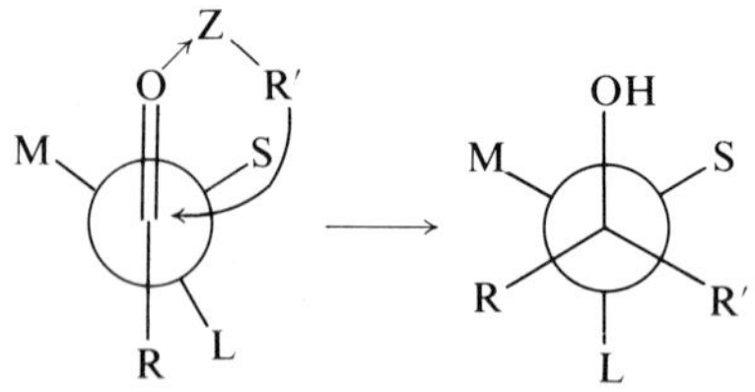

Z
O
R′
M
S
R
L
3
OH
M
S
R
R′
L
4

Figure 2. Stereochemical course of open chain model.

Predictions using this open chain model (**3**) (Fig. 2) have been successfully demonstrated in many examples (3–8), some of which are shown in Table 1.

2. *Cyclic Model (Rigid Model)*

Table 2 represents some reaction results with aldehydes and ketones (**1**) whose asymmetric carbon atom carries one functional group such as OH, OR, and NH_2, capable of complexing with reagents. In these cases, a five-membered cyclic model (**5**) was proposed as the explanation of stereochemical results, based on the following considerations (9a). The transition state of the lowest energy is postulated to be the one in which the relatively rigid, five-membered cyclic ring, as **5** (Fig. 3), is produced due to the complexing of the reagent with both the oxygen of the carbonyl group and the functional group attached to the asymmetric carbon atom. The approach of the nucleophile from the less hindered side is expected to give a diastereoisomer (**6**), which is considered to be produced in excess. This model (**5**) interprets the major diastereoisomers correctly in many examples (9–11), some of which are

TABLE 2

$$\underset{\mathbf{1}}{RCO{-}CR_1R_2R_3} \xrightarrow{R'-Z} \underset{\mathbf{2}}{RR'C(OH){-}CR_1R_2R_3}$$

	1					Product		
No.	R_1	R_2	R_3	R	R′ − Z	Threo : erythro	Prediction[a]	Ref.
10	Ph	Me	OH	Ph	MeLi	1 : 11.5	Erythro	9a
11	Ph	Me	OH	Me	PhMgCl	1 : 3.33	Threo	9b
12	Ph	Me	OMe	Ph	MeLi	1 : 2	Erythro	9a
13	Ph	OH	H	*p*-Tolyl	$LiAlH_4$	1 : 5.5	Erythro	9b
14	*p*-Tolyl	OH	H	Ph	$LiAlH_4$	1 : 4.9	Erythro	9b
15	Me	OMe	H	Ph	$NaBH_4$	27 : 73	Erythro	10
16	Me	NH_2	H	Ph	*p*-TolylMgBr	1 : 50	Erythro	9c

[a] Prediction by five-membered cyclic model (**5**).

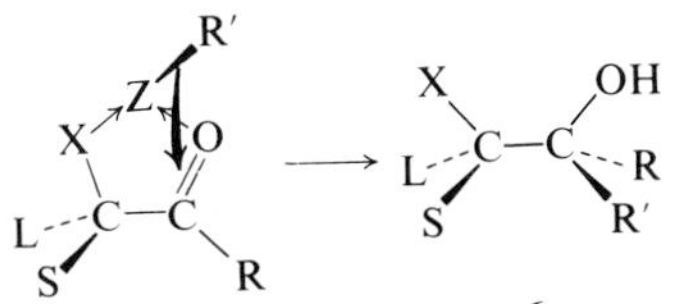

5 **6**

X = OH, OR, NH_2, etc.

Figure 3. Stereochemical course of five-membered cyclic model.

shown in Table 2. However, exceptions to the prediction, as No. 11 (9b), and criticism claiming oversimplification (11) have also been reported.

Deviations from the prediction by the model (**5**) are also observed when the functional group attached to the asymmetric carbon atom becomes bulky (No. 19 and No. 20). This phenomenon may be explained (10) by the steric consideration that the bulky functional group and the bulky oxygen of the carbonyl group are forced to orient themselves *trans* to each other in the transition state [similar to the open chain model (**3**) or the dipolar model (**9**)] due to their repulsive forces. On the application of the cyclic model (**5**), therefore, it is important to pay attention to the bulkiness of the functional group attached to the asymmetric carbon atom.

The successful application of the five-membered cyclic model (**5**) in many examples suggests the possibility of a six-membered analog (**7**) in the stereochemical prediction of 1,2-asymmetric induction in the reactions of ketones having such a functional group as NH_2, OR, or OH at β-position to the carbonyl group (10,12). The results of the reactions of such ketones are shown in Table 3 (No. 21 and No. 22). The direction of 1,2-asymmetric

TABLE 3

$$\underset{\mathbf{1}}{RCO{-}CR_1R_2R_3} \xrightarrow{R'{-}Z} \underset{\mathbf{2}}{RR'C(OH){-}CR_1R_2R_3}$$

	1					Product		
No.	R_1	R_2	R_3	R	$R'-Z$	Threo : erythro	Prediction	Ref.
17	Me	NH_2HCl	H	Ph	$NaBH_4$	1 : 11.5	Erythro[a]	10
18	Me	NHCOPh	H	Ph	$NaBH_4$	1 : 3.3	Erythro[a]	10
19	Me	NMe_2HCl	H	Ph	$NaBH_4$	1.1–1.3 : 1	Erythro[a]	10
20	Me	N(Me)(CH_2Ph)	H	Ph	$NaBH_4$	7–8 : 1	Erythro[a]	10
21	CH_2NH_2HCl	Me	H	Ph	$NaBH_4$	9 : 1	Threo[b]	10
22	CH_2OH	Me	H	Ph	$NaBH_4$	2.5 : 1	Threo[b]	10
23	CH_2OH	OMe	H	Ph	$NaBH_4$	1 : 1.4	—[c]	10
24	CH_2NH_2HCl	OMe	H	Ph	$NaBH_4$	1.5 : 1	—[c]	10

[a] Prediction by five-membered cyclic model (**5**).
[b] Prediction by six-membered cyclic model (**7**).
[c] See text.

induction is in agreement with the prediction by a six-membered cyclic model (**7**) to give diastereoisomer (**8**) predominantly (10,13). The six-membered cyclic model (**7**) is, thus, applicable for predicting the stereochemical course in the hydride reductions of ketones having a functional group at β-position to the carbonyl group (10), as shown in Figure 4.

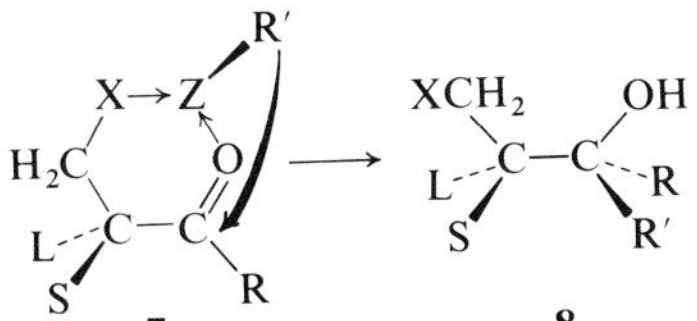

$X = OH, OR, NH_2$, etc.

Figure 4. Stereochemical course of six-membered cyclic model.

On the other hand, stereoselectivity is decreased in the reductions of ketones having such functional groups both at α- and β-positions to the carbonyl group (No. 23 and No. 24). This phenomenon is reasonable because the effect of the functional group at α-position on the direction of 1,2-asymmetric induction is opposite to that of the group at β-position, thus resulting in a cancellation of these effects (10).

3. Dipolar Model

The introduction of a halogen substituent at α-position of the carbonyl group is known to affect the conformation and the reactivity of the parent carbonyl compounds (14). Dipolar model (**9**) is devised in these cases assuming dipolar interaction between the halogen and the oxygen of the carbonyl group, as shown in Figure 5 (15). Approach of the reagent from

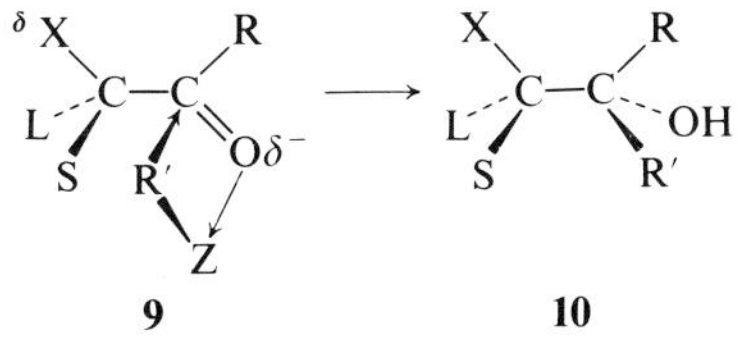

X = halogen

Figure 5. Stereochemical course of dipolar model.

the less hindered side will give diastereoisomer **10** predominantly. As shown in Table 4, the direction of 1,2-asymmetric induction in the reactions of ketones having a halogen at the asymmetric carbon atom is predictable by the model (**9**).

TABLE 4

$$\underset{\mathbf{1}}{RCO{-}CR_1R_2R_3} \xrightarrow{R'-Z} \underset{\mathbf{2}}{RR'C(OH){-}CR_1R_2R_3}$$

	1					Product		
No.	R_1	R_2	R_3	R	$R' - Z$	Threo:erythro	Prediction[a]	Ref.
25	Et	Cl	H	H	BuLi	1:2.3	Erythro	15a
26	Et	Cl	H	Bu	$NaBH_4$	4–6:1	Threo	15a

[a] Prediction by dipolar model (**9**).

B. Models of Staggered Conformations

Felkin and his co-workers (16a,16b) proposed alternate models for the interpretation of the stereochemical course of 1,2-asymmetric induction in the reactions of hydrides and Grignard reagents with acyclic aldehydes and ketones. Their proposal is based on the following considerations. The reaction occurs via reactant-like transition states, in which the nucleophilic part of the reagent (R′) lies in the π-axis of the carbonyl carbon. For torsional strain involving partial bonds the transition states are considered to be in staggered conformations, in which R′ is approximately skew to two of the groups attached to the asymmetric carbon atom. Furthermore, the important steric interactions are thought to be caused by the nucleophilic entity R′ and the substituent R. The direction and the degree of 1,2-asymmetric induction are thus considered to be determined by the relative stabilities of these staggered transition states. It follows from these premises that, in the absence of a polar substituent (such as halogen) at the asymmetric carbon atom, the least strained of the six possible transition states is considered to be **11**, followed by **12**, **13**, etc., as shown in Figure 6. These models were also applied to the interpretation of the stereochemical course of 1,3-asymmetric induction (16c).

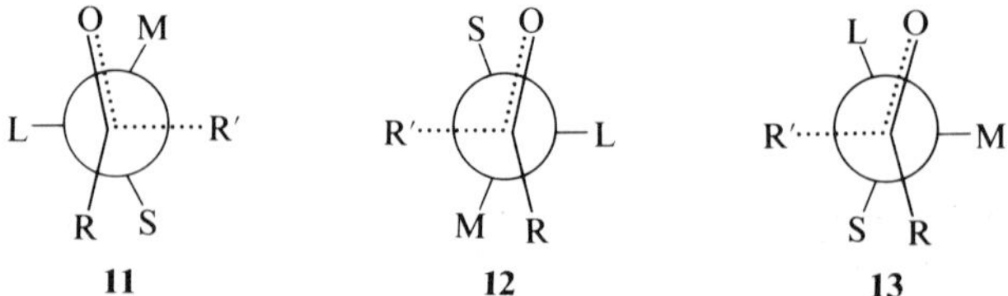

Figure 6. Models of staggered transition states.

In the presence of a polar substituent at the asymmetric carbon atom, those transition states in which the separation between R′ and the polar substituent is greatest are considered to become stable.

The same concept has been also applied to the interpretation of the stereochemical course of the reactions of cyclohexanone derivatives (see Section IV-A).

C. An Approach to Semiquantitative Prediction

Although the models so far discussed are useful for the prediction of the direction of 1,2-asymmetric induction, they cannot be used in a quantitative sense. An approach to the semiquantitative prediction has been made recently by Karabatsos (17) with alternative models, which are discussed below.

From the Curtin-Hammett principle, which is considered to be applicable to reactions of carbonyl compounds with hydrides and Grignard reagents, the diastereoisomeric ratio of 1,2-asymmetric induction in these reactions is considered to be dependent solely on the free energy difference ($\Delta G_A^{\ddagger} - \Delta G_B^{\ddagger}$) between the corresponding diastereoisomeric transition states. Quantitative prediction of the product stereoselectivities, therefore, requires the knowledge of the structures and the energy contents of their diastereoisomeric transition states. Judging from the rapidity and exothermicity of the reactions, as well as product stereospecificities that are independent on the relative stabilities of the diastereoisomers produced, it is reasonable to assume that these transition states are reactant-like, without any appreciable bond making and breaking having occurred.

On the other hand, investigations on the conformations of unsaturated molecules have revealed that a single bond is eclipsed with the double bond (C=C or C=O) as shown in **14** (18). Based on these findings, the populations of rotational isomers of acyclic aldehydes have been calculated by analyses of their NMR coupling constants (19). The results in neat states are shown in Table 5. These results indicate that the conformation **16** (in which R eclipses with oxygen) is preferred to **15** (in which hydrogen eclipses with oxygen) in the equilibrium mixture of **15** $\rightleftharpoons$ **16** when R is methyl, ethyl, *n*-propyl, *n*-amyl, *i*-propyl, or phenyl. Although the conformations of these

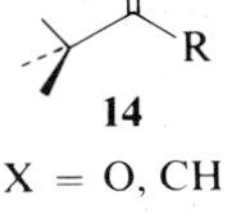

X
R
14
X = O, CH_2

Figure 7. Conformation of unsaturated molecules.

TABLE 5[a]

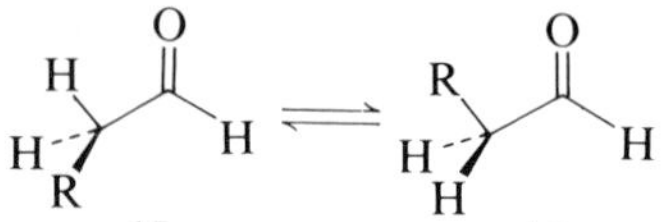

15 ⇌ 16

R	ΔH°
Me	−800 cal/mole
Et	−700
n-Pr	−600
n-Am	−500
i-Pr	−400
Ph	−300[b]
t-Bu	+250

[a] Data taken from ref. 19.
[b] Variable with solvents.

transition states are clearly the complexed rather than the free carbonyl, there seems to be no reason to expect that the stable conformations in the transition states are other than the eclipsed. For instance, aldehyde oxime *O*-methyl ethers and *N*-methylphenylhydrazones were also found to be consistent with the eclipsed conformations such as **17** and **18**, respectively (20).

Figure 8. Conformations of oxime *O*-methyl ethers and *N*-methylphenylhydrazones.

Based on the eclipsed conformations discussed above, the stable transition states in the reaction of an acyclic carbonyl compound (**19**) having substituents L, M, and S attached to the adjacent asymmetric carbon atom may be visualized as shown in Figure 9. The transition states **20**, **21**, and **22** are considered to give diastereoisomer A, while **23**, **24**, and **25** to the other diastereoisomer B. Among them, **20** and **23** should be the most stable pair, because the incoming bulky group R′ is closest to the smallest group S. The diastereoisomeric ratio A/B is, therefore, expected to be predictable by the

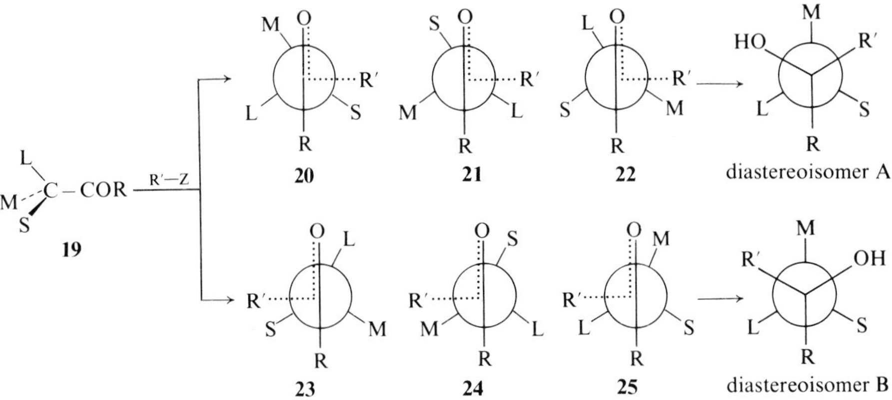

Figure 9. Models for 1,2-asymmetric induction proposed by Karabatsos (17).

difference in the relative stabilities of **20** and **23**. Since group interactions in **20** and **23** may be evaluated as

$$\mathbf{20}: (\mathrm{M} \leftrightarrow \mathrm{O}) + (\mathrm{R}' \leftrightarrow \mathrm{S}) + (\mathrm{R} \leftrightarrow \mathrm{S}) + (\mathrm{R} \leftrightarrow \mathrm{L}) + (\mathrm{R}' \leftrightarrow \mathrm{M}) \quad (1)$$

$$\mathbf{23}: (\mathrm{L} \leftrightarrow \mathrm{O}) + (\mathrm{R}' \leftrightarrow \mathrm{S}) + (\mathrm{R} \leftrightarrow \mathrm{S}) + (\mathrm{R} \leftrightarrow \mathrm{M}) + (\mathrm{R}' \leftrightarrow \mathrm{L}) \quad (2)$$

we have

$$\begin{aligned} \Delta G^{\ddagger}_{\mathrm{A}} - \Delta G^{\ddagger}_{\mathrm{B}} &= -RT \ln (\mathrm{A/B}) \\ &= (\mathrm{M} \leftrightarrow \mathrm{O}) - (\mathrm{L} \leftrightarrow \mathrm{O}) + [(\mathrm{R} \leftrightarrow \mathrm{L}) + (\mathrm{R}' \leftrightarrow \mathrm{M}) \\ &\quad - (\mathrm{R} \leftrightarrow \mathrm{M}) - (\mathrm{R}' \leftrightarrow \mathrm{L})] \end{aligned} \quad (3)$$

Assuming that the term $[(\mathrm{R} \leftrightarrow \mathrm{L}) + (\mathrm{R}' \leftrightarrow \mathrm{M}) - (\mathrm{R} \leftrightarrow \mathrm{M}) - (\mathrm{R}' \leftrightarrow \mathrm{L})]$ is negligible due to the cancelling, we get

$$\begin{aligned} \Delta G^{\ddagger}_{\mathrm{A}} - \Delta G^{\ddagger}_{\mathrm{B}} &= -RT \ln (\mathrm{A/B}) \\ &\fallingdotseq (\mathrm{M} \leftrightarrow \mathrm{O}) - (\mathrm{L} \leftrightarrow \mathrm{O}) \end{aligned} \quad (4)$$

The magnitudes of the terms $(\mathrm{M} \leftrightarrow \mathrm{O})$ and $(\mathrm{L} \leftrightarrow \mathrm{O})$ are available from the data in Table 5. Therefore the diastereoisomeric ratio A/B is expected to be predictable by eq. 4 (17).

This method of prediction requires that the transition states of the reactions be reactant-like, as shown in Figure 9. A probe on this point was checked by the reactions which give isotopically related diastereoisomers, whose relative stabilities are considered to be almost equal (21). In these cases the diastereoisomeric ratios are considered to approach unity as the transition states of the reactions become product-like. The results in Table 6 indicate that extensive bond breaking and making have occurred in the transition states of hydride reductions.

TABLE 6[a]

$$\underset{\mathbf{1}}{\mathrm{RCO{-}CR_1R_2R_3}} \xrightarrow{\mathrm{R'{-}Z}} \underset{\mathbf{2}}{\mathrm{RR'C(OH){-}CR_1R_2R_3}}$$

	1					Product	$\Delta\Delta G^{\ddagger}$ (kcal/mole)	
No.	R_1	R_2	R_3	R	R′ − Z	A:B[b]	Found	Calcd.
27	Ph	Me	D	CD_3	MeMgI[c]	75:25	−0.65	−0.60
28	Ph	Me	D	H	$LiAlD_4$[c]	58:42	−0.19	−0.60
29	Ph	Me	D	H	$NaBD_4$[d]	55:45	−0.14	−0.60
30	Ph	Me	D	H	$NaBD_4$[e]	63:37	−0.32	−0.50

[a] Data taken from ref. 21.
[b] See Figure 9.
[c] In ether at 25°C.
[d] In tetrahydrofuran at 66°C.
[e] In *i*-propanol at 25°C.

In spite of appreciable approximations and deviations from ideal conditions, the validity of this method of prediction by eq. (4) has proven to be fairly satisfactory in many cases (17), not only in the direction but also in the degree of 1,2-asymmetric induction. Some examples of predictability by eq. 4 are shown in Table 1.

Finally it must be noted that the various types of empirical models so far described should be applied only to kinetically controlled reactions. The results of catalytic reductions or thermodynamically controlled reactions (i.e., prolonged reactions under Meerwein-Ponndorf reduction conditions) are not necessarily those predictable by the above models.

III. HYDRIDE REDUCTIONS OF α-KETO ACID ESTERS OF OPTICALLY ACTIVE ALCOHOLS (PRELOG'S RULE)

Extensive studies by McKenzie and his co-workers originally showed that addition reactions to the keto-carbonyl groups of α-keto acid esters of optically active alcohols (**26**) proceeded stereoselectively to give, after hydrolysis, the corresponding optically active α-hydroxy acids (**27**) (Fig. 10).

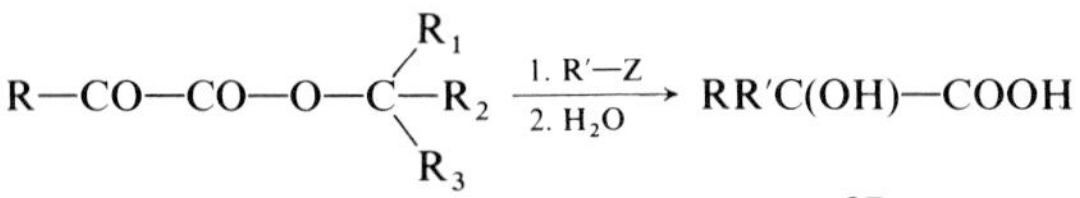

$$\underset{\mathbf{26}}{\text{R—CO—CO—O—C}(R_1)(R_2)(R_3)} \xrightarrow[\text{2. } H_2O]{\text{1. R}'\text{—Z}} \underset{\mathbf{27}}{\text{RR}'\text{C(OH)—COOH}}$$

Figure 10

Prelog and his co-workers (22–25) re-examined the addition reactions of this type of compounds mainly with Grignard reagents, and proposed an empirical rule to correlate the configurations of the optically active alcohols used with those of the α-hydroxy acids produced in excess. This rule, usually called Prelog's rule, is summarized in Figure 11 (22). The most stable conformations

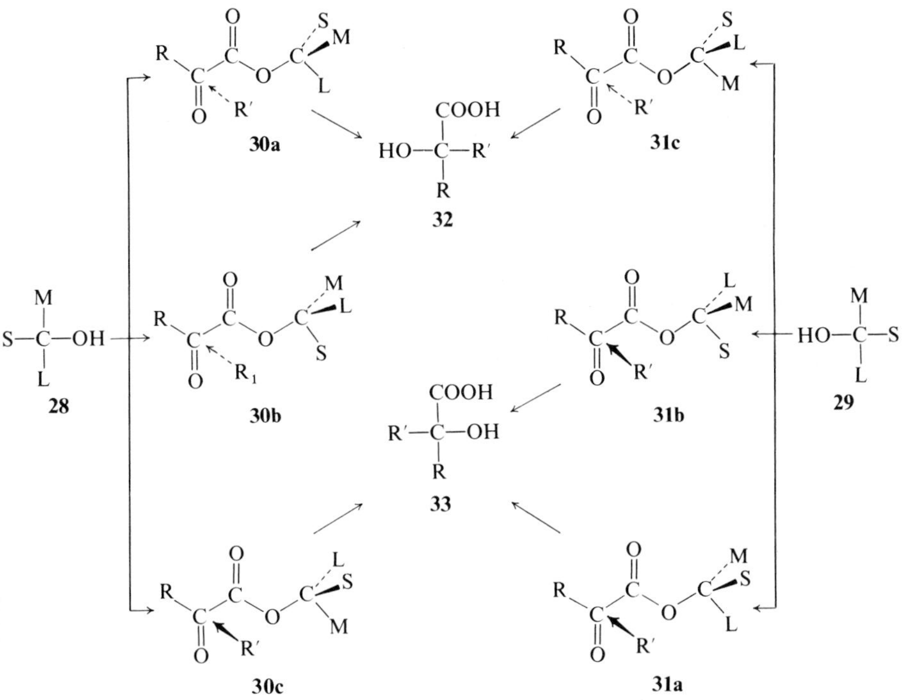

Figure 11. Prelog's rule.

of α-keto acid esters are thought to be such that the skeleton of the molecule, R—CO—CO—O—C—, lies in one plane, and both the two adjacent carbonyl groups as well as RCO and —C— orient themselves in *trans* position. In this conformation three rotamers, due to the rotation around C—O bond of the alcohol part, are possible. As shown in Figure 11, they are **30a**, **30b**, and **30c**, or **31a**, **31b**, and **31c**, depending on the configuration of the alcohol used. The approach of the reagent to the keto-carbonyl group will be favored from the less hindered side, which will be decided by the bulkiness of the substituents attached to the asymmetric carbon atom. Therefore, when an optically active alcohol of the **28**-type is used, rotamers **30a**, **30b**, and **30c** will be produced. Among them, **30a** and **30b** are expected to give α-hydroxy acid of the **32**-type, while **30c** is expected to yield α-hydroxy acid of the **33**-type, on hydrolysis. It is assumed, then, that the use of an alcohol of the **28**-type will produce α-hydroxy acid of the **32**-type predominantly, if the populations of each rotamer are equal. This conclusion becomes more probable if rotamer **30a** is more stable than the others (23). The same consideration leads to the prediction that α-hydroxy acid of the **33**-type will be produced in excess when an alcohol of the **29**-type is used. Prelog and his co-workers showed the validity of this rule in many examples (22–25).

This rule is also useful for the determination of the absolute configuration of a given alcohol. Thus, the alcohol is converted to the corresponding phenylglyoxylate, which is reacted with methyl magnesium halide. From the sign of rotation of atrolactic acid obtained by complete hydrolysis of the reaction product, the configuration of the alcohol is deduced according to the rule, as shown in Figure 11 (24). This method is called "atrolactic acid method."

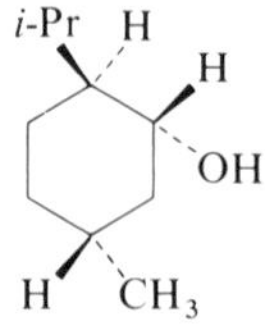

34 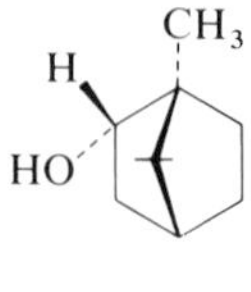 **35** 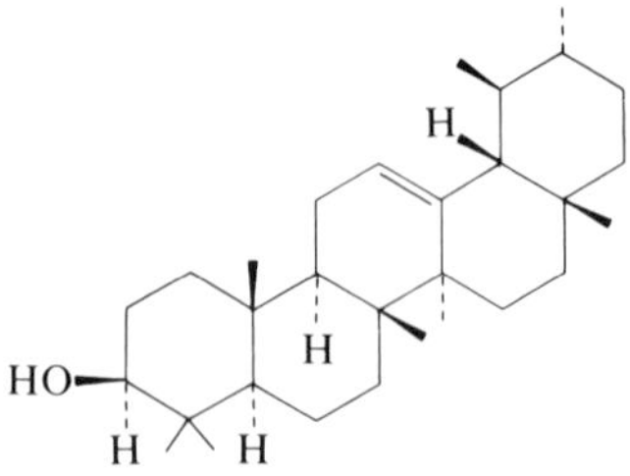 **36**

Figure 12

Prelog and his co-workers (25) also examined lithium aluminum hydride reductions of the esters of phenylglyoxylic acid with (−)-menthol (**34**), (+)-borneol (**35**), and α-amyrin (**36**), in order to test the applicability of this rule. The reaction of (−)-menthyl phenylglyoxylate (**37**) is shown in Figure 13. As the configuration of (−)-menthol is known to be the **29**-type, the ester

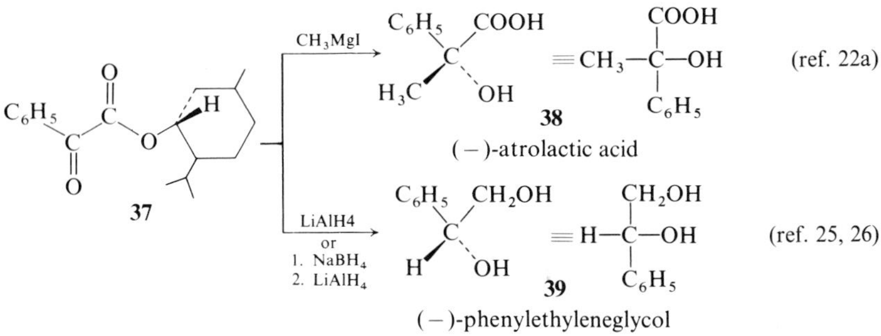

Figure 13. Reactions of (−)-menthyl phenylglyoxylate.

37 written in Figure 13 corresponds to **31a** in Figure 11. The reaction of the ester (**37**) with methyl magnesium iodide and lithium aluminum hydride gave, after hydrolysis, (−)-atrolactic acid (**38**) and (−)-phenylethyleneglycol (**39**), respectively. These results are in accordance with the prediction using this model. The reduction of the ester (**37**) with sodium borohydride was also reported (26) to be consistent with the rule. On the other hand, (+)-borneol (**35**) and α-amyrin (**36**) are known to be the **28**-type. Lithium aluminum hydride reductions of their corresponding phenylglyoxylates gave (+)-phenylethyleneglycol in both cases, also in accordance with the rule. The results (26) of the reductions of cholesteryl phenylglyoxylate with lithium aluminum hydride, sodium borohydride, and aluminum isopropoxide are also explainable by supposing that the configuration of cholesterol is the **29**-type.

In the reaction of phenyldihydrothebaine phenylglyoxylate (**40**), however, an anomaly to the rule was reported with hydride reductions. Thus although the reaction of the ester (**40**) with methyl magnesium iodide afforded (−)-atrolactic acid (**38**) of 70% optical purity (27), reduction with lithium aluminum hydride, or sodium borohydride gave (+)-phenylethyleneglycol (**41**) (28), as shown in Figure 14.

The Grignard reaction is postulated (27) to proceed in conformation **42** and **43**, in which the two carbonyl groups are coplanar transoid, perpendicular to ring A and at an angle of 60° to ring B, as shown in Figure 15. The entering reagent will approach the keto-carbonyl group from the direction

CH_3O
C_6H_5—CO—COO
C_6H_5
N—CH_3
CH_3O
40

CH_3MgI →
COOH
CH_3—C—OH
C_6H_5
38
(−)-atrolactic acid
(ref. 27)

$LiAlH_4$ or $NaBH_4$ →
CH_2OH
HO—C—H
C_6H_5
41
(+)-phenylethyleneglycol
(ref. 28)

Figure 14. Reactions of phenyldihydrothebaine phenylglyoxylate.

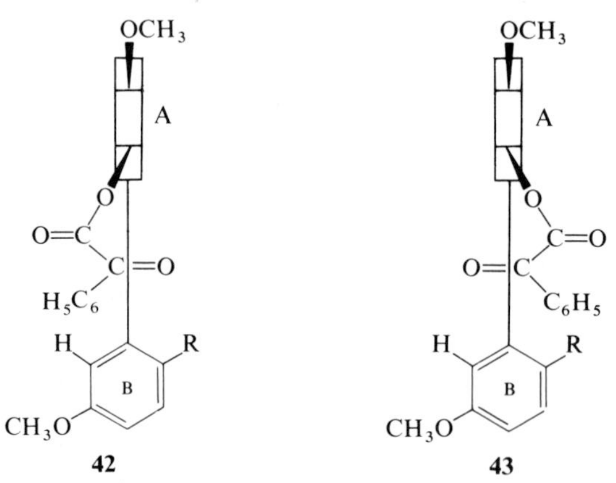

Figure 15

of ring A, because the approach of the reagent from the other side is considered to be blocked by ring B. In the course of these two transition states, the compression effects between the phenylglyoxylate chain and hydrogen in **42** and the phenylglyoxylate chain and R in **43**, come into play as the keto-carbonyl group begins to assume a tetrahedral geometry. Thus the reaction is postulated to occur from the direction of ring A on **42**, predominantly, to give (−)-atrolactic acid in excess. It must be noted that this asymmetric induction is effected by the asymmetry of the biphenyl system itself.

On the other hand, reduction of phenyldihydrothebaine phenylglyoxylate (**40**) with lithium aluminum hydride or sodium borohydride gave (+)-phenylethyleneglycol (**41**), whose absolute configuration is in enantiomeric relationship with that of (−)-atrolactic acid produced by the addition of methyl magnesium iodide mentioned above (28). Attempts to obtain phenyldihydrothebaine mandelate by sodium borohydride reduction were unsuccessful. The anomaly of the stereochemical course of the hydride reductions compared with that of the Grignard reactions may be attributed (28) to the abnormal reactivity order of the carbonyl groups; as the ester carbonyl group is reduced faster than the keto-carbonyl group with sodium borohydride, contrary to the results mentioned previously. These findings demonstrate the need for extreme caution in the assignments of the absolute configurations by hydride reductions of the phenylglyoxylates.

IV. HYDRIDE REDUCTIONS OF CYCLOHEXANONE DERIVATIVES

Numerous investigations have hitherto been reported on the stereochemistry of nucleophilic additions to the carbonyl group of cyclohexanone derivatives (29). In 1953 Barton (30) proposed the general rule. The rule states that the reduction of cyclohexanone derivatives with lithium aluminum hydride and with sodium borohydride, in general, affords the equatorial epimer if the ketone is not hindered, the axial epimer if it is hindered or very hindered. As will be seen in the following, Barton's proposal is considered to be basically correct.

A. The Concept of Steric Approach Control and Product Development Control

Dauben and his co-workers (31) studied the factors which control the ratios of diastereoisomers produced in the hydride reductions of alkylcyclohexanones and steroidal ketones. In these reductions the reaction is expected to involve the primary formation of a metallo-organic complex between the carbonyl group and the hydride reagent, followed by hydride transfer. Thus two factors are supposed to be operative. The first is a steric consideration involving competitive attacks of a reagent from a favored (unhindered) or an unfavored (hindered) side in the formation of the initial metallo-organic complex. This factor is called "steric approach control." The second is a consideration of relative energetics in the formation of the products once the complex is formed. This factor, therefore, is dependent on the relative stabilities of the product and is called "product development control." In general, it is said that hydride reductions of unhindered ketones are basically product-development controlled, while steric approach control becomes progressively more important in the hydride reductions of hindered ketones.

Hydride reductions of 4-tert-butylcyclohexanone (**44**) give diastereoisomeric mixtures enriched in *trans*-isomer (**45**) as shown in Figure 16. The

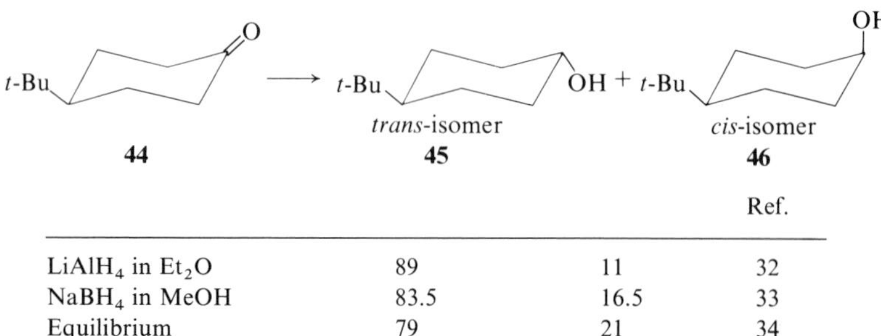

			Ref.
$LiAlH_4$ in Et_2O	89	11	32
$NaBH_4$ in MeOH	83.5	16.5	33
Equilibrium	79	21	34

Figure 16. Stereochemical course of hydride reductions of 4-tert-butylcyclohexanone.

results are explained by the above concept as follows. The tert-butyl group is considered to orient itself exclusively in an equatorial position due to its effective bulkiness and is remote from the carbonyl group. Hence the approach of the reducing species to the carbonyl group is equally favored from either side (from the axial side or from the equatorial side). Therefore steric approach control is not considered to be a dominant factor. On the other hand, *trans*-isomer (**45**) is more stable than *cis*-isomer (**46**), as is evident from the equilibrium position shown in Figure 16. Accordingly, the stereochemical course of the reduction is said to be product-development controlled. The results of hydride reductions of the so-called unhindered ketones, such as 4-methylcyclohexanone (84% *trans*-isomer by lithium aluminum hydride (31c), 82% *trans*-isomer by sodium borohydride (35)), and 3-cholestanone (90% 3β-hydroxy isomer by lithium aluminum hydride (31b, 36), 85% 3β-hydroxy isomer by sodium borohydride (37a)) are considered to be product-development controlled.

Hydride reductions of 11-keto steroids (**47**) are reported to give preferentially 11β-hydroxy isomers (**48**) (37). As shown in Figure 17,

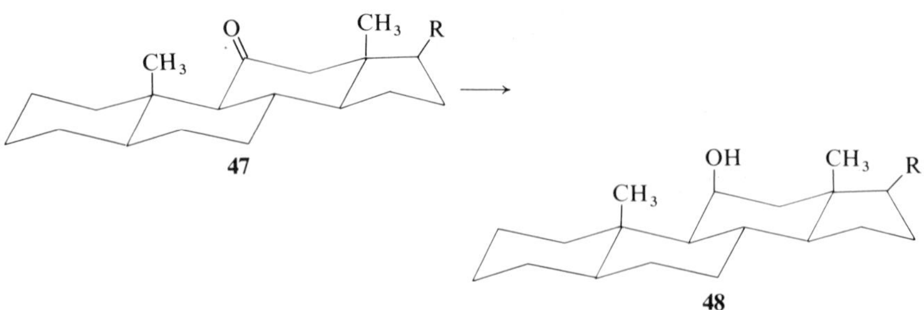

Figure 17. Stereochemical course of hydride reductions of 11-keto steroids.

approach of the reagent from the β-side to the carbonyl group at the 11-position of the steroidal nucleus is strongly inhibited by the two angular methyl groups of C-18 and C-19. In these cases, therefore, approach of the reagent is considered to occur preferentially from the α-side. On the other hand, severe 1,3-diaxial methyl–hydroxyl interactions in 11β-hydroxy isomers will make them less stable than the corresponding 11α-hydroxy isomers. Thus the results of hydride reductions of 11-keto steroids are interpreted as the consequences of steric approach control. Hydride reductions of rigid bicyclic ketones are also known to be steric approach controlled. These are shown in Figure 18. Hydride reductions of norcamphor (**49**) and

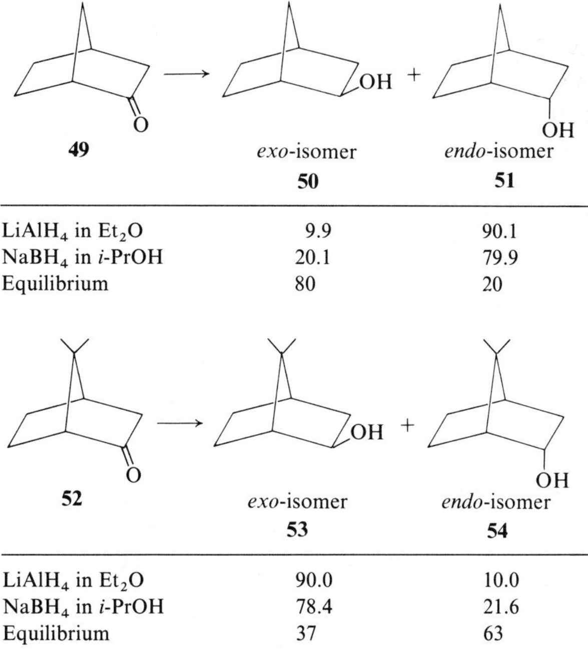

	exo-isomer **50**	*endo*-isomer **51**
$LiAlH_4$ in Et_2O	9.9	90.1
$NaBH_4$ in *i*-PrOH	20.1	79.9
Equilibrium	80	20

	exo-isomer **53**	*endo*-isomer **54**
$LiAlH_4$ in Et_2O	90.0	10.0
$NaBH_4$ in *i*-PrOH	78.4	21.6
Equilibrium	37	63

Figure 18. Stereochemical course of hydride reductions of norcamphor and apocamphor (data taken from ref. 38).

apocamphor (**52**) always give less stable isomers predominantly (38). The results of hydride reductions of the so-called hindered ketones are interpreted to be steric-approach controlled.

It is interesting to note that the hydride reductions of 3,3,5-trimethylcyclohexanone (**55**) show results which are intermediate to the above two cases. The ketone (**55**) is considered to be conformationally homogeneous as shown in Figure 19, because the alternative chair conformation would have a

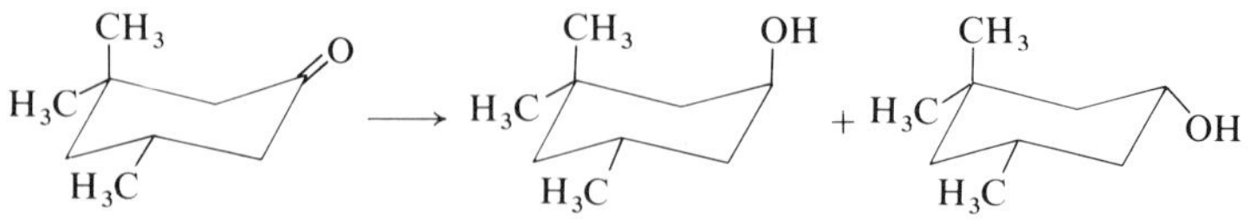

	trans-isomer **56**	*cis*-isomer **57**	Ref.
$LiAlH_4$ in Et_2O	55	45	39a
$NaBH_4$ in *i*-PrOH	55	45	39b
Equilibrium	6	94	39c

Figure 19. Stereochemical course of hydride reductions of 3,3,5-trimethylcyclohexanone.

severe 1,3-diaxial methyl–methyl interaction. Thus approach of the reagent to the carbonyl group from axial direction would suffer some hindrance by the axial methyl group at C-3. This effect would give more *trans*-isomer (**56**). On the other hand, the effect of the relative stabilities of the products favors the *cis*-isomer (**57**). The results in Figure 19 are, therefore, considered to be both steric-approach controlled and product-development controlled. Similar results were obtained in 2-cholestanone (52% 2β-hydroxy isomer and 37% 2α-hydroxy isomer with lithium aluminum hydride (31b)).

It can therefore be said that the stereochemical course of hydride reductions of cyclohexanone derivatives is explained by the concept of steric approach control and product development control.

On the other hand, Richer (40) proposed an interpretation based solely on steric factors, considering that the axial hydrogens in positions 2 and 6 as well as 3 and 5 of cyclohexanone derivatives would interfere with a species approaching the carbonyl group. Thus attacks of reagents small enough not to interfere with the axial hydrogens in positions 3 and 5 can be expected to be directed exclusively from the axial side of the carbonyl group, due to the interference with the axial hydrogens in positions 2 and 6. For example, the results in Figure 16 may be interpreted as a preferential attack of the reagent from the axial direction. As the size of the entering group becomes larger, more interactions with axial hydrogens in positions 3 and 5 are felt, favoring to a greater extent of the equatorial attack on the unhindered ketone. The same effect should also be expected by the introduction of an axial methyl group in positions 3 and 5, as the results in Figure 19 show.

Felkin and his co-workers (16a,16b) proposed the interpretation which is in line with that in the reactions of acyclic ketones (Section II-B). Thus reactions all proceed via reactant-like transition states. Steric strain between the nucleophilic entity and the axial substituents at positions 3 and 5 may be important in the axial attack, whereas torsional strain between the nucleophilic entity and the axial substituents at positions 2 and 6 may be important in the equatorial attack. The stereochemical course of the reactions is expected to be determined by the relative magnitudes of steric strain and torsional strain in the corresponding transition states.

B. Factors that Affect Stereochemistry

1. Reagent

Changing the reducing species sometimes shows a pronounced effect on the ratios of diastereoisomers obtained by hydride reductions. The results of reductions of 2-methylcyclohexanone (**58**) with various hydride reagents are shown in Figure 20. Quite large differences in stereoselectivity are observed.

58	*trans*-isomer **59**	*cis*-isomer **60**	Ref.
$LiAlH_4$ in THF	76	24	41
$LiAlH(OMe)_3$ in THF	31	69	41
$LiAlH(OEt)_3$ in THF	73	27	41
$LiAlH(O\text{-}t\text{-}Bu)_3$ in THF	70	30	41
$NaBH_4$ in *i*-PrOH	69	31	42
B_2H_6 in DG	65	35	43
Diisoamylborane in DG	23	77	43
Diisopinocamphenylborane in DG	8	92	43
Equilibrium	85	15	31c

Figure 20. Stereochemical course of reductions of 2-methylcyclohexanone (THF = tetrahydrofuran, DG = diglyme).

This phenomenon is the result of increasing steric approach control as the bulkiness of the reducing species increases. Similar results were obtained in the case of 2-methylcyclopentanone (41,43). It is interesting to note that the use of lithium trimethoxyaluminohydride gives an exceptional result (41).

As described above, reductions of 3,3,5-trimethylcyclohexanone (**55**) with lithium aluminum hydride and sodium borohydride give a roughly 1 : 1 mixture of diastereoisomers, due to the effect of both steric approach control and product development control. As shown in Figure 21, the less stable isomer (**56**) becomes the predominant product when lithium aluminum hydride reduction is performed with the addition of methanol or aluminum chloride under kinetically controlled conditions. This is probably due to the increased bulkiness in the reducing species. However, when reduction with

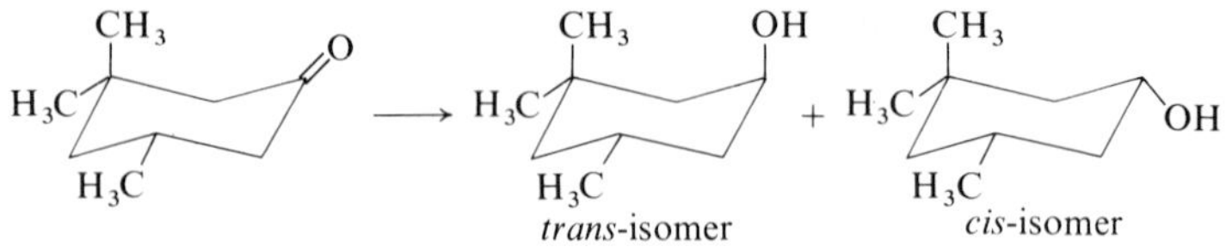

	56	57
$LiAlH_4$ in Et_2O	55	45
$LiAlH(OMe)_3$ in Et_2O	75	25
$LiAlH_4$-$AlCl_3$ (1:3) in Et_2O (kinetically controlled)	85	15
$LiAlH_4$-$AlCl_3$ (1:3) in Et_2O (thermodynamically controlled)	~0	~100

Figure 21. Stereochemical course of reductions of 3,3,5-trimethylcyclohexanone (data taken from ref. 39a).

lithium aluminum hydride–aluminum chloride (1:3) is performed in the presence of a small excess of ketone, the more stable isomer (**57**) results, almost to the exclusion of its epimer (**56**). This reaction is considered to be a thermodynamically controlled one, in which the equilibrium between the aluminum alkoxides (**61** and **62**) lies exclusively to the equatorial epimer (**62**) (Fig. 22) due to the bulkiness of the dichloroaluminoxy group (39a). Application of this method to acyclic ketones has also been reported (44).

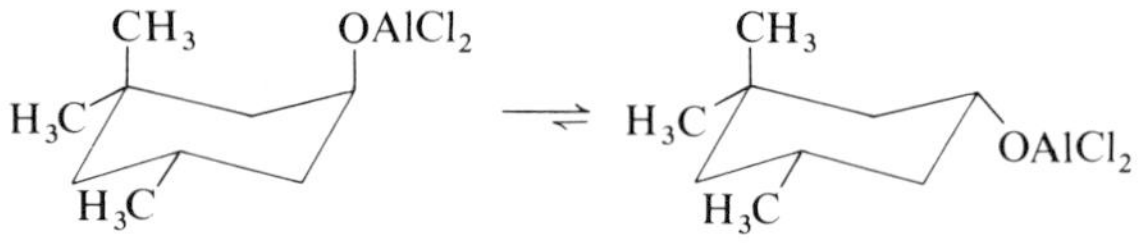

Figure 22. Equilibrium of alkoxyaluminum dichlorides.

The addition of boron trifluoride in amine borane reduction of 4-tert-butylcyclohexanone (**44**) is also reported to appreciably increase the ratio of the less stable isomer (**46**) (45).

From a comparison of the ratios of diastereoisomers produced by the reduction of sodium borohydride and lithium aluminum hydride, it is said that the former is the more bulky reagent (31a, 31b). Although some doubts exist as to the exact determination of these ratios (30c, 45), the reason is attributed to a difference in the degree of solvation (30a,30b), or to a difference in the reaction mechanism (47).

2. *Solvent and Temperature*

It is known that the stereochemistry of hydride reductions is also dependent on the solvent used. For example, lithium aluminum hydride reduction of 3,3,5-trimethylcyclohexanone (**55**) gives 55% *trans*-isomer (**56**) in ether and 74% *trans*-isomer (**56**) in tetrahydrofuran (39a). The results of sodium borohydride reduction of the ketone (**55**) in various solvents are shown in Figure 23. The fact that the ratios of diastereoisomers vary with the

$\xrightarrow{NaBH_4}$

	trans-isomer **56**	*cis*-isomer **57**
In *i*-PrOH	55	45
In diglyme	55	45
In 65% aqueous *i*-PrOH	59.5	40.5
In EtOH	67	33
In 71% aqueous MeOH	73	27

Figure 23. Stereochemical course of sodium borohydride reductions of 3,3,5-trimethylcyclohexanone in various solvents (data taken from ref. 39b).

solvent used supports the proposal that both lithium aluminum hydride and sodium borohydride are solvated in these media (39b). The molar ratios of ketone and reducing agent show little effect on the stereochemical course in this reduction (39b).

Reaction temperature also affects the stereochemistry of the reaction. Examinations on 3,3,5-trimethylcyclohexanone (**55**) (33), 4-tert-butylcyclohexanone (**44**) (33), and 3-cholestanone (47b) revealed, however, that variations in temperature caused minor changes in the relative amounts of diastereoisomers. In general, it is said that the greatest stereoselectivity in the reductions of hindered ketones to the less stable isomers can be achieved in good solvating media at low temperature (35).

3. *Neighboring Group Participation*

Stevens and his co-workers (48) reported that sodium borohydride reduction of various 2-alkylamino-2-phenylcyclohexanones (**63**) gave *trans*-isomer (**64**) predominantly. For example, 2-methylamino-2-phenylcyclohexanone (**63**, R = methyl) gave *trans*-isomer (**64**, R = methyl) in at least 80% yield. The results indicate that hydride ion attacks from the direction of the alkylamino group.

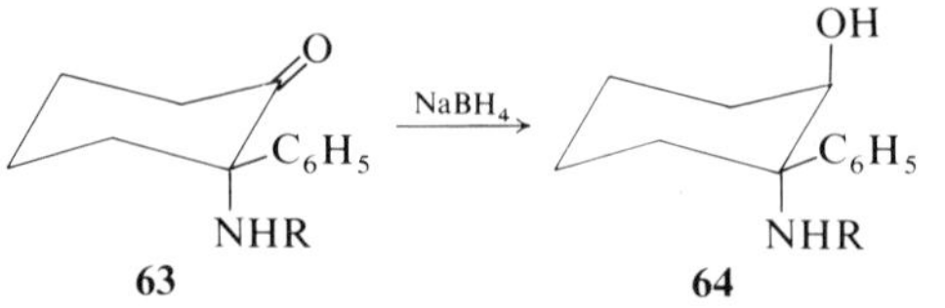

Figure 24. Sodium borohydride reduction of 2-alkylamino-2-phenylcyclohexanone.

Reduction of the keto-acid (**65**) with sodium borohydride gives predominantly the alcohol (**66**), which is the isomer produced by the attack of the reagent from the direction of the carboxyl group (49a). This result is

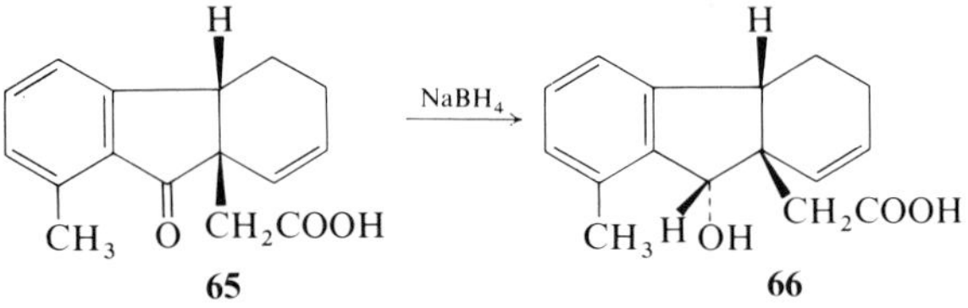

Figure 25

envisioned as an intramolecular solvation of the borohydride ion by the carboxylate ion. From the examinations on the model compounds, however, it is reported that the results are in accordance with the postulate that the major product is the more stable isomer and offers no support for the possibility of such an assistance of hydride ion transfer (49b). On the other hand, results of sodium borohydride reduction of *cis*- and *trans*-8-oxodecahydronaphthoic acid are consistent with the postulate that carboxylate ion inhibits the attack of borohydride ion (50).

The stereochemistry of hydride reductions of cyclohexenone derivatives has been recently reviewed by Toromanoff (51).

V. REDUCTIONS WITH ASYMMETRIC HYDRIDE REAGENTS

Many examples of asymmetric induction through optically active reagents have hitherto been reported (2a). It is interesting to note that this type of asymmetric induction has a similarity with enzyme reactions in that the asymmetry is induced intermolecularly.

A. Reductions with Optically Active Grignard Reagents

As shown in Figure 26, Grignard reagent having hydrogen at β-position (**68**) acts as a reducing agent by the transfer of hydride ion to the carbonyl

$$\underset{\mathbf{67}}{R_2C{=}O} + \underset{\mathbf{68}}{R'CH_2CH_2MgX} \longrightarrow \underset{\mathbf{69}}{\left[R_2C\cdots O\cdots Mg(Cl)\cdots CH_2\cdots CH(R')\cdots H\right]} \longrightarrow \underset{\mathbf{70}}{R_2CH{-}OH} + \underset{\mathbf{71}}{R'CH{=}CH_2}$$

Figure 26. Reduction of ketone with Grignard reagent.

carbon via a six-membered ring complex (**69**) (52). Asymmetric induction is also known to occur when this reduction is performed with unsymmetric ketones and optically active Grignard reagents (53). For example, Vavon and his co-workers (53b) reported that the reduction of sec-butyl phenyl ketone with isobornylmagnesium chloride, which is obtained from pinene hydrochloride, gave sec-butylphenylcarbinol of 72% optical purity. Extensive investigations on the asymmetric induction in the reductions of unsymmetric ketones with optically active Grignard reagents having one asymmetric carbon atom were made by Mosher and his co-workers (54,55) in order to relate the configuration of the optically active alcohol produced in excess with that of the reagent. Table 7 represents some results of the reductions of ketones (**72**) with optically active Grignard reagent (**73**) from (+)-1-chloro-2-methylbutane.

TABLE 7[a]

$$\underset{\mathbf{72}}{RR'C{=}O} + \underset{\mathbf{73}}{Me{-}C(H)(Et){-}CH_2MgCl} \longrightarrow \underset{\mathbf{74}}{H{-}C(R)(R'){-}OH} + \underset{\mathbf{75}}{HO{-}C(R)(R'){-}H}$$

R' \ R	Me	Et	*n*-Pr	*i*-Pr	C_6H_{11}	C_6H_5	*t*-Bu	Ref.
t-Bu	+13	+11	+11	+5	+2.5	−16	—	54a
C_6H_5	+3.9	+5.7	+5.9	+24	+25	—	+16	54b
C_6H_{11}	+3.6	+8.8	+8.9	+2.1	—	−25	−2.5	54c

[a] Positive signs refer to the production of **74** in excess. Negative signs refer to the production of **75** in excess. Values indicate the extent of asymmetric induction.

These data are interpreted as follows (54c). The transition states of the reaction may be represented as **76** and **77**, in which the asymmetric carbon atom of the reagent constitutes a member of the postulated ring. The stereochemical course of the reaction is then decided by the relative stabilities of **76** and **77**. In these transition states steric interferences are considered to be caused by the groups both lying in the same side of the ring; that is, R versus

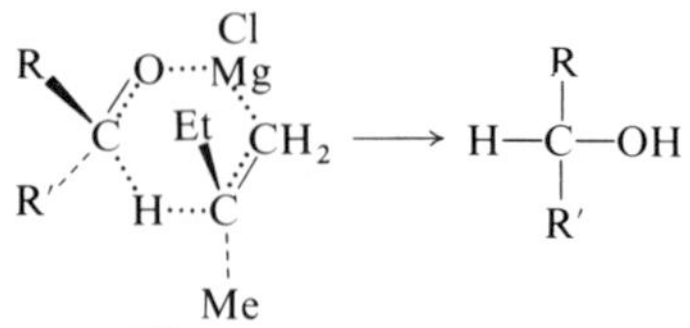

Figure 27. Stereochemical course of asymmetric Grignard reduction.

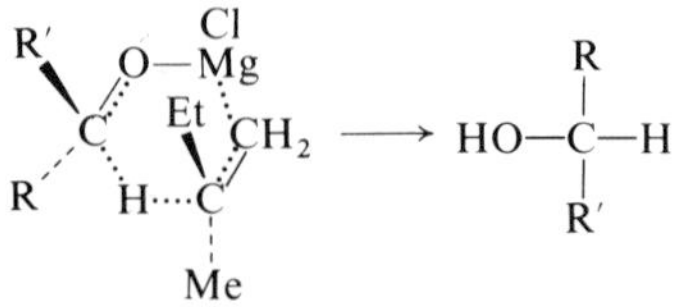

ethyl and R′ versus methyl in **76**, R′ versus ethyl and R versus methyl in **77**. Therefore, when R′ is bulkier than R, **74** should predominate over **75** in the reaction products.

As shown in Table 7, this interpretation is quite satisfactory for the reductions of alkyl tert-butyl ketones (**72**, R = alkyl, R′ = tert-butyl). In these cases the configurations of the predominant enantiomers are in accordance with the prediction. Moreover, stereospecificity decreases as the alkyl group becomes larger. Mosher and his co-workers (54a) examined various alkyl tert-butyl ketones and concluded that the steric bulkiness of the alkyl groups in the reactions was increased by branching at α-position to the carbonyl group.

However, such a clearcut interpretation does not necessarily hold in the cases of alkyl phenyl ketones and alkyl cyclohexyl ketones, as shown in Table 7. For example, in alkyl phenyl ketones, stereospecificity of the reductions increases as the alkyl group becomes bulkier. Moreover, the greatest stereospecificity is observed in the reduction of cyclohexyl phenyl ketone contrary to the prediction of low stereospecificity because approximately the same stereospecificity is found in the asymmetric reduction of methyl phenyl ketone (3.9%) and methyl cyclohexyl ketone (3.6%). These inconsistencies in the prediction seem to suggest that interpretation of the direction and degree of asymmetric induction by analyzing only the steric interferences of opposing groups in the postulated cyclic transition states (**76** and **77**) is an oversimplification. It seems necessary to evaluate more complicated factors such as coordinated solvent, and slight structural and electronic variations in the transition states (54c). It is also recognized that the stereospecificity is affected by the rate of reduction (54d).

Of special interest in this type of asymmetric induction is the reduction of isopropyl phenyl ketone with optically active Grignard reagent from (+)-1-chloro-2-phenylbutane to give (−)-isopropylphenylcarbinol of 82% optical purity (55). It is striking to note that the reducing agent with a single asymmetric center is capable of inducing such a high degree of asymmetry in the intermolecular reaction.

Originally it was considered necessary, for asymmetric induction by Grignard reagents, that the asymmetric center should be at β-position as in **78** (Fig. 28) and should constitute a member of the postulated ring in the

```
     CH2MgCl              CH2MgCl
        |                    |
   R—C—H                 CH2
        |                    |
        R'              R—C—H
                             |
                             R'
       78                   79
```

Figure 28

transition states as shown in **76** and **77** (54e). However, it was revealed that reductions of isobutyrophenone with γ-asymmetric Grignard reagents from (−)-1-chloro-3-phenylbutane (**79**, R = methyl, R′ = phenyl) and (−)-1-chloro-3-phenylpentane (**79**, R = ethyl, R′ = phenyl) showed stereospecificity giving (+)-isopropylphenylcarbinol of 25% and 29% optical purities, respectively (56).

B. Reductions with Optically Active Alkoxides

Asymmetric reductions of unsymmetric ketones in optically active alcohols under Meerwein-Ponndorf reduction conditions are considered to be theoretically possible. Doering and Young (57) showed that 6-methyl-2-heptanone was reduced in (+)-2-butanol to 5.9% optically active (+)-6-methyl-2-heptanol (**82**, R′ = *n*-hexyl) in the presence of aluminum 2-butoxide, and also that methyl cyclohexyl ketone was reduced in (+)-3-methyl-2-butanol to 21.8% optically active (+)-methylcyclohexylcarbinol (**82**′, R = cyclohexyl) in the presence of aluminum 3-methyl-2-butoxide. These results are interpreted by a mechanism similar to that described in Section V-A. Thus the transition states of this reaction may be represented as **80** and **81**, which give **82** and **83**, respectively (Fig. 29). From a consideration of the steric interactions between opposing groups, **80** is expected to be of lower energy when R and R′ are bulkier than methyl. The configuration of the predominant enantiomers produced in the reactions described above are in accordance with those predicted by **80** (57).

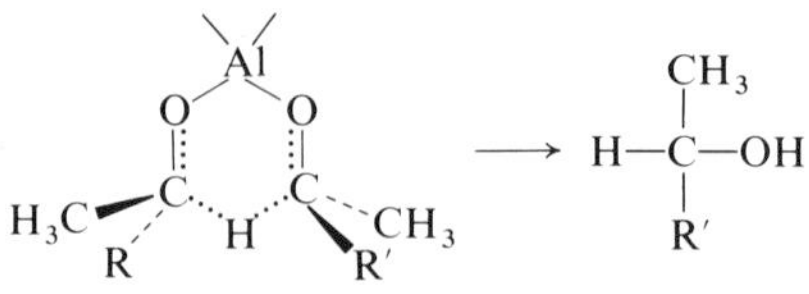

Figure 29. Asymmetric Meerwein-Ponndorf reduction.

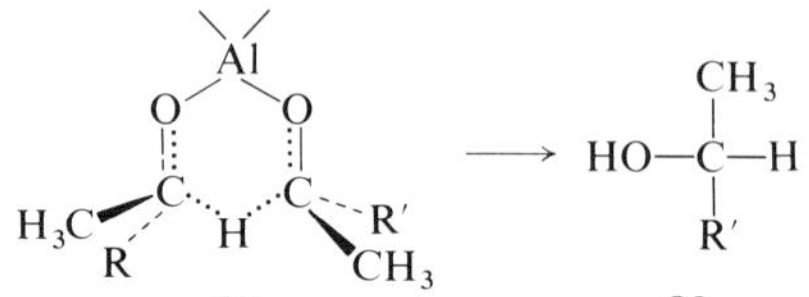

Assignment of the absolute configuration of optically active 6,6′-dinitro-2,2′-diphenic acid (**84** and **85**) using the asymmetric Meerwein-Ponndorf reduction was made by Mislow and his co-workers (58), as shown in Figure 30. Racemic biphenyl ketone (a mixture of **88** and **89**) was prepared from racemic 6,6′-dinitro-2,2′-diphenic acid. Incomplete reduction of this racemic ketone in S(+)-pinacolyl alcohol (**90**, R = tert-butyl) or S(+)-2-octanol (**90**, R = *n*-hexyl) in the presence of aluminum tert-butoxide afforded a levorotatory reduction product and a dextrorotatory starting ketone. The stereochemical course of this reaction is interpreted as follows (58). **91** and **92** are the transition states of the reaction and their relative stabilities are expected to be decided by the steric interactions; that is, R versus jutting phenyl and methyl versus hydrogen in **91**, R versus hydrogen and methyl versus jutting phenyl in **92**. Since R is bulkier than methyl in these cases, **92** is considered to be more stable and, therefore, to be a preferred transition state. Thus, the reaction via **92** must be faster than that via **91**. As a result, when the reaction is stopped before completion, **93** is expected to predominate over **94** in the reduction product, while **88** is expected to predominate over **89** in the recovered ketone. Therefore (−)-biphenylalcohol and (+)-biphenylketone are represented as **93** and **88**, respectively. As (+)-biphenylketone (**88**) is obtained from (−)6,6′-dinitro-2,2′-diphenic acid, the latter is represented as **84**, or S configuration. Recent determination of the absolute configuration by X-ray diffraction demonstrated that this assignment was correct (59).

Examples of asymmetric reductions through magnesiobromide salt of optically active alcohols are also known (60).

Among the alkoxyaluminum dichlorides (RR′CHOAlCl_2), (−)-isobornyloxyaluminum dichloride (**95**) has proven to be a fairly stereoselective

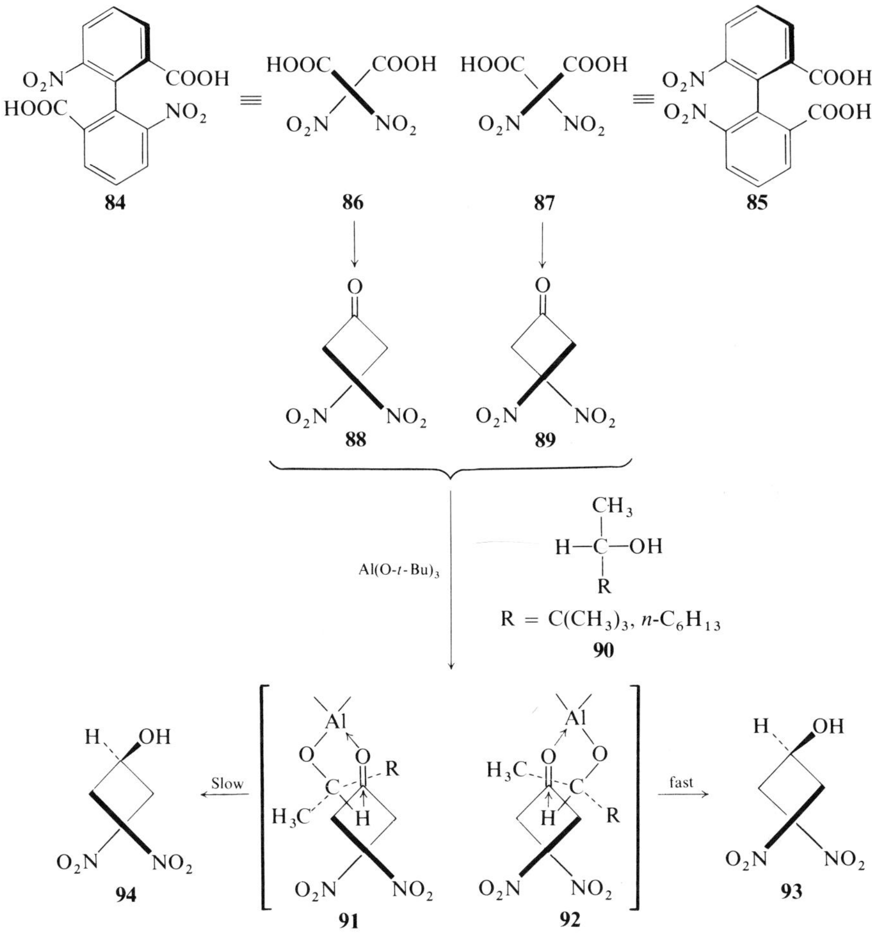

Figure 30. Assignment of the absolute configuration by asymmetric reduction (Mislow and his co-workers).

reducing agent in the reduction of ketones to the corresponding alcohols by hydride transfer (61). Although the direction of asymmetric induction is predictable by analyses of the postulated six-membered transition states (**96** and **97**) (62), these analyses are not useful in the quantitative sense. Thus stereochemical explanation of these results has been attempted (63) with models proposed by Karabatsos (see Section I-C).

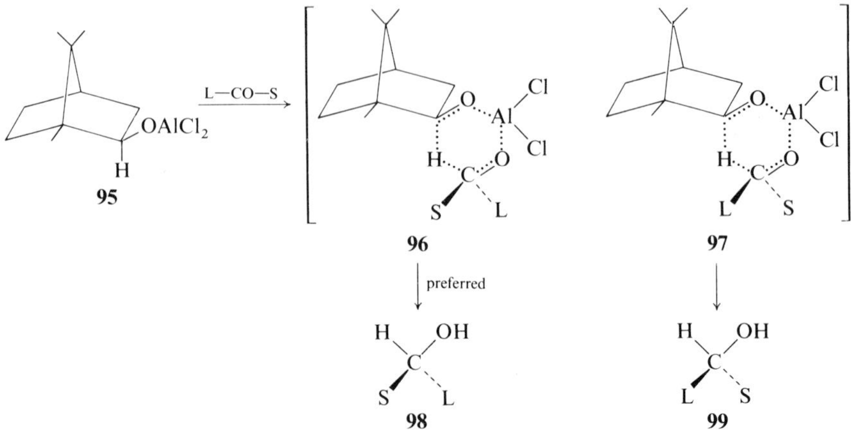

Figure 31. Stereochemical course of asymmetric reduction with (−)-isobornyloxyaluminum dichloride.

C. Reductions with Optically Active Hydrides

Although initial attempts (64) to reduce 2-butanone and pinacolone asymmetrically with lithium aluminum hydride in the presence of (+)-camphor was proved to be unsuccessful (65), intermolecular asymmetric reductions based on the same consideration did prove successful with lithium aluminum hydride modified with optically active substances such as alcohols, sugars, and alkaloids (66,67). For example, ethanol modified lithium aluminum hydride-3-*O*-benzyl-1,2-*O*-cyclohexylidene-α-D-glucofuranose complex (molar ratio of ethanol:$LiAlH_4$:sugar derivative = 1:1:1) was shown to reduce acetophenone to α-phenethylalcohol of 70.9% optical purity. In spite of their highly complexed systems approaches to interpretation of these stereochemical results have also been reported (66,67).

Finally diisopinocamphenylborane, known as a useful reagent for the asymmetric hydroboration of olefins (68), is also effective in the asymmetric reductions of unsymmetric ketones (43).

References

1. E. L. Eliel, *Stereochemistry of Carbon Compounds*, McGraw-Hill, New York, 1962, p. 68.
2. (a) J. I. Klabunowski, *Asymmetrische Synthese*, Deutscher Verlag der Wissenschaften, Berlin, 1963; (b) L. Velluz, J. Valls, and J. Mathieu, *Angew. Chem. Intern. Ed.*, **6**, 778 (1967).
3. D. J. Cram and F. A. Abd Elhafez, *J. Am. Chem. Soc.*, **74**, 5828 (1952).
4. D. J. Cram and J. Allinger, *J. Am. Chem. Soc.*, **76**, 4516 (1954).
5. Y. Gault and H. Felkin, *Bull. Soc. Chim. France*, **1960**, 1342.
6. D. J. Cram, F. A. Abd Elhafez, and H. Weingarten, *J. Am. Chem. Soc.*, **75**, 2293 (1953).

7. D. J. Cram and F. D. Greene, *J. Am. Chem. Soc.*, **75**, 6005 (1953).
8. D. J. Cram, F. A. Abd Elhafez, and H. L. Nyquist, *J. Am. Chem. Soc.*, **76**, 22 (1954).
9. (a) D. J. Cram and K. R. Kopecky, *J. Am. Chem. Soc.*, **81**, 2748 (1959); (b) J. H. Stocker, P. Sidisunthorn, B. M. Benjamin, and C. J. Collins, *J. Am. Chem. Soc.*, **82**, 3913 (1960); (c) B. M. Benjamin, H. J. Schaeffer, and C. J. Collins, *J. Am. Chem. Soc.*, **79**, 6160 (1957).
10. S. Yamada and K. Koga, *Tetrahedron Letters*, **1967**, 1711.
11. J. H. Stocker, *J. Org. Chem.*, **29**, 3593 (1964).
12. (a) J. Sicher, M. Svoboda, M. Hrdá, J. Rudinger, and F. Šorm, *Collection Czech. Chem. Commun.*, **18**, 487 (1953); (b) T. Matsumoto, T. Nishida, and H. Shirahama, *J. Org. Chem.*, **27**, 79 (1962).
13. (a) G. Drefahl and H. H. Hoerhold, *Chem. Ber.*, **94**, 1641 (1961); (b) A. Pohland, L. R. Peters, and H. R. Sullivan, *J. Org. Chem.*, **28**, 2483 (1963); (c) P. S. Portoghese and D. A. Williams, *Tetrahedron Letters*, **1966**, 6299.
14. (a) E. J. Corey, *J. Am. Chem. Soc.*, **75**, 2301 (1953); (b) L. J. Bellamy and R. L. Williams, *J. Chem. Soc.*, **1957**, 4294.
15. (a) J. W. Cornforth, R. H. Cornforth, and K. K. Mathew, *J. Chem. Soc.*, **1959**, 112; (b) D. J. Cram and D. R. Wilson, *J. Am. Chem. Soc.*, **85**, 1245 (1963).
16. (a) M. Chérest, H. Felkin, and N. Prudent, *Tetrahedron Letters*, **1968**, 2199; (b) M. Chérest and H. Felkin, *Tetrahedron Letters*, **1968**, 2205; (c) M. J. Brienne, C. Ouannès, and J. Jacques, *Bull. Soc. Chim. France*, **1968**, 1036.
17. G. J. Karabatsos, *J. Am. Chem. Soc.*, **89**, 1367 (1967).
18. E. L. Eliel, N. L. Allinger, S. J. Angyal, and G. A. Morrison, *Conformational Analysis*, Interscience, New York, 1965, p. 19.
19. G. J. Karabatsos and N. Hsi, *J. Am. Chem. Soc.*, **87**, 2864 (1965).
20. (a) G. J. Karabatsos and N. Hsi, *Tetrahedron*, **23**, 1079 (1967); (b) G. J. Karabatsos and K. L. Krumel, *ibid.*, **23**, 1097 (1967).
21. G. J. Karabatsos and T. H. Althuis, *Tetrahedron Letters*, **1967**, 4911.
22. (a) V. Prelog, *Helv. Chim. Acta*, **36**, 308 (1953); (b) V. Prelog, *Bull. Soc. Chim. France*, **1956**, 987.
23. V. Prelog, O. Ceder, and M. Wilhelm, *Helv. Chim. Acta*, **38**, 303 (1955).
24. (a) W. G. Dauben, D. F. Dickel, O. Jeger, and V. Prelog, *Helv. Chim. Acta*, **36**, 325 (1953); (b) V. Prelog and G. Tsatsas, *Helv. Chim. Acta*, **36**, 1178 (1953); (c) W. R. Feldman and V. Prelog, *Helv. Chim. Acta*, **41**, 2396 (1958); (d) K. Mislow, V. Prelog, and H. Scherrer, *Helv. Chim. Acta*, **41**, 1410 (1958).
25. V. Prelog, M. Wilhelm, and D. B. Bright, *Helv. Chim. Acta*, **37**, 221 (1954).
26. S. P. Bakshi and E. E. Turner, *J. Chem. Soc.*, **1961**, 168.
27. J. A. Berson and M. A. Greenbaum, *J. Am. Chem. Soc.*, **80**, 445 (1958).
28. J. A. Berson and M. A. Greenbaum, *J. Am. Chem. Soc.*, **81**, 6456 (1959).
29. A. V. Kamernitzky and A. A. Akhrem, *Tetrahedron*, **18**, 705 (1962).
30. D. H. R. Barton, *J. Chem. Soc.*, **1953**, 1027.
31. (a) W. G. Dauben, G. J. Fonken, and D. S. Noyce, *J. Am. Chem. Soc.*, **78**, 2579 (1956); (b) W. G. Dauben, E. J. Blanz, Jr., J. Jiu, and R. A. Micheli, *J. Am. Chem. Soc.*, **78**, 3752 (1956); (c) W. G. Dauben and R. E. Bozak, *J. Org. Chem.*, **24**, 1596 (1959).
32. E. L. Eliel and M. N. Rerick, *J. Am. Chem. Soc.*, **82**, 1367 (1960).
33. P. T. Lansbury and R. E. Macleay, *J. Org. Chem.*, **28**, 1940 (1963).
34. E. L. Eliel and R. S. Ro, *J. Am. Chem. Soc.*, **79**, 5992 (1957).
35. M. G. Combe and H. B. Henbest, *Tetrahedron Letters*, **1961**, 404.
36. (a) C. W. Shoppee and G. H. R. Summers, *J. Chem. Soc.*, **1950**, 687; (b) H. R. Nace and G. L. O'Connor, *J. Am. Chem. Soc.*, **73**, 5824 (1951); (c) O. H. Wheeler and J. L. Mateos, *Can. J. Chem.*, **36**, 1049 (1958).

37. (a) L. H. Sarett, M. Feurer, and K. Folkers, *J. Am. Chem. Soc.*, **73**, 1777 (1951); (b) S. Bernstein, R. H. Lenhard, and J. H. Williams, *J. Org. Chem.*, **18**, 1166 (1953).
38. R. Howe, E. C. Friedrich, and S. Winstein, *J. Am. Chem. Soc.*, **87**, 379 (1965).
39. (a) H. Haubenstock and E. L. Eliel, *J. Am. Chem. Soc.*, **84**, 2363 (1962); (b) H. Haubenstock and E. L. Eliel, *J. Am. Chem. Soc.*, **84**, 2368 (1962); (c) E. L. Eliel and H. Haubenstock, *J. Org. Chem.*, **26**, 3504 (1961).
40. J. C. Richer, *J. Org. Chem.*, **30**, 324 (1965).
41. H. C. Brown and H. R. Deck, *J. Am. Chem. Soc.*, **87**, 5620 (1965).
42. H. C. Brown and J. Muzzio, *J. Am. Chem. Soc.*, **88**, 2811 (1966).
43. H. C. Brown and D. B. Bigley, *J. Am. Chem. Soc.*, **83**, 3166 (1961).
44. Ref. 17, p. 26.
45. W. M. Jones, *J. Am. Chem. Soc.*, **82**, 2528 (1960).
46. K. D. Hardy and R. J. Wicker, *J. Am. Chem. Soc.*, **80**, 640 (1958).
47. (a) D. M. S. Wheeler and J. W. Huffman, *Experientia*, **16**, 516 (1960); (b) O. R. Vail and D. M. S. Wheeler, *J. Org. Chem.*, **27**, 3803 (1962).
48. C. L. Stevens, A. B. Ash, A. Thuillier, A. J. Amin, A. Balys, W. E. Dennis, J. P. Dickerson, R. P. Glinski, H. T. Hanson, M. D. Pillai, and J. W. Stoddard, *J. Org. Chem.*, **31**, 2593 (1966).
49. (a) H. O. House, R. G. Carlson, H. Müller, A. W. Noltes, and C. D. Slater, *J. Am. Chem. Soc.*, **84**, 2614 (1962); (b) H. O. House, H. Babad, R. B. Toothill, and A. W. Noltes, *J. Org. Chem.*, **27**, 4141 (1962).
50. D. M. S. Wheeler and M. M. Wheeler, *J. Org. Chem.*, **27**, 3796 (1962).
51. E. Toromanoff, *Topics in Stereochemistry*, Vol. 2, Interscience, New York, 1967, p. 157.
52. J. D. Roberts and M. C. Caserio, *Basic Principles of Organic Chemistry*, Benjamin, New York, 1964, p. 361.
53. (a) G. Vavon, C. Rivière, and B. Angelo, *Compt. Rend.*, **222**, 959 (1946); (b) G. Vavon and B. Angelo, *Compt. Rend.*, **224**, 1435 (1947); (c) F. Tatibouët, *Bull. Soc. Chim. France*, **1951**, 867.
54. (a) W. M. Folay, F. J. Welch, E. M. LaCombe, and H. S. Mosher, *J. Am. Chem. Soc.*, **81**, 2779 (1959); (b) R. MacLeod, F. J. Welch, and H. S. Mosher, *J. Am. Chem. Soc.*, **82**, 876 (1960); (c) E. P. Burrows, F. J. Welch, and H. S. Mosher, *J. Am. Chem. Soc.*, **82**, 880 (1960); (d) V. E. Althouse, E. Kaufmann, P. Loeffler, K. Ueda, and H. S. Mosher, *J. Am. Chem. Soc.*, **83**, 3138 (1961); (e) H. S. Mosher and E. LaCombe, *J. Am. Chem. Soc.*, **72**, 4991 (1950).
55. J. S. Birtwistle, K. Lee, J. D. Morrison, W. A. Sanderson, and H. S. Mosher, *J. Org. Chem.*, **29**, 37 (1964).
56. J. D. Morrison, D. L. Black, and R. W. Ridgway, *Tetrahedron Letters*, **1968**, 985.
57. W. von E. Doering and R. W. Young, *J. Am. Chem. Soc.*, **72**, 631 (1950).
58. P. Newman, P. Rutkin, and K. Mislow, *J. Am. Chem. Soc.*, **80**, 465 (1958).
59. H. Akimoto, T. Shioiri, Y. Iitaka, and S. Yamada, *Tetrahedron Letters*, **1968**, 97.
60. (a) G. Vavon and A. Antonini, *Compt. Rend.*, **230**, 1870 (1950); (b) G. Vavon and A. Antonini, *Compt. Rend.*, **232**, 1120 (1951); (c) A. Streitwieser, Jr., *J. Am. Chem. Soc.*, **75**, 5014 (1953).
61. E. L. Eliel and D. Nasipuri, *J. Org. Chem.*, **30**, 3809 (1965).
62. (a) D. Nasipuri and G. Sarkar, *J. Indian Chem. Soc.*, **44**, 165, 425 (1967); (b) D. Nasipuri and C. K. Ghosh, *J. Indian Chem. Soc.*, **44**, 556 (1967).
63. D. Nasipuri, G. Sarkar, and C. K. Ghosh, *Tetrahedron Letters*, **1967**, 5189.
64. A. A. Bothner-By, *J. Am. Chem. Soc.*, **73**, 846 (1951).
65. P. S. Portoghese, *J. Org. Chem.*, **27**, 3359 (1962).
66. S. R. Landor, B. J. Miller, and A. R. Tatchell, *J. Chem. Soc. (C)*, **1967**, 197.

67. O. Červinka and O. Bchovcky, *Collection Czech. Chem. Commun.*, **32**, 3897 (1967) and references cited therein.

68. (a) H. C. Brown, N. R. Ayyangar, and G. Zweifel, *J. Am. Chem. Soc.*, **86**, 397 (1964); (b) G. Zweifel, N. R. Ayyangar, T. Munekata, and H. C. Brown, *J. Am. Chem. Soc.*, **86**, 1076 (1964); (c) H. C. Brown, *Hydroboration*, Benjamin, New York, 1962; (d) G. Zweifel and H. C. Brown, *Organic Reactions*, Vol. 13, Wiley, New York, 1963, p. 1.

Steric Selectivity in the Addition of Carbenes to Olefins

ROBERT A. MOSS

Wright Laboratory, School of Chemistry, Rutgers University, The State University of New Jersey, New Brunswick

I. INTRODUCTION

In 1954 Doering and Hoffmann (1) showed that dichlorocarbene could add to an olefinic double bond to yield a dichlorocyclopropane. In 1956, Skell and Garner (2) reported analogous reactions of dibromocarbene and, by studying competitive reactions with structurally different olefins, demonstrated that dibromocarbene behaved as an electrophilic reagent during its addition to the olefinic π bond. Shortly thereafter, Doering and Henderson (3) announced a similar study of dichlorocarbene. These papers form a nucleus from which there has developed, over the past decade, an ever-increasing interest in the carbene–olefin cyclopropanation reaction.

In much of this work steric effects go unmentioned. This omission, which has recently begun to be corrected, can be traced to the fact that dichlorocarbene and dibromocarbene, the first two carbenes subjected to relative addition rate studies, are (in an electrophilic sense) among the most selective carbenes known. This is particularly true of dichlorocarbene; and even though attention was called to the fact that steric hindrance seemed to retard the addition of dibromocarbene (relative to dichlorocarbene) to tetramethylethylene (3), the striking parallel of the carbene–olefin addition rates with those of olefin bromination and epoxidation (2b) led to an emphasis on electronic effects in carbene–olefin addition reactions.

The purpose of this chapter is to call attention to the importance of steric effects in these reactions. The chapter is not intended to be an exhaustive catalog of all carbene–olefin reactions in which steric effects are noticeable. Rather, it is intended to be illustrative of the kinds and magnitudes of steric effects which have been observed, and which therefore may be expected in other systems.

In the following, the term "carbene," when used with reference to a specific reaction, is usually intended to refer to a divalent carbon species. The terms "carbenoid" or "carbene-complex" are intended to refer to species which, while capable of cyclopropanating olefins are not, in fact, divalent carbon species.

II. STERIC EFFECTS IN CARBENE ADDITIONS TO SPECIAL OLEFINS

A. Bicyclic or Tricyclic Olefins

Olefinic substrates may be so constructed that only one mode of approach of the carbene (with respect to the plane of the olefinic carbons and their α substituents) is possible. A major class of sterically demanding substrates is constituted by the bicyclic alkenes. Here advantage is taken of the known reluctance toward *endo* attack by electrophilic reagents.

Thus the addition of the Simmons-Smith reagent, $(ICH_2)_2Zn \cdot ZnI_2$, to norbornene gives only the *exo*-tricyclooctane, **1** (4). Addition of the methylene

$$+ \ CH_2I_2 \xrightarrow{\text{Zn—Cu}} \quad (1)$$

H
H
1

species (from diazomethane and cuprous chloride) to *anti*-7-norbornenol proceeds in the same sense (5).

The methylene-transfer reagent of the Simmons-Smith reaction is well known to have great steric sensitivity toward substrate double bonds. In norbornene the opposition of the *syn* proton of the methano bridge provides a smaller barrier to *exo* attack than do the oppositions of the *endo* protons of the ethano bridge to *endo* attack (6). Addition of the reagent to norbornadiene affords *both exo* and *endo* monoadducts in a 5.7:1 distribution (4).

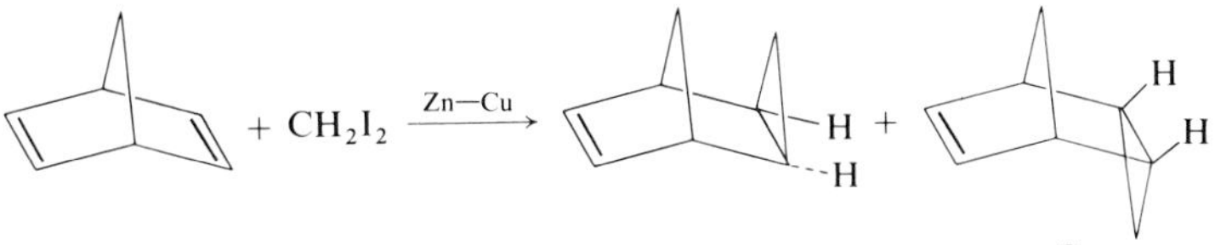

+ CH_2I_2 —Zn—Cu→ H H + H H (2)

2

It is not clear whether the formation of some **2** reflects a greater ease of *endo* attack in norbornadiene, as opposed to norbornene, or whether a prior complexing of the carbenoid intermediate to the underside of the diene is involved. A parallel observation of exclusive *exo* attack on norbornene but concomitant *exo* and *endo* attack on norbornadiene has been reported for carboethoxycarbene (see below, Section IV-B-1).

The general preference for *exo* attack of carbenes and carbenoids on bicyclic and tricyclic olefins is observed in the cyclopropanation of *endo*- and *exo*-dicyclopentadiene, **3** and **4** (4).

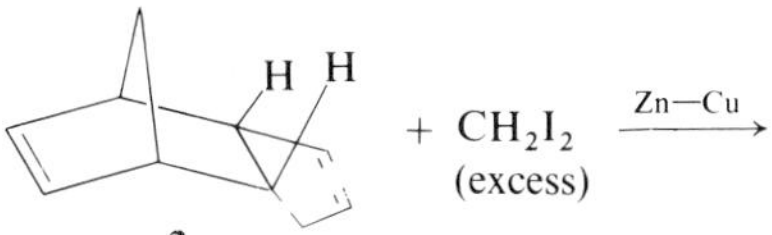

3 + CH_2I_2 (excess) —Zn—Cu→

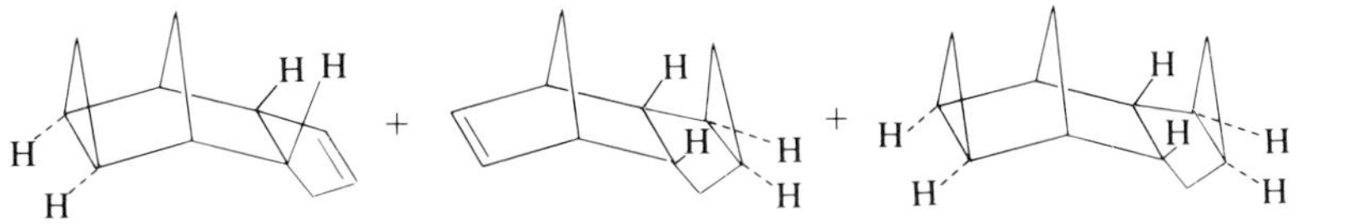

+ + (3)

4 + CH_2I_2 (excess) —Zn—Cu→

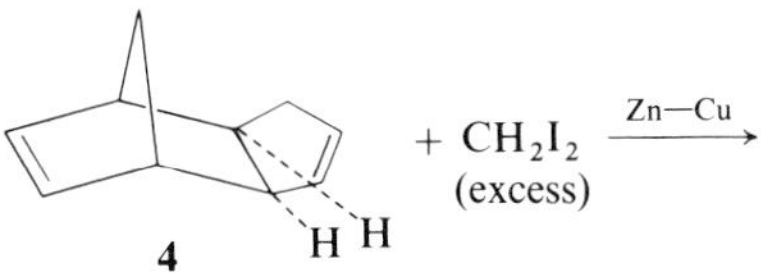

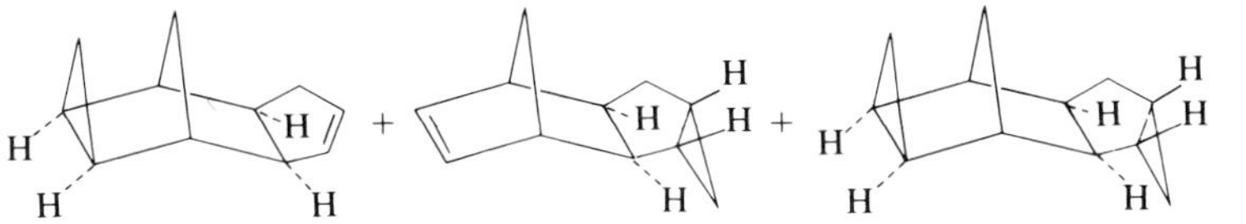

+ + (4)

A related observation is found in the cyclopropanation of hexamethyl Dewar benzene with diazomethane-cuprous bromide. This reaction, eq. 5a, leads to *exo* mono and diadducts (7a). Addition of carbomethoxycarbenoid

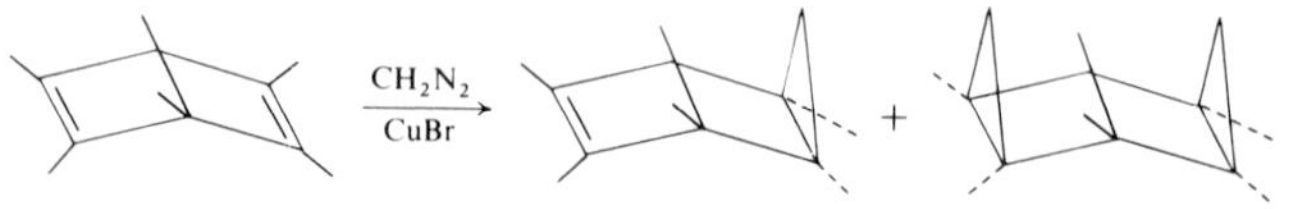

CH_2N_2 / CuBr → + (5a)

to this substrate follows a similar pattern, eq. 5b (7b). Related steric effects

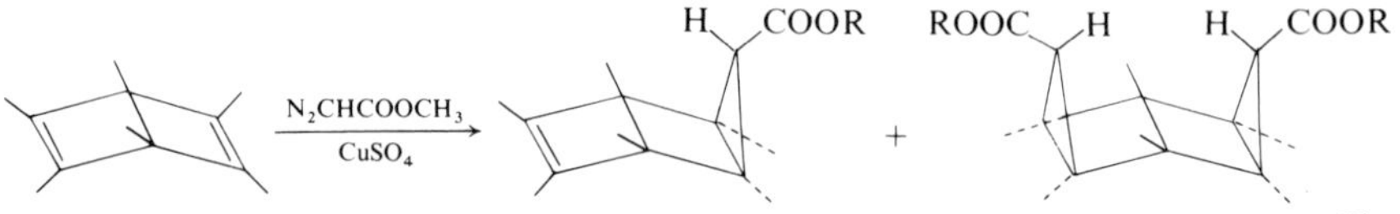

$N_2CHCOOCH_3$ / $CuSO_4$ → H COOR + ROOC H H COOR (5b)

operate in the additions of dibromocarbene (7c) and dichlorocarbene (7d) to Δ^3-carene, eq. 5c.

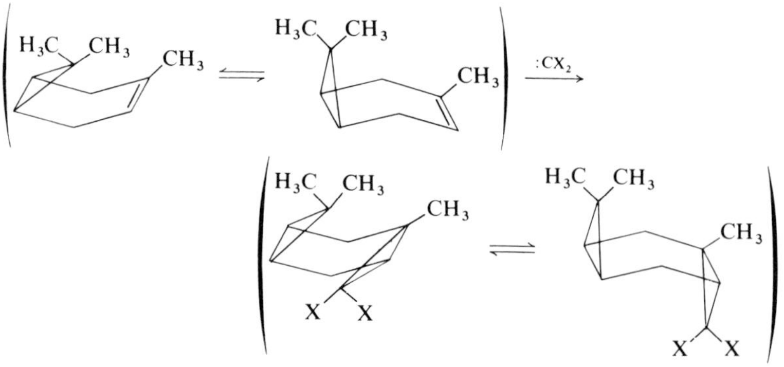

H_3C CH_3 CH_3 ⇌ H_3C CH_3 CH_3 —:CX_2→ H_3C CH_3 CH_3 X X ⇌ H_3C CH_3 CH_3 X X (5c)

Dichlorocarbene also adds to norbornene (8,9) and to 7-oxanorbornene (10) only from the *exo* direction. The addition of this carbene to *endo*-dicyclopentadiene, occurs almost exclusively (*exo*) at the cyclopentene double bond, eq. 6 (11). Adduct **5** was formed in at least 96% purity when

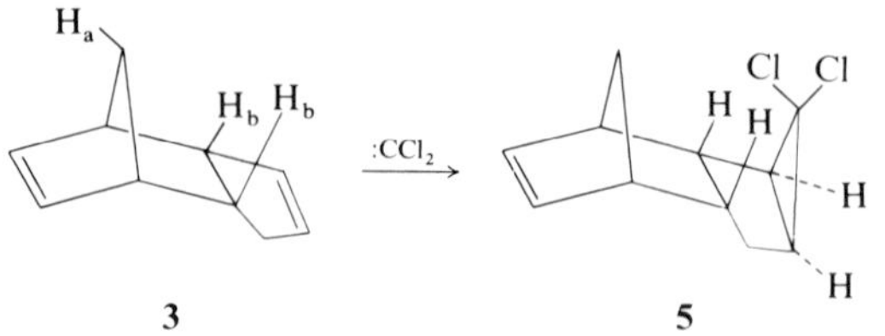

H_a H_b H_b **3** —:CCl_2→ Cl Cl H H H H **5** (6)

the dichlorocarbene was generated in a tenfold excess of diene **3**. Since it is known that CCl_2 can attack the norbornene double bond (8,9), it is clear that eq. 6 represents the outcome of a competitive process, in which there is demonstrated a much higher reactivity of the cyclopentene over the norbornene double bond of **3**, toward CCl_2. The origin of the reactivity difference is presumably the greater steric hindrance toward incoming CCl_2 provided by H_a, relative to H_b, in **3**.

The steric influence of the *syn*-7-proton of norbornene is clearly seen in the attempted addition of dibromocarbene, eq. 7 (8). The anticipated

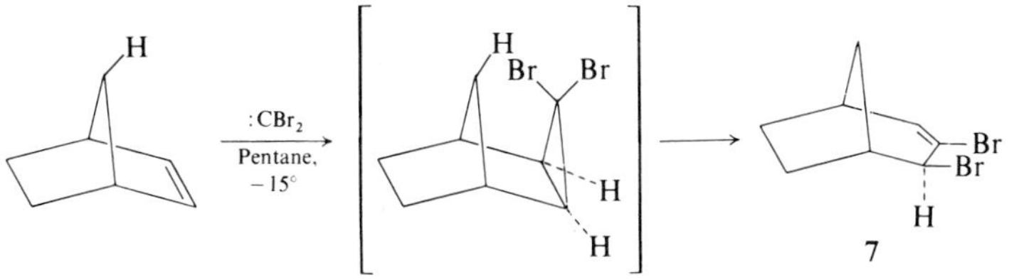

:CBr$_2$ / Pentane, −15° → [**6**] → **7** (7)

product (**6**) could not be isolated. Instead, bicyclooctene **7** was found, free of isomers. Product **6** is less stable than the corresponding dichloride, at least partly because of the increased strain due to the *syn*-7-proton–halogen interaction, which is more serious in the dibromide than in the dichloride. The failure to observe **7**-*endo*-Br, together with the absence of any *endo*-**6**, suggests that the attack of CBr_2 on norbornene was exclusively *exo* (8).

The addition of CBr_2 to *norbornadiene* may involve some *endo* attack. The instability of the cyclopropyl products clouds the results (8).

It has also been shown that addition of CCl_2 to the bicyclic system of the furanose derivative (**8**) is exclusively *exo* (12).

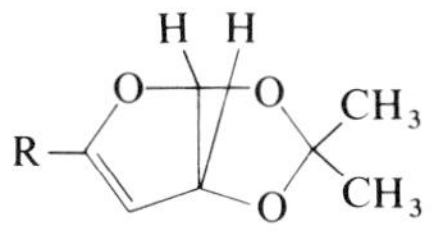

8

An interesting study of monohalocarbenoid additions to norbornene has been carried out by Jefford and co-workers. The addition of a monohalocarbene to norbornene could give four products, eq. 8. Despite an early confusion of product identities (13), it is now clear that both bromocarbenoids and chlorocarbenoids add to norbornene with formation only of the *exo* cyclopropanes, **9** and **10** (14). The *exo*-*syn*-halocyclopropane (**9**) was not, in fact, isolated. Rather (**13**) its rearrangement product was obtained. The differential

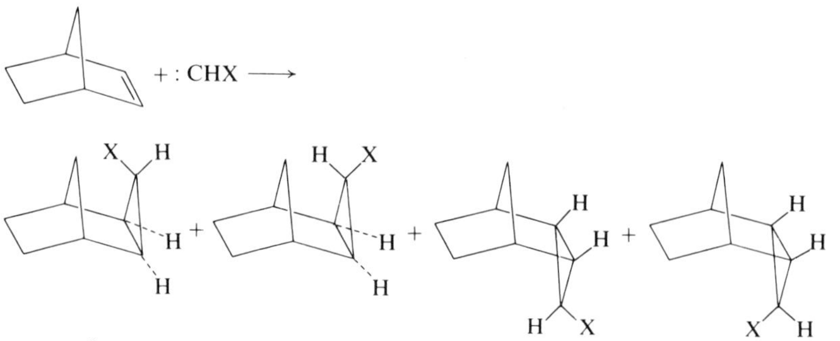

$+ :CHX \longrightarrow$

9 + 10 + 11 + 12 (8)

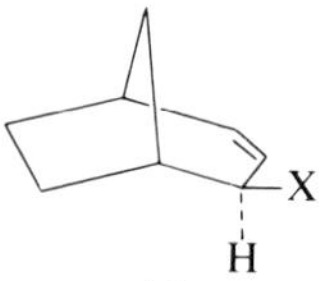

13

stability of **9** and **10** is understandable in terms of the nonbonded interaction of the *syn*-bridge proton and the halogen atom in **9**, and the application of the Woodward-Hoffmann rules to cyclopropyl systems of this type (15). Assuming that the observed yield of **13** accurately mirrored the initial yield of **9**, the **9**:**10** distribution was about 2:5 for both the chlorocarbenoid and bromocarbenoid (CH_2Cl_2 + CH_3Li + LiBr) reactions. The chlorocarbenoid generated with bromide-free methyllithium gave a **9**:**10** distribution of 1:6 (14). These observations of halocarbenoid additions with *anti*-halo stereoselectivity are of great interest, in view of the fact that halocarbenoid additions to such olefins as *cis*-butene or cyclopentene occur with *syn*-halo stereoselectivity; i.e., with formation mainly of the cyclopropane isomer having halogen *cis* to the larger number of alkyl groups (see below, Section IV-B-4). As will be discussed later (Section IV-A), the particular stereoselectivity observed during the addition of an unsymmetrically substituted carbene or carbenoid to a simple olefin is thought to reflect the balance of the attractive and repulsive nonbonded interactions of the carbenic and olefinic substituents in the addition transition state. In additions of the halocarbenoids to norbornene the dominant factor controlling stereochemistry is the steric hindrance afforded by the *syn*-7 proton, which, although not sufficient to force *endo* addition, clearly mitigates in favor of the *anti*-halo mode of *exo* addition.

Other bicyclic olefins, 1- and 2-methylnorbornene and bicyclo[2.2.2]octene, were also reacted with a chlorocarbenoid (16). In the bicyclooctene case symmetry removes the *exo–endo* problem, but the cyclopropanation was again found to exhibit *anti*-chloro stereoselectivity (*syn*-Cl:*anti*-Cl = 1:5). With the methylnorbornenes only *exo* additions were observed, but the *syn*:*anti* partition was sensitive to the site of the methyl group, being 1:1.5 for 1-methylnorbornene (**14**) and 1:9.8 for 2-methylnorbornene (**15**). These

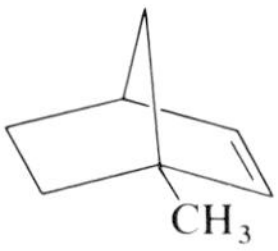

CH_3

14

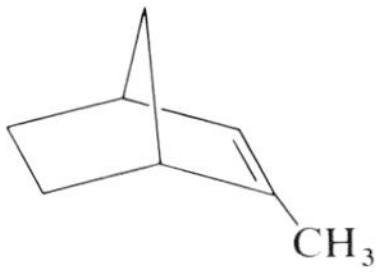

CH_3

15

results have been rationalized (16) with recourse to the arguments developed below (Section IV-B-4).

The exclusive *exo* addition of the halocarbenoids to the norbornenes is interesting, in view of the ready addition of chlorocarbenoid to bicyclo[2.2.2]octene, which serves as a model for the *endo* face of norbornene.

The stereochemistry of addition of fluorochlorocarbene to norbornene is understandable on the basis of our previous discussion of the monohalocarbenoid additions. As indicated in eq. 9, the addition is purely *exo*, and

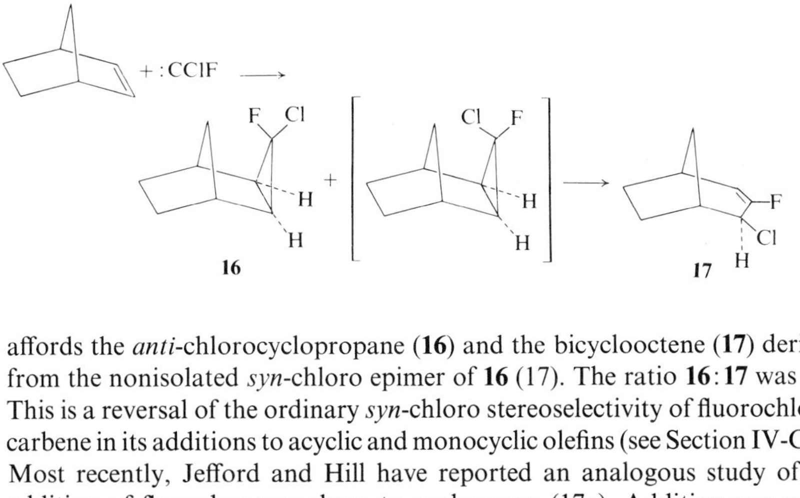

+ :CClF ⟶

F Cl, H, H — **16** + [Cl F, H, H] ⟶ F, Cl, H — **17** (9)

affords the *anti*-chlorocyclopropane (**16**) and the bicyclooctene (**17**) derived from the nonisolated *syn*-chloro epimer of **16** (17). The ratio **16**:**17** was 1.1. This is a reversal of the ordinary *syn*-chloro stereoselectivity of fluorochlorocarbene in its additions to acyclic and monocyclic olefins (see Section IV-C-1). Most recently, Jefford and Hill have reported an analogous study of the addition of fluorobromocarbene to norbornene (17a). Addition was again *exo*, but the bromo analog of **17** was formed in excess (1.3:1) of the bromo analog of **16**. The disparate stereoselectivity exhibited by CFCl and CFBr in *exo* addition to norbornene is unexplained, and offers a challenge to the simple steric argument which we have discussed. Further work on these and related rigid olefinic substrates is needed and promised (17a).

B. Steroidal Olefins

The steroidal olefins constitute another class of sterically demanding substrates. In the B ring, addition of CF_2 to the Δ^5 unsaturation in the androstene (**18**) and the pregnadiene (**19**) gave only the $5\beta,6\beta$ cyclopropanes, eq. 10 (18). The apparently exclusive β attack has been attributed to a stereoelectronic effect [see below (19)]. A clear steric effect is recognized in the

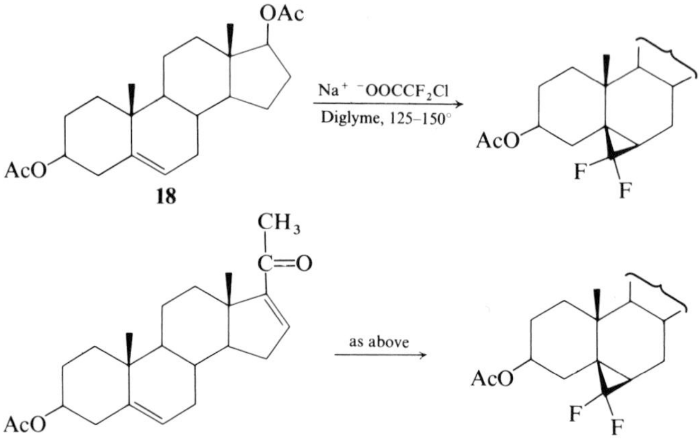

OAc; AcO; **18**; $Na^+\ {}^-OOCCF_2Cl$ / Diglyme, 125–150°; AcO; F; F

CH_3; C=O; AcO; **19**; as above; AcO; F; F (10)

failure of CCl_2 to attack **18** and **19**. The necessary β approach is possible only for CF_2; the β approach of the larger CCl_2 is too strongly hindered by the β methyl substituent at C_{10}. This argument gains weight when it is observed that the 19-*nor*-steroids (**20** and **21**), which have no methyl substituent at C_{10}, do add CCl_2 affording *5β,6β* cyclopropanes (18).

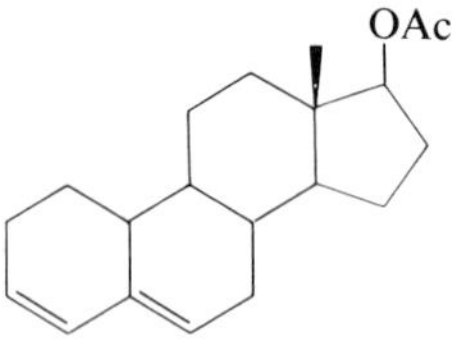

OAc; **20**; OAc; C≡CH; **21**

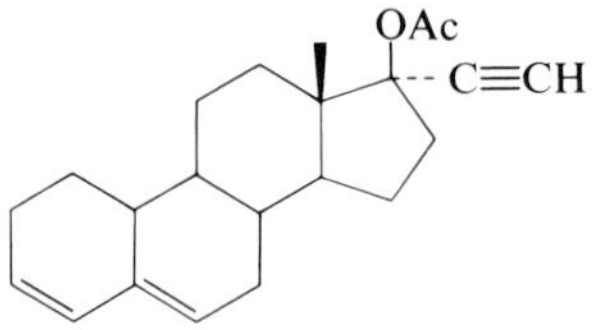

Further differential steric effects in CF_2 and CCl_2 additions appear in the attack of CCl_2 on $\Delta^{3,5}$ dienes (**22** and **23**) only at the Δ^3 position (which afforded $3\alpha,4\alpha$ adducts), while CF_2 was capable of attacking **22** at both Δ^3 (α attack) *and* Δ^5 (β attack) positions.

In a summary of the above results the Δ^5 steroid system is attacked from the β direction by dihalocarbenes. This attack is opposed by the steric influence of a β substituent at C_{10}. As exceptions, it has been reported that

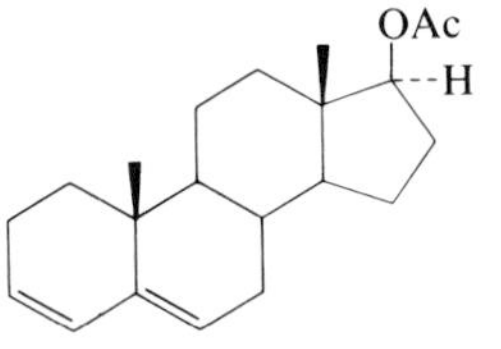

22

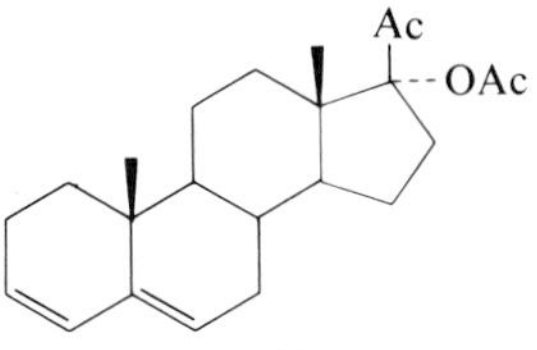

23

17β-acetoxy-3-ethoxy-$\Delta^{3,5}$-estradiene (a 19-norsteroid) added CBr_2 at Δ^5 to give *both* α and β cyclopropanes (20). Also, B-norcholesterol and its acetate (**24**) are cyclopropanated by the Simmons-Smith reagent almost exclusively from the α side (21).

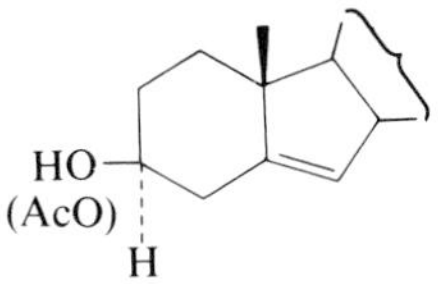

24

It is interesting to note that the 3β, homoallylic, hydroxy group of B-norchlolesterol fails to direct Simmons-Smith methylenation from the β direction, as might have been expected from studies of the methylenation of monocyclic homoallylic ene-ols (Section II-C). The demands of the *steroid structure itself* are paramount in **24** (21, Joska et al.).

Nazer has shown that the additions of CCl_2 and CBr_2 to ergosteryl tetrahydropyranyl ether (**25**) afford 5α,6α cyclopropanes (19). Thus carbene

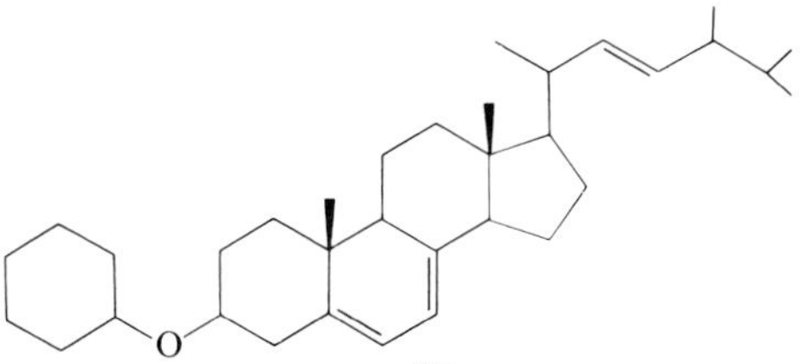

25

additions to Δ^5 or $\Delta^{3,5}$ steroids usually lead to 5β,6β cyclopropanes, while carbene addition to the $\Delta^{5,7}$ steroid, **25**, leads to a 5α,6α derivative. This has been rationalized as a stereoelectronic effect (19). Assuming that the transition state for carbene addition involves substantial positive charge, unsymmetrically placed on the olefinic carbons, transition state bonding of the carbene to the Δ^5 or $\Delta^{3,5}$ steroid will be most advanced at C_6, in order to take advantage of the tertiary or allylic character of C_5. Moreover, in order to have maximum orbital overlap in that transition state, the electrophile will attack C_6 from the β (axial direction). On the other hand, in order to utilize the allylic character of C_6, a $\Delta^{5,7}$ diene will be bonded to an attacking carbene (in the transition state) mainly at C_5, and, for maximum orbital overlap, this attack must come from the α (axial) direction (18,19). Nazer's conclusions (19) receive support from the elegant studies of Bond and Cornelia (19a). These workers showed that the additions of CF_2, CCl_2, and CBr_2 to the $\Delta^{5,7}$ diene, 7-dehydrocholesteryl benzoate, were exclusively 5α, 6α. Moreover, although they confirmed the inertness of the Δ^5 double bond of β C_{19} steroids to CCl_2 (18), it was found that CCl_2 would add to a tetrasubstituted (6-methyl substituent) Δ^5 steroid, *and from the α direction.* In this case, it was suggested (19a) that attack may occur α at C_5 since in the transition state, partial positive character can now reside on a *tertiary* C_6.

There have been several reports of carbene additions to Δ^6 steroids. The addition of difluorocarbene to dienones such as **26** can lead to both α (22) and β (23) 6,7-cyclopropanation. Additions of CCl_2 to Δ^6-, and of both CF_2 and CCl_2 to Δ^7-cholestene benzoates have been reported to lead to α cyclopropanations. Stereoelectronic explanations were offered (19a).

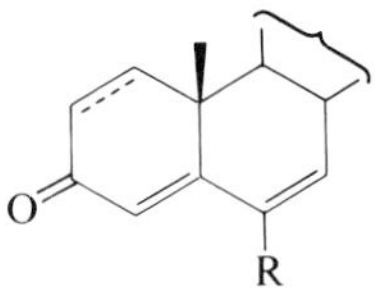

O R

(R = H, CH_3, or halogen)

26

Addition reactions have also been reported for ring A unsaturated steroids. Additions of CF_2 at Δ^1 and Δ^4 positions (24), and of CBr_2 at the Δ^2 and Δ^3 positions (19), all led to α cyclopropanation. Moreover, CCl_2 addition to the $\Delta^{3,5}$ dienes (**22** and **23**) occurred in the α sense at Δ^3, as did the addition of CF_2 (see above) (18). It is tempting to ascribe these α additions to steric hindrance toward β addition originating at the β methyl substituent at C_{10}. However, it is to be noted that the foregoing substrates were all conjugated ene-ones or dienes, and stereoelectronic effects may also have been involved in some cases.

Additions of CBr_2 (25) and the carboethoxycarbenoid from diazoacetic ester and copper sulfate (26) to monounsaturated Δ^2 substrates also afforded only α cyclopropanated products. The addition of CF_2 to steroid **27**, however, gave both 2α,3α and 2β,3β cyclopropanes, whereas the addition of the larger CCl_2 led only to the 2α,3α derivative (18). The studies of the Δ^2 steroids would again seem to implicate the β methyl substituent at C_{10} as a principal seat of steric hindrance during the carbene additions. With the removal of this group steric control of carbene additions at the Δ^2 position becomes less

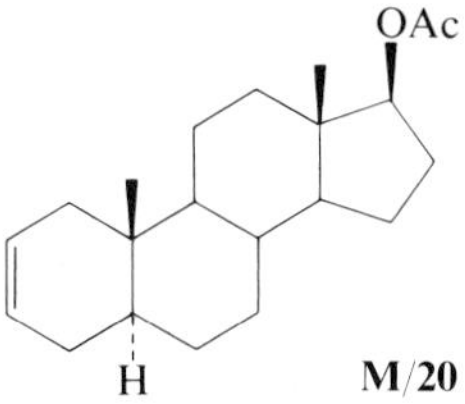

M/20

27

complete. Thus addition of the Simmons-Smith carbenoid to **28** gave epimeric 2,3-cyclopropanes (26), as did the addition of CBr_2 to **29** (27).

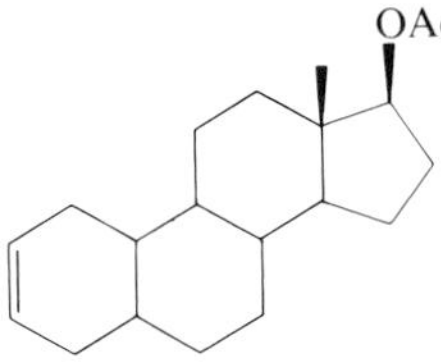

28

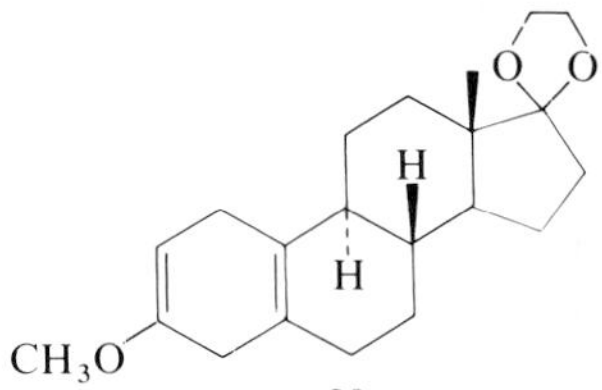

29

Additions of CBr_2 to 5,10 double bonds, as in the normal estrone derivative (**29**) occurred mainly from the β direction, but α addition was shown to compete successfully in the 9β-estrone derivative **30** (28a). α-Addition of "CH_2" to the 5,10 double bond of 3α-hydroxy steroids has been achieved, making use of the *cis* directive ability of the hydroxy function on the Simmons-Smith reagent (28b), as discussed in Section II-C below.

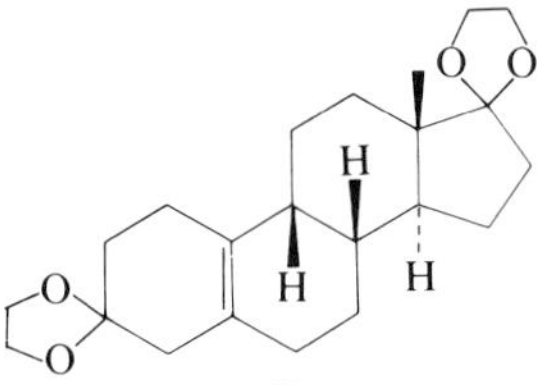

30

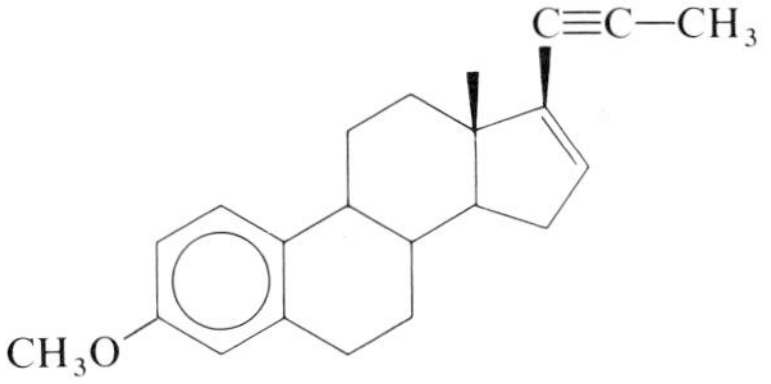

31

Attack of CF_2 on the unsaturated D ring of **31** led to α cyclopropanation (along with products of the carbene attack on the acetylenic linkage) (29).

C. Miscellaneous Olefins

There are also some miscellaneous substrates in which significant stereochemical selectivity has been observed upon carbene addition. It is well known, for example, that the addition of the Simmons-Smith reagent to allylic or homoallylic alcohols involves a prior complexation of the carbenoid and the hydroxyl oxygen, such that the stereochemistry of cyclopropanation will, generally, be *cis* relative to the hydroxyl group.* Cyclopropanation of cyclopent-3-ene-1-ol serves as one example, eq. 11 (4,30). Analogous

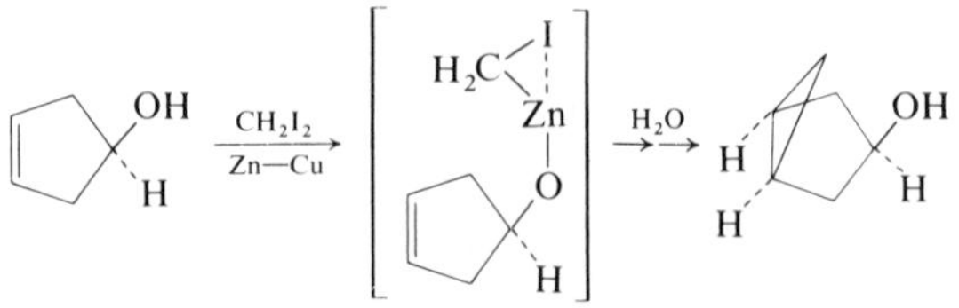

$$\xrightarrow[\text{Zn—Cu}]{CH_2I_2} \left[H_2C \cdots I \cdots Zn - O \right] \xrightarrow{H_2O} \tag{11}$$

reactions have been reported (31), and a detailed study of the stereochemistry of Simmons-Smith methylenation of cyclic allylic alcohols has been carried out (32). There it was discovered (32) that a methoxyl substituent can function as well as a hydroxyl substituent as the directing group. Acetoxy substituents fail to exhibit the *cis* directive ability. In fact, the Simmons-Smith cyclopropanation of a cyclic, allylic acetate has been shown to occur *trans* to the acetoxy group (33a). Not only did the acetoxy group fail to direct *cis*, allylic cyclopropanation by effective prior complexation of the carbenoid, but it probably resulted in steric hindrance to *cis* addition. More recently it has been reported that the carbomethoxy substituent exerts a *cis* directive influence on Simmons-Smith cyclopropanation, similar to, but not so strong as, the influence of the hydroxy group (33b). There has also appeared a detailed study of the dependence of the rate and stereochemistry of Simmons-Smith addition to various cyclohexenols, as a function of the spatial disposition of the functional groups (33c).

Tris methylenation of *cis-cis-cis*-1,4,7-cyclononatriene (**32**) with the Simmons-Smith reagent led to a major product (80–90%) in which the three cyclopropyl units were mutually *cis* (34). Inspection of models of the proposed, rather rigid, "crown" conformation of the triene (35) suggests that the initial methylenation should occur from the more accessible "top-side"

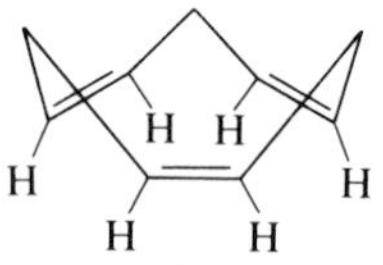

32

of the crown. The resultant diene-cyclopropane will be rather similar to **32** in conformation and steric accessibility of the remaining double bonds, and so the second, and then the third methylenation will also occur from the "top-side," leading to an all-*cis* product.

Additions of carbenic reagents to *cis,trans,trans*-1,5,9-cyclododecatriene (**33**) follow an interesting pattern. With CCl_2 a *trans* double bond is preferentially attacked (36), but the degree of preference, 91:9 (37), is much greater

* See however, C. D. Poulper, E. C. Friedrich and F. Winstein, *J. Am. Chem. Soc*, **91**, 6892 (1969).

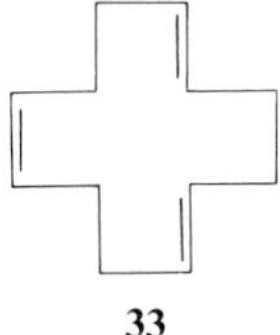

33

than the statistically expected 2:1. Addition of CBr_2 to **33** also leads to initial attack at a *trans* double bond [preference 89:11 (37, Nozaki et al.) or 72–75:28–26 (37a)] and addition of the Simmons-Smith carbenoid leads to an even greater selectivity, 97:3 (37, Nozaki et al.; 38). Addition of a second mole of a carbenic reagent to the *trans*-monocyclopropane derived from **33** involves preferential attack at the *cis* double bond. Preferences for the formation of *cis–trans* rather than *trans–trans* dicyclopropyl derivatives are reported to be 84:16 for CBr_2, 87:13 for CCl_2, and 81:19 for the Simmons-Smith carbenoid (37, Nozaki et al.). Attempts to rationalize these effects have been made (39), but the author believes that a complete explanation has not yet appeared.

1-*cis*-5-*trans*-Cyclododecadiene behaves similarly as a substrate for CCl_2 and CBr_2 (39a). Again, initial carbene attack occurs mainly at the *trans* double bond. The *trans*/*cis* preferences are ca. 65:35 (CCl_2) and 56:44 (CBr_2).

III. STERIC EFFECTS IN CARBENE ADDITIONS TO ACYCLIC OR MONOCYCLIC OLEFINS

Steric effects operating during carbenic additions to acyclic or monocyclic olefins can be detected by examination of the relative rates of addition of the carbene to a *series* of olefins. Several rather clear demonstrations of steric effects in CCl_2 additions have been reported. Nefedov and co-workers list the indicated relative addition rates for the reaction of CCl_2 with various l-substituted cyclohexenes (k_{rel} for cyclohexene = 1.00) (40). The results

	CH_3			
k_{rel}	8.43	1.14	6.19	0.40

appear to correlate with expectations based on the relative ease of approach of CCl_2 to the double bond. Another study considered the additions of CBr_2 and CCl_2 to a series of l-alkyl-1-methylallenes. Relative rates for some of the additions, which occur at the more highly alkylated double bond, eq. 12, are shown in Table 1 (41).

$$(H_3C)(R)C{=}C{=}CH_2 + :CX_2 \longrightarrow \text{1-}H_3C\text{-1-}R\text{-2,2-}X_2\text{-3-methylenecyclopropane} \quad (12)$$

TABLE 1

Relative Rates of Addition of CX_2 to $(R)(CH_3)C{=}C{=}CH_2$[a]

R	CCl_2	CBr_2
CH_3	1.00	1.00
CH_3CH_2	0.64	0.56
$(CH_3)_2CH$	0.11	0.21
$(CH_3)_3C$	0.00	0.00
$(CH_3)_3CCH_2$	0.28	0.37

[a] Data of ref. 41.

The trends for both CCl_2 and CBr_2 additions are clear and readily explained by the steric hindrance to carbene attack provided by R. However, the larger CBr_2 might have been expected to be more discriminating, in a steric sense, than CCl_2, and this is not reflected in the data. The expected differential steric demand of these carbenes was observed, however, in a study of the competition of styrene and vinylmesitylene for CCl_2 or CBr_2 (41a). Although CCl_2 added to both olefins at about the same rate, CBr_2 added to vinylmesitylene 3.3 times more slowly than to styrene. Rationalization (41a) was based on differential steric hindrance, provided by the 2- and 6-methyl substituents of the mesitylene, which more seriously affected the addition of the bulkier CBr_2 (41a).

Similar conclusions about the importance of steric effects in CCl_2 additions to alkenes have been drawn by Seyferth and co-workers (42) and also by Moss and Mamantov, who studied CCl_2 additions to a series of 1-alkylethylenes, Table 2 (43a).

The first three olefins in Table 2 form a series corresponding to successive α-methylation of the olefinic alkyl substituent. The relative CCl_2 addition rates along this series are correlated by the Taft steric substituent constants, $\log(k/k_0) = \delta E_s$, with $\delta \sim 1.0$. In contrast, along the series of the

TABLE 2

Relative Rates of Addition of CCl_2 to $RCH{=}CH_2$[a]

R	Relative rate
CH_3CH_2	1.00
$(CH_3)_2CH$	0.43
$(CH_3)_3C$	0.029
$CH_3CH_2CH_2CH_2$	1.24
$(CH_3)_2CHCH_2$	0.99
$(CH_3)_3CCH_2$	0.97

[a] Data of ref. 43a.

last three olefins corresponding to successive β methylations of the olefinic alkyl substituent essentially no steric effects were observed.

Similar conclusions have been reached by Seyferth and Dertouzos, who found that neopentylethylene was 0.78 times as reactive as 1-heptene toward CCl_2 (at 80°), whereas the *t*-butyl-type olefin, 3,3-dimethylpentene-1, was only 0.043 times as reactive. Interestingly, CF_2, with its smaller steric demand, was found to add relative to heptene-1 in each case, about three times faster than CCl_2 to the *t*-butyl-type olefin (43b).

A study of the additions of *p*-tolylcarbenoid (from alkyl-lithiums and *p*-methylbenzal bromide or iodide) to ethylethylene, *i*-propylethylene, and *t*-butylethylene, also revealed the rate-retarding influence of substrate α-methylation in carbenic addition reactions (see Section IV-B-2 and ref. 44).

Steric hindrance toward carbene addition can also be inferred from studies of the selectivity of a carbene toward a series of olefins, each of which differs in degree or pattern of alkylation. It is necessary, however, to know what kind of selectivity is to be expected for a given carbene in the absence of dominant differential steric effects along such an olefin series. Doering (3) and Skell (2b) have demonstrated the electrophilic nature of the attack of CCl_2 and CBr_2 on olefins, which was pictured as occurring via the charge separated transition state (**34**).

$$\underset{\mathbf{34}}{\left[\;\mathrm{C{-}C^{+}}\cdots\mathrm{{}^{-}CX_2}\;\longleftrightarrow\;\mathrm{{}^{+}C{-}C}\cdots\mathrm{{}^{-}CX_2}\;\longleftrightarrow\;\overset{\delta^+}{\mathrm{C{=}C}}\cdots\overset{\delta^-}{\mathrm{CX_2}}\;\right]}$$

In **34** the olefinic carbon atoms have become positive, relative to the ground-state olefin. The substitution of alkyl groups for protons on these carbons will, by electron-releasing induction and/or hyperconjugation,

Table 3
Relative Rates of Addition of Some Dihalocarbenes to Olefins

Olefin	Relative rates		
	CFCl[a]	CCl_2[b]	CBr_2[c]
$(CH_3)_2C{=}C(CH_3)_2$	31	6.5	3.5
$(CH_3)_2C{=}C(CH_3)H$	6.5	2.8	3.2
$(CH_3)_2C{=}CH_2$	1.00	1.00	1.00
cis-$CH_3CH{=}CHCH_3$	0.14	0.23	—
trans-$CH_3CH{=}CHCH_3$	0.097	0.15	—
$C_2H_5(H)C{=}CH_2$	0.0087	0.011	0.07[d]

[a] Data of ref. 45, −10°.
[b] Data of refs. 3 and 45, −10° to −20°.
[c] Data of ref. 2b, 0–3°.
[d] 1-Hexene.

stabilize the transition state. If transition state **34** has substantial charge-separated character, then the energy-lowering effect of successive alkylations will be more important in this transition state than in the ground state olefin; and the activation energy for carbene addition to the more heavily alkylated olefins will be smaller than that for carbene addition to the less alkylated olefins. In Table 3 relative addition rate data are collected for CFCl, CCl_2, and CBr_2, which demonstrate the electrophilic nature of these carbenes' attacks on olefins. CFCl is one of the most selective carbenes studied in the olefin addition reaction (45).

Selectivity trends such as those exhibited by CFCl, CCl_2, and CBr_2 are readily explained by the simple electronic argument just outlined. The sensitivity of CFCl and CCl_2 to the degree of olefinic alkylation is very apparent. The maximum rate ratios are 3560 for CFCl and 590 for CCl_2. The greater selectivity of CFCl has been explained by noting that it should be a more stable carbene than CCl_2. The dihalocarbene ground state may be represented by resonance hybrid (**35**) (46).

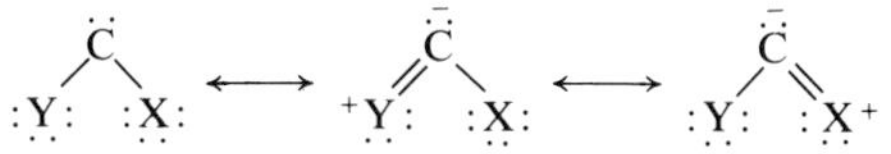

$$\ddot{C}(:\ddot{Y}:)(:\ddot{X}:) \longleftrightarrow \bar{\ddot{C}}(={}^{+}\ddot{Y}:)(:\ddot{X}:) \longleftrightarrow \bar{\ddot{C}}(:\ddot{Y}:)(=:\ddot{X}^{+})$$

35

The significance of the resonance interactions, hence the stability of the carbene, is predicted to increase in the order F ≫ Cl > Br > I. The resonance interactions of the carbene center and its halogen substituents are lost in **34**, and if the nature of the halogens is more important in determining the free energy of the carbene itself than that of transition state (**34**), the differential activation energy for the selective additions of any two dihalocarbenes to a given pair of olefins will mainly reflect the relative "internal stabilization" (3) of the carbenes.

It is clear, however, from the CBr_2 data of Table 3, that the differing steric demands of various carbenes can seriously undermine the simple arguments just presented. Note that there is essentially no rate enhancement when tetramethylethylene is substituted for trimethylethylene in the CBr_2 addition reaction. This has been rationalized (3) as being due to greater steric hindrance toward addition to the more heavily substituted olefin. With the smaller carbenes, CFCl and CCl_2, steric discrimination between olefins is not readily apparent (though it no doubt contributes) and, at least superficially, electronic arguments can account for the data (however, see below).

From the foregoing considerations, we can suggest several circumstances in which we are likely to see steric discrimination in carbene–olefin additions. If the carbene is very reactive, the transition state for its addition to an olefin will be reactantlike. Such a transition state will be relatively loose, and, also, will involve little charge separation. In this case both electronic and steric discrimination on the part of the carbene will be slight and neither will strongly dominate. If the carbene is of moderate reactivity, but carries bulky substituents, we might expect to observe typical electrophilic selectivity toward the lesser substituted olefins, but increasing steric hindrance toward addition to tetrasubstituted ethylenes. If the carbene is highly stabilized but carries small substituents, electrophilic selectivity may not be obscured in addition reactions to highly alkylated olefins. But, as the substrates are made less reactive (less alkylated) and the addition transition state more productlike ("tighter"), steric hindrance to addition will become more important and, perhaps, observable.

On the basis of its behavior toward the C—H bond, cyclopentadienylidene (**36**) is a very reactive carbene (47). Relative rates of addition of **36** to olefins appear in Table 4.

36

TABLE 4
Relative Rates of Addition of Cyclopentadienylidene to Olefins[a]

Olefin	Relative rate
Tetramethylethylene	0.74
Trimethylethylene	0.75
Cyclohexene	1.00
1-Hexene	0.94
Styrene	0.94
t-Butylethylene	0.70

[a] Data of ref. 47.

Selectivity of **36** is minimal. A mild steric selectivity is reflected in the lower rates of addition to the tetra- and trisubstituted olefins, as well as in the low rate of addition to *t*-butylethylene. Comparison of the selectivity data for **36** with that for CFCl (Table 3) shows just how dramatic the electrophilic discrimination of carbenes can be, when it is operative. Steric effects which modify the addition reactions of triphenyl- (47a) and tetraphenylcyclopentadienylidene (47b) have also been reported.

Related to **36** is the cyclohexadienylidene, **37**. Somewhat more selective than **36**, **37** also shows the effects of steric hindrance in its addition to heavily substituted olefins, adding to tetramethylethylene and trimethylethylene at nearly the same rate (48). Also similar to **36** in its selectivity is dicarbo-

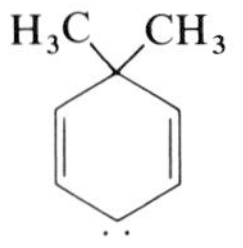

37

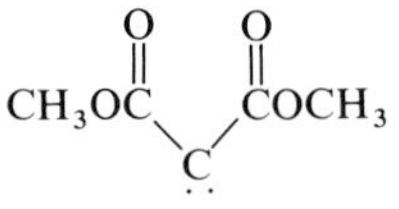

38

methoxycarbene (**38**), which adds to tetramethylethylene at about 0.9 times the rate of its addition to trimethylethylene (49).

An interesting pattern of steric discrimination is displayed by 2,2-diphenylcyclopropylidene (**39**), Table 5 (50). Geometry **40** (from ref. 50) was

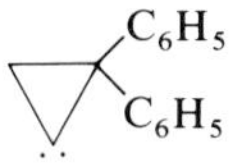

39

TABLE 5
Relative Rates of Addition of 2,2-Diphenylcyclopropylidene to Olefins[a]

Olefin	Relative rate
Tetramethylethylene	0.41
Isobutene	1.00
cis-Butene	1.15
Cyclohexene	1.23
trans-Butene	0.42
1-Butene	0.22

[a] Data of ref. 50.

drawn to account for the slow rate of addition of **39** to tetramethylethylene, relative to isobutene, and for the rapid rates of addition to *cis*-butene and cyclohexene.

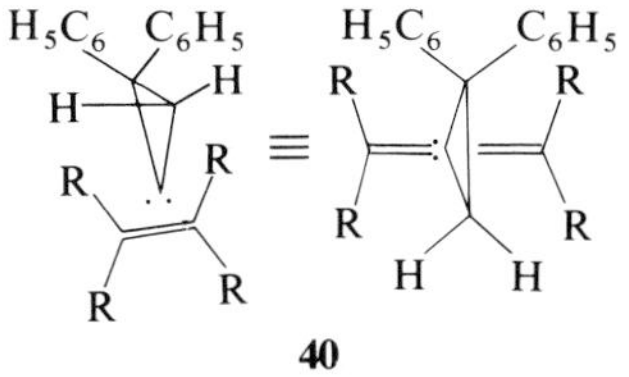

40

In **40**, both phenyl groups are opposed by methyl groups during addition to tetramethylethylene. During additions to *cis*-butene or cyclohexene both phenyl groups could be opposed by protons, whereas, during addition to isobutene only one phenyl group would have to oppose an olefinic methyl group. It may be noted that the proposed geometry of carbene–olefin interaction (**40**) is very productlike (i.e., has the geometry of the product spiropentanes). Examination of models suggests that the data (Table 5) can also be rationalized with geometries based upon more reactantlike transition states, analogous to those discussed below (Section IV-A).

There are many examples of carbenic species that are of moderate reactivity because of substituent interactions with the carbenic center or of carbenoid complexation of the carbenic center to a metal or metal salt. These species often show moderate electrophilic selectivity toward the lesser alkylated olefins, but a dominant steric selectivity toward the more heavily alkylated olefins. This phenomenon has already been illustrated for CBr_2 (see above). Similar behavior has been observed with the phenylbromocarbene species generated from benzal bromide and potassium *t*-butoxide (51), which, although fairly selective toward lesser substituted olefins, adds to tetramethylethylene only 1.3 times faster than to isobutene. This species is probably a weak complex of phenylbromocarbene and potassium *t*-butoxide, since the photolysis of phenylbromodiazirine yields an intermediate with a differing olefin selectivity (52). This latter species is probably a free phenylbromocarbene. Interestingly, although it is more selective than the base-generated species, steric effects seem to be mitigated. It adds to tetramethylethylene 4.4 times faster than to isobutene, and 1.7 times faster than to trimethylethylene. Presumably, though more charge separated, the addition transition states of the photolytic species are somewhat more reactantlike or "looser" than are those of the base-generated carbenoid, and steric effects are more moderate.

The zinc carbenoid intermediate in the Simmons-Smith reaction is a bulky species which exhibits considerable steric discrimination in its addition to bicyclic olefins (Section II-A). Steric discrimination is also apparent in its additions to simple olefins, as is shown by the data in Table 6 (53).

TABLE 6
Relative Addition Rates of the Simmons-Smith Reagent to Olefins[a]

Olefin	Relative rate
Tetramethylethylene	1.29
Trimethylethylene	2.18
α-Methyl-α-ethylethylene	2.53
Cyclohexene	1.00
cis-3-Hexene	0.83
trans-3-Hexene	0.42
1-Hexene	0.36
t-Butylethylene	0.14

[a] Data of ref. 53.

Steric discrimination has also been observed in the addition reactions of phenoxycarbenoid (ϕOCHLiHal), which added to tetramethylethylene more slowly than to other, lesser alkylated olefins (54). Steric effects have been held accountable for the failures of the carbenoid (**41**) to add to tetramethylethylene (55), and of the carbene (**42**) to add to *cis*-stilbene, triphenylethylene, and tetraphenylethylene (56). Both species add readily to less hindered olefins.

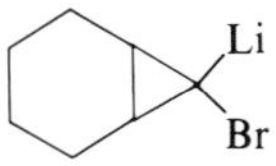

Li Br

41

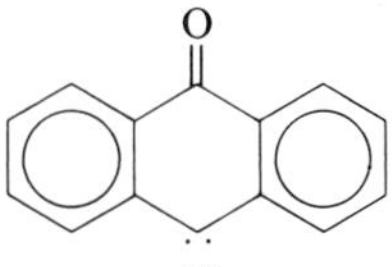

O

42

Examples of highly selective (and presumably highly stabilized) carbenes are CFCl and CCl_2. Examination of Table 3 shows that the selectivity of CFCl, relative to that of CCl_2, falls off as progressively less reactive olefins are used as substrates. A detailed discussion is given in reference 45, in which it is shown that CFCl is less discriminating than CCl_2 toward mono and dialkylated olefins. This behavior was attributed to increasingly dominant steric discrimination by the larger CCl_2 in the tighter transition states for the additions to the less reactive mono and dialkylated olefins. The strong steric selectivity exhibited by CCl_2 in its addition to monosubstituted olefins has been separately demonstrated (see above, Table 2).

IV. STEREOSELECTIVITY OF UNSYMMETRICALLY SUBSTITUTED CARBENES AND CARBENOIDS

A. Qualitative Models

The addition of an unsymmetrically substituted carbene or carbenoid to an olefin which lacks both a center of symmetry, and a twofold rotational axis coincident with the double bond, will afford isomeric cyclopropane products. A general case is illustrated in eq. 13.

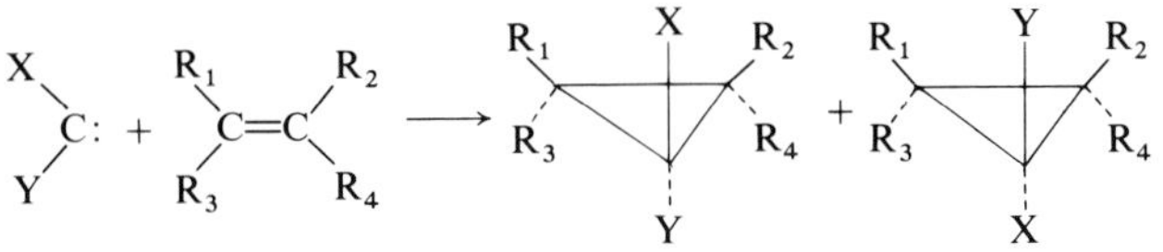

$$XYC: + R_1R_3C{=}CR_2R_4 \longrightarrow \text{(cyclopropane: } R_1, R_2, X \text{ syn; } R_3, R_4, Y\text{)} + \text{(cyclopropane: } R_1, R_2, Y \text{ syn; } R_3, R_4, X\text{)} \tag{13}$$

When the products are stable to the reaction conditions, their distribution is kinetically controlled, and reflects the relative energies of the alternative transition states involved in the addition reaction. When the product distribution is appreciably different from 1.0, the carbene is said to display stereoselectivity in its addition reaction.

In the following, unless otherwise noted, we discuss only carbenes that are known to be highly *stereospecific* in their additions to such olefins as *cis* and *trans*-butene. (That is, in these additions, the geometrical relationships of the substituents of the starting olefins are preserved in the product cyclopropanes.) Such carbenes are presumably of singlet multiplicity in the reactive state under consideration (57). If the carbene depicted in eq. 13 were of triplet multiplicity, intermediates such as **43** would intervene in its addition reaction. Not only will rotation A result in the well-known phenomenon of

43

nonstereospecificity in triplet carbene addition reactions, but rotation B will contribute to the distribution of the two "stereospecific" cyclopropanes formed in eq. 13. The factors which determine product distribution in two-step carbene additions proceeding via diradical intermediates such as **43** would, at least in part, be distinct from those operative in the presumably one-step, stereospecific, singlet carbene additions.

Several authors have discussed possible transition states for the singlet carbene olefin addition reaction (2,3,8,58,59). The various suggestions were similar, but a detailed, generally accepted picture has yet to emerge. A frequently cited model suggests that, in the transition state, the carbene lies over the olefin, in a plane roughly parallel to the plane defined by the olefinic carbons and their α substituents. In this arrangement, there is overlap between the carbene's vacant p orbital and the filled olefinic π orbital, that is, electrophilic attack of the carbene on the olefin. A side-view representation of this transition state is given in **44a** (from ref. 8).

It should be noted that **44a** is a reactantlike transition state. Structure **44b** represents a somewhat more productlike transition state, in which the

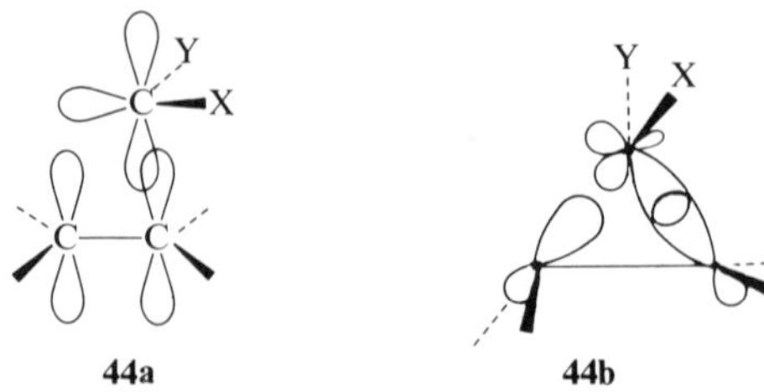

44a **44b**

geometry has begun to approach that of the product cyclopropane. The more stabilized the carbene, the more selective it will be in its olefin addition reactions, and the more the transition states for those reactions will resemble **44b**. Furthermore, as will be seen, there is a general parallel of high selectivity of a carbene in its additions to members of a set of olefins and appreciable stereoselectivity in its addition to any given olefin. The factors which determine stereoselectivity will therefore be discussed in terms of model **44b**.

Top views of the two transition states which lead to the two products of eq. 13 are shown in **45a** and **45b**. In **45a** there may be strong nonbonded

45a **45b**

interactions between the groups R_1, R_2, and X and R_3, R_4, and Y. In alternative **45b** the interactions would be between R_1, R_2, and Y and R_3, R_4, and X. It was early assumed that all such interactions were repulsive (steric hindrance), and that the product distribution in eq. 13 could be predicted merely by comparing the number and relative strengths of these repulsive interactions in each alternative transition state. However, with the failure of this approach for chlorocarbenoid addition reactions (60,61) it became necessary to allow for attractive nonbonded interactions in these transition states (58,61).

Consider, for example, the addition of a monosubstituted carbene to *cis*-butene, eq. 14. Transition states **46a** and **46b** may be drawn for this

(14)

Syn *Anti*

reaction. If only repulsive nonbonded interactions of carbene and olefin substituents had to be considered, then *anti* cyclopropane, corresponding to **46b**, would be observed as the major reaction product. The experimental

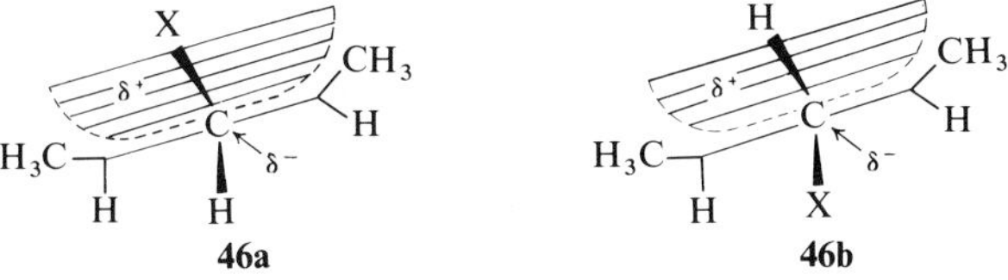

46a **46b**

facts in many cases are quite the reverse, and *syn* cyclopropane, corresponding to transition state **46a**, is found to be the major reaction product. In such cases it has been concluded that the sum of all nonbonded interactions in **46a**, relative to **46b**, is attractive and energy-lowering, rather than repulsive and energy-increasing. The following rationale has been suggested (58,61,62).

The attack of the carbene on the olefin is electrophilic. In the transition state, the olefinic carbons have developed some positive charge. Their alkyl substituents donate electron density inductively and/or hyperconjugatively toward the olefinic carbon atoms, becoming somewhat positive themselves. In **46a**, where carbene substituent X lies over the region of induced positive charge, we may well expect strong Coulombic interactions between carbenic and olefinic substituents. These will be especially important when X carries a charged center or a permanent dipole. In **46b** such interactions will be minimal.

In theory, it should be possible to construct a curve representing the contribution of the nonbonded and Coulombic interactions in **46a** to the total energy of **46a**, as a function of a "reaction coordinate" for the addition reaction. Such a curve might have the general shape of curve *A* in Figure 1.

Toward the reactantlike end of the "reaction coordinate," Coulombic interactions between the carbenic and olefinic substituents could be attractive, and the curve *A* would describe a shallow minimum. At the productlike end

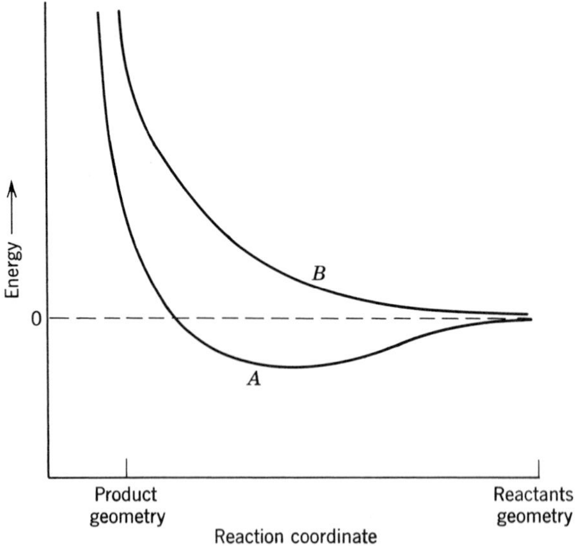

Figure 1. Energy of nonbonded interactions between carbenic and olefinic substituents as a function of a "reaction coordinate" for the carbene-olefin addition reaction.

of the "reaction coordinate" the curve would rise steeply. Here the interactions would be repulsive, and of the type generally described as "steric hindrance." On the other hand, depending on the nature of X (e.g., $X = -\overset{+}{N}R_3$), the curve could be repulsive and everywhere positive, as in curve *B*, Figure 1. Assuming that curve *A* applies to reaction (14) and, depending on the positioning of the transition state along the "reaction coordinate" axis, the interactions in **46a** would be either zero, stabilizing, or destabilizing. If it is assumed that these interactions are unimportant in alternative transition state **46b**, then the stereoselectivity observed in eq. 14 could be predicted. It is apparent that this kind of approach could be generalized to the olefin addition reactions of disubstituted carbenes.

Since a different curve would have to be drawn for every set of carbenic and olefinic substituents, and since the shape of these curves as well as the internuclear separation of the various atoms and the exact geometry at the transition state are unknown, a detailed theoretical approach is presently impractical. The most that can be done is to qualitatively predict whether a given set of interactions will be attractive or repulsive, based upon several orienting experiments, and reasonable guesses as to the effect of changing the size or electronic character of carbenic and olefinic substituents. As will be seen, much of the data can be accounted for in such primitive terms.

B. Stereoselectivity of Monosubstituted Carbenes and Carbenoids

1. Carboalkoxycarbenes and Related Species

The general observation for the addition of carboalkoxycarbenes to simple olefins is one of *anti* stereoselectivity. This is true both for the very reactive species produced by the photolysis of an alkyl diazoester [which is capable of facile insertion into saturated C—H bonds (63) and is only moderately discriminating in addition to olefins which differ in degree of alkylation (64)] and also for the species produced by the catalytic decomposition of an alkyl diazoester [which does not give the insertion reaction, though it also adds to olefins with little discrimination (65), and is presumably a carbene-complex or carbenoid].

Early examples of carboalkoxycarbene additions have been reviewed (66). The addition of carbomethoxycarbene, produced either photolytically or catalytically (anhydrous $CuSO_4$), to *cis*-butene is reported to yield isomeric products **47a** and **47b** in the ratio 2:5 (67). In similar reactions of carboethoxycarbene the analogous *syn*:*anti* ratios are identical, but minor differences are noted with change of the catalyst to copper chloride, or change in the light source (64).

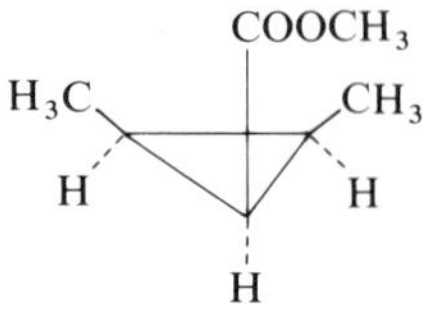

47a 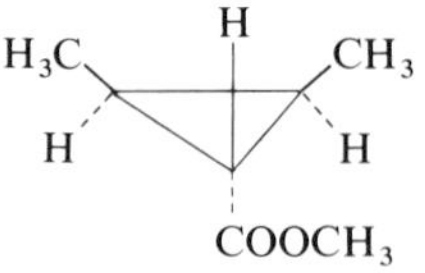 **47b**

Additions of carboethoxycarbene to styrene (68), ethylacrylate, and vinyl bromide, and of carbomethoxycarbene to vinyl-*t*-butylether (69), have all been reported to afford mainly the respective *anti*-carboethoxycyclopropanes.

Photolytically produced carboethoxycarbene added to cyclohexene with *syn*:*anti* (*endo*:*exo*) selectivity of 1:1.6. A copper–bronze catalyzed addition was much more selective; the *syn*:*anti* distribution of 7-carboethoxynorcaranes was 4:69 (70). A similar effect was noted in additions to cyclopentadiene. Photolytically generated carbene afforded *syn* and *anti* adducts in the distribution 1:3; catalytically generated carbene afforded these adducts in a 1:5 distribution (71). This pattern was repeated in carboethoxycarbene addition to 1,4-cyclohexadiene: photochemical generation, *syn*:*anti* = 1:4; copper powder catalytic generation, *syn*:*anti* = 1:60 (72). In a similar study Berson and Hand observed a ratio of about 1:7 for the catalytic addition of carboethoxycarbene to 1,4-cyclohexadiene, rather than 1:60 (73). The origin of this discrepancy is not known. Copper catalyzed addition of carboethoxycarbene to 1,3-cyclohexadiene afforded *syn* and *anti* cyclopropanes in a 1:5 distribution (73).

Addition of carboethoxycarbene (copper catalysis) to cyclooctatetraene is reported to yield *anti* or *exo* ester (**48**) (74). A small amount of the *endo* epimer is probably also formed (75). A recent study of the addition of

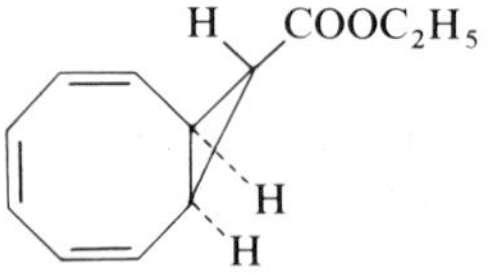

48

carbomethoxycarbene to a cyclopropene gave the results outlined in eq. 15 (76). The preference of the intermediate for *anti* addition is marked by the dominance (77.6%) of the "*exo–exo*" product. The "*exo–endo*" product (17.1%) could have come from either a *syn* or *anti* addition (or, more likely, both). Only the "*endo–endo*" minor product must have come from a *syn* addition. An analogous reaction of 1,2-diphenyl-3-carbomethoxycyclopropene with carbomethoxycarbene (copper–bronze catalyst) led to *exo–exo*

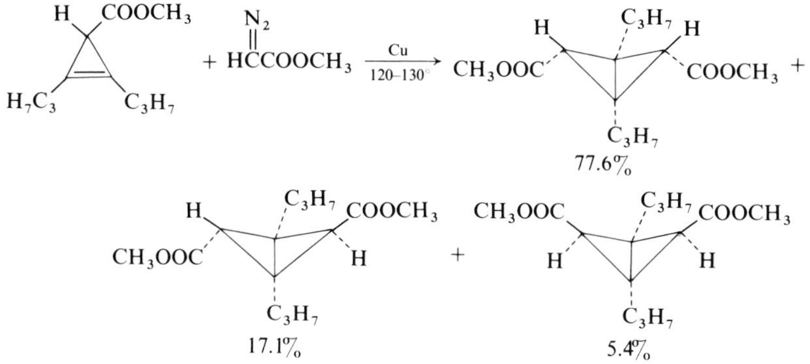

$$\text{(15)}$$

(major) and *exo–endo* (minor) products. *Endo–endo* epimer was not observed (77).

Reactions of carboethoxycarbene (CuCN catalysis) with vinylene carbonate (**49**) and its monomethyl derivative were reported to afford mainly the *anti* carboethoxycyclopropanes (78). A similarly catalyzed addition to norbornene afforded the *anti–exo* product, **50** (79), though a minor amount of the *syn–exo* epimer (**51**) is also formed (75,80). In line with the discussion

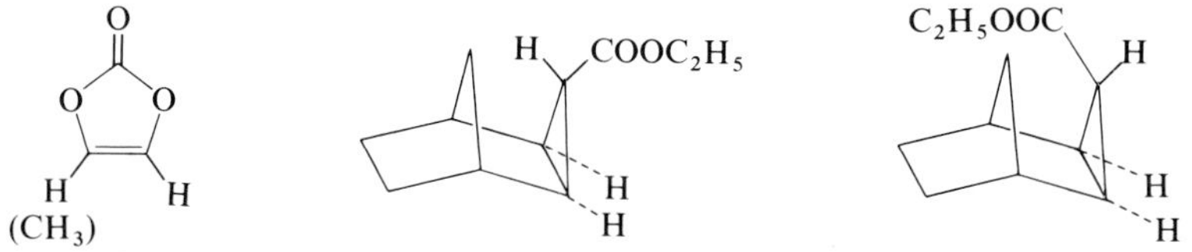

49 **50** **51**

in Section II-A *endo* addition was not observed in this reaction (79). Also, in analogy to previous observations for CH_2 (4) and CBr_2 (8), the CuCN-catalyzed addition of carboethoxycarbene to *norbornadiene* was reported to give both *anti–exo* and *anti–endo* products, eq. 16 (79), *exo* addition predominating, 2:1. In a recent study of the copper-catalyzed addition of

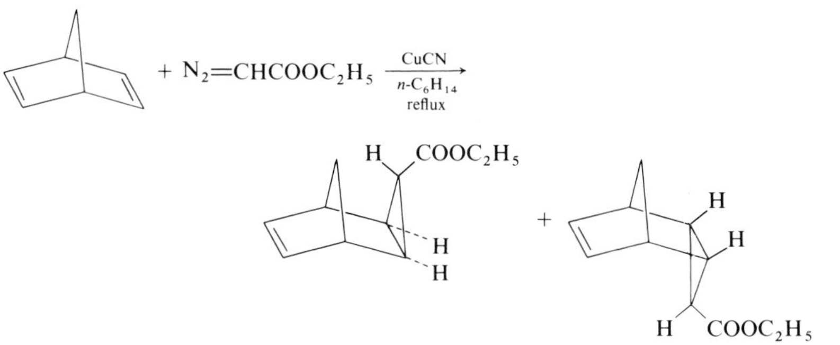

$$+ \; N_2{=}CHCOOC_2H_5 \xrightarrow[n\text{-}C_6H_{14}\ \text{reflux}]{CuCN} \qquad \text{(16)}$$

carbomethoxycarbene to norbornadiene the isolated diadducts were reported to both be *exo*, *exo*. One isomer was *anti*, *anti* and the other was *anti*, *syn* (81). It is not known whether any *exo*, *endo* diadducts were formed in this reaction, though *endo* monoadduct was observed.

There have been studies of carboalkoxycarbene additions to several condensed aromatic systems in which the product cyclopropanes were reasonably stable toward valence tautomerism. For example, various additions of carboalkoxycarbenes to naphthalene (attack at the 1,2 position) (82); anthracene (attack at the 1,2 position) (83); and phenanthrene (attack at the 9,10 position) (84) afforded mainly *anti* carboalkoxycyclopropane derivatives.

Two aspects of the foregoing data require comment. First, we must rationalize the dominant *anti* addition stereoselectivity invariably observed for carboalkoxycarbenes over a wide variety of substrates. Second, we should account for the quantitative differences in stereoselectivity as a function of the method of generation of the carbene (i.e., photolytic or catalytic decomposition of an alkyl diazoester).

In terms of the model described in Section IV-A, consider the generalized *anti* and *syn* addition transition states, **52a** and **52b**, respectively.

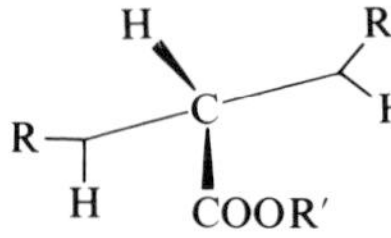

52a 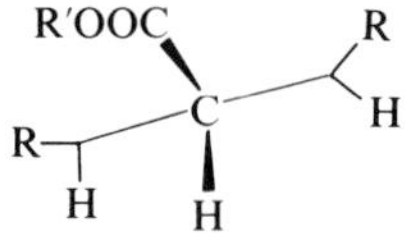 **52b** 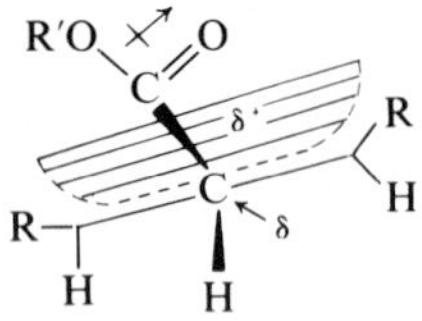 **52c**

Steric hindrance to addition will certainly be of less importance in **52a**, the *anti* transition state. This biasing factor will not be offset by electrostatic factors favoring **52b**, because (*1*) the poor olefinic discrimination of both photolytically (64), and catalytically (65) generated carboalkoxycarbenes suggests that there is not much charge separation in the addition transition state; and (*2*) the carbenic substituent which lies over the olefinic substituents in **52b** carries a strong permanent dipole, involving its α carbon as the positive end, which may be expected to interact in a destabilizing manner with whatever positive charge does reside on the olefinic centers and their substituents, **52c**. In other terms, a hypothetical curve of the energy of the nonbonded interactions in the *syn* transition state, **52a** or **52c**, versus a reaction coordinate would probably resemble Figure 1, curve *b*.

In view of the foregoing analysis, it is worth observing that, under conditions where its addition reaction is reasonably stereospecific, trifluoromethylcarbene adds to *cis*-butene with *anti* stereoselectivity, eq. 17 (85).

Predominant *anti* stereoselectivity is also shown in the addition of trifluoromethylcarbene to propene, *anti*:*syn* = 1.2 (86). Trifluoromethylcarbene

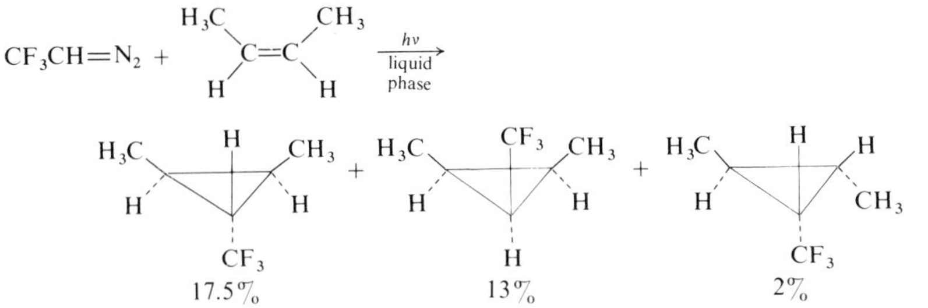

$$CF_3CH{=}N_2 + (H_3C)(H)C{=}C(CH_3)(H) \xrightarrow[\text{liquid phase}]{h\nu}$$

H₃C, H, CH₃, H, CF₃ — 17.5% + H₃C, CF₃, CH₃, H, H — 13% + H₃C, H, H, CH₃, CF₃ — 2% (17)

would be expected to experience similar differential steric and (in view of the strong permanent dipoles in the trifluoromethyl substituent) Coulombic nonbonded interactions in transition-state orientations analogous to those of **52**. [In this context note that methylcarbene, produced by photolysis of diazoethane, adds to propene with *syn* stereoselectivity (see Section IV-B-5).]

Similar considerations are suggested by studies of the photochemical decomposition of arylsulfonyldiazomethanes. Addition of phenylsulfonylcarbene (**53**) to cyclohexene gave the expected norcaranes with a *syn*:*anti* distribution of 1:2.1 (87). A similar, though smaller, preference for *anti* stereoselectivity was observed with the *p*-methoxy derivative of **53**, which added to *cis*-butene to afford a *syn*:*anti* product distribution of 1:1.3 (88).

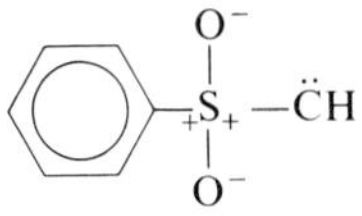

$$C_6H_5-\overset{O^-}{\underset{O^-}{S^{++}}}-\ddot{C}H$$

53

Finally it was shown above, that changing from a photolytic generation of a carboalkoxycarbene to a catalytic generation almost always results in an *increased anti* stereoselectivity toward a given olefinic substrate. Since the photolytic intermediate is probably a free carbene, whereas the catalytic intermediate is doubtless a carbene–catalyst complex (carbenoid), it has been suggested that the latter species, being bulkier, will suffer the greater steric hindrance in addition to any olefin, therefore giving the greater *anti* stereoselectivity (70). This argument is necessarily somewhat vague, because the precise structure of the carbenoid is unknown, and the differential steric factors operating during its *syn* or *anti* addition to olefins cannot, therefore, be assessed. However, whatever the exact structure of the carbenoid, it is a *less reactive* species than the photolytic carbene (65,70). Transition states for the carbenoid's addition to olefins will be more productlike ("tighter") than those for the photolytic, or free carbene. Hence we should expect that the unfavorable nonbonded interactions which lead to *anti* stereoselectivity will operate most strongly in the *carbenoid* addition reaction. This argument, which stresses the relationship of carbene reactivity to stereoselectivity, will be seen to have considerable generality. It is particularly useful in discussing the stereoselectivity differences between formally identical "carbenes" generated in different ways.

2. *Arylcarbenes*

Extensive studies of the stereoselectivity of arylcarbenes and carbenoids have been carried out by Closs and co-workers (58,61,89). Additions of "free" arylcarbenes, from photolyses of aryldiazomethanes, eq. 18 (58,61), or of arylcarbenoids, derived either from the action of alkyllithiums on benzal halides, eq. 19 (58,61), or from the metal halide catalyzed decomposition of aryldiazomethanes, eq. 20 (89), have been reported. In these cases additions

$$\mathrm{ArCH{=}N_2} \xrightarrow[-\mathrm{N_2}]{h\nu} [\mathrm{Ar{-}\ddot{C}H}] \xrightarrow{\text{alkene}} \text{cyclopropane (H, Ar)} \tag{18}$$

$$\mathrm{ArCHBr_2} \xrightarrow[-\mathrm{RBr}]{\mathrm{RLi}} \left[\mathrm{ArCH}\begin{matrix}\mathrm{Br}\\ \mathrm{Li}\end{matrix}\right] \xrightarrow[-\mathrm{LiBr}]{\text{alkene}} \text{cyclopropane (H, Ar)} \tag{19}$$

$$\mathrm{ArCH{=}N_2} \xrightarrow[-\mathrm{N_2}]{\mathrm{MX}} [\mathrm{ArCH\cdot MX}] \xrightarrow[-\mathrm{MX}]{\text{alkene}} \text{cyclopropane (H, Ar)} \tag{20}$$

to alkenes such as *cis*-butene are generally *syn* stereoselective. That is, the major product cyclopropane is the isomer with the aryl group *cis* to the largest possible number of alkyl groups. Data from studies of eqs. 18 and 19 appear in Table 7.

The following general trends are observed: (*a*) In any given case, the carbenoid exhibits a greater *syn* stereoselectivity than the free (photolytic) carbene; (*b*) Aryl substituents enhance the *syn* stereoselectivity in the order p-OCH_3 > p-CH_3 > m-Cl > p-Cl > H; (*c*) *Syn* stereoselectivity for additions to acyclic alkenes is most pronounced with *cis*-butene.

Also relevant was the observation that, at least in the one case investigated, thermodynamic factors strongly favored *anti*-1-phenyl-*cis*-2,3-dimethylcyclopropane over its *syn*-phenyl epimer, demonstrating that the *syn* stereoselectivity of the carbenic additions is a kinetically controlled phenomenon (58). Finally it was demonstrated that small quantities of

TABLE 7
Syn:*Anti* Distributions of Cyclopropanes Formed by Arylcarbene (eq. 18) and Arylcarbenoid (eq. 19) Additions to Alkenes[a,b]

Starting olefin	Phenyl substituent				
	H	*p*-Cl	*m*-Cl	*p*-CH_3	*p*-OCH_3
1-Butene	2.1 (1.0)	2.1 (1.1)	2.5 (1.2)	2.7 (1.2)	3.0 (1.4)
cis-2-Butene	2.4 (1.1)	2.9 (1.1)	3.7 (1.2)	4.5 (1.7)	8.1 (2.8)
2-Methyl-2-butene	1.3 (1.1)	1.4 (1.1)		1.4 (1.2)	
Cyclohexene[c]	3.0[d] (1.1)				

[a] Data of ref. 58, determined at $-10°$ in hydrocarbon media.

[b] Data in parentheses are for the photolytic reaction, eq. 18; other data are for the carbenoid reaction, eq. 19.

[c] Data of ref. 90.

[d] Ether was present in this reaction. The value would probably have been higher in its absence; see text.

diethyl ether caused a lowering of the *syn* stereoselectivity in the carbenoid addition reactions, eq. 19. For example, the *syn*:*anti* distribution for the addition of *p*-tolylcarbenoid to *cis*-butene was ca. 2.9 in the presence of ether, but it was 4.5 in pure hydrocarbon (58).

Transition states **54a** and **54b** have been drawn for the carbenoid and carbene *syn* additions to *cis*-butene (from ref. 58).

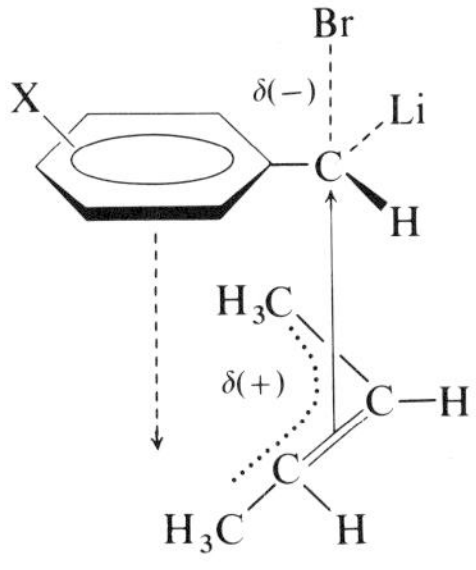

54a
Carbenoid

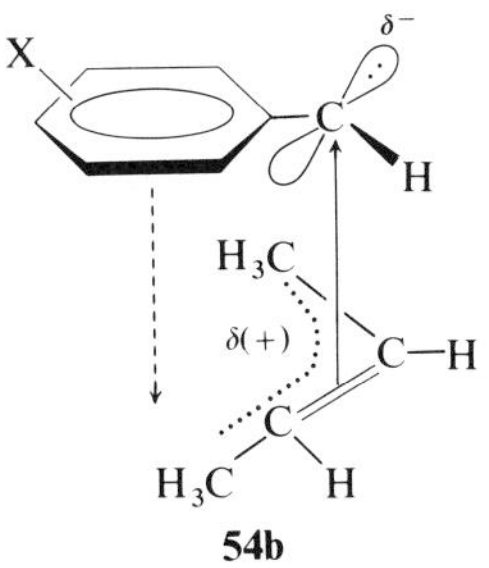

54b
Free Carbene

In these transition states, it was suggested, the positive charge induced on the olefinic substituents during the electrophilic carbenoid or carbene attack is mitigated by interaction with the aromatic π electrons. This energetically *favorable* interaction is possible only with a *syn* arrangement of carbenic and olefinic substituents and more than offsets accompanying unfavorable steric interactions and entropy effects (e.g., from the necessary damping of rotation about the aryl-carbenic carbon bond). *Syn* addition is thus favored.

The trends noted above can be accommodated by **54**, which in turn derives from the stereoselectivity model discussed in Section IV-A. (*a*) The free carbene is a more reactive species than the carbenoid; **54b** is a "looser" transition state than **54a**, and all of the interactions which determine stereoselectivity, as well as the positive charge placed on the olefinic substituents, will be less important in **54b**. *Syn* stereoselectivity will therefore be less pronounced in the free carbene additions. [With regard to the freeness of the arylcarbenes produced in eq. 18, note that photolysis of *trans*-stilbene oxide leads to a phenylcarbene with stereoselectivity and C—H insertion selectivity essentially identical to those of the phenylcarbene generated by the photolysis of phenyldiazomethane (91).] (*b*) Donating substituents such as *p*-CH_3 and *p*-OCH_3 should buttress the polarizability of the aryl π electrons. Greater *syn* stereoselectivity of arylcarbenes and carbenoids substituted with these groups is expected in terms of **54**. That *m*-Cl and *p*-Cl substituents also enhance the *syn* stereoselectivity of arylcarbenoids may indicate that the unshared electron pairs on the chlorine atom play a direct role in determining the stereoselectivity. This is reasonable in view of the fact that chlorocarbenoids exhibit *syn*-chloro stereoselectivity (see below, Section IV-B-4). (*c*) With *cis*-butene, as compared with a mono- or trisubstituted acyclic olefin, the difference in nonbonded interactions between carbenic and olefinic substituents for *syn* and *anti* transition states will be maximized. A more pronounced stereoselectivity for carbene additions to *cis*-butene might then be expected. Finally, addition of the Lewis base, ether, to the carbenoid reaction medium could serve to solvate the LiBr leaving group, thus "activating" the carbenoid, making transition state **54a** more reactantlike, and lowering the *syn* stereoselective tendency.

In a recent probe of steric hindrance in reaction 19 Moss (44) studied stereoselectivity in the additions of *p*-tolylcarbenoid to ethylethylene, *i*-propylethylene, and *t*-butylethylene. The carbenoid was generated by the action of butyllithium in pentane on either *p*-methylbenzal bromide or iodide, or by the action of methyllithium in ether on *p*-methylbenzal bromide. In agreement with a carbenoid mechanism, the observed stereoselectivity depended on the generative conditions. The most discriminating intermediate was produced from the *p*-methylbenzal iodide. Data for this species appear in Table 8. The data have been discussed in terms of postulated transition states **55a** and **55b** (from ref. 44). In **55a** the balance of nonbonded interactions favoring *syn* stereoselectivity can be reversed by enlarging R. The observation of preferential *anti* addition to *t*-butylethylene was the first report of *anti* stereoselectivity for an arylcarbene or carbenoid in addition to a simple olefin. The relative addition rates of *p*-tolylcarbenoid in the *syn* and

TABLE 8

Selectivity in Addition to Olefins of *p*-Tolylcarbenoid from *p*-Methylbenzal Iodide and Butyllithium in Pentane, $-10°$[a]

R in $RCH{=}CH_2$	Stereoselectivity *syn*:*anti*	Rate of carbenoid addition to $RCH{=}CH_2$[b]	
		Syn mode	*Anti* mode
C_2H_5	2.7	0.99	0.37
i-C_3H_7	1.9	0.46	0.24
t-C_4H_9	0.42	0.071	0.17

[a] Data of ref. 44.

[b] Relative to *trans*-butene. All rates have been normalized to the standard by multiplication by 2.

anti modes were calculated from the overall rates of addition, relative to *trans*-butene, and the stereoselectivity data (Table 8). As would be expected, retardation of the carbenoid additions, as a function of enlarging R, was more severe in the *syn* mode of addition (**55a**), where the interaction was between the tolyl and R groups, than in the *anti* mode of addition (**55b**), where only the carbenoid's proton opposed the R group.

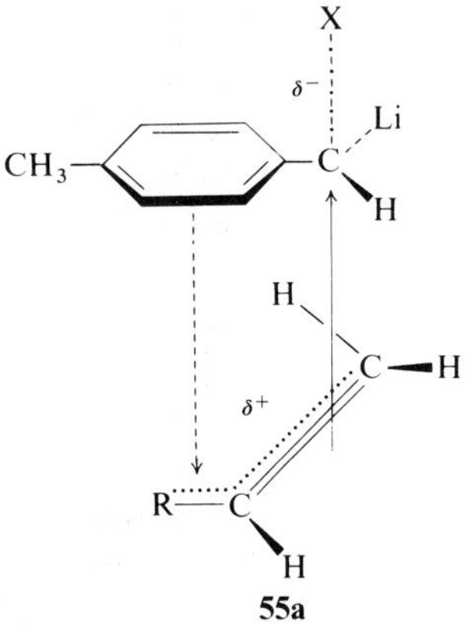

55a *Syn*

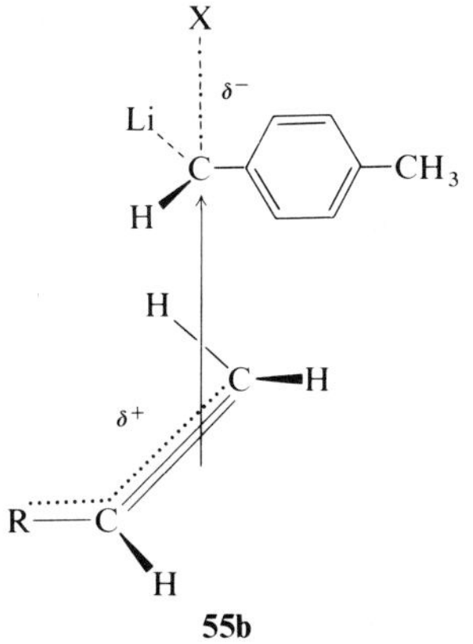

55b *Anti*

Much stereoselectivity data has been reported for reaction 20 (89). Aryldiazomethanes employed included the *p*-CH_3, *p*-CH_3O, and unsubstituted compounds. Catalysts included $ZnCl_2$, $ZnBr_2$, ZnI_2, LiBr, $LiClO_4$, $MgBr_2$, and $CoBr_2$. Substrates included butadiene, cyclohexene, *cis*-butene, trimethylethylene, 1-butene, and divinyl ether. Each catalyst led to a different stereoselectivity with any given aryldiazomethane and olefin (although *syn* stereoselectivity was the general qualitative result). Thus each catalyst afforded a different carbenoid. Even reaction (19) (RLi = CH_3Li) was found to involve, at least partially, different intermediates than reaction (20) (MX = LiBr); the latter reaction affording the more discriminating carbenoid. In spite of the complexity of the data, the trends observed in previous studies (see the above discussion of ref. 58) were still found to hold. For any given substrate and catalyst, the *syn* stereoselectivity of the arylcarbenoids was in the order *p*-OCH_3 > *p*-CH_3 > *p*-H. For any given carbenoid, *cis*-butene afforded the most pronounced *syn* stereoselectivity.

Some of the *syn* addition preferences were amazingly high. With a *cis*-butene substrate, *p*-anisylcarbenoid ($ZnBr_2$ catalyst) gave *syn*:*anti* = 53.0; the analogous addition of *p*-tolylcarbenoid (ZnI_2 catalyst) gave 21.0. With cyclohexene, *p*-anisylcarbenoid ($ZnCl_2$ catalyst) gave *syn*:*anti* = 34.0, whereas under similar conditions the parent phenylcarbenoid gave a *syn* preference of only 3.0. It is interesting to note that the carbenoid produced from benzal iodide and diethyl zinc adds to cyclohexene with a *syn* stereoselective preference of 17.0 (92).

It was pointed out (89) that many of these high-yield, highly *syn* stereoselective arylcarbenoid additions are synthetically useful, since the *anti*-arylcyclopropane can always be obtained from the kinetically favored *syn*-arylcyclopropane by base-catalyzed isomerization.

As simplest representations of some of the carbenoid intermediates in these metal-halide-catalyzed aryldiazomethane decompositions, **56**, **57**, and **58** were suggested (89). Clearly, although the precise structures of the

$$\underset{\mathbf{56}}{Ar{-}\overset{+}{C}HLi\ X^-} \qquad \underset{\mathbf{57}}{ArCHLiX} \qquad \underset{\mathbf{58}}{Ar{-}C(ZnX)(H){-}X}$$

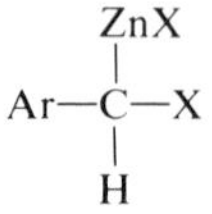

carbenoids in these reactions are unknown, the *syn* stereoselectivity can be rationalized by the transition state models previously discussed. Furthermore, the change in stereoselectivity as a function of the catalyst could be rationalized by considering the expected degree of electrophilicity of the derived carbenoid (89).

Finally, in contrast to the *syn* stereoselectivity of phenylcarbene in its additions to simple olefins, it has been reported that phenylcarbene, produced by the thermolysis of a benzaldehyde tosylhydrazone, sodium methoxide mixture, added to naphthalene and anthracene at the 1,2 positions, with exclusive, low-yield formation of the *anti* cyclopropyl products (93). However, the stability of the *syn* epimers under the reaction conditions is unknown.

3. *Alkoxy, Aryloxy, and Related Carbenes*

Schöllkopf and co-workers have generated a number of alkoxy and aryloxycarbenic species. The generative techniques involved the reaction of an alkyllithium with a chloromethyl ether ($ROCH_2Cl$) or of methyllithium–lithium iodide with a dichloromethyl ether ($ROCHCl_2$). The exact nature of the intermediates involved in these reactions is unclear, but because different stereoselectivities were observed when these two methods were employed to add "methoxycarbene" to cyclohexene (see Table 9) it seems likely that these reactions occur largely through the intermediacy of lithium carbenoids (ROCHLiX). The stereoselectivity exhibited by these species upon their addition to olefins can be either *syn* or *anti*, depending on the olefin. A summary of much of the data is given in Table 9.

Table 9 demonstrates the strong dependence of the stereoselectivity of addition of these carbenoids on the nature of the substrate. Additions to the acyclic olefins, *cis*-butene and trimethylethylene are predominantly *syn*. In terms of our working model (Section IV-A), this suggests that in the electrophilic additions of these carbenoids, stabilizing interactions between the lone electron pairs on the oxygen atom and the partial positive charges induced on the *syn* olefinic alkyl groups are the dominant factors, and make the *syn* addition transition state of lower energy than the *anti* alternative. This situation is very delicate, and can be reversed by increasing the steric demands of the substrate, so that repulsive nonbonded interactions dominate in the *syn* addition mode. Thus the cyclic alkenes, cyclopentene, cyclohexene, and cycloheptene add the oxycarbenoids with *anti* stereoselectivity. With cyclopentadiene and cyclooctene, the stereoselectivity reverts back to a dominant *syn* character. The former olefin does not have the steric problems associated with cyclopentene, whereas the latter may be large enough and flexible enough so that its steric demands *vis-à-vis syn* carbenoid addition may be closer to those of *cis*-butene than to those of cyclohexene.

Schöllkopf has recently discussed an alternative transition state model, in which the lithium atom of the carbenoid, rather than the alkoxy or aryloxy group is the source of the steric hindrance experienced by the carbenoid in its addition to olefins (54). In terms of this model, cases in which *syn* aryloxycarbenoid addition predominated were explained as resulting from steric hindrance to *anti* addition (in which the carbenoid's lithium atom was depicted as being *syn* to the largest number of olefinic substituents). Recently, methoxycarbene has been generated by pyrolysis (125°) of (dimethoxymethyl) trimethoxysilane (54a). *Anti*:*syn* ratios for addition to *cis*-butene and cyclohexene were 1.52 and 7.0, respectively. Note the change from the net *syn* addition of methoxycarbenoid to *cis*-butene (Table 9). For both substrates, the pyrolytic methoxycarbene exhibits enhanced *anti* selectivity, *vis-à-vis* the

TABLE 9
Stereoselectivity of Alkoxy and Aryloxycarbenoids

Olefin	"Carbene"[a]	Conditions[b]	Stereoselectivity[c]	Ref.
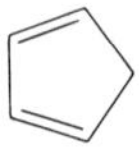	$CH_3O\ddot{C}H$	B	*Syn* (3.8)	95
	$n\text{-}C_6H_{13}O\ddot{C}H$	B	*Syn* (3.6)	95
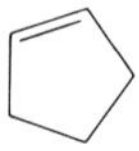	$CH_3O\ddot{C}H$	B	*Anti* (2.0)	95
	$ClCH_2CH_2O\ddot{C}H$	B	*Anti* (2.3)	96
	$n\text{-}C_6H_{13}O\ddot{C}H$	B	*Anti* (1.9)	95
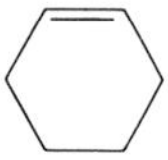	$CH_3O\ddot{C}H$	A	*Anti* (4–4.5)	94
	$CH_3O\ddot{C}H$	B	*Anti* (1.7)	95
	$n\text{-}C_4H_9O\ddot{C}H$	A	*Anti* (4.5)	94
	$n\text{-}C_6H_{13}O\ddot{C}H$	B	*Anti* (3.2)	95
	$ClCH_2CH_2O\ddot{C}H$	B	*Anti* (4.)	96
	$i\text{-}C_3H_7O\ddot{C}H$	A	*Anti* (7.)	94
	$\phi O\ddot{C}H$	A	*Anti* (1.8)[d]	94,97
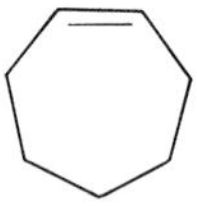	$ClCH_2CH_2O\ddot{C}H$	B	*Anti* (1.5)	96
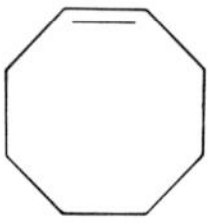	$CH_3O\ddot{C}H$	B	*Syn* (2.0)	95
	$ClCH_2CH_2O\ddot{C}H$	B	*Syn* (1.8)	96
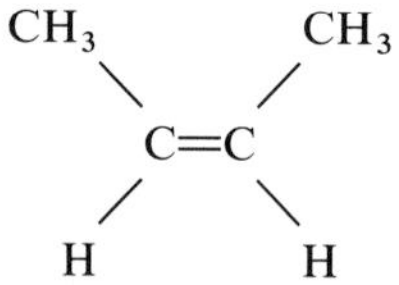 CH_3, CH_3 / C=C / H, H	$CH_3O\ddot{C}H$	B	*Syn* (7.1)	95
	$ClCH_2CH_2O\ddot{C}H$	B	*Syn* (3.)	96
	$\phi O\ddot{C}H$	A	*Syn* (3.7)	97
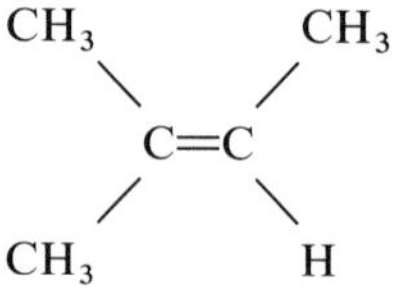 CH_3, CH_3 / C=C / CH_3, H	$ClCH_2CH_2O\ddot{C}H$	B	*Syn* (1.4)	96
	$\phi O\ddot{C}H$	A	*Syn* (3.2)	54

[a] For simplicity, the formal carbenes are listed.

[b] Conditions: A, $ROCH_2Cl + RLi$; B, $ROCHCl_2 + CH_3Li \cdot LiI$. The reaction temperatures vary.

[c] The dominant cyclopropane product is listed, together with the factor by which it exceeds the epimeric minor product.

[d] See, however, the values in ref. 54.

lithium carbenoid. While this may constitute some support for Schöllkopf's alternative transition state model (see above), comprehensive analysis of alkoxy and aryloxycarbenoid stereoselectivity waits upon further research.

A recent study of "phenoxycarbene" reveals a strong dependence of stereoselectivity on the nature of the halogen of the halomethylphenyl ether from which the species was derived (54). Thus the *exo*:*endo* distributions of the 7-phenoxynorcaranes produced by the action of salt-free butyllithium in ether on $C_6H_5OCH_2X$ in cyclohexene were 6.00, 4.20, and 2.10, where X was F, Cl, and Br, respectively. With butyllithium–lithium bromide in ether the analogous ratios were 2.15, 2.45, and 2.10, respectively. The present author believes that these data are strongly suggestive of carbenoid ($C_6H_5OCHLiX$) intermediacy in these reactions, though it has been noted that a true carbene intermediate might explain the data if the lithium halide molecule had not become solvent-separated from the carbene–olefin moiety at the transition state (54).

Stereoselectivity data have also been reported for phenylthio (98) and phenylselenocarbenes (or carbenoids) (99) (**59** and **60**). Phenylthiocarbene

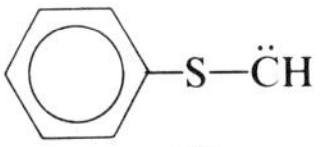

C_6H_5—S—$\ddot{C}$H

59

C_6H_5—Se—$\ddot{C}$H

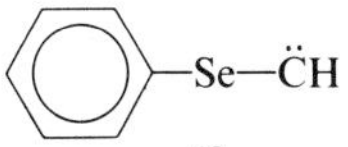

60

could be produced either by the action of potassium *t*-butoxide or butyllithium on chloromethylphenyl sulfide. Additions to cyclohexene, 1,3-cyclohexadiene, *cis*-butene, and *cis*-1,2-*bis*-phenylthioethylene were all *syn* stereoselective. Additions of phenylselenocarbene, from potassium *t*-butoxide and chloromethylphenyl selenide, to cyclohexene and 1,3-cyclohexadiene were also *syn* stereoselective. Note that phenoxycarbenoid adds with *anti* stereoselectivity to cyclohexene (Table 9), whereas its phenylthio and phenylseleno analogs add to this olefin with *syn* stereoselectivity. Thus, if the phenoxy species is presumed to favor *anti* addition to cyclohexene because the nonbonded carbene–olefin interactions in the *syn* transition state are repulsive, then the change of oxygen to sulfur or selenium could, at relatively little steric cost, change the character of these interactions to attractive, because of the greater polarizability of the sulfur and selenium atoms. A parallel effect is seen in the stereoselectivities of the dihalocarbenes (Section IV-C-1).

4. *Halocarbenes*

The stereoselectivity of halocarbenes depends strongly on their mode of generation. Thus free halocarbenes show little or no stereochemical preference in their olefin addition reactions, while halocarbenoids exhibit a pronounced *syn* stereoselectivity.

The gas phase reactions of fluoro(tritio)carbene with propene, butene-1, *cis*-butene, and trimethylethylene led, in each case, to the formation of equal amounts of *syn*-fluoro and *anti*-fluorocyclopropane isomers (100). Similarly, no stereoselectivity was observed in the additions of chlorocarbene or bromocarbene (from the halodiazomethanes) to 1-butene, *cis*-butene, cyclohexene, and trimethylethylene (101). Monohalocarbenes have also been generated by photolyses of polyhalomethanes (102). The stereoselectivity of the postulated intermediates appears to depend on the presence and nature of solvent and, as yet, other undefined experimental parameters.

The above experiments presumably involved free halocarbenes. A different picture emerges when the species are generated by the action of alkyllithiums on di- and trihalomethanes. The carbenoid produced from butyllithium and methylene chloride, which is probably lithium dichloromethide (103), adds to alkenes with marked *syn*-chloro stereoselectivity. Additions to various substrates, with *syn*:*anti* distributions, are (60,61,101, 104): 1-butene (3.4), *cis*-butene (5.5), cyclohexene (3.2), and trimethylethylene (1.6). Addition of the chlorocarbenoid ($CH_3Li + CH_2Cl_2$) to cyclopentene in ether led to about a threefold excess of *endo* over *exo*-6-chlorobicyclo[3.1.0]hexane (105). The chlorocarbenoid generated from CH_2Cl_2 and CH_3Li-LiBr is also *syn* stereoselective in its addition to cyclooctatetraene (106,107), the *syn*:*anti* distribution was 3–4:1. *Syn* stereoselectivity in chlorocarbenoid addition to styrene has been reported (108), and it has been shown that the *syn*:*anti* distribution in this reaction depends on the origin of the methyllithium used to generate the chlorocarbenoid from dichloromethane (109). The distribution was 1.7, 2.3, and 1.6 when the methyllithium was prepared from methyl bromide, chloride, and iodide, respectively. These results indicate the importance of halocarbenoids in these reactions.

Closs has rationalized the *syn* stereoselectivity of the halocarbenoids in the manner previously used to account for the *syn* stereoselectivity of the arylcarbenoids, Section IV-B-2 (90,101); that is, in halocarbenoid additions energetically favorable electrostatic interactions between the halogen substituent's lone pairs and the olefinic alkyl groups (which have become somewhat positive in the transition state) stabilize the *syn* relative to the *anti* transition state. The *syn* transition state has been rendered as **61** (from ref. 101).

In the addition reactions of free halocarbenes [which, in contrast to the halocarbenoids, show little ability to discriminate between differently alkylated olefinic π bonds (101,104)], it seems likely that the transition states are reactantlike. The relatively great spatial separation of carbenic and olefinic substituents, together with the lack of substantial induced positive character on the olefinic substituents, negate the nonbonded interactions responsible for stereoselectivity. At most, only a mild steric effect is seen with the larger halogens (102).

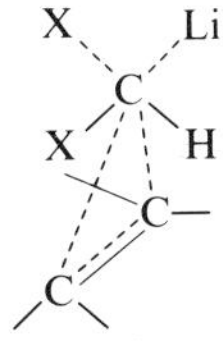

X Li C X H C— C

61

Although the reaction of methyllithium–lithium bromide with methylene chloride in cyclooctatetraene leads mainly to *syn*-9-chlorobicyclo[6.1.0] nonatriene, **62a** (106,107), the reaction of dipotassium cyclooctatetraenide with chloroform leads to essentially pure *anti*-chloro product **62b**, eq. 21 (110).

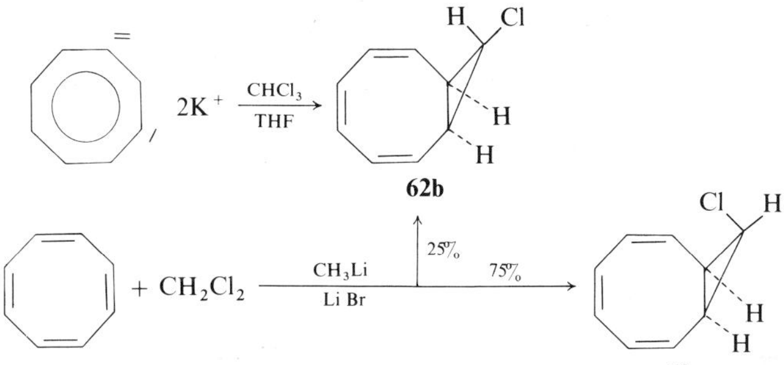

$2K^+$ $\xrightarrow[\text{THF}]{CHCl_3}$ H Cl H H **62b**

+ CH_2Cl_2 $\xrightarrow[\text{Li Br}]{CH_3Li}$ 25% 75% Cl H H H **62a** (21)

The substitution of lithium for the potassium counterions does not alter the stereochemistry. The hypothesis that lithium cyclooctatetraenide reacts with chloroform to produce a lithium chlorocarbenoid and cyclooctatetraene, which then recombine, is clearly not tenable. The mechanism of the cyclooctatetraenide reaction is still uncertain, and it is not possible to account for the nearly complete *anti*-chloro stereoselectivity at this time. It may be noted that, with dichloromethylmethyl ether, cyclooctatetraenide leads to *anti*-9-methoxybicyclo[6.1.0]nonatriene (110).

Another rather exceptional reaction is that of lithium dichloromethide and *N*-benzylideneaniline, which appeared to yield only aziridine **63**, and none of its *anti*-chloro isomer, eq. 22 (111).

A complicated area of carbenoid chemistry involves "salt effects." Although the reaction of methylene chloride and methyllithium yields the carbenoid, lithium dichloromethide, in the presence of other halide ions, such as bromide or iodide, the initial carbenoid may be diverted to any one of the

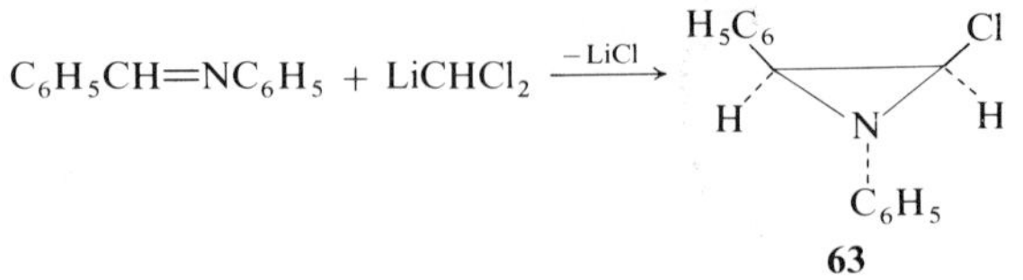

$$C_6H_5CH{=}NC_6H_5 + LiCHCl_2 \xrightarrow{-LiCl} \mathbf{63} \quad (22)$$

possible lithium dihalomethides. Ultimately, in the presence of an olefin and bromide and/or iodide ions lithium dichloromethide may yield chloro, bromo, and/or iodocyclopropanes. An indication of the complexity of these reactions is given by a composite scheme (eq. 23) which "illustrates a number of possible reaction pathways to account for the observed products from the reaction of methylene chloride and methyllithium in the presence of cyclohexene" (112).

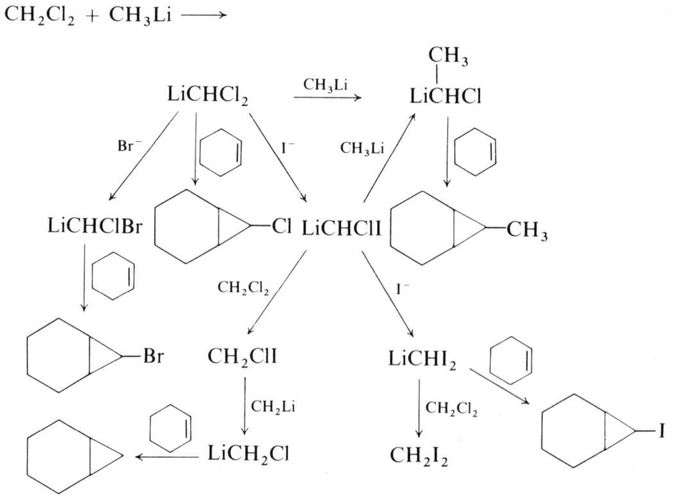

$$CH_2Cl_2 + CH_3Li \longrightarrow \quad (23)$$

As indicated in eq. 23 (from ref. 112), methylcarbenoid may also be formed and lead to methylnorcaranes. Although not reported in ref. 112, iodonorcaranes are also products in eq. 23 (113) and could arise from either chloroiodomethyllithium or more probably diiodomethyllithium. The iodonorcaranes have been shown not to arise by iodide displacement on the 7-chloronorcaranes. It was suggested that the methylnorcaranes might have arisen via stereospecific reactions of methyllithium with the iodonorcaranes (113), but this has been shown not to occur under the actual reaction conditions of eq. 23 (114). Therefore it appears that all of the halonorcaranes, as well as the methylnorcaranes are primary reaction products. Related "salt effects" have been found to operate in the reaction of styrene with methylene chloride and methyllithium in the presence of various halide ions (108,109).

Although it is beyond the scope of this review to discuss the factors that determine the relative rates of the competing processes outlined in eq.23, it is worthwhile to note the stereoselectivity of the various halocarbenoid intermediates. These data are presented in Table 10.

TABLE 10

Stereoselectivity of Halocarbenoid Additions to Cyclohexene. *Syn*:*Anti* Distribution of 7-Halonorcaranes[a]

Carbenoid source	Chloro-norcaranes	Bromo-norcaranes	Iodo-norcaranes
$CH_2Cl_2 + CH_3Li$ (CH_3Cl + Li)	2.1	—	—
(CH_3Br + Li)	2.2	~1.1	—
(CH_3I + Li)	2.0–2.3	—	1.6[b]
(CH_3Cl + Li + LiI)	2.5	—	—
$CH_2Br_2 + CH_3Li$ (CH_3Br + Li)	—	1.2–1.4	—
$CHBr_3 + CH_3Li$ (CH_3Cl + Li)	—	2.0	—
(CH_3I + Li)	—	~2.9	—

[a] Data of ref. 112.
[b] Data of ref. 113.

It is seen that all of the halocarbenoids add with *syn* stereoselectivity to cyclohexene. The data are similar to those found for the reactions of lithium dichloromethide, discussed above, and quite different from the nil or *anti* stereoselectivity exhibited by the free halocarbenes (100–102). Interpretation of the data in Table 10 is subject to the same limitation which applies to other lithium carbenoid systems; namely, lack of knowledge of the actual structure of the carbenoids. The lithium carbenoids are most likely aggregates in solution, and the species shown in eq. 23 and related schemes and transition states are, no doubt, rather simplistic. Subject to this limitation, the observed *syn* stereoselectivity of the halocarbenoids generally fits the model we have been discussing throughout (see above, the discussion of **61**).

5. *Alkylcarbenes*

In general, alkylcarbenes do not undergo olefin addition reactions from the singlet state, because intramolecular reactions intervene. There are, however, some reports of alkylcarbene or carbenoid additions to olefins and also some stereoselectivity data.

Diazoethane yielded methylcarbene upon photolysis in the gas phase. The carbene added to propene, affording *cis*- and *trans*-1,2-dimethylcyclopropane (115). A slight *syn* stereoselectivity was observed; the product distribution was 1.4:1, independent of pressure over the range 400–1400 torr.

Interestingly, methylcarbene addition to *trans*- or *cis*-butene was apparently slower than to propene and only traces of possible addition products were observed. This could have been an example of steric hindrance to carbene addition. A slowing of the intermolecular addition reaction would permit virtually all of the carbene (lifetime estimated to be ca. 7×10^{-10} sec) to react by hydride shift to ethylene, or by attack on its diazoethane precursor.

A number of methylcarbenoid additions have been reported; however, the stereoselectivity data are somewhat erratic. In the zinc carbenoid series the reaction of ethylidene iodide with cyclohexene (zinc–copper couple) afforded a small yield of essentially pure *anti* adduct, **64b** (4). On the other hand, the treatment of ethylidene iodide with diethylzinc in cyclohexene yielded both

H_3C H H H **64a** H CH_3 H H **64b** C_6H_5—C(=O)—O—CH(I)—CH_3 **65**

epimers of 7-methylnorcarane, **64a**:**64b** = 1.5:1 (92). Treatment of iodide **65** with a zinc–copper couple in boiling ether, followed by refluxing cyclohexene, afforded a 25% yield of **64a** and **64b** in a 1:1 distribution (116). Presumably different carbenoids are involved in each of these reactions, and the balance of nonbonded interactions shifts from stabilizing to destabilizing from one *syn* transition state to another.

In the lithium carbenoid series, in which alkylcarbenoids are apparently generated by the reactions of halocarbenoids with alkyllithium reagents (see eq. 23), care must be taken to ascertain that any alkylcyclopropane products are really *primary* reaction products, and did not arise by reaction of an alkyllithium with a halocyclopropane. The latter mechanism has been shown to operate during the reaction of 7,7-dibromonorcarane and methyllithium (117).

The action of methyllithium (lithium iodide) on methylene chloride in styrene was reported to produce only the *syn* methylcarbenoid adduct, *cis*-1-phenyl-2-methylcyclopropane (44,108). However, in a repeat of this reaction it was observed that both the *cis* and *trans* epimers were formed in 1:1 distribution (109). It was suggested that the products may have been formed via reaction of phenyliodocyclopropanes with methyllithium.

The reaction of methylene chloride, methyllithium (lithium iodide) and cyclooctatetraene produced a low yield of only the *syn* addition product, **66** (107).

In the absence of iodide ion the major products were the *syn* and *anti*

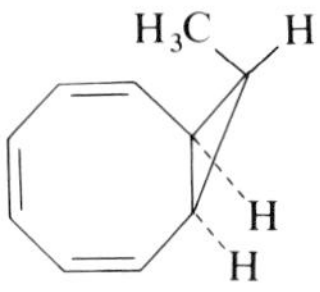

H_3C H H H

66

chloro analogs of **66**, but some **66** also formed. It was shown that reaction of **67** with methyllithium *did not* lead to **68**, eq. 24 (107), and, presumably, **66** did not arise in such a manner either. The reaction of dilithium cyclooctatetraenide with 1,1-dichloroethane afforded **66** in 20% yield, without formation of its *anti* epimer (107).

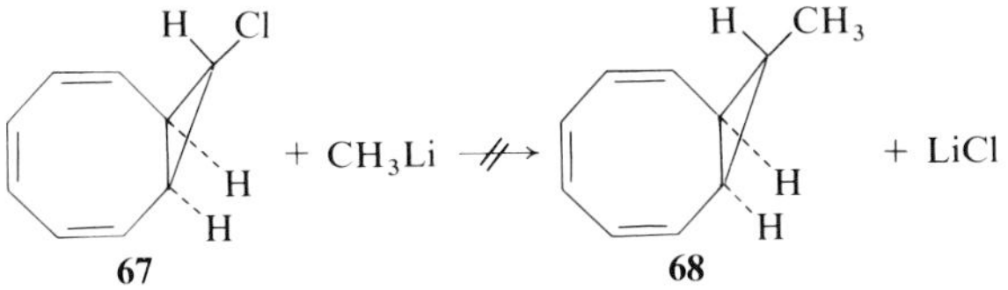

$$\mathbf{67}\ (\text{H, Cl; H, H}) + CH_3Li \not\rightarrow \mathbf{68}\ (\text{H, } CH_3\text{; H, H}) + LiCl \tag{24}$$

Various methylcarbenoid addition reactions have been studied with cyclohexene as substrate (see eq. 23). The data of Table 11 shows that the 7-methylnorcaranes, **64a** and **64b** are generally formed with **64a** predominating, i.e., with *syn* stereoselectivity.

TABLE 11

7-Methylnorcarane Distribution in Methylcarbenoid Additions to Cyclohexene[a]

Carbenoid source	**64a**:**64b** (*syn*:*anti*)
$CH_2Cl_2 + CH_3Li$ (CH_3Cl + Li)	~1.0
(CH_3Br + Li)	~1.7
(CH_3I + Li)	1.6–1.8
(CH_3Cl + Li + LiI)	~1.8

[a] Data of ref. 112.

It was suggested that the methylnorcarane products could have arisen from precursory iodocyclopropanes (last two entries in Table 11) (113), but this possibility has been ruled out for the particular reaction conditions in question (114).

At present it seems presumptuous to attempt analysis of the above data. Clearly, methylcarbenoid stereoselectivity is very dependent on carbenoid source. In some cases it is not known with certainty that the methylcyclopropane products are primary products. The situation seems less clear than that of the species previously discussed and more research is needed.

C. Stereoselectivity of Disubstituted Carbenes and Carbenoids

1. Dihalocarbenes

Fluorochlorocarbene has received considerable attention. It has been generated by the basic decomposition of methyl dichlorofluoroacetate and added to cyclohexene (118) and dihydropyran (119). In both cases the carbene added with *syn*-chloro stereoselectivity. Adducts **69** and **70** were formed in excess of their isomers by factors of 1.2–1.5 and 1.5–1.6, respectively.

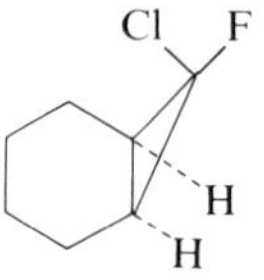

69 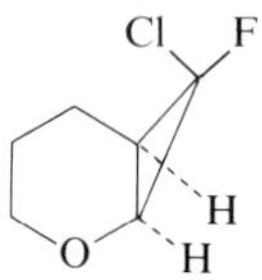 **70**

With acyclic substrates, *syn*-chloro stereoselectivity was also observed. Thus CFCl, generated by reaction of potassium *t*-butoxide with *sym*-difluorotetrachloroacetone at $-10°$, added to *cis*-butene, trimethylethylene, and 1-butene with preferential formation of the *syn*-chlorocyclopropane by factors of 3.1, 2.4, and 1.5, respectively (45,120). Addition to styrene was *syn*-chloro stereoselective by the small factor of 1.2 (121).

Fluorobromocarbene was generated by the action of potassium *t*-butoxide on fluorodibromomethane at -10 to $-15°$ and was added to cyclohexene, trimethylethylene, and 1-hexene with *syn*-bromo stereoselectivity. The *syn*-bromo:*anti*-bromo preferences were 1.4–1.7, 1.4–1.5, and 1.5, respectively. With styrene, *anti*-bromo, or *syn*-fluoro stereoselectivity was observed, the preference was 1.1:1 (121).

It is not yet clear whether these carbenic species are free carbenes or carbene-base complexes. The demonstration of similar reactivities for the CCl_2 species produced by the action of potassium *t*-butoxide on chloroform and by the action of sodium iodide on phenyl(bromodichloromethyl) mercury (42) suggests that the former method produces essentially free CCl_2. By implication CFCl and CFBr, produced in *t*-butoxide reactions, may also be true divalent carbon species.

Considering the addition of CFX (X = Cl, Br) to *cis*-butene, for example, in terms of our model (Section IV-A) two transition states, **71a** and **71b**, can be drawn. The former leads to *syn*-fluoro product; the latter leads to *syn*-X product.

The data indicate that **71b** is of lower free energy. This has been attributed to stabilizing interactions of the 3*p* or 4*p* lone pair electrons of X

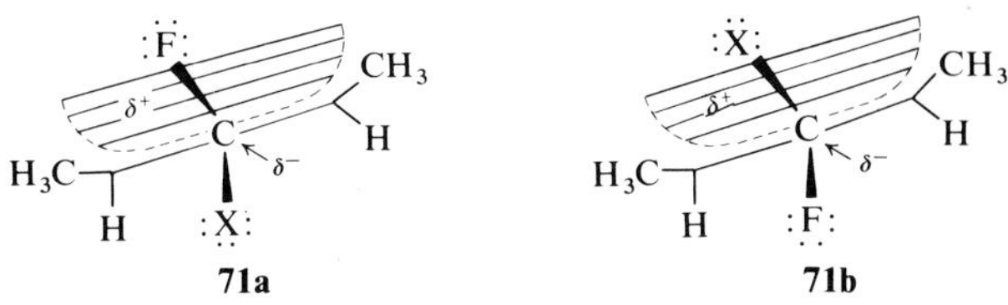

71a **71b**

with the partially positive olefinic substituents. Such interactions are more favorable than the analogous interaction of the less polarizable fluorine $2p$ electrons in transition state **71a**.

In a tighter or otherwise altered set of transition states in which greater steric demands were present *syn*-Fluoro addition rather than *syn*-X addition could be expected. An example is the addition of fluorochlorocarbene to norbornene, in which adduct **16** is slightly preferred over its *anti*-Fluoro epimer (see Section II-A). The departure from *syn*-chloro stereoselectivity is

F Cl H H +

16

due to steric hindrance originating in repulsive nonbonded interactions between the chlorine atom and the *syn*-bridge proton.

2. *Phenylhalocarbenes*

Stereoselectivity has been reported for the species produced by the action of potassium *t*-butoxide on benzal bromide (51) and by the photolysis of phenylbromodiazirine (52). Comparison of the abilities of the two species to discriminate between different olefins suggested that the photolytic species was probably a free carbene but that the base-generated species was likely to be a carbenoid (52). However, the stereoselectivity displayed by each of the two species was very similar. Additions to *cis*-butene and trimethylethylene led to preferential formation of the *syn*-bromocyclopropane. The *syn*-bromo:*anti*-bromo distributions were: *cis*-butene, 1.55 (photolysis) and 1.35 (base); trimethylethylene, 1.31 (photolysis), and 1.28 (base).

Phenylchlorocarbene species have been produced by several methods, and a good deal of stereoselectivity data is available. One generative method involved the action of methyllithium on benzal chloride (122). The reactive intermediate here probably was the carbenoid, **72** (123). A second generative method involved the reaction of potassium *t*-butoxide with benzal chloride

(90,124); while a third method involved the photolysis of phenylchlorodiazirine (125). Stereoselectivity data for various olefinic substrates, employing each of these methods is gathered in Table 12.

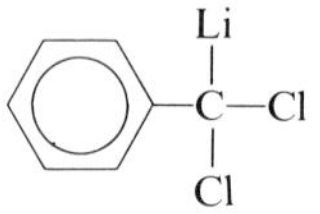

$$C_6H_5-C(Li)(Cl)-Cl$$

72

The trends in these data are quite clear. With the exception of the cyclopentene case, over a variety of generative procedures and temperatures, the phenylchlorocarbene species add to the olefins so as to produce an excess of that product cyclopropane having the chlorine atom *cis* to the larger number of alkyl substituents (*syn*-chloro addition).

TABLE 12

Stereoselectivity of Phenylchlorocarbene Additions to Various Olefins, *Syn*-Cl : *Anti*-Cl Distributions

Olefin	Generative method		
	$\phi CHCl_2 + CH_3Li(LiBr)$[a]	$\phi CHCl_2 + K^+\bar{O}\text{-}t\text{-Bu}$	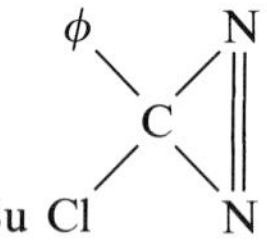 $\phi(Cl)C<(N=N)$ (phenylchlorodiazirine) $+ h\nu$[b]
cis-Butene	3.0	3.0[c], 2.2[d]	2.0
Cyclohexene	—	2.0[c] —	—
Trimethylethylene	1.6	1.5[c], 1.2[d]	1.3
1-Butene	1.8	1.7[c] —	—
Cyclopentene	—	0.7[e] —	—

[a] At −40°, data of ref. 90.
[b] At ca. 25°, data of refs. 125 and 126.
[c] At 60–75°, data of ref. 90.
[d] At ca. 25°, data of ref. 126.
[e] At 60–75°, ref. 127.

Comparison of the species produced by the butoxide and photolytic methods in their abilities to distinguish between different olefinic substrates suggests that, as in the phenylbromocarbene case, the butoxide method does not lead to a free carbene (125,126). The intermediate(s) of each reaction summarized in Table 12 is probably different from the others. Nevertheless, the stereoselectivity hardly varies from one generative procedure and temperature to the next.

Phenylfluorocarbene has been generated by the action of potassium *t*-butoxide on either α-bromo-α-fluorotoluene (128) or on α-chloro-α-fluorotoluene (129). The stereoselectivity of this species is summarized in Table 13. It is not yet known whether free phenylfluorocarbene is produced in these reactions.

TABLE 13
Stereoselectivity of Phenylfluorocarbene Additions to Various Olefins, *Syn*-F : *Anti*-F Distributions

Olefin	*Syn*-F : *Anti*-F Distribution	
cis-Butene	2.0[a]	1.23[b]
Cyclohexene	2.1[a]	—
Trimethylethylene	0.77[a]	0.76[b]
2-Methylpentene-2	0.85[a]	—

[a] Data of ref. 129, 60–80°.
[b] Data of ref. 126, ca. 25°.

Table 14 collects some of the stereoselectivity data for three phenylhalocarbenes. Comparisons within this table seem justified because, for the cases shown, the species were generated under common conditions of base and temperature. Furthermore, in the cases of phenylbromo and phenylchlorocarbenes, it is known that the base-generated and the photolytically-generated intermediates exhibit very similar stereoselectivity.

TABLE 14
Comparative Stereoselectivity Data for Phenylhalocarbenes[a]

	Stereoselectivity : *syn*-halo : *anti*-halo		
Olefin	ϕ—C—Br[b]	ϕ—C—Cl[c]	ϕ—C—F[c]
cis-Butene	1.35	2.2	1.23
Trimethylethylene	1.28	1.2	0.76

[a] Produced by the action of potassium *t*-butoxide on benzal halides at ca. 25°.
[b] Ref. 51.
[c] Ref. 126.

In all cases, however, the product distributions are not far from 1:1, and the thermodynamically controlled product distribution is unknown.

The phenyl group being more sterically demanding than the halogen atoms, a prediction of *anti*-phenyl or *syn*-halo stereoselectivity would not be unreasonable. This is, in fact, generally observed. Of interest is the slight decrease in *syn*-halo stereoselectivity (*cis*-butene data), accompanying the change from phenylchloro to phenylbromocarbene. This suggests that whatever increase in *syn*-halo addition could *a priori* be expected as a result of exchanging the chlorine for the more polarizable bromine atom is, in fact, cancelled by the greater size and steric demand of the bromine atom (51).

The stereoselectivity reversal in the addition of phenylfluorocarbene to trimethylethylene is also of interest. In the reaction of either fluorochlorocarbene (120) or fluorobromocarbene (121) with trimethylethylene, the *anti*-fluoro addition mode is favored to a *greater extent* than is the *anti*-phenyl addition mode in the reactions of phenylbromocarbene (51) or phenylchlorocarbene (126) with the same olefin. Preferential *anti*-fluoro (*syn*-phenyl) addition of phenylfluorocarbene to trimethylethylene is therefore expected, and points to the dominance of attractive nonbonded interactions between the phenyl and olefinic alkyl groups in determining the stereoselectivity of this reaction. On the other hand, an analogous argument also predicts predominant *syn*-phenyl addition of phenylfluorocarbene to *cis*-butene. This is not observed. A possible explanation notes that *cis*-butene is a ten times poorer acceptor of phenylfluorocarbene than is trimethylethylene (126). The addition of phenylfluorocarbene to *cis*-butene would proceed through a "tighter," more productlike transition state than does the addition to trimethylethylene. In such a transition state the nonbonded interactions associated with *syn*-phenyl addition have become repulsive, rather than attractive (126).

Attempts have been made to treat the above ideas in a more quantitative manner. Thus one can inquire whether the stereoselectivity of e.g., phenylfluorocarbene addition to olefins can be "exactly" predicted from an expression of the following kind:

$$\left(\frac{syn\text{-F}}{syn\text{-Cl}}\right)_{\text{F—C—Cl}}\left(\frac{syn\text{-Cl}}{syn\text{-Ph}}\right)_{\text{Ph—C—Cl}} = \left(\frac{syn\text{-F}}{syn\text{-Ph}}\right)_{\text{Ph—C—F}}$$

Here the input data consist of the stereoselectivity results for CFCl and for phenylchlorocarbene. When this calculation is carried out, with all carbenes generated in the same manner and at 25°, the predicted stereoselectivities of phenylfluorocarbene are 0.88 (*cis*-butene) and 0.54 (trimethylethylene). The observed values (Table 14) were 1.23 and 0.76, respectively (125). It has been pointed out (125) that this type of treatment should only succeed when the three carbenes are of comparable kinetic selectivity. CFCl is about 10 times more selective than either of the phenylhalocarbenes and the transition states of its olefin addition reactions are likely to be considerably "tighter" than those of the other two carbenes (126). If the CFCl addition transition states were "looser," the *syn*-F : *syn*-Cl ratios, which presumably depend on the differential polarizabilities of Cl and F, would increase, and the predicted *syn*-F/*syn*-Ph ratios would more closely approach the observed values.

This point was not recognized by Schlosser and Heinz (126a) who concluded, from a study of stereoselectivity in fluoro, chloro, and fluorochloro *carbenoid* addition reactions, that since an expression of the above type was not followed, there must be something wrong with present thinking on the origin of stereoselectivity effects. In the light of the foregoing, and since the dihalocarbenoid is likely to be a much more selective species than the monohalocarbenoids, the conclusions of Schlosser and Heinz (126a) cannot be regarded as cogent.

3. *Other Disubstituted Carbenes*

It has been reported that chlorotrimethylsilylcarbene, **73**, adds to cyclohexene with production of only one cyclopropane, presumably **74** (130).

$(CH_3)_3Si—\ddot{C}—Cl$

73

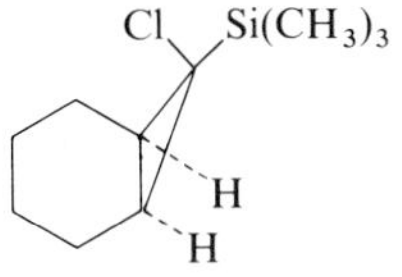

Cl Si(CH$_3$)$_3$ H H

74

The addition of methoxyphenylcarbene to 1-decene (158°) afforded a low yield of the expected cyclopropanes, with the *cis*-methoxy isomer predominating, 1.8:1 (131). Addition of phenylcyanocarbene to trimethylethylene was *syn*-cyano stereoselective by a factor of 1.86:1 (132). It should be noted that it is not yet known whether the addition reactions of these three carbenes are stereospecific. Although subject to reservations because of the paucity of data, the above species exhibit stereoselectivity consonant with predictions based on relative steric hindrance in the two modes of addition.

The reaction of benzal chloride with methyllithium (lithium bromide) produces phenylmethylcarbenoids as well as phenylchlorocarbenoids, and 1-phenyl-1-methylcyclopropanes are formed from olefinic substrates. At −40°, phenylmethylcarbenoid added to trimethylethylene, *cis*-butene, and 1-butene with *syn*-phenyl stereoselectivity. The *syn*-phenyl : *anti*-phenyl distributions were 1.4, 2.3, and 1.8, respectively (90). Comparison of stereoselectivity data for the addition of methylcarbenoid to cyclohexene (*syn*-methyl addition preferred by a factor of ca. 1.8, see Table 11) and for the addition of phenylcarbenoid to cyclohexene (*syn*-phenyl addition preferred by a factor of 3.0, see Table 7), suggests that the addition of phenylmethylcarbenoid to cyclohexene would be mainly *syn*-phenyl. Although the cyclohexene addition has not been studied, the three reported cases all exhibit *syn*-phenyl stereoselectivity.

In contrast to the above, photolysis of 1-phenyldiazoethane in *cis*-butene ($-40°$), showed *anti*-phenyl stereoselectivity. In this reaction, which should involve free phenylmethylcarbene, the *syn*-phenyl:*anti*-phenyl distribution was 1:1.3 (90). However, decomposition of the diazo compound with $ZnCl_2$ in *cis*-butene gave *syn*-phenyl stereoselectivity again, 2.0:1 (89).

The phenylmethylcarbenoids add with predominant *syn*-phenyl stereoselectivity, and it seems likely that this is not the stereochemistry expected from steric hindrance considerations. Attractive nonbonded interactions between the phenyl and olefinic alkyl groups, as discussed for phenylcarbenoid additions (Section IV-B-2) may be invoked to explain the observations. The free phenylmethylcarbene shows a slight *anti*-phenyl stereoselectivity, which may reflect the most probable approach geometry of the carbene and olefin in a "loose" transition state.

References

1. W. von E. Doering and A. K. Hoffmann, *J. Am. Chem. Soc.*, **76**, 6162 (1954).
2a. P. S. Skell and A. Y. Garner, *J. Am. Chem. Soc.*, **78**, 3409 (1956).
2b. P. S. Skell and A. Y. Garner, *J. Am. Chem. Soc.*, **78**, 5430 (1956).
3. W. von E. Doering and W. A. Henderson, Jr., *J. Am. Chem. Soc.*, **80**, 5274 (1958).
4. H. E. Simmons, E. P. Blanchard, and R. D. Smith, *J. Am. Chem. Soc.*, **86**, 1347 (1964), and references therein.
5. R. E. Pincock and J. I. Wells, *J. Org. Chem.*, **29**, 965 (1964).
6. See P. von R. Schleyer, *J. Am. Chem. Soc.*, **89**, 701 (1967), for another possible explanation.
7a. E. Müller and H. Kessler, *Tetrahedron Letters*, **1968**, 3037.
7b. H. Prinzbach and E. Druckrey, *Tetrahedron Letters*, **1968**, 4285.
7c. M. Mühlstädt, H. van Phiet, J. Graefe, and H. Frischleder, *Tetrahedron*, **24**, 6075 (1968).
7d. H. Frischleder, J. Graefe, H. van Phiet, and M. Mühlstädt, *Tetrahedron*, **25**, 2081 (1969).
8. W. R. Moore, W. R. Moser, and J. E. LaPrade, *J. Org. Chem.*, **28**, 2200 (1963).
9. R. C. De Selms and C. M. Combs, *J. Org. Chem.*, **28**, 2206 (1963); E. Bergman, *J. Org. Chem.*, **28**, 2210 (1963).
10. L. Ghosez, P. Laroche, and G. Slinckx, *Tetrahedron Letters*, **1967**, 2767.
11. L. Ghosez, P. Laroche, and L. Bastens, *Tetrahedron Letters*, **1964**, 3745.
12. J. S. Brimacombe, P. A. Gent, and T. A. Hamor, *Chem. Commun.*, **1967**, 1305.
13. C. W. Jefford and R. Medary, *Tetrahedron Letters*, **1966**, 2069.
14. C. W. Jefford, E. H. Yen, and R. Medary, *Tetrahedron Letters*, **1966**, 6317; C. W. Jefford and R. Medary, *Tetrahedron*, **23**, 4123 (1967).
15. R. Hoffmann and R. B. Woodward, *Accts. Chem. Res.*, **1**, 17 (1968); C. H. DePuy, *Accts. Chem. Res.*, **1**, 33 (1968).
16. C. W. Jefford and W. Wojnarowski, *Tetrahedron Letters*, **1968**, 193; *Tetrahedron*, **25**, 2089 (1969).
17. L. Ghosez, G. Slinckx, M. Glineur, P. Hoet, and P. Laroche, *Tetrahedron Letters*, **1967**, 2773.

17a. C. W. Jefford and D. T. Hill, *Tetrahedron Letters*, **1969**, 1957.
18. L. H. Knox, E. Velarde, S. Berger, D. Cuadriello, P. W. Landis, and A. D. Cross, *J. Am. Chem. Soc.*, **85**, 1851 (1963).
19. M. Z. Nazer, *J. Org. Chem.*, **30**, 1737 (1965).
19a. F. T. Bond and R. H. Cornelia, *Chem. Commun.*, **1968**, 1189; R. H. Cornelia, Ph.D. thesis, Oregon State University, 1968.
20. A. B. Font, *Bull. Soc. Chim. France*, **1964**, 419.
21. P. G. Gassman and W. E. Hymans, *Chem. Commun.*, **1967**, 795; P. G. Gassman and W. E. Hymans, *Tetrahedron*, **24**, 4437 (1968); J. Joska, J. Fajkos, and F. Sorm, *Coll. Czech. Chem. Commun.*, **33**, 2049 (1968).
22. T. L. Popper, F. E. Carlon, H. M. Marigliano, and M. D. Yudis, *Chem. Commun.*, **1968**, 277; C. Beard, I. T. Harrison, L. Kirkham, and J. H. Fried, *Tetrahedron Letters*, **1966**, 3287.
23. G. Tarzia, N. H. Dyson, I. T. Harrison, J. A. Edwards, and J. H. Fried, *Steroids*, **9**, 387 (1967); I. T. Harrison, C. Beard, L. Kirkham, B. Lewis, I. M. Jamieson, W. Rooks, and J. H. Fried, *J. Med. Chem.*, **11**, 868 (1968); see also; C. Beard, B. Berkoz, N. H. Dyson, I. T. Harrison, P. Hodge, L. H. Kirkham, G. S. Lewis, D. Giannini, B. Lewis, J. A. Edwards, and J. H. Fried, *Tetrahedron*, **25**, 1219 (1969).
24. C. Beard, N. H. Dyson, and J. H. Fried, *Tetrahedron Letters*, **1966**, 3281.
25. R. C. Cookson, D. P. G. Hamon, and J. Hudec, *J. Chem. Soc.*, **1963**, 5782.
26. M. E. Wolff, S.-Y. Cheng, and W. Ho, *J. Med. Chem.*, **11**, 864 (1968).
27. A. J. Birch and G. S. R. Subba Rao, *Tetrahedron, Suppl. 7*, 391 (1966); A. J. Birch, J. M. H. Graves, and J. B. Siddall, *J. Chem. Soc.*, **1963**, 4234.
28a. A. J. Birch and G. S. R. Subba Rao, *J. Chem. Soc.*, (C), **1967**, 2509; A. J. Birch, J. M. Brown, and G. S. R. Subba Rao, *J. Chem. Soc.*, **1964**, 3309.
28b. R. Ginsig and A. D. Cross, *J. Am. Chem. Soc.*, **87**, 4629 (1965).
29. P. Crabbé, P. Anderson, and E. Velarde, *J. Am. Chem. Soc.*, **90**, 2998 (1968).
30. S. Winstein and J. Sonnenberg, *J. Am. Chem. Soc.*, **83**, 3235 (1961).
31. For examples, P. Radlick and S. Winstein, *J. Am. Chem. Soc.*, **86**, 1866 (1964); A. C. Cope, S. Moon, and C. H. Park, *J. Am. Chem. Soc.*, **84**, 4843 (1962); H. D. Berndt and R. Wiechert, *Angew. Chem., Intern. Ed.*, **8**, 376 (1969).
32. W. G. Dauben and G. H. Berezin, *J. Am. Chem. Soc.*, **85**, 468 (1963); also, W. G. Dauben and A. C. Ashcraft, *J. Am. Chem. Soc.*, **85**, 3673 (1963).
33a. A. C. Cope, S. Moon, and P. E. Peterson, *J. Am. Chem. Soc.*, **84**, 1935 (1962).
33b. J. J. Sims, *J. Am. Chem. Soc.*, **87**, 3511 (1965); T. Hanafusa, L. Birdladeanu, and S. Winstein, *J. Am. Chem. Soc.*, **87**, 3510 (1965).
33c. J. H-H. Chan and B. Rickborn, *J. Am. Chem. Soc.*, **90**, 6407 (1968).
34. R. S. Boikess and S. Winstein, *J. Am. Chem. Soc.*, **85**, 343 (1963).
35. P. Radlick and S. Winstein, *J. Am. Chem. Soc.*, **85**, 344 (1963); K. G. Untch, *J. Am. Chem. Soc.*, **85**, 345 (1963).
36. J. M. Locke and E. W. Duck, *Chem. Ind.*, **1965**, 1727.
37. H. Nozaki, S. Katô, and R. Noyori, *Can. J. Chem.*, **44**, 1021 (1966); M. Mühlstädt and J. Graefe, *Chem. Ber.*, **99**, 1192 (1966).
37a. M. Mühlstädt and J. Graefe, *Z. Chem.*, **8**, 299 (1968).
38. H. Nozaki, M. Kawanisi, and R. Noyori, *J. Org. Chem.*, **30**, 2216 (1965).
39. M. Ohno, M. Okamoto, and N. Naruse, *Tetrahedron Letters*, **1965**, 1971.
39a. J. Graefe and M. Mühlstädt, *Z. Chem.*, **9**, 23 (1969).
40. O. M. Nefedov, M. N. Manakov, and A. A. Ivashenko, *Izvest. Akad. Nauk. SSSR, Otd. Khim. Nauk.*, **1962**, 1242; *Chem. Abstr.*, **58**, 5528 (1963).
41. A. Bezaguet and M. Bertrand, *Compt. Rend. Acad. Sci. Paris* (C), **262**, 428 (1966).

41a. O. M. Nefedov and R. N. Shafran, *Zhur. Obsch. Khim.*, **37**, 1561 (1967).
42. D. Seyferth, M. E. Gordon, J. Y.-P. Mui, and J. M. Burlitch, *J. Am. Chem. Soc.*, **89**, 959 (1967).
43a. R. A. Moss and A. Mamantov, *Tetrahedron Letters*, **1968**, 3425.
43b. D. Seyferth and H. Dertouzos, *J. Organometallic Chem.*, **11**, 263 (1968).
44. R. A. Moss, *J. Org. Chem.*, **30**, 3261 (1965).
45. R. A. Moss and R. Gerstl, *J. Org. Chem.*, **32**, 2268 (1967).
46. J. Hine and S. J. Ehrenson, *J. Am. Chem. Soc.*, **80**, 824 (1958); J. Hine, *Divalent Carbon*, Ronald Press Co., New York, 1964.
47. R. A. Moss, *J. Org. Chem.*, **31**, 3296 (1966); *Chem. Commun.*, **1965**, 622.
47a. H. Dürr and L. Schrader, *Chem. Ber.*, **102**, 2026 (1969).
47b. H. Dürr and G. Scheppers, *Chem. Ber.*, **100**, 3236 (1967).
48. M. Jones, Jr., A. M. Harrison and K. R. Rettig, *J. Am. Chem. Soc.*, **91**, 7462 (1969).
49. M. Jones, Jr., A. Kulczycki, Jr., and K. F. Hummel, *Tetrahedron Letters*, **1967**, 183.
50. W. M. Jones, M. H. Grasley, and W. S. Brey, Jr., *J. Am. Chem. Soc.*, **85**, 2754 (1963).
51. R. A. Moss and R. Gerstl, *Tetrahedron*, **22**, 2637 (1966).
52. R. A. Moss, *Tetrahedron Letters*, **1967**, 4905.
53. E. P. Blanchard and H. E. Simmons, *J. Am. Chem. Soc.*, **86**, 1337 (1964); B. Rickborn and J. H.-H. Chan, *J. Org. Chem.*, **32**, 3576 (1967).
54. U. Schöllkopf and H. Görth, *Annalen*, **709**, 97 (1967).
54a. W. H. Atwell, D. R. Weyenberg, and J. G. Uhlmann, *J. Am. Chem. Soc.*, **91**, 2025 (1969).
55. W. R. Moore, H. R. Ward, and R. F. Merritt, *J. Am. Chem. Soc.*, **83**, 2019 (1961).
56. G. Cauquis and G. Reverdy, *Tetrahedron Letters*, **1968**, 1085.
57. P. S. Skell and R. C. Woodworth, *J. Am. Chem. Soc.*, **78**, 4496 (1956); P. P. Gaspar and G. S. Hammond, in *Carbene Chemistry*, W. Kirmse, Ed., Academic Press, New York, 1964, pp. 235 ff; R. A. Moss and J. R. Przybyla, *J. Org. Chem.*, **33**, 3816 (1968).
58. G. L. Closs and R. A. Moss, *J. Am. Chem. Soc.*, **86**, 4042 (1964).
59. R. Hoffmann, *J. Am. Chem. Soc.*, **90**, 1475 (1968).
60. G. L. Closs and L. E. Closs, *J. Am. Chem. Soc.*, **82**, 5723 (1960).
61. G. L. Closs, R. A. Moss, and J. J. Coyle, *J. Am. Chem. Soc.*, **84**, 4985 (1962).
62. U. Schöllkopf, *Chem. Eng. News*, Aug. 5, 1963, p. 42.
63. W. von E. Doering and L. H. Knox, *J. Am. Chem. Soc.*, **83**, 1989 (1961).
64. T. C. Neil, Ph.D. Thesis, The Pennsylvania State University, 1964; *Diss. Abstr.*, **25**, 6239 (1965).
65. P. S. Skell and R. M. Etter, *Chem. Ind.*, **1958**, 624.
66. W. Kirmse, *Carbene Chemistry*, Academic Press, New York, 1964, pp. 97 ff.
67. W. von E. Doering and T. Mole, *Tetrahedron*, **10**, 65 (1960).
68. E. N. Trachtenberg and G. Odian, *J. Am. Chem. Soc.*, **80**, 4015 (1958).
69. K. B. Wiberg, R. K. Barnes, and J. Albin, *J. Am. Chem. Soc.*, **79**, 4994 (1957).
70. P. S. Skell and R. M. Etter, *Proc. Chem. Soc.*, **1961**, 443; see, also, the elegant work of W. R. Moser, *J. Am. Chem. Soc.*, **91**, 1135, 1141 (1969), which appeared too late to be incorporated into this chapter in detail, but which bears importantly on catalyzed carboethoxycarbene additions.
71. J. Warkentin, E. Singleton, and J. F. Edgar, *Can. J. Chem.*, **43**, 3456 (1965).
72. H. Musso and U. Biethan, *Chem. Ber.*, **97**, 2282 (1964).
73. J. A. Berson and E. S. Hand, *J. Am. Chem. Soc.*, **86**, 1978 (1964).
74. K. F. Bangert and V. Boekelheide, *J. Am. Chem. Soc.*, **86**, 905 (1964).
75. R. A. Moss, unpublished work.
76. J. H. Leftin, E. Gil-Av, and A. Pines, *Chem. Commun.*, **1968**, 396.
77. I. A. D'yakonov, V. V. Razin, and M. I. Komendantov, *Tetrahedron Letters*, **1966**, 1127.

78. F. W. Breitbeil, III, J. J. McDonnell, T. A. Marolewski, and D. T. Dennerlein, *Tetrahedron Letters*, **1965**, 4627; see also, F. W. Breitbeil, III, D. T. Dennerlein, A. E. Fiebig, and R. E. Kuznicki, *J. Org. Chem.*, **33**, 3389 (1968).
79. R. R. Sauers and P. E. Sonnet, *Tetrahedron*, **20**, 1029 (1964).
80. R. R. Sauers, S. B. Schlosberg, and P. E. Pfeffer, *J. Org. Chem.*, **33**, 2175 (1968).
81. S. C. Clarke and B. L. Johnson, *Tetrahedron*, **24**, 5067 (1968).
82. R. Huisgen and G. Juppe, *Chem. Ber.*, **94**, 2332 (1961).
83. E. Müller and H. Kessler, *Annalen*, **692**, 58 (1966).
84. G. E. Hall and J. P. Ward, *Tetrahedron Letters*, **1965**, 437.
85. J. H. Atherton and R. Fields, *J. Chem. Soc. (C)*, **1967**, 1450.
86. J. H. Atherton and R. Fields, *J. Chem. Soc. (C)*, **1968**, 1507.
87. R. A. Abramovitch and J. Roy, *Chem. Commun.*, **1965**, 542.
88. A. M. Van Leusen, R. J. Mulder, and J. Strating, *Rec. Trav. Chim.*, **86**, 225 (1967).
89. S. H. Goh, L. E. Closs, and G. L. Closs, *J. Org. Chem.*, **34**, 25 (1969).
90. G. L. Closs and J. J. Coyle, *J. Org. Chem.*, **31**, 2759 (1966).
91. H. Dietrich and G. W. Griffen, *Tetrahedron Letters*, **1968**, 153, and references therein, especially footnote 4.
92. J. Furukawa, N. Kawabata, and J. Nishimura, *Tetrahedron Letters*, **1968**, 3495.
93. H. Nozaki, M. Yamabe, and R. Noyori, *Tetrahedron*, **21**, 1657 (1965).
94. U. Schöllkopf, A. Lerch, and W. Pitteroff, *Tetrahedron Letters*, **1962**, 241; U. Schöllkopf and W. Pitteroff, *Chem. Ber.*, **97**, 636 (1964).
95. U. Schöllkopf and J. Paust, *Chem. Ber.*, **98**, 2221 (1965).
96. U. Schöllkopf, J. Paust, A. Al-Azrak, and H. Schumacher, *Chem. Ber.*, **99**, 3391 (1960).
97. U. Schöllkopf, A. Lerch, and J. Paust, *Chem. Ber.*, **96**, 2266 (1963).
98. U. Schöllkopf, G. J. Lehmann, J. Paust, and H-D. Härtl, *Chem. Ber.*, **97**, 1527 (1964); U. Schöllkopf and G. J. Lehmann, *Tetrahedron Letters*, **1962**, 165.
99. U. Schöllkopf and H. Küppers, *Tetrahedron Letters*, **1963**, 105.
100. Y.-N. Tang and F. S. Rowland, *J. Am. Chem. Soc.*, **89**, 6420 (1967); Y.-N. Tang and F. S. Rowland, *J. Am. Chem. Soc.*, **88**, 626 (1966).
101. G. L. Closs and J. J. Coyle, *J. Am. Chem. Soc.*, **87**, 4270 (1965); G. L. Closs and J. J. Coyle, *J. Am. Chem. Soc.*, **84**, 4350 (1962).
102. T. Marolewski and N. C. Yang, *Chem. Commun.*, **1967**, 1225; compare with, N. C. Yang and T. A. Marolewski, *J. Am. Chem. Soc.* **90**, 5644 (1968).
103. Reviews: G. Köbrich, *Angew. Chem., Intern. Ed.*, **6**, 41 (1967); *Bull. Soc. Chim. France*, **1969**, 2712; W. Kirmse, *Angew. Chem., Intern. Ed.*, **4**, 1 (1965).
104. G. L. Closs and G. M. Schwartz, *J. Am. Chem. Soc.*, **82**, 5729 (1960).
105. M. S. Baird and C. B. Reese, *Tetrahedron Letters*, **1967**, 1379.
106. E. A. LaLancette and R. E. Benson, *J. Am. Chem. Soc.*, **85**, 2853 (1963).
107. T. J. Katz and P. J. Garratt, *J. Am. Chem. Soc.*, **86**, 4876 (1964).
108. W. L. Dilling, *J. Org. Chem.*, **29**, 960 (1964).
109. R. M. Magid and J. G. Welch, *Tetrahedron Letters*, **1967**, 2619.
110. T. J. Katz and P. J. Garratt, *J. Am. Chem. Soc.*, **86**, 5194 (1964); T. J. Katz and P. J. Garratt, *J. Am. Chem. Soc.*, **85**, 2852 (1963).
111. J. A. Deyrup and R. B. Greenwald, *Tetrahedron Letters*, **1965**, 321.
112. W. L. Dilling and F. Y. Edamura, *J. Org. Chem.*, **32**, 3492 (1967); W. L. Dilling and F. Y. Edamura, *Tetrahedron Letters*, **1967**, 587; W. L. Dilling and F. Y. Edamura, *Chem. Commun.*, **1967**, 183.
113. R. M. Magid and J. G. Welch, *Chem. Commun.*, **1967**, 518.
114. R. L. Hatch and P. D. Gardner, *Chem. Commun.*, **1967**, 1019.
115. H. M. Frey, *J. Chem. Soc.*, **1962**, 2293; H. M. Frey, *Chem Ind.*, **1962**, 218.

116. G. Wittig and M. Jautelat, *Annalen*, **702**, 24 (1967).
117. E. T. Marquis and P. D. Gardner, *Chem. Commun.*, **1966**, 726.
118. T. Ando, H. Yamanaka, S. Terabe, A. Horike, and W. Funasaka, *Tetrahedron Letters*, **1967**, 1123.
119. T. Ando, H. Yamanaka, and W. Funasaka, *Tetrahedron Letters*, **1967**, 2587.
120. R. A. Moss and R. Gerstl, *Tetrahedron*, **23**, 2549 (1967).
121. W. Funaska, T. Ando, H. Yamanaka, H. Kanehira, and Y. Shimokawa, "Symposium on Organic Halogen Compounds," Tokyo, Nov. 29–30, 1967; Abstracts, pp. 25 ff.
122. R. A. Moss, *J. Org. Chem.*, **27**, 2683 (1962).
123. D. F. Hoeg, D. I. Lusk, and A. L. Crumbliss, *J. Am. Chem. Soc.*, **87**, 4147 (1965).
124. J. E. Hodgkins, J. D. Woodyard, and D. L. Stephenson, *J. Am. Chem. Soc.*, **86**, 4080 (1964); F. R. Jensen and D. B. Patterson, *Tetrahedron Letters*, **1966**, 3837.
125. R. A. Moss, J. R. Whittle, and P. Freidenreich, *J. Org. Chem.*, **34**, 2220 (1969).
126. R. A. Moss and J. R. Przybyla, *Tetrahedron*, **25**, 647 (1969).
126a. M. Schlosser and G. Heinz, *Angew. Chem., Intern. Ed.*, **7**, 820 (1968).
127. G. L. Closs, private communication.
128. R. A. Moss, *Tetrahedron Letters*, **1968**, 1961.
129. T. Ando, Y. Kôtoku, H. Yamanaka, and W. Funasaka, *Tetrahedron Letters*, **1968**, 2479.
130. D. Seyferth and E. M. Hanson, *J. Am. Chem. Soc.*, **90**, 2438 (1968).
131. R. J. Crawford and R. Raap, *Can. J. Chem.*, **43**, 356 (1965).
132. P. C. Petrellis, H. Dietrich, E. Meyer, and G. W. Griffen, *J. Am. Chem. Soc.*, **89**, 1967 (1967).

Stereoselectivity in Quaternizations of Tertiary Amines

A. T. BOTTINI

Department of Chemistry, University of California, Davis, California

I. INTRODUCTION

A. Historical

Early in this century, Scholtz and his co-workers (1–4) described the first studies concerned with stereoselectivity in amine quaternizations. These studies, mainly of reactions of 1,2-disubstituted piperidines with alkyl halides, were followed—nearly 25 years later—by those of Mills, Parkin, and Ward (5), who examined alkylations of 1,4-disubstituted piperidines, and of Polonovski and Polonovski (6,7), who examined *N*-oxidations of a number of tertiary amines including tropine, atropine, and scopolamine. Another quarter century passed before interest in stereoselectivity of amine quaternizations was renewed by Fodor and his co-workers (8) with their admirable studies of the stereochemistry of tropane alkaloids. In 1959 Closs (9) showed how NMR spectroscopy could be applied so productively to this field, and since 1960 widespread and intensive efforts, depending to a considerable degree on NMR spectroscopy, have been devoted to the study of stereoselectivity in amine quaternizations, particularly reactions of tertiary cyclic amines with alkyl halides and sulfonate esters.

B. Scope

The main purpose of this article is to review research concerned with the stereochemistry of *N*-alkylations and *N*-oxidations of tertiary amines. With the exception of particular findings that are pertinent to discussion of these classes of reactions, studies of protonated and deuteriated amines are not reviewed here.

C. The Nature of Stereoselectivity in Quaternization Reactions

Detailed consideration of the reaction of *trans*-1,2-dimethyl-3-isopropylaziridine (**1**) with deuteriomethyl benzenesulfonate (eq. 1) will serve to illustrate the problems and goals that provide the motivation for study of stereoselectivity in amine quaternizations.

Because trivalent nitrogen inverts rapidly (10,11), **1** exists as a mixture of energetically different conformers (invertomers) **1a** and **1b** in rapid equilibrium. As the steric requirements of the isopropyl group are greater than those of the methyl group (12), which in turn are greater than those of hydrogen or the electron pair on nitrogen (13), **1b** is the more stable conformer. When **1** is quaternized with deuteriomethyl benzenesulfonate, the configuration at nitrogen becomes fixed, and two diasteromeric salts, **2a** and **2b**, are formed. If the salts are formed in unequal amounts, the reaction can be described as stereoselective (14), and the more one salt predominates, the greater the degree of stereoselectivity. Determination of the degree and manner of stereoselectivity of this reaction requires estimation of the relative amounts of **2a** and **2b** and their identification. It is also necessary to establish whether or not the products are equilibrated under the conditions used for their preparation.

i-C_3H_7, H_3C, N, CH_3 — **1a** ⇌ i-C_3H_7, H_3C, N, H_3C — **1b**

↓ $CD_3OSO_2\phi$ ↓ $CD_3OSO_2\phi$ (1)

i-C_3H_7, H_3C, N^+, ϕSO_3^-, D_3C, CH_3 — **2a**; i-C_3H_7, H_3C, N^+, ϕSO_3^-, H_3C, CD_3 — **2b**

Theoretical considerations and experimental observations strongly support the point of view that transition states for quaternizations of aliphatic amines involve tetrahedral nitrogen, with the large lobe of the sp^3 or nearly sp^3 orbital used for bond development to the quaternizing agent, as depicted in **3** (15). Thus, during and after the activation process, the configuration of the amine moiety is well approximated by that of the corresponding amine conformer. As a consequence fruitful collision of deuteriomethyl benzenesulfonate with either **1a** or **1b** gives only the corresponding quaternary salt **2a** or **2b**. Alternative transition states, e.g., **4** and **5**, which would provide pathways for conversion of **1a** to **2b**, or **1b** to **2a**, without intermediacy of the other conformer, are unlikely for steric and electronic reasons. Formation of **4** would require an inordinate amount of energy because of the nonbonded interactions introduced. Consideration of a transition state represented by **5** is less subject to criticism on steric grounds; but not only is the population of an aliphatic amine with trigonal nitrogen exceedingly small, a *p* orbital has poorer directional quality for bond formation than an sp^3-like orbital. Therefore the rate constant, as well as rate, for quaternization of an amine with trigonal nitrogen can be expected to be less than that for the same amine with tetrahedral-like nitrogen.

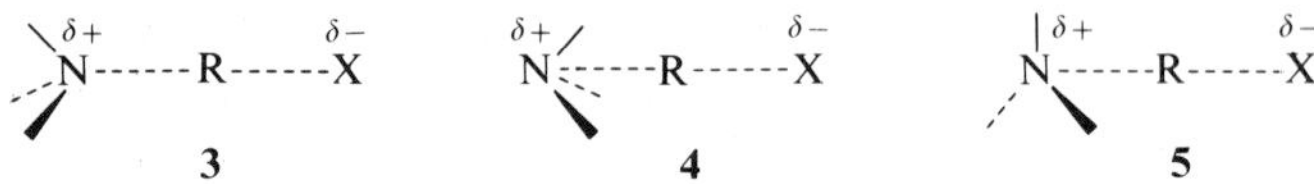

If quaternization reactions occurred more rapidly than nitrogen inversion, the ratio of products formed in the reaction of **1** with deuteriomethyl benzenesulfonate would be the same as the conformational equilibrium constant; that is, **2b**/**2a** ≡ **1b**/**1a**. In addition, reaction of **1** with any other quaternizing agent, under otherwise identical conditions, would give the same product ratio. In fact, the degree and manner of stereoselectivity in amine quaternizations depends on the structure of the incoming or quaternizing group and the leaving group, as well as the structure of the amine, the solvent, and the temperature.

For the reactions discussed in this article, knowledge of nitrogen inversion frequencies and rate constants of quaternization allows one to estimate that the amines involved undergo 10^3–10^7 inversions per quaternization. Under these actual conditions, the Curtin-Hammett principle (16) applies, and the product composition is determined by the difference in free energies of the transition states leading to the diasteromeric quaternary products; the larger this difference, the more one product will predominate. Thus, determination of the kinetically controlled product composition from the reaction of **1** with deuteriomethyl benzenesulfonate provides a measure of the free energy difference of the diasteromeric transition states leading to **2a** and **2b**, and identification of **2a** and **2b** establishes the relative free energies of the respective transition states. Determination of the product ratio, **2a**/**2b** or **2b**/**2a**, is a relatively easy matter; identification is not.

In a transition state for quaternization of an amine, the new bond between carbon and nitrogen is certainly longer than that between fully bonded carbon and nitrogen. Therefore, in considering quaternization of **1** with deuteriomethyl benzenesulfonate, one might expect that the transition state leading to **2b** would have less energy because the fully bonded methyl group is in the less-hindered position. However, unlike fully bonded carbon, the carbon undergoing attack does not have the tetrahedral configuration at the transition state; it is effectively coplanar with the atoms fully bonded to it, and this tends to bring these atoms closer to the amine moiety. Consequently the contribution to the free energy of the transition state resulting from nonbonded interactions involving groups on incoming carbon might be greater than that for the same groups on fully bonded carbon. So although one can expect that the product ratio **2b/2a** will be less than the conformational equilibrium constant for **1**, that is, **1b/1a**, one cannot deduce whether this ratio will be greater than unity.

Another way of reaching the same conclusion is to note that the kinetically controlled product ratio **2b/2a** is equal to the conformational equilibrium constant *times* the ratio of rate constants for the respective conformers (17), that is, (**2b/2a**) $\equiv k_{1b}/k_{1a}$ (**1b/1a**). As the incoming group assumes the more hindered position during quaternization of the more stable conformer **1b**, it can be anticipated that **1b** will have the smaller rate constant.

For the vast majority of amines (**1** is a notable exception; see Section III-G), estimation of the conformational equilibrium constant is difficult and subject to considerable uncertainty. Conceivably, as more is learned about how variables in quaternizations of tertiary amines affect stereoselectivity, these results will allow increasingly accurate inferences concerning conformational equilibria of the amines.

II. DETERMINATION OF THE DEGREE OF STEREOSELECTIVITY IN QUATERNIZATIONS

A. Fractional Crystallization

The oldest and still most commonly used method for separating diastereomeric quaternary salts is fractional crystallization. In the earliest studies concerned with stereoselectivity in amine quaternizations, the degree of stereoselectivity was estimated by fractional crystallization and examination of the melting behavior of the individual salts and their mixtures. Today, fractional crystallization, when used at all in stereoselectivity studies, is nearly always combined with a spectroscopic method, which provides the means of quantitative analysis (e.g., 18,19).

Conclusions or inferences based on analyses made by means of fractional crystallization, with or without the aid of melting-point composition curves, must be accepted with reservation. This is because samples of constant melting point obtained by recrystallization of quaternary mixtures have subsequently been shown to contain two stereoisomers, each in substantial proportion. The isomeric *N*-ethyltropinium iodides (**6** and **7**) provide a particularly noteworthy example of a pair of stereoisomeric quaternary salts that defied analysis by means of fractional crystallization. In 1953 Findlay (20) reported that reactions of tropine (**8a**) with ethyl iodide and *N*-ethylnortropine (**8b**) with methyl iodide gave products with the same melting point (319°), which was not depressed on mixing. As reactions of **8b** with *n*-propyl iodide and *N*-*n*-propylnortropine (**8c**) with ethyl iodide gave products with different melting points, Findlay concluded that the same *N*-ethyltropinium iodide was obtained from both routes. Soon after Findlay's report, examination of X-ray powder patterns (21) and IR spectra (22) of the two products showed that they were not identical, but apparently because of their sharp melting points they were still regarded as pure substances. Recent analysis by means of NMR spectroscopy (23,24) has revealed that both reaction products are mixtures containing at least 20% of the minor product.

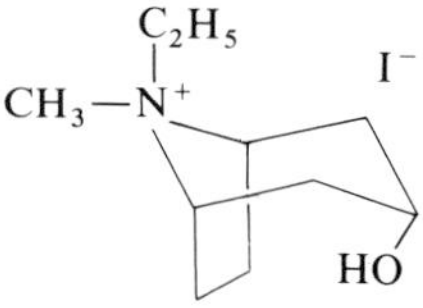

C_2H_5 / CH_3—N^+ I^- HO

6

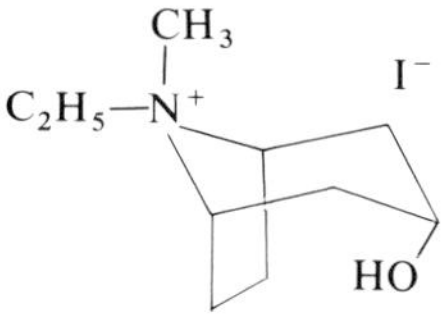

CH_3 / C_2H_5—N^+ I^- HO

7

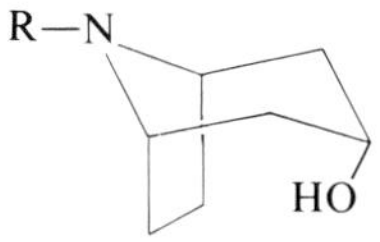

R—N HO

8a: R = CH_3;
b: R = C_2H_5; **c**: R = *n*-C_3H_7

It is important to recognize that a number of important studies on the stereochemistry of quaternizations were carried out by investigators whose primary concern was identification of major products rather than degree of stereoselectivity, and who routinely removed minor products by fractional crystallization. Where minor products were not investigated, only limiting conclusions can be made concerning the degree of stereoselectivity.

B. NMR Spectroscopy

Following the example of Closs (9), numerous investigators have applied NMR spectroscopy to the analysis of stereoisomeric quaternary salts obtained from alkylation (15,18,23–60), *N*-oxidation (50,61), and protonation or deuteriation (18,26,27,29,45,48–51,53,56,62–69) of tertiary amines. This method has been used most commonly to analyze mixtures of salts possessing *N*-methyl groups or other groups that give singlets in the NMR spectrum; however, it has also been possible to obtain satisfactory analyses using more complicated signals such as AB patterns due to nonequivalent benzylic hydrogens (28, 36, 38). (Also see refs. 128–131 and 133.)

NMR analysis of quaternization products has its limitations and pitfalls, but these can usually be circumvented. The problem most frequently encountered, first recognized by Closs, is the coincidence of otherwise diagnostic bands due to similar groups in diastereomeric salts. Lack of discernible differences in spectra taken in one solvent does not necessarily preclude successful analysis by NMR because spectra of diastereomeric quaternary salts, especially bands due to α-hydrogens of *N*-alkyl groups, are dependent on solvent (34,36,38) as well as anion (34,45) and concentration (37).* For example, the *N*-methyl bands of the diastereomeric 1-ethyl-1-methyl-4-phenylpiperidinium iodides, which are coincident in spectra taken in deuterium oxide, are separated by 0.08 ppm in spectra taken in deuteriochloroform (34,36); also, exchange of benzenesulfonate for iodide increases the chemical shift in deuterium oxide from nil to 0.06 ppm (34). Significantly, failure to recognize that the *N*-methyl bands of these isomeric piperidinium iodides, in water (27,48), or the isomeric *N*-ethyltropinium iodides (**6** and **7**), in wet (43) but not dry (23, 24) dimethyl sulfoxide-d_6, are coincident, coupled with misassignment of another band as the "second" *N*-methyl band, led to incorrect conclusions concerning the stereoselectivity of reactions leading to these mixtures. It seems likely that some other analyses, particularly the claims of high stereoselectivity, based solely on NMR examination that uses a single solvent or anion may also be in error.

In principle, use of double resonance or isotopic labeling could allow simplification of spectra of diastereomeric quaternary salts that would otherwise be too complex for straightforward NMR analysis. Also, computer averaging could be used to assist in recording spectra of salts with low solubility. It appears that no use of these techniques or of carbon-13 NMR has been made in the determination of the degree of stereoselectivity in amine quaternizations.

C. Other Methods

The importance of NMR spectroscopy in this field can be appreciated when one considers that, although diastereomeric quaternary salts or other mixtures have been distinguished by infrared spectroscopy (e.g., 18,19,22,37, 53,70), paper chromatography (18,54,71), thin-layer chromatography (15), and their optical rotation (e.g., 4,70,72,73), there is only one paper describing the use of any of these methods to determine quantitatively the degree of stereoselectivity in quaternizing reactions. Koczka and Bernáth (19) analyzed mixtures of diastereomeric codeine and morphine salts by examination of their infrared spectra and optical rotations; NMR spectroscopy was not used because of the low solubility of the salts.

* This writer's opinion is that the observed solvent, anion, and concentration dependence of these spectra reflects differences in the degree or nature of ion pairing rather than dramatic differences in conformer populations (cf. 63).

III. ASSIGNMENT OF STEREOCHEMISTRY TO QUATERNARY AMMONIUM SALTS

A. NMR Spectroscopy

NMR spectroscopy has been widely applied, directly and indirectly, in assigning configurations to quaternary ammonium salts. Direct applications have been based on empirical relationships of varying degrees of reliability as well as sound theoretical considerations. Indirect applications, where NMR spectroscopy is used solely as an analytical tool and the configurational assignments are based on another principle, are discussed in following sections.

Moynehan, Schofield, Jones, and Katritzky (55,56) were able to assign configurations to the *cis*- and *trans*-*N*-methylquinolizidinium iodides (**9** and **10**, respectively) as well as several *C*,*N*-dimethylquinolizidinium iodides, and they noted that the *N*-methyl resonances of the *cis* fused salts occurred at δ3.08–3.16 ppm while those of the *trans* fused salts occurred at δ2.94–2.99 ppm. The same order of *N*-methyl resonances has been observed for the methiodides of *cis–trans*- (**11**) and *trans–trans*-hexahydrojulolidine (**12**) (49) and several stereoisomeric indole alkaloid methiodides (25,58,59) in that the *N*-methyl resonances of those salts in which the quinolizidine fusion is *cis* occur at lower field.

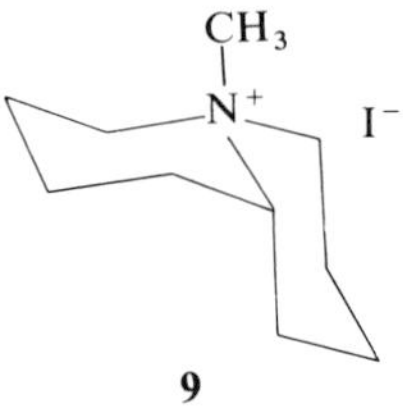

9 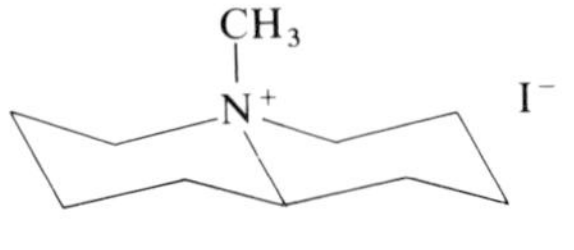 **10**

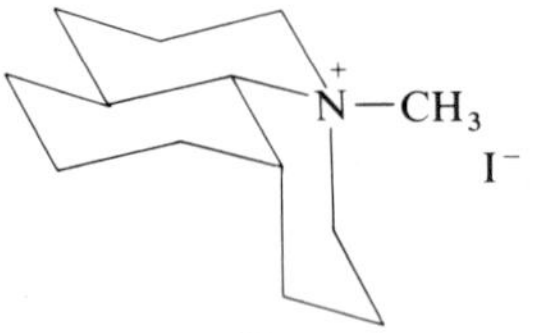

11 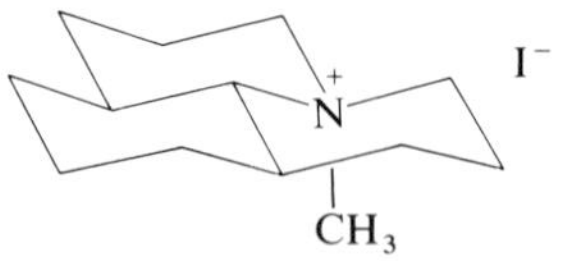 **12**

Another noteworthy difference in the spectra of **9** and **10** is the significantly greater *N*-methyl bandwidth of **10** (74). There appear to be no configurational assignments to diastereomeric quaternary salts based on differences in *N*-methyl or other band widths.

Note that in the *cis* fused quinolizidinium salts the *N*-methyl group is axial with respect to one piperidine ring and equatorial to the other, while in the *trans* fused isomers the *N*-methyl group is axial with respect to both piperidine rings. Indeed, for most pairs of diastereomeric quaternary *N*-methylpiperidinium salts (34,45–47, cf. 37) that have different *N*-methyl resonances, the diastereomer with the axial *N*-methyl group has the higher field *N*-methyl resonance. This rule must be qualified because at least one example is known where the order of axial and equatorial *N*-methyl resonances is solvent dependent (34). Interestingly, for a number of 4-substituted 1,1-dialkylpiperidinium salts, change of solvent from deuteriochloroform to deuterium oxide invariably causes the band due to equatorial α-hydrogens to undergo a greater upfield chemical shift than that of the similar axial hydrogens (34,37).

As with the isomeric *N*-methylquinolizidinium iodides, the *N*-methyl resonance of *cis*-fused *N*-methylindolizidinium iodide (**13**) occurs at lower field than that of its *trans*-fused isomer (54); this is in accord with calculations that, admittedly, have no more than qualitative significance.

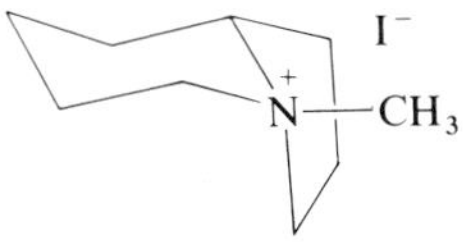

13 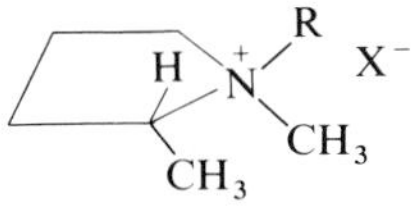 **14** 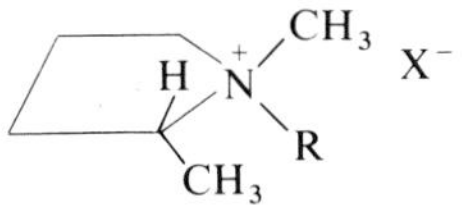 **15**

a: R = H, X = Cl; **b**: R = C_2H_5, X = I
c: R = *n*-C_3H_7, X = I; **d**: R = ϕCH_2; X = I

McKenna et al. (18,26,27,37,53), using the reverse of a method employed earlier by Closs (9), assigned configurations to a number of *N*-methyl quaternary ammonium salts on the basis of conformational arguments and empirical relationships between the NMR spectra of the salts and the corresponding *N*-methylamine hydrochlorides. For example, Becconsall, Jones, and McKenna (26,27) argued that *cis*-1,2-dimethylpyrrolidine hydrochloride (**14a**) is less stable than its *trans* isomer (**15a**), and accordingly they assigned the less intense *N*-methyl band at δ2.73 ppm in the spectrum of the chloroform solution to **14a** and the more intense band at δ3.03 ppm to **15a**. For a series of diastereomeric quaternary iodides (**14b–d** and **15b–d**) prepared from 1,2-dimethylpyrrolidine, they observed *N*-methyl signals at δ2.94–3.02 or 3.23–3.35 ppm (26,27). They assigned configurations to these salts on the assumption that differential shielding of the *N*-methyl groups would be similar to that for the hydrochlorides, i.e., the high and low field *N*-methyl signals were ascribed to **14** and **15**, respectively. It was noted that results of dubious value would be obtained where there was a marked conformational change between the proton and quaternary salts (27,37), as is believed to exist for derivatives of 3-methylazabicyclo[3.3.1]nonane (51), or where the preferred conformation of the proton salts or quaternary salts was unclear, as with *cis*-1,2-dimethylpiperidine hydrochloride (37,48). This method cannot be applied when only one *N*-methyl signal can be seen for the major and minor diastereomers of the proton salt, and many investigators have been frustrated by such an observation (9,15,18,26,27,40,48). A much more serious failing is the possibility that the summed bond anisotropies might give reversed orders of *N*-methyl signals for the two types of salts (37).

McKenna and his co-workers (36,38,75) have also used NMR spectroscopy in a more limited but theoretically sounder manner to assign configurations to quaternary products from *N*-benzyl cyclic amines and deuteriobenzyl iodide. By use of either the Johnson-Bovey (76) or the point-dipole (77) model, they calculated that the effects of the benzene ring anisotrophies in the most important conformer of 1,1-dibenzyl-2-methylpiperidinium iodide, that is, **16**, would lead to a greater degree of magnetic nonequivalence for the equatorial methylene hydrogens. Their calculated chemical shift differences were in very good agreement with the experimental results. A similar procedure applied to corresponding salts in the 2-methylpyrrolidine and decahydroquinoline systems indicated that the methylene hydrogens *trans* to the methyl group and the equatorial methylene hydrogens were the more nonequivalent. They were also able to calculate that the upfield singlets in the spectra of 1,1-dibenzyl-3-methyl- and 4-phenylpiperidinium iodides were due to the equatorial methylene groups.

ϕ C H H N$^+$ C H H ϕ I^- — **16**

i-C_3H_7 H_3C X^- N^+ ϕCH_2 CH_3 — **17**

i-C_3H_7 H_3C X^- N^+ H_3C $CH_2\phi$ — **18**

Bernáth et al. (28,29) have also made configurational assignments based in part on differing degrees of magnetic nonequivalence in *N*-benzyl quaternary salts.

Other examples where the predictable effect of the benzene ring on chemical shifts of neighboring groups allows configurational assignments are provided by *N*-benzyl salts derived from 1,2,3-trialkylaziridines (31); the preferred conformation of the phenyl groups in, for example, **17** and **18** results in a downfield shift of the signals due to the *cis*-*C*-alkyl hydrogens.

B. IR Spectroscopy

McKenna et al. (15,18,37,53) confirmed and extended observations of earlier workers (22,70) who had noted characteristic differences between epimeric pairs of tropine and camphidine quaternary salts in the 850–900 cm^{-1} region. They proposed the following rule as a subsidiary criterion for assigning configurations to certain *N*-methyl quaternary salts (15,37,53): "in the approximate region 840–900 cm^{-1} there is a diagnostic band at high ($>$ca. 885 cm^{-1}) or at lower frequency for *N*-methyl equatorial or axial respectively (or, in the case 2-methylpyrrolidine derivatives, for *N*-methyl respectively *trans* or *cis* to *C*-methyl.") The rule is not generally applicable (37,43,48), and it can provide only supporting evidence when applied.

C. X-Ray Crystallography

The structures of hunterburnine α- and β-methochlorides (78,79), one of the diastereomeric *N*-ethyltropinium bromides (80), the major products from reactions of ($\pm$)-*N*-ethylcamphidine with methyl iodide (81), and pseudotropine with ethyl bromide (60b), and ($\pm$)-1-acetyl-16-methylaspidospermidine-4-methiodide (82) have been determined by X-ray crystallography. That the *N*-ethyltropinium bromide whose structure was determined corresponds to the major product from *N*-ethyltropine and methyl iodide has been questioned (37), but subsequent work (60b,83b) has confirmed the original structural assignment.

D. Lactone Formation

Although necessarily limited in applicability, lactone formation has often been used to establish the configuration of quaternization products and was essential to Fodor's classical work (8) in this field. Two methods have been used. The older, first tried unsuccessfully by Mills, Parkin, and Ward (5), involves lactonization of an *N*-carboxymethyl quaternary salt that bears a hydroxyl group (41,46,84–88). The other method involves conversion of a tertiary amino alcohol to its α-halo-acetate ester, which can undergo intramolecular quaternization to the lactone (47,90,91); the lactone must then be converted to the *N*-alkoxycarbonylmethyl quaternary salt or other suitable derivative for comparison with the salt whose structure is to be established. These methods work well for formation of six-membered lactones (47,85–87,89–91) but sometimes fail (5,47), give a poor yield (84), or give oligomeric lactone (60,83) when used in attempts to prepare seven-membered lactones. Examples of the two methods, due respectively to Fodor, Tóth; and Vincze (85), and House, Tefertiller, and Pitt (47), are given in eqs. 2 and 3.

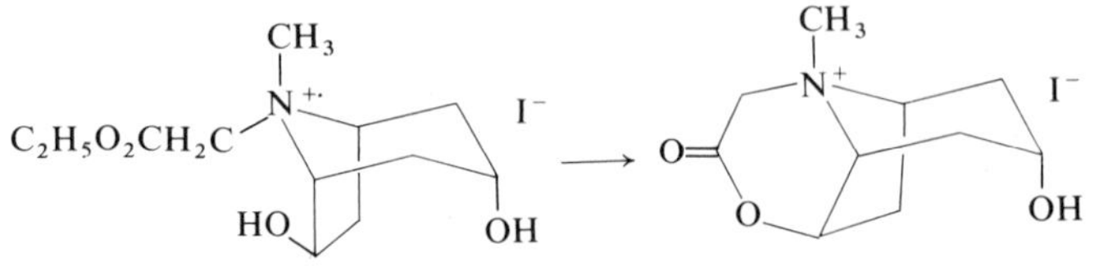

(2)

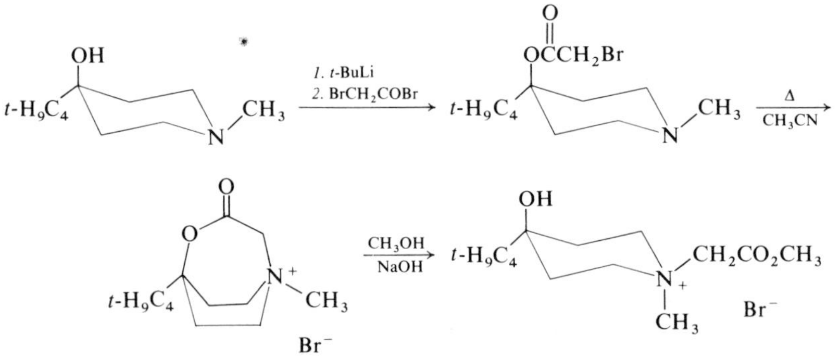

(3)

E. Equilibration and Partial Equilibration

In principle, configurational assignments can be made with confidence to members of any pair of diastereomeric quaternization products that differ significantly in free energy in a predictable manner *if* they can be equilibrated cleanly in solution. Diastereomeric 1,1,2,3-tetraalkylaziridinium salts such as **17** and **18** are equilibrated by halide ion in methanol (31) or acetonitrile (30) at room temperature via the corresponding β-haloalkylamine, **19** (eq. 4), and configurational assignments to the aziridinium salts so far studied, except 1-deuteriomethyl-1-methyl salts including **2a** and **2b**, are based on equilibration studies.

$$\mathbf{17} \underset{-I^-}{\overset{I^-}{\rightleftharpoons}} CH_3CHICH(i\text{-}C_3H_7)N(CH_3)CH_2\phi \underset{I^-}{\overset{-I^-}{\rightleftharpoons}} \mathbf{18} \qquad (4)$$

19

Equilibration of quaternization products less strained than aziridinium salts requires more vigorous conditions, and assignments of configuration to such systems based on equilibration or partial equilibration studies (15,19,28,

29,37,40,51,52) have been limited to *N*-benzyl- or *N*-allyl-*N*-methyl quaternary salts. Unlike the aziridinium salts, these are equilibrated by initial attack of (usually) iodide on an *N*-alkyl group, which gives a tertiary cyclic amine and alkyl iodide, followed by requaternization. Provided that the difference in reactivity at methyl and benzyl or allyl is quite large, equilibration will occur relatively cleanly. However, a small difference in reactivity will lead to significant amounts of four salts rather than two. The latter situation could result in incorrect conclusions if the users of this method examine changes in composition only by NMR or IR spectroscopy and make no serious effort to determine the extent of these side reactions. Although methyl appears to be displaced from nitrogen much less readily than benzyl (cf. refs. 52 and 92), there is insufficient quantitative data to exclude the possibility of complications due to these side reactions.

McKenna et al. (15,37,51,52), who have applied this method most widely, have called attention to other limitations and drawbacks. The method does not work with all *N*-benzyl-*N*-methyl quaternary salts as they were unable to interconvert the diastereomeric *N*-benzyltropanium iodides (**20** and **21**). These salts are insoluble in chloroform, normally the solvent of choice, and use of ethanol or acetonitrile led to recovery of the starting mixture, even though the corresponding diastereomeric camphidinium iodides could be equilibrated in acetonitrile. Partial destruction of the quaternary salts may occur, and it is therefore sometimes necessary to determine conditions that minimize this attrition. Another difficulty is encountered when the starting mixture for the equilibration studies is a mixture of diastereomers, each in substantial amount, so that the observed degree of change is small.

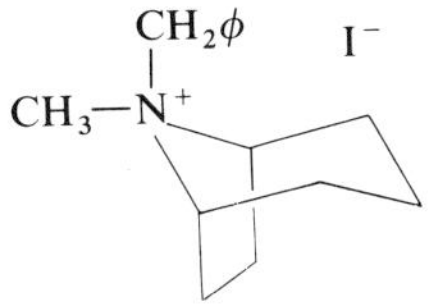

$CH_2\phi$ I^-; CH_3-N^+

20

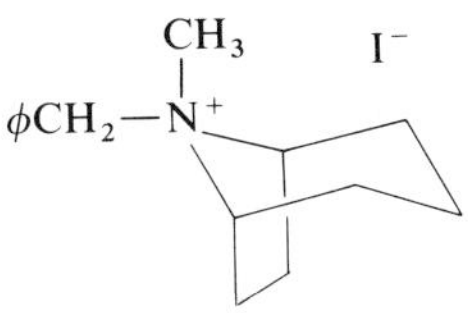

CH_3 I^-; ϕCH_2-N^+

21

It should be noted that quaternizations in nonprotic solvents at elevated temperatures that give quaternary iodides with an aryl group (93) or an easily displaced group such as benzyl, allyl, or alkoxycarbonylmethyl on nitrogen might well be accompanied by partial or complete equilibration of the diastereomeric products (37). Aziridinium halides equilibrate even more readily, and failure to recognize this led to incorrect conclusions in the first study of the stereoselectivity of their formation (cf. 31 and 35).

Note also that conversion of a lower melting diastereomeric salt to its higher melting isomer by heating at a temperature between the melting points of the two salts does not establish that the higher melting is the more stable diastereomer in solution (cf. refs. 3–5,19).

F. Comparison of Stereoselectivity in Related Reactions

Dependence of stereoselectivity on solvent has been observed for quaternizations of several aziridines (30,32) and piperidines (23,33,50). For reactions of *trans*-1,2-dimethyl-3-isopropylaziridine (**1**) and its 3-*t*-butyl homolog with alkyl benzenesulfonates, change of solvent from benzene to methanol increases the proportion of product formed by attack of alkyl group *cis* to the larger *C*-alkyl group. Particularly dramatic is the influence of solvent on the stereochemistry of reactions of 1-methyl-4-*t*-butylpiperidine with benzyl chloride (33). As the solvent is changed from benzene to methanol at 30°, the product ratio changes from 1.6:1 to 0.74:1; in benzene the major product is formed by reaction of benzyl chloride with the less stable ax.-methyl conformer, while in methanol the major product is formed by reaction with the more stable eq.-methyl conformer.

These and results of other studies on solvent dependence of stereoselectivity follow a consistent pattern that can be stated as follows: under otherwise constant reaction conditions, change of solvent from benzene to methanol results in formation of a greater proportion of product that corresponds to the more stable amine conformer. More limited data (23,30,33; however, see ref. 42b) indicate that use of acetonitrile gives stereoselectivities intermediate between those obtained in benzene and methanol. [The implication (33) that benzyl chloride reacts more rapidly with 1-methyl-4-*t*-butylpiperidine in methanol than in acetonitrile is incorrect.] Significantly, changes in product composition with solvent are greater with alkyl chlorides or sulfonates than for quaternizations with the much more commonly used iodides (33).

The relationship between stereochemistry of quaternization and leaving group has not been studied as extensively as that with solvent, but data obtained with 4-substituted piperidines, tropine, and pseudotropine (23) indicate that change of leaving group from chloride or bromide to iodide favors formation of the diastereomeric salt that corresponds to the more stable amine conformer.

McKenna, McKenna, and White (52) proposed an empirical criterion for assigning configurations to *N*-methyl-*N*-alkyl cyclic quaternary iodides based on the difference in degrees of stereoselectivity shown in reactions of the *N*-alkyl base with methyl iodide and the *N*-methyl base with the alkyl iodide. They felt that if methylation gave a significantly greater degree of stereoselectivity than its inverse reaction, that is, the reaction of *N*-methyl base and alkyl iodide, this would correspond to predominant attack by the quaternizing agent on the more hindered nitrogen flank, for example, axial attack for a piperidine. On the other hand, significantly higher stereoselectivity in the reaction with alkyl iodide would indicate the predominant attack on the less hindered flank. Although the reasoning that led to this criterion was not stated, it seems reasonable to argue for quaternizations leading to the 1-methyl-1-*n*-propyl-4-*t*-butylpiperidinium iodides, for example, that *if* transition state **22** has less free energy than its diastereomer **23**, and transition state **24** has less than **25**, the difference in free energy between **22** and **23** will be greater than that for **24** and **25**. This criterion has been applied by McKenna et al. to 3-azabicyclo[3.3.1]nonanes (37,51), 4-aza-5α-cholestanes (15,37), camphidines (18,37), *trans*-decahydroquinolines (18,37,39), 2-methylpiperidines (18,37,39), 2-methylpyrrolidines (18, 37), 4-phenylpiperidines (37), and nortropanes (37), and by Casy, Beckett, and Iorio (40) to 4-aryl-5-methyl-1,2,5,6-tetrahydropyridines.

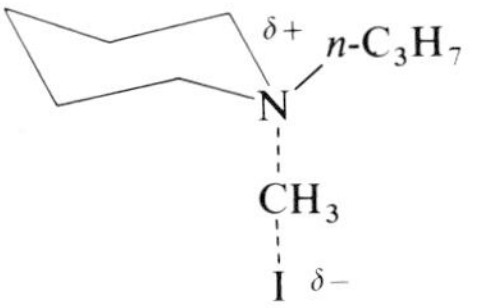

22 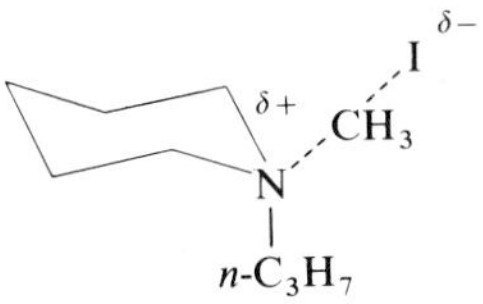 **23**

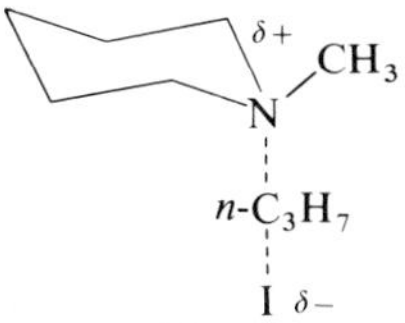

24 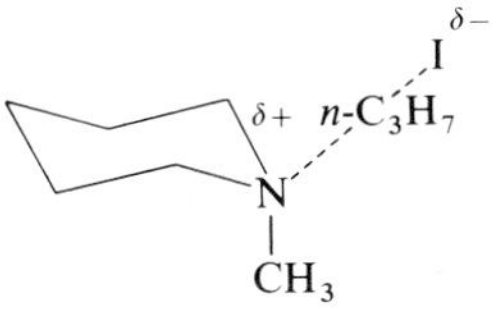 **25**

Although the predominant *N*-methyl-*N*-alkyl cyclic quaternary salt formed in a given quaternization is generally the minor product when the exocyclic groups are introduced in the opposite order, reactions of methyl iodide with 1-isopropyl-*trans*-decahydroquinoline and 1-isopropyl-2-methylpiperidine give the same major product as their inverse reactions (37,39), and all four reactions show marked stereoselectivity. It was pointed out that each of the salts with an ax.-isopropyl group would have significantly greater energy than its respective diastereomer because of the eq.-2-alkyl substituent. Although a severe interaction involving at least one of the isopropyl-methyl groups cannot be avoided in any conformation of the ax.-isopropyl salts, each of their diastereomers possesses a low-energy conformation (e.g., **26**). It was suggested that this special factor affects the free energies of the diastereomeric transition states so as to cause predominant formation of the salt with the eq.-isopropyl group.

These three empirical criteria, based on differences in degrees of stereoselectivity, are capable of wide and fairly easy application. However, further research, particularly with systems possessing special structural features, seems called for in order to test the limits of their reliability.

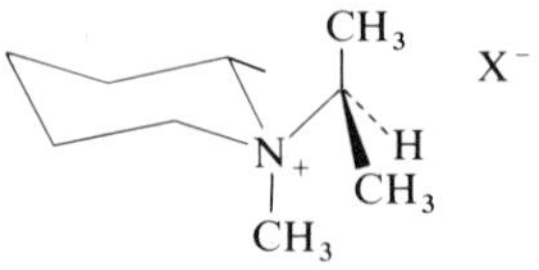

26

G. Analysis of Quaternization Rates

Kinetic data has been used to assign configurations to diastereomeric quaternary salts from the reaction of *trans*-1,2-dimethyl-3-isopropylaziridine (**1**) with deuteriomethyl benzenesulfonate (30). Because nitrogen inversion in **1** is slow on the NMR time scale, and because the *N*-methyl resonances of its conformers are resolvable, it is possible to determine the conformational equilibrium constant by direct examination of its NMR spectrum (94); in methanol at 30°, **1b**/**1a** = 3.8. Quaternization of **1** with deuteriomethyl benzenesulfonate in methanol at ca. 30° gave a product ratio of 2.9:1 (32), and determination of the rate constant for quaternization (1.76×10^{-3} M^{-1} sec^{-1}) in 90% methanol at 30° allowed calculation of two sets of rate constants for the conformers corresponding to formation of the major product from either **1a** or **1b**. When **1a**, the less stable conformer, is quaternized, the benzyl group approaches *cis* to *C*-methyl. Comparison of the calculated rate constants for **1a** (6.3 and 2.2×10^{-3} M^{-1} sec^{-1}, respectively) with that for *trans*-1,2,3-trimethylaziridine (2.19×10^{-3} M^{-1} sec^{-1}) under the same conditions indicated that the major product was formed from the more stable conformer **1b**.

An ambitious attempt by Imbach, Katritzky, and Kolinski (48) to assign configurations to products of 1-alkyl-4-phenylpiperidines and 1-alkyl-4-*t*-butylpiperazines by comparison of their rates of formation with quaternization rates of model bi- and polycyclic amines was seriously undermined by, among other things (cf. ref. 37), inaccurate determination of the degree of stereoselectivity for the reaction of 1-ethyl-4-phenylpiperidine with methyl iodide (33,37). The limited results with 1,2,3-trialkylaziridines, however, indicate that their approach has some merit and deserves further investigation.

In discussing their conclusions concerning preferred direction of *N*-alkylation of 4-*t*-butylpiperidines and 3-azabicyclo[3.*n*.1]alkanes, that is, axial and equatorial attack, respectively, House, Tefertiller, and Pitt (47) noted that the monocyclic amines quaternized much more rapidly. The speed of reaction was related to lesser degree of new C—N bond formation and, thus, to the different steric requirements of the incoming group relative to the fully bonded group in transition states for quaternizations of the different systems. Interestingly, quaternization of methyl benzenesulfonate with 4-cyclohexyl-1-ethylpiperidine gives a greater proportion of product corresponding to the more stable eq.-ethyl conformer than does its slower reaction with the 4-phenyl analog (33). The argument relating manner of stereoselectivity with degree of new C—N bond formation, which may be an oversimplification or even fundamentally incorrect (37), provides a convenient rationale for differences in degree of stereoselectivity in reactions of closely related compounds, such as 4-cyclohexyl- and 4-phenyl-1-ethylpiperidine (33), and for dependence of stereoselectivity on solvent (23,30,32) and leaving group (23).

The reverse of this method, that is, assignment of configuration or preferred conformation to amines based on comparison of quaternization rates, has been applied to a wide range of tertiary cyclic amines (49,59, 95–102).

H. Ring Closure of Tertiary Amino Halides or Sulfonates

Conformational arguments have been used by Katritzky (103), Meyer and Sapianchiay (54), and Brown and McKenna (104) to assign configurations to the predominant diastereomeric quaternary salt formed by cyclization of tertiary amino halides or sulfonates. At the heart of these arguments is the assumption that the sign of the free-energy difference between the possible transition states is the same as that of the corresponding products. Katritzky, for example, argued that *cis-c*-3-methylquinolizidinium iodide

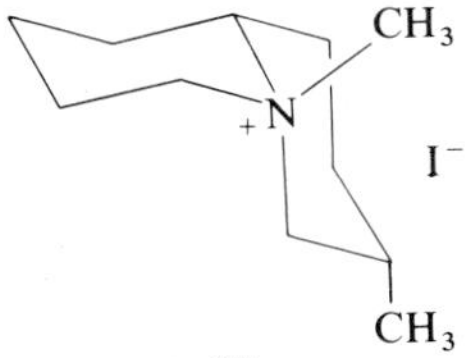

CH_3 ^+N I^- CH_3

27

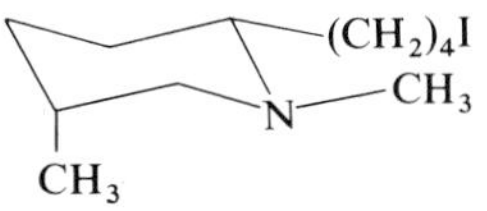

$(CH_2)_4I$ N CH_3 CH_3

28

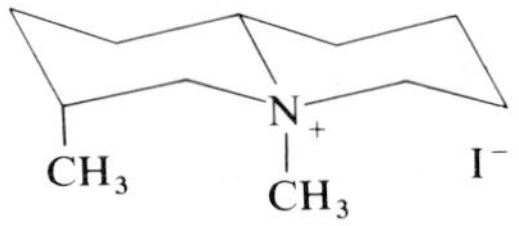

N^+ CH_3 CH_3 I^-

29

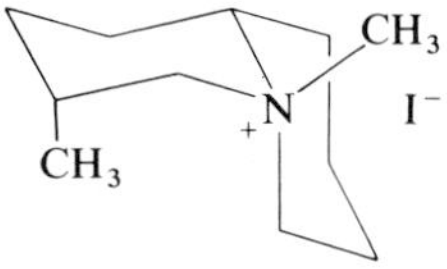

CH_3 ^+N I^- CH_3

30

(**27**) would be the preferred quaternization product of *cis*-1,5-dimethyl-2-(4′-iodobutyl) piperidine (**28**) because formation of **29** and **30**, the latter a conformer of **27**, would require generation of unfavorable 1,3-diaxial interactions of two methyl groups and methyl and methylene, respectively. This method has been applied most generally by Brown and McKenna (104), and "many of the relevant results obtained so far accord with expectation." The potential of this method depends on the correctness of the basic assumption and the soundness of the conformational arguments used. (See also ref. 130.)

I. Other Methods

In addition to the foregoing methods, which have had or are capable of widespread application, a number of other methods of more limited applicability have been used to assign configurations to quaternization products.

Crow (73) noted that the change in $[M]_D$ in passing from (−)-lupinene (**31**) to its methiodide is of similar magnitude but opposite sign to that seen with (+)-epilupinene (**32**). From this he concluded that the two methiodides had different ring junctures, and considerations similar to those used later (55,56,103) for *C*-methyl-substituted quinolizidines led him to assign the *cis*-ring juncture to lupinine methiodide. Assignments of configuration have also been made to quaternary salts with the codeine (72,105) and morphine (72,106) skeletons and to methiodides of lycorine-type alkaloids (107) on the basis of comparisons of their optical rotations with those of the corresponding bases.

Support for the stereochemical assignments to the diastereomeric 1-benzyl-1-methyl-4-*t*-butylpiperidinium chlorides was obtained (33) by comparison of their benzyl-methylene and 1-methyl resonance frequencies with the corresponding time-averaged resonance frequencies in the spectrum of 1-benzyl-1-methylpiperidinium chloride (**33**). This comparison and the assumption that the 4-*t*-butyl group has no effect on the chemical shifts of exo-α-hydrogens of the groups in the l-position allowed calculation of values of ca. 2.2 for the conformational equilibrium constant of **33**, that is, **33b**/**33a**. Although the results of equilibration studies (37) agree with the magnitude of the calculated values for **33b**/**33a**, it should be noted that the assumption made seems tenuous for piperidinium salts (108) as well as cyclohexanes (109).

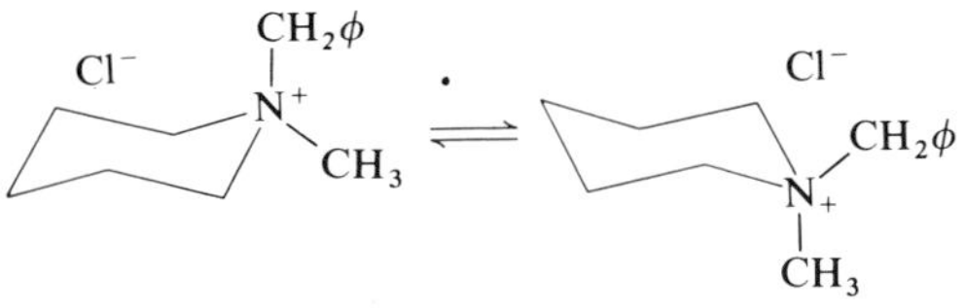

Polonovski and Polonovski (7) claimed that one of the scopolamine *N*-oxides (**34** and **35**) undergoes a transannular reaction when treated with hydrogen bromide to give scopinium bromide (**36**), which, in turn, was reduced to pseudoscopine (**37**). This sequence may be used to assign configuration **35** to the *N*-oxide. Although the melting points reported for **37** and its picrate were later found to be in good agreement with those obtained for **37** prepared by potassium borohydride reduction of scopinone (**38**), and its picrate (110), attempts (111,112) to repeat the transannular reaction have failed.

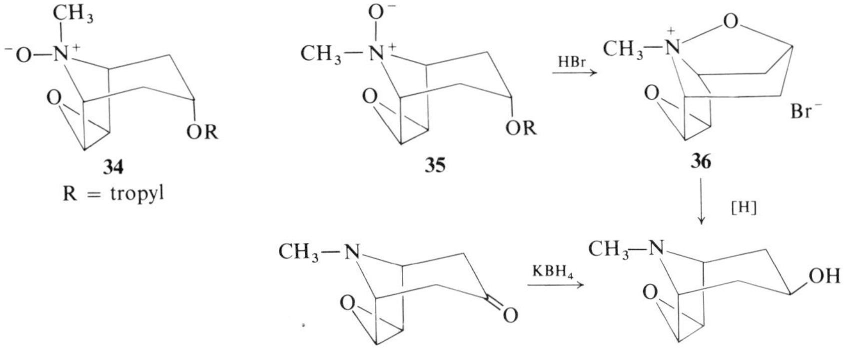

Schofield and Wells (57) found that Hoffman degradation of the quinolizidine methiodide that had been assigned the *cis* fused configuration (**9**) gave 2-(3′-butenyl)-1-methylpiperidine (**39**) as the major product, together with small amounts of quinolizidine and 6-methylazacyclodecene (**40**). In contrast, **40** was the major product from similar treatment of the diastereomeric quinolizidine methiodide (**10**), and only minor amounts of **39** and quinolizidine were formed. Consideration of β-elimination pathways available to the quinolizidine methiodides provides added support to the assigned structures.

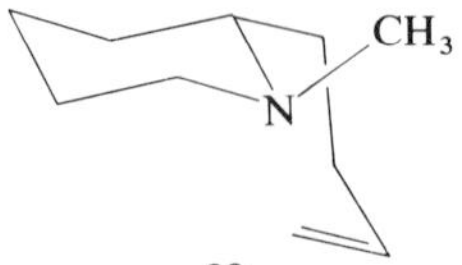

House and Tefertiller (46) were able to make unambiguous configurational assignments to the products (**41** and **42**) from 3-methylaza-9-bicyclo[3.3.1]nonanone and methyl bromoacetate when they found that one of the diastereomeric salts undergoes an intramolecular aldol condensation to give **43**. They also established configurations of quaternization products (e.g., **44**) from methyl bromoacetate and the corresponding epimeric alcohols by oxidizing them to either **41** or **42**.

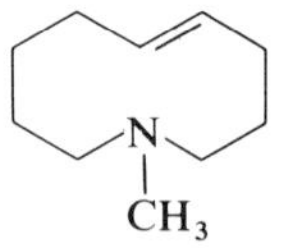

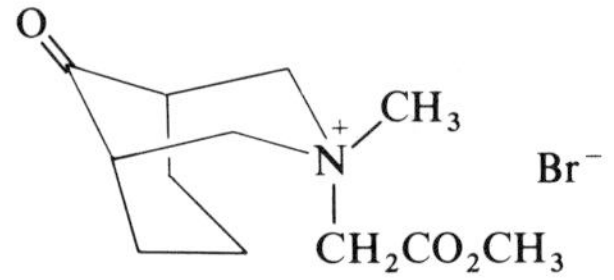

IV. STEREOCHEMISTRY OF QUATERNIZATIONS

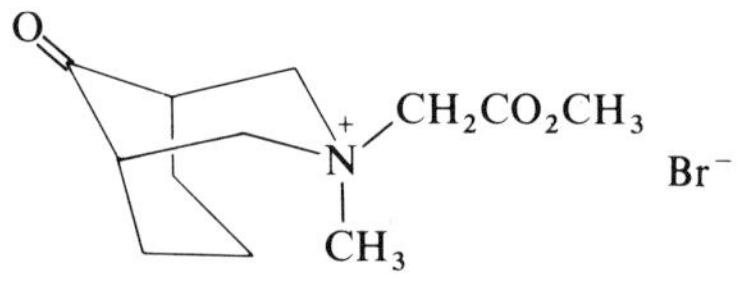

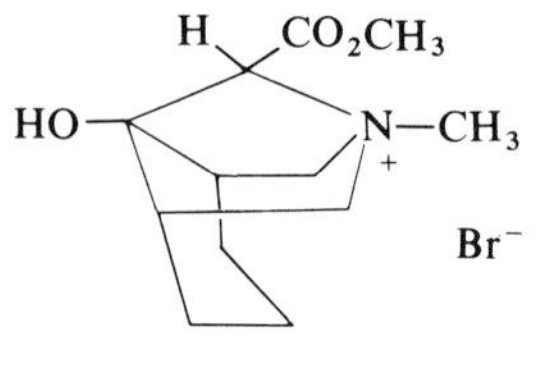

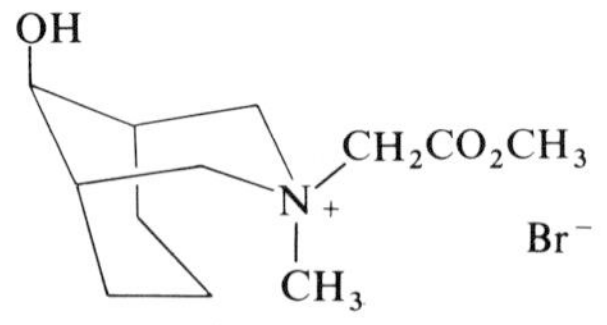

A. Piperidines

1. 4-Substituted Piperidines

Several groups of workers (5,18,27,33,36–38,41,47,48,52,53) have studied stereoselectivity in quaternizations of 4-substituted 1-alkylpiperidines (**45**), and their results are summarized in Table 1. It is unlikely that any of the products is equilibrated under the reaction conditions, and care seems to have been taken to avoid fractional crystallization prior to analysis. There was some controversy concerning the quaternizations of 1-ethyl- and 1-isopropyl-4-phenylpiperidine with methyl iodide, but this has been resolved (37), and there now appears to be general agreement on the stereochemical assignments to the products. (See also ref. 127,129.)

The most important transition states for these quaternizations are certainly those in which the piperidine ring is in the chair conformation with the 4-substituent equatorial. Therefore, the product distributions reflect free energy differences of transition states corresponding to the products **46** and **47**. With the exception of some reactions of benzyl chloride and ethyl bromoacetate, the major product is formed by axial attack, that is, from the more stable amine conformer. Greatest stereoselectivities are observed for methylations and deuteriomethylations, and greatest variations in stereoselectivity, achieved by change of solvent or leaving group, are observed for benzylations. Variation of the 4-substituent has only a slight effect on stereoselectivity.

2. *4,4-Disubstituted Piperidines*

In Table 2 are summarized the results observed for three quaternizations of 4-substituted 1-methyl-4-hydroxypiperidines (**48**). The results are similar to those found for like reactions of the corresponding 4-substituted 1-methylpiperidines, and this indicates that the important transition states are those in which the hydroxyl group is axial.

3. *3-Substituted Piperidines*

The only quaternization of a 3-substituted piperidine for which stereoselectivity has been determined (38) is that of 1-dideuteriobenzyl-3-methylpiperidine with benzyl iodide in acetone. As in the reaction of the 4-phenyl analog (Table 1), the major product is formed by axial attack, and the observed product ratio was 3:1.

4. *2-Substituted Piperidines*

In addition to the early studies of Scholtz (1,3,4), stereoselectivity in quaternizations of 2-substituted piperidines (**51**) has been studied by McKenna et al. (18,26,27,36–38,52,53) and by Kawazoe and Tsuda (50). Several features of their results, which are summarized in Table 3, are noteworthy. As with 4-substituted piperidines, change from methanol to a nonhydroxylic solvent for deuteriomethylation of 1,2-dimethylpiperidine results in formation of a greater percentage of product corresponding to the less stable amine conformer. Compared with similar reactions in the 4-substituted piperidine series, methylations and deuteriomethylations give somewhat less stereoselectivity. This can be taken to mean that a 2-substituent

TABLE 1
Stereoselectivity in Quaternizations of 4-Substituted Piperidines

$$\mathbf{45} \xrightarrow{R''X} \mathbf{46} + \mathbf{47}$$

R	R′	R″X	Solvent	Temp., °C	% **46**[a]	% **47**[a]	Method[b]	Ref.
CH_3	t-C_4H_9	CD_3OTs	$(CH_3)_2C{=}O$	R.T.	87	13	A	47
CH_3	t-C_4H_9	C_2H_5OTs	$(CH_3)_2C{=}O$	25	83	17	A	47
CH_3	t-C_4H_9	ϕCH_2OTs	$(CH_3)_2C{=}O$	25	69	31	A	47
CH_3	t-C_4H_9	ϕCH_2Cl	CH_3OH	30	*58*	*42*	F, I	33
CH_3	t-C_4H_9	ϕCH_2Cl	CH_3CN	30	*49*	*51*	F, I	33
CH_3	t-C_4H_9	ϕCH_2Cl	ϕH	30	*39*	*61*	F, I	33
CH_3	ϕ	CD_3I	$(C_2H_5)_2O$	R.T.	*67–75*	*25–33*	F	37
CH_3	ϕ	C_2H_5I	$(CH_3)_2C{=}O$	Reflux	*50–60*	*40–50*	F	37
CH_3	ϕ	i-C_3H_7I	$(CH_3)_2C{=}O$	Reflux	*50–60*	*40–50*	F	37
CH_3	ϕ	ϕCH_2I	$(CH_3)_2C{=}O$	R.T.	*50*	*50*	E, F	37
CH_3	ϕ	$C_2H_5O_2CCH_2Br$	—	—	48 (52)	52 (48)	—	41
CH_3	OH	$C_2H_5O_2CCH_2Br$	ϕH	R.T.	42[c]	58[c]	—	5
C_2H_5	t-C_4H_9	CH_3OTs	$(CH_3)_2C{=}O$	25	80	20	A	47
C_2H_5	c-C_6H_{11}	$CH_3OSO_2\phi$	CH_3CN	30	*81*	*19*	F	33
C_2H_5	c-C_6H_{11}	$CH_3OSO_2\phi$	ϕH	30	*78*	*22*	F	33
C_2H_5	c-C_6H_{11}	CH_3I	CH_3CN	30	*77*	*23*	F	33

C_2H_5	c-C_6H_{11}	CH_3I	ϕH	30	*77*	*23*	F	33
C_2H_5	ϕ	CH_3I	$(CH_3)_2C{=}O$	Reflux	*67–75*	*25–33*	F	37
C_2H_5	ϕ	ϕCH_2Cl	None	100	38[c]	62[c]	—	5
i-C_3H_7	ϕ	CH_3I	$(CH_3)_2C{=}O$	Reflux	*67–75*	*25–33*	F	37
i-C_3H_7	ϕ	CH_3I	CH_3CN	R.T.	67[d]	33[d]	—	48
t-C_4H_9	ϕ	CH_3I	CH_3CN	R.T.	>95	<5	—	48
ϕCH_2	t-C_4H_9	CH_3OTs	$(CH_3)_2C{=}O$	25	85	15	A	47
ϕCH_2	ϕ	CH_3I	$(CH_3)_2C{=}O$	R.T.	*78–88*	*12–22*	E, F	37
ϕCD_2	ϕ	ϕCH_2I	$(CH_3)_2C{=}O$	R.T.	*75*	*25*	A	38

[a] Unless noted otherwise, product compositions were determined using NMR spectroscopy, and stereochemical assignments are those given in the reference. Percentages in italics were calculated from data given in the reference.

[b] The method used for assigning configurations is one noted in the indicated part of Section III.

[c] Analyzed by comparison of melting point with those of known mixtures. Configurational assignments made by the writer.

[d] Reverse of the assignments made in the reference.

reduces the difference in free energy of the transition states corresponding to the products **52** and **53** as shown, that is, with the group at C-2 equatorial.*

TABLE 2
Stereoselectivity in Quaternizations of 4-Substituted 4-Hydroxy-1-methylpiperidines

$$\underset{\mathbf{48}}{\text{4-R-4-OH-1-}CH_3\text{-piperidine}} \xrightarrow[(CH_3)_2C{=}O]{R'X} \underset{\mathbf{49}}{\text{(N-}R'\text{ axial)}\ X^-} + \underset{\mathbf{50}}{\text{(N-}CH_3\text{ axial)}\ X^-}$$

R	R′X	Temp., °C	% **49**[a]	% **50**[a]	Method[b]	Ref.
t-C_4H_9	CD_3OTs	25–30	83	17	A	47
t-C_4H_9	$CH_3O_2CCH_2Br$	25	46	54	D	47
ϕ	$C_2H_5O_2CCH_2Br$	R.T. and reflux	45	55	D	41

[a] Analysis by NMR spectroscopy. Stereochemical assignments are those given in the reference.

[b] The method used for assigning configurations is one noted in the indicated part of Section III.

Significantly, increase in size of the 2-substituent favors equatorial attack. In contrast to the 4-substituted piperidines, quaternizations of 2-substituted 1-methylpiperidines with ethyl, n-propyl, isopropyl, or benzyl iodide occur mainly by equatorial attack. Isopropylation of 1,2-dimethylpiperidine is highly stereoselective, as is its inverse reaction, which gives the same principal product, and this was rationalized on the basis of a conformational argument noted in Section III-F. Interestingly, benzyl iodide gives significantly less axial attack in its reaction with 1,2-dimethylpiperidine than it does in its reaction with the 1-dideuteriobenzyl analog. Finally, N-oxidations with 30% hydrogen peroxide give stereochemical results intermediate between those obtained with methyl iodide and ethyl iodide. (Also see ref. 129.)

* Note that axial attack on an amine conformer with an ax.-2-substituent, which allows the incoming group to avoid one gauchelike interaction, results in formation of a conformer of the quaternary product that is formed by equatorial attack on a conformer of the amine in which the 2-substituent is equatorial. An argument against the importance of transition states with the alkyl group at C-2 axial is provided by comparison of data in Table 3 with that in Table 5 obtained for *trans*-decahydroquinolines (**57**), which can be regarded as piperidines with fixed eq.-2 substituents.

5. *2,6-Disubstituted Piperidines*

The only investigation of stereoselectivity in quaternizations of 2,6-disubstituted piperidines is that by Kawazoe and Tsuda (50), whose results with *cis*-1,2,6-trimethylpiperidine (**54**, R = CH_3) and its 6-ethyl analog (**54**, R = C_2H_5) are summarized in Table 4. As with 2-substituted piperidines, change of the 6-substituent or alkylating group from methyl to ethyl increases the importance of equatorial attack, and the stereochemical course of *N*-oxidation is intermediate between that of methylation and ethylation.

6. *trans-Decahydroquinolines*

McKenna et al. (18,26,27,36–38,52,53) have studied stereoselectivity in quaternizations of *trans*-decahydroquinolines (**57**), which can be regarded as piperidines with alkyl substituents at C-2 and C-3 fixed in the equatorial conformation. Of these, the 2-substituent can be expected to have more effect on stereoselectivity, and comparison of the results summarized in Table 5 with those observed for 2-methylpiperidines (Table 3) shows that the stereochemical results are virtually identical. (Also see ref. 129.)

7. *Codeines and Morphines*

In 1958, Koczka and Bernáth (105) showed that the product isolated from the reaction of codeine (**60a**) and benzyl iodide was diastereomeric with that obtained from *N*-benzylnorcodeine (**60b**) and methyl iodide. Later, Bognár and Szabó (106) recorded similar findings for reactions of O_3,O_6-diacetylmorphine (**60c**) with allyl bromide and *n*-propyl bromide and their inverse reactions. Configurational assignments, which were based on differences in optical rotation (72,105,106), indicated that the products isolated were formed by axial attack. Although no second product from any of these reactions, or that of dihydrocodeine and benzyl iodide, was reported, it seems unlikely that any is more than 90% stereoselective.

More recently, Koczka and Bernáth (19) described the details of their study of stereoselectivity in quaternizations of morphine (**60d**) with allyl iodide and *N*-allylnormorphine (**60e**) with methyl iodide. The greatest degree of stereoselectivity (<15% of minor product) was seen in reactions of **60e** with methyl iodide in benzene at room temperature or in chloroform at 4°. Reactions were also carried out in ethanol and in methanol. Products from reactions carried out at reflux temperatures contained 15–25% of the minor quaternary salt. As the major product from morphine and allyl iodide was isomerized to its diastereomer by heating in chloroform at 80° for 80 hr, it

TABLE 3

Stereoselectivity in Quaternizations of 2-Substituted Piperidines

51 $\xrightarrow{R''X}$ **52** + **53**

R	R′	R″X	Solvent	Temp., °C	% **52**[a]	% **53**[a]	Method[b]	Ref.
CH_3	CH_3	CD_3I	75% aq. CH_3OH	23	71	29	A	50
CH_3	CH_3	CD_3I	CH_3OH	23	69	31	A	50
CH_3	CH_3	CD_3I	CH_3OH	70	*65*	*35*	A	50
CH_3	CH_3	CD_3I	$HCCl_3$	23	69	31	A	50
CH_3	CH_3	CD_3I	$(CH_3)_2C{=}O$	23	66	34	A	50
CH_3	CH_3	CD_3I	$(C_2H_5)_2O$	23	66	34	A	50
CH_3	CH_3	CD_3I	$(C_2H_5)_2O$	R.T.	*50–60*	*40–50*	A, F	18,37
CH_3	CH_3	CD_3I	Dioxane	23	63	37	A	50
CH_3	CH_3	CD_3I	ϕH	23	63	37	A	50
CH_3	CH_3	CD_3I	c-C_6H_{12}	23	64	36	A	50
CH_3	CH_3	C_2H_5I	$(CH_3)_2C{=}O$	23	43	57	A	50
CH_3	CH_3	C_2H_5I	$(CH_3)_2C{=}O$	R.T.	*50*	*50*	F	37
CH_3	CH_3	n-C_3H_7I	$(CH_3)_2C{=}O$	R.T.	*40–50*	*50–60*	A, F	18,37
CH_3	CH_3	i-C_3H_7I	$(CH_3)_2C{=}O$	Reflux	*<8*	*>92*	F	37
CH_3	CH_3	ϕCH_2I	$(CH_3)_2C{=}O$	Reflux	*40–50*[c]	*50–60*[c]	E, F	18,37,52
CH_3	CH_3	$O{-}OH_2$	30% H_2O_2	23	64	36	A	50
CH_3	CH_3	$O{-}OH_2$	30% H_2O_2	70	*55*	*45*	A	50
CH_3	C_2H_5	CD_3I	$(CH_3)_2C{=}O$	23	56	44	A	50
CH_3	C_2H_5	C_2H_5I	$(CH_3)_2C{=}O$	23	36	64	A	50
CH_3	C_2H_5	C_2H_5I	$(CH_3)_2C{=}O$	70	*40*	*60*	A	50
CH_3	C_2H_5	$O{-}OH_2$	30% H_2O_2	23	50	50	A	50

C_2H_5	CH_3	CH_3I	$(CH_3)_2C{=}O$	R.T.	*67–75*	*25–33*	F	37
C_2H_5	n-C_3H_7	C_3H_5I	None	> R.T.	—[d]		—	3
C_2H_5	n-C_3H_7	ϕCH_2I	None	> R.T.	—[d]		—	1
C_2H_5	C_2H_5CHOH	ϕCH_2I	None	> R.T.	*75 (25)*[e]	*25 (75)*[e]	—	4
n-C_3H_7	CH_3	CH_3I	$(CH_3)_2C{=}O$	R.T.	*67–75*	*25–33*	A, F	37
n-C_3H_7	n-C_3H_7	ϕCH_2I	None	> R.T.	—[d]		—	3
n-C_3H_7	C_2H_5CHOH	ϕCH_2I	—[f]	—[f]	*75 (25)*[e]	*25 (75)*[e]	—	4
i-C_3H_7	CH_3	CH_3I	$(CH_3)_2C{=}O$	Reflux	*92*	*<8*	F	37
n-C_4H_9	n-C_3H_7	ϕCH_2I	None	> R.T.	—[d]		—	3
i-C_5H_{12}	n-C_3H_7	ϕCH_2I	None	R.T.	—[d]		—	1
i-C_5H_{12}	C_2H_5CHOH	ϕCH_2I	—[f]	—[f]	—[d]		—	4
ϕCH_2	CH_3	CH_3I	$(CH_3)_2C{=}O$	Reflux	*67–75*[c]	*25–33*[c]	E, F	18,37,52
ϕCH_2	n-C_3H_7	CH_3I	None	R.T.	—[d]		—	1
ϕCH_2	n-C_3H_7	C_2H_5I	None	120	—[d]		—	1
ϕCH_2	n-C_3H_7	$CH_3O_2CCH_2I$	None	> R.T.	—[d]		—	1
ϕCD_2	CH_3	ϕCH_2I	$(CH_3)_2C{=}O$	R.T.	*75*	*25*	A	38
$CH_2CO_2CH_3$	n-C_3H_7	ϕCH_2I	None	R.T.	—[d]		—	1

[a] Unless noted otherwise, product compositions were determined using NMR spectroscopy, and stereochemical assignments are those given in the reference. Percentages in italics were calculated from data given in the reference.

[b] The method used for assigning configurations is one noted in the indicated part of Section III.

[c] A slightly greater stereoselectivity was noted for the reaction carried out at room temperature.

[d] Two products isolated.

[e] Analysis by fractional crystallization.

[f] Presumably as for the reaction of the *N*-ethyl analog.

TABLE 4
Stereoselectivity in Quaternizations of *cis*-6-Substituted 1,2-Dimethylpiperidines[a]

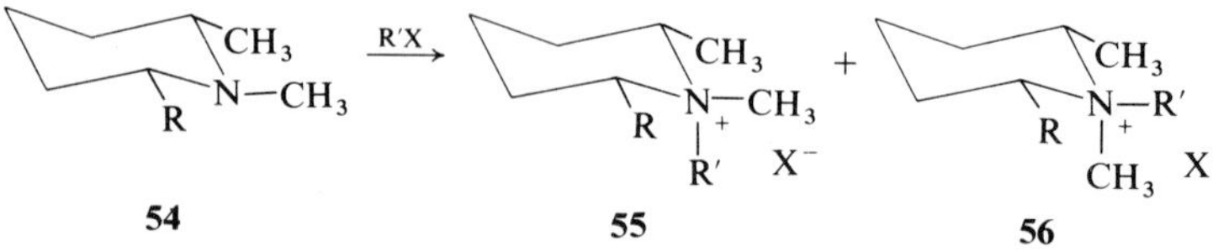

54 → (R′X) 55 + 56

R	R′X	Solvent	Temp., °C	% **55**	% **56**
CH_3	CD_3I	$(CH_3)_2C{=}O$	23	44	56
CH_3	C_2H_5I	$(CH_3)_2C{=}O$	23	11	89
CH_3	C_2H_5I	$(CH_3)_2C{=}O$	70	12	88
CH_3	$O{-}OH_2$	30% H_2O_2	23	39	61
C_2H_5	CD_3I	$(CH_3)_2C{=}O$	23	34	66
C_2H_5	C_2H_5I	$(CH_3)_2C{=}O$	23	9	91
C_2H_5	$O{-}OH_2$	30% H_2O_2	23	19	81

[a] Data from ref. 50. Configurational assignments were based on empirical NMR correlations.

TABLE 5
Stereoselectivity in Quaternizations of *trans*-Decahydroquinolines[a]

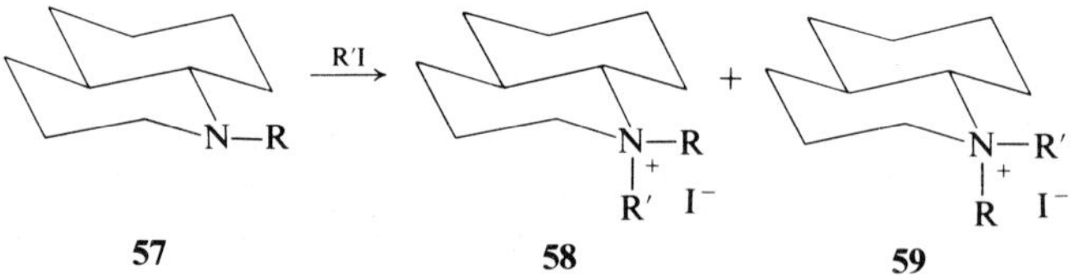

57 → (R′I) 58 + 59

R	R′	% **58**[b]	% **59**[b]	Method[c]	Ref.
CH_3	CD_3[d]	67–75	25–33	F	37
CH_3	C_2H_5[e]	<50	>50	F	18,37
CH_3	n-C_3H_7	40–50	50–60	F	37
CH_3	i-C_3H_7	<8	>92	F	37
CH_3	ϕCH_2	40–50	50–60	E	37
C_2H_5	CH_3[e]	67–75	25–33	F	37
n-C_3H_7	CH_3	67–75	25–33	F	37
i-C_3H_7	CH_3	>92	<8	F	37
ϕCH_2	CH_3	67–75	25–33	E	37
ϕCD_2	ϕCH_2[f]	ca. 75	ca. 25	A	38

[a] Unless noted otherwise, reactions were carried out in refluxing acetone.
[b] Calculated from product ratio reported in the reference. Analysis by NMR spectroscopy.
[c] The method used for assigning configurations is one noted in the indicated part of Section III.
[d] Carried out in ether at room temperature.
[e] Essentially the same stereoselectivity was observed for the reaction carried out in acetone at room temperature (18).
[f] Carried out at room temperature.

was concluded that both quaternizations took place predominantly by axial attack. Similar equilibration experiments with the major product from codeine (**60a**) and benzyl iodide indicated that it was the less stable diastereomer, thereby confirming the previous configurational assignments based on optical rotation data.

Significantly, the degree and manner of stereoselectivity observed in quaternizations of compounds containing the morphine skeleton, which can be thought of as a piperidine with equatorial substituents at C-3 and C-4 and axial substituents at C-2 and C-4 (cf. **61**), appear to be similar to what is observed in quaternizations of 4-substituted piperidines.

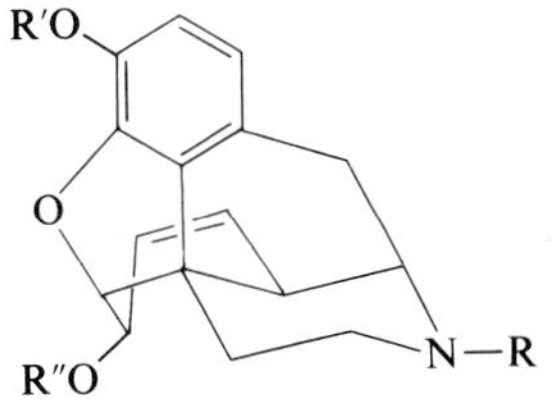

60a: R = R′ = CH_3, R″ = H
b: R = ϕCH_2, R′ = CH_3, R″ = H
c: R = CH_3, R′ = R″ = CH_3CO
d: R = CH_3, R′ = R″ = H
e: R = C_3H_5, R′ = R″ = H

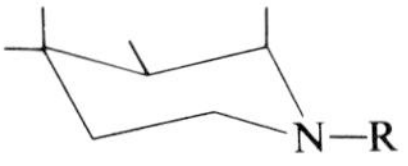

61

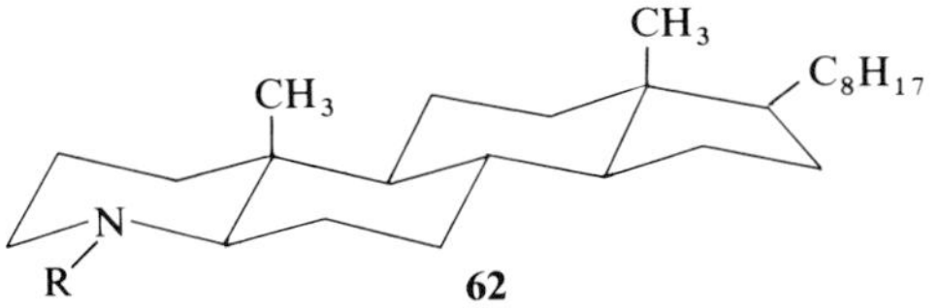

62

8. *4-Aza-5α-cholestanes*

4-Aza-5α-cholestanes (**62**) can be regarded as piperidines with fixed equatorial substituents at C-2 and C-3 and a fixed ax.-methyl group at C-3. Quaternization with the piperidine ring in the chair conformation therefore causes a 1,3-diaxial interaction of the *N*-methyl group with either the incoming group (axial attack) or the fully bonded *N*-alkyl group (equatorial attack). Examination of Table 6, which is a summary of results obtained by McKenna et al. (15,37,38), shows that little stereoselectivity was observed in methylations and deuteriomethylation, and that alkylations of the *N*-methyl base occur mainly by axial attack. Interestingly, Brown, McKenna, and McKenna (38) could not obtain the *N*-benzyl benziodide from attempted reactions of the *N*-benzyl base in acetone at room temperature for periods up to 15 days or at reflux temperatures for shorter times.

TABLE 6
Stereoselectivity in Quaternizations of 4-Aza-5α-cholestanes[a]

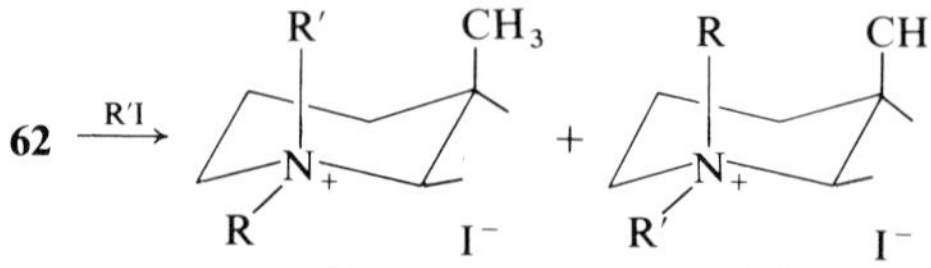

$$\mathbf{62} \xrightarrow{R'I} \mathbf{63} + \mathbf{64}$$

R	R′	Solvent	% **63**[b]	% **64**[b]	Method[c]
CH_3	CD_3	$(CH_3)_2C{=}O$	50–60	40–50	F
CH_3	C_2H_5	C_2H_5OH	25–33	67–75	F
CH_3	*n*-C_3H_7	$(CH_3)_2C{=}O$	25–33	67–75	F
CH_3	ϕCH_2	$(CH_3)_2C{=}O$	12–22	78–88	E
C_2H_5	CH_3	CH_3OH	ca. 50	ca. 50	F
n-C_3H_7	CH_3	$(CH_3)_2C{=}O$	ca. 50	ca. 50	F
ϕCH_2	CH_3	$(CH_3)_2C{=}O$	50–60	40–50	F

[a] Data from refs. 15 and 37. Reactions were carried out at reflux temperatures.
[b] Calculated from the reported product ratio.
[c] The method used for assigning configurations is one noted in the indicated part of Section III.

9. *3-Azabicyclo[3.n.1]alkanes*

In 1961 Trojánek et al. (70) described quaternizations of a number of *N*-alkylcamphidines (3-alkylaza-1,8,8-trimethylbicyclo[3.2.1]octanes; **65**) derived from (+)-camphorimide, and reported the isolation of only one isomer from each reaction. In the years immediately following McKenna et al. (18,26,27,37,38,52,53) examined, with one exception, the same reactions of racemic *N*-alkylcamphidines and showed that mixtures were obtained in all cases. Although different conditions were used by the two groups, this would not account for the different observations. It seems most likely that the earlier investigators had failed to identify a second salt in reaction mixtures where it was a minor component. (Also see ref. 129.)

As a first approximation, a camphidine can be regarded as a piperidine with axial substituents at C-3 and C-5. Thus one might expect that the degree and manner of stereoselectivity observed in camphidine quaternizations would be more closely related to that for piperidines with an ax.-methyl at

C-3, i.e., 4-aza-5α-cholestanes, than 4-substituted piperidines. In fact, examination of the data in Table 7, which are those of McKenna et al., shows that the stereoselectivity pattern is similar to that of 4-substituted piperidines and quite different from that of 4-aza-5α-cholestanes. Reactions of *N*-alkylcamphidines with methyl iodide or deuteriomethyl iodide are more stereoselective than alkylations of the *N*-methyl base. In all quaternizations which show stereoselectivity the preferred direction of attack is *syn* to the ethylene bridge, which corresponds to axial attack.

TABLE 7
Stereoselectivity in Quaternizations of Camphidines

N—R (**65**) —R′I, $(CH_3)_2C{=}O$, Reflux→ N^+—R, R′ I^- (**66**) + N^+—R′, R I^- (**67**)

R	R′	% **66**[a]	% **67**[a]	Method[b]	Ref.
CH_3	CD_3	>89	<11	A, F	37
CH_3	C_2H_5	67–75	25–33	C	81
CH_3	n-C_3H_7	67–75	25–33	A	18,37
CH_3	i-C_3H_7	50	50	—	18,37
CH_3	ϕCH_2	50–60[c]	40–50[c]	A, E	38,52
C_2H_5	CH_3	>89	<11	C	81
C_2H_5	n-C_3H_7	67–83[d]	17–33[d]	F	18
n-C_3H_7	CH_3	>89	<11	A	18,37
n-C_3H_7	C_2H_5	>95[d,e]	<5[d,e]	F	18
i-C_3H_7	CH_3	>89	<11	A	18,37
ϕCH_2	CH_3	>89	<11	A, E	38,52

[a] Analysis by NMR spectroscopy unless noted otherwise. Calculated from product ratio reported in the reference.

[b] The method used for assigning configurations is one noted in the indicated part of Section III.

[c] Partial equilibration may have occurred.

[d] Analysis by fractional crystallization.

[e] Only one isomer found.

One rationale for the stereoselectivity pattern seen in camphidine quaternizations is that the conformation of the piperidine ring in transition states is poorly approximated by the regular chair form. McKenna, McKenna, Tulley, and White (18) have pointed out that a regular chair or regular boat conformation would be unlikely for camphidinium salts because

Table 8

Stereoselectivity in Quaternizations of 3-Azabicyclo[3.*n*.1]alkanes Other than Camphidines[a]

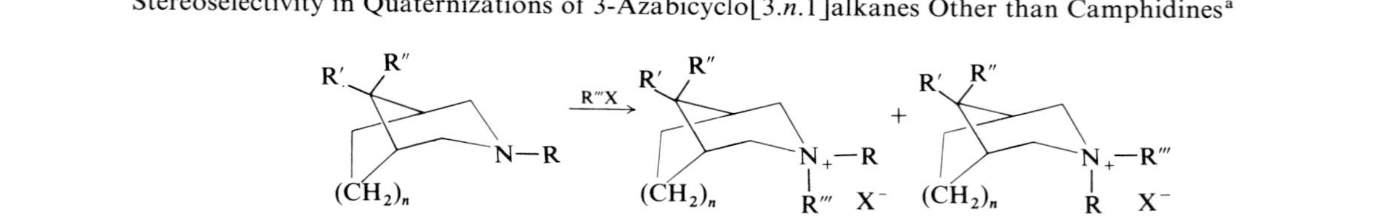

n	R	R′	R″	R‴ X	Temp., °C	% **69**[b]	% **70**[b]	Method[c]	Ref.
1	CH_3	O		CD_3OTs	56	29	71	A	45
1	CH_3	OH	H	CD_3OTs	30	9	91	I	45
1	CH_3	H	OH	CD_3OTs	30	11	89	I	45
2	CH_3	H	H	C_2H_5I	Reflux	*25–33*	*67–75*	F	37,51
2	CH_3	H	H	ϕCH_2I	Reflux	*25–33*	*67–75*	E	37,51
2	C_2H_5	H	H	CH_3I	Reflux	*9–11*	*89–91*	F	37,51
2	ϕCH_2	H	H	CH_3I	Reflux	*9–11*	*89–91*	E	37,51
2	CH_3	O		CD_3I	30	14	86	A	45
2	CH_3	O		CD_3OTs	30	18	82	A	45
2	CH_3	O		$CD_3—S^+—(CD_3)_2O^-$	130[d]	18	82	A	45
2	CH_3	O		$CH_3O_2CCH_2Br$	25	50	50[e]	I	46
2	CH_3	OH	H	CD_3OTs	56	25	75	I	45
2	CH_3	OH	H	$CH_3O_2CCH_2Br$	70[f]	—[g]		I	46
2	CH_3	H	OH	CD_3OTs	30	29	71	I	45

2	CH_3	H	OH	CD_3OTs	56	33	67	I	45
2	CH_3	H	OH	CD_3OTs	70 [f]	32	68	I	45
2	CH_3	H	OH	$CH_3O_2CCH_2Br$	70 [f]	25	75 [h]	D	46
3	CH_3	O		CD_3OTs	30	37	63	A	45
3	CH_3	OH	H	CD_3OTs	56	— [i]			45
3	CH_3	H	OH	CD_3OTs	30	25	75	I	45

[a] Solvent was acetone unless noted otherwise.

[b] Analysis by NMR spectroscopy. Percentages in italics were calculated from product ratio given in the reference.

[c] The method used for assigning configurations is one noted in the indicated part of Section III.

[d] No solvent specified.

[e] Included 30% of aldol condensation product **43**.

[f] No solvent.

[g] Although it was believed that the crude reaction product was a mixture, only the salt corresponding to **70** could be isolated.

[h] Isolated as the lactone.

[i] Composition could not be determined.

of high internal strain. They considered that a deformed chair conformation, that is, partway deformed toward the regular boat form, is more likely than the twisted boat conformation suggested earlier by Trojánek et al. (70), and an X-ray study (81) of the main product from *N*-ethylcamphidine and methyl iodide (**66**, R = C_2H_5, R′ = CH_3) indicates that the deformed chair conformation is preferred, at least in the crystalline state.

Stereoselectivities reported by House (45,46) and McKenna (37,51) and their co-workers for quaternizations of 3-azabicyclo[3.*n*.1]alkanes (**68**) other than camphidines are summarized in Table 8.

The stereochemistry of the products obtained from reactions of methyl bromoacetate was established by chemical means. The configurations of the three pairs of deuteriomethylation products from each ring system were related chemically, but the stereochemical assignments rest on the assumptions that the preferred direction of *N*-alkylation of the 3-azabicyclo[3.3.1]-nonanes (**68**, $n = 2$) is the same with deuteriomethyl tosylate and methyl bromoacetate, and that the preferred direction of deuteriomethylation is the same for all three ring systems. (Note that the conclusion concerning the stereochemistry of quaternization of 3-azabicyclo[3.2.1]octanes (**68**, $n = 1$) is opposite to that reached for camphidines.) Equilibration experiments established that the major product from methyl iodide and 3-benzylazabicyclo[3.3.1]nonane is the more stable diastereomer; the assigned stereochemistry is compatible with this stability order if the piperidinium ring is in the boat conformation, but incompatible if it is in the chair conformation.

The only bicyclic system studied by both groups is the 3-azabicyclo-[3.3.1]nonane system. Although they agreed that the preferred direction of alkylation is *syn* to the smaller carbon bridge, they disagreed as to the preferred conformation of the salts and the transition states leading to them. Lygo, McKenna, and Sutherland (51) interpreted the NMR spectrum of 3-methylazabicyclo[3.3.1]nonane methiodide (**69**, $n = 2$, R = R‴ = CH_3) as indicating that the preferred form of the piperidinium ring is the twist-boat conformation, and they argued that the stereoselectivity pattern shown by the 3-azabicyclononanes is best rationalized in terms of preferred axial attack on twist-boat piperidine. House et al. expressed the opinion that the preferred conformation of the transition states and salts is that with the heterocyclic ring in a chairlike conformation, and similar stereoselectivities in quaternizations of the epimeric 3-azabicyclo[3.3.1]nonanols seem to support this view.

10. Tropane and Related Compounds

This group of compounds (**71–75**) has figured prominently in investigations of the stereochemistry of amine quaternizations. The first unequivocal configurational assignments to a pair of diastereomeric quaternary salts were made to the products of a reaction of a tropane derivative, many of the studies of stereoselectivity in amine quaternizations followed from or gained inspiration from the pioneering work of Fodor and his co-workers (8) on the configuration of quaternization products of tropane alkaloids, and NMR spectroscopy was first applied in this field (9) to salts of tropane derivatives. Investigators of salts of tropane and its derivatives have also encountered more than their share of experimental difficulties, some of which have been noted earlier (see Sections II-A, II-B, III-A, and III-D).

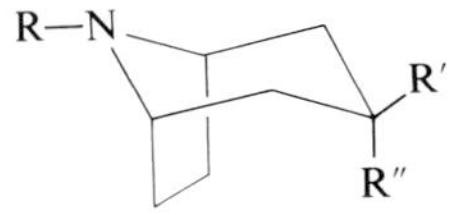

71

a: R′ = R″ = H; **b**: R′ = OH, R″ = H
c: R′ = H, R″ = OH; **d**: R′ = H, R″ = O-Tropyl

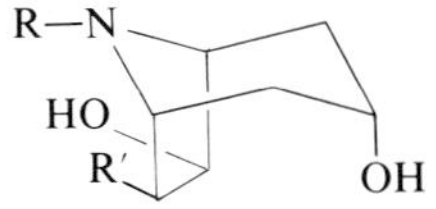

72a: R′ = H
b: R′ = OH

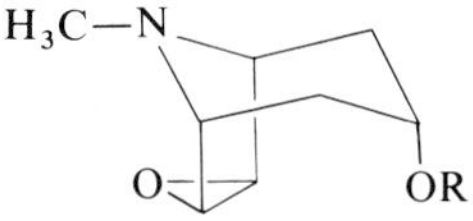

73a: R = Tropyl
b: R = H

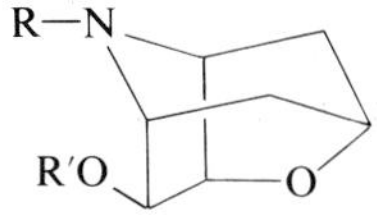

74a: R = H
b: R = CH_3CO

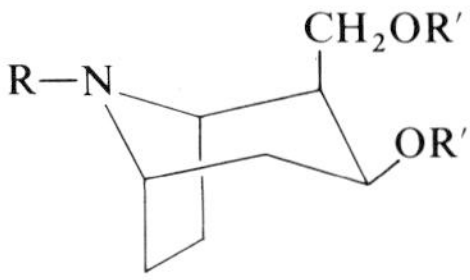

75a: R′ = H
b: R′ = CH_3CO

Although structural assignments to alkylation products of most of the di- and tri-substituted tropanes (**72**–**75**) were well established, the stereochemistry of quaternization of tropane (**71a**, R=CH_3) and the 3-tropanols (**71b** and **71c**, R=CH_3) has been a subject of controversy. Nearly two decades ago, Fodor, Koczka, and Lestyán (84) reported that the product from *N*-ethoxycarbonylnorpseudotropine (**71b**, R=$C_2H_5O_2CCH_2$) and methyl iodide could be converted to a lactone in poor yield by heating the corresponding acid at 200°. Like treatment of the product from pseudotropine (**71b**, R=CH_3) and ethyliodoacetate gave no evidence for lactone formation. These results were interpreted as indicating that the quaternizations occurred by equatorial attack. In 1964, results of an X-ray diffraction study of an *N*-ethyltropinium bromide were reported (80) which indicated that the major product from tropine (**71c**, R=CH_3) and ethyl iodide is also formed by equatorial attack. In 1967, McKenna et al. (37) questioned these structural assignments and, citing empirical correlations, suggested that quaternizations of tropanes occur mainly by axial attack. The following year, two groups (60a, 83a) obtained a lactone from pseudotropine bromoacetate, which gave on hydrolysis the betaine corresponding to the major product

TABLE 9

Stereoselectivity in Quaternizations of Tropane and Related Compounds

$$\mathbf{71, 72, 73, 74,}\text{ or }\mathbf{75} \xrightarrow{R'''X}$$

Compound	R	R‴X	Solvent	Temp., °C	% eq. R‴[a]	% ax. R‴[a]	Method[b]	Ref.
71a	CH_3	CD_3I	$(C_2H_5)_2O$	R.T.	*89–93*	*7–11*	—	37[c]
71a	CH_3	CD_3OBs	CD_3CN	30	68	32	A	60[b]
71a	CH_3	CD_3I	DMSO-d_6	R.T.	82[d]	18[d]	A	43
71a	CH_3	C_2H_5I	$(CH_3)_2C{=}O$	Reflux	*67–75*	*25–33*	F	37[c]
71a	CH_3	i-C_3H_7I	$(CH_3)_2C{=}O$	Reflux	*67–75*	*25–33*	F	37[c]
71a	CH_3	ϕCH_2I	$(CH_3)_2C{=}O$	R.T.	*89–93*[e]	*7–11*[e]	F	37[c]
71a	CH_3	$C_2H_5O_2CCH_2I$	CH_3CN	30	90	10	A,I	23
71a	C_2H_5	CH_3I	$(CH_3)_2C{=}O$	Reflux	*89–93*	*7–11*	F	37[c]
71a	i-C_3H_7	CH_3I	$(CH_3)_2C{=}O$	Reflux	*>93*	*<7*	F	37[c]
71a	ϕCH_2	CH_3I	$(CH_3)_2C{=}O$	R.T.	*67–75*[e]	*25–33*[e]	F	37[c]
71a	ϕCD_2	ϕCH_2I	$(CH_3)_2C{=}O$	R.T.	*80*	*20*	—	38
71b	CH_3	CD_3I	DMSO-d_6	R.T.	84	16	A	43
71b	CH_3	C_2H_5I	Abs.C_2H_5OH	R.T.	~75[f]	~25[f]	A	9
71b	CH_3	C_2H_5I	CH_3CN	30	78[f]	22[f]	C	60b
71b	CH_3	C_2H_5OBs	CH_3OH	30	83	17	C	60b
71b	CH_3	C_2H_5OBs	CH_3CN	30	80	20	C	60b
71b	CH_3	C_2H_5OBs	ϕH	30	80	20	C	60b
71b	CH_3	ϕCH_2Br	CH_3OH	30	89	11	A	23
71b	CH_3	$C_2H_5O_2CCH_2I$	CH_3CN	30	92	8	C,I	60
71b	C_2H_5	CH_3I	C_2H_5OH	R.T.	85	15	A	9
71b	$C_2H_5O_2CCH_2$	CH_3I	ϕH	30	92	8	I	60, 83

71c	CH_3	CD_3I	DMSO-d_6	R.T.	85	15	A	43
71c	CH_3	C_2H_5I	CH_3CN	30	72[g]	28[g]	C	60b, 84
71c	CH_3	C_2H_5OBs	CH_3CN	30	86	14	C	60b, 84
71c	CH_3	$C_2H_5O_2CCH_2Br$	C_2H_5OH	30	87	13	I	60, 83
71c	CH_3	$C_2H_5O_2CCH_2Br$	CH_3CN	30	86	14	I	60, 83
71c	CH_3	$C_2H_5O_2CCH_2Br$	ϕH	30	77	23	I	60, 83
71c	CH_3	$C_2H_5O_2CCH_2I$	C_2H_5OH	30	89	11	I	60, 83
71c	CH_3	$C_2H_5O_2CCH_2I$	CH_3CN	30	88	12	I	60, 83
71c	CH_3	$C_2H_5O_2CCH_2I$	ϕH	30	85	15	I	60, 83
71c	CH_3	O—OH_2	C_2H_5OH	R.T.		100[d]	A	61
71c	CO_3	CH_3I	DMSO-d_6	R.T.	83	17	A	43
71c	C_2H_5	CH_3I	DMSO-d_6	R.T.	50[d]	50[d]	A	43
71c	$C_2H_5O_2CCH_2$	CH_3I	CH_3CN	30	86	14	I	60, 83
71d	CH_3	—[h]	$(C_2H_5)_2O$	R.T.		100[d]	A	61
72a	CH_3	CD_3I	DMSO-d_6	R.T.	87[d]	13[d]	A	43
72a	CH_3	C_2H_5	DMSO-d_6	R.T.	83	17	A	43
73a	CH_3	O—OH_2	C_2H_5OH	R.T.		100[d]	A	61
73b	CH_3	CD_3I	DMSO-d_6	R.T.	90–93[d]	7–10[d]	A	43
74a	CH_3	CD_3I	DMSO-d_6	R.T.	74[d]	26[d]	A	43
75a	CH_3	CD_3I	DMSO-d_6	R.T.	38	62	A	43

[a] Analysis by NMR spectroscopy. Percentages in italics were calculated from data given in the reference.
[b] The method used for assigning configurations is one noted in the indicated part of Section III.
[c] In ref. 37, the configurational assignments indicated here were given with considerable qualification.
[d] Analysis questionable because of possible interference by hydroxyl resonance with an *N*-methyl band.
[e] Attempts to equilibrate the diastereomeric salts failed.
[f] The claim in ref. 43 that this reaction in DMSO-d_6 gives 100% axial attack appears to be incorrect.
[g] The analysis given in ref. 43 for the mixture obtained in DMSO-d_6 is incorrect because of failure to resolve the two *N*-methyl bands.
[h] The epoxidizing agent was monoperphthalic acid.

TABLE 10
Stereoselectivity in Quaternizations of Quinolizidine and Substituted Quinolizidines with Methyl Iodide[a,b]

76 $\xrightarrow{CH_3I}$ 77 + 78a ⇌ 78b

R	% 77[c]	% 78[c]	Ref.
H	~100	~0	56
c-1-CH_3	~0	~100	56
c-1-OH[d]	~0	~100	42b
c-1-OH, *t*-1-CH_3[d]	~0	~100	42b
c-1-OH, *t*-1-ϕ[e]	~100	~0	42b
c-1-O_2CCH_3[d,f]	~0	~100	42a
c-1-$O_2CC_2H_5$[d,f]	~0	~100	42a
c-1-CH_2OH	~0	~100[g]	73
t-1-CH_3	>95	<5	56
t-1-OH[d,h]	>99	<1	42b
t-1-OH[e]	50	50	42b
t-1-OH, *c*-1-CH_3[e]	50	50	42b
t-1-OH, *c*-1-ϕ[e]	50	50	42b
t-1-O_2CCH_3[d]	50	50	42a
t-1-$O_2CC_2H_5$[d]	50	50	42a
t-1-CH_2OH	~100[g]	~0	73
c-2-CH_3	~100	~0	56
t-2-CH_3[i]	~50	~50	56
c-3-CH_3	~0	~100	56
t-3-CH_3	>90	<10	56
c-4-CH_3	~0	~100	56
t-4-CH_3	~100	~0	56

[a] The prefixes *c* and *t* are used to indicate the *cis* or *trans* relationship of the hydrogen atom at C-10 and that of the hydrogen or other group at the substituted carbon atom. The ring conformation is indicated by the prefixes *cis* and *trans*.

from alkoxycarbonylmethylations of pseudotropine. This led to the incorrect conclusions that the original structural assignments of Fodor et al. (84) were in error. Subsequent work (60b,83b) has shown that this lactone was formed by inter- rather than intramolecular quaternization. The structures of the major products from quaternizations of pseudotropine with ethyl bromide (60b) and *N*-ethyl nortropine and methyl iodide (80,83b) have been determined by X-ray crystallography. Both of these products are formed by equatorial attack, that is, from the less stable amine conformer. That reactions of these amines with ethyl bromoacetate also occur predominantly by equatorial attack has been shown by conversion of the principal products, via the 2-hydroxyethyl and 2-chloroethyl salts, to the main products obtained with ethyl bromide. Comparison of stereo selectivities observed for other quaternizations of tropane, the 3-tropanols, and their *N*-homologs **(71a-c)** with that seen in other systems (see Section V-A) indicates that these other reactions also occur mainly by equatorial attack. It should also be noted here that exo-α-hydrogens of substituents in the axial conformation on cyclohexanes and piperidines are, generally, more shielded than when in the equatorial conformation (Section III-A); the opposite is generally the case for tropanes. Empirical correlations of NMR data led to incorrect inferences concerning the major pathway of certain tropane quaternizations, for example, pseudotropine with ethyl iodide (43).

Stereoselectivities in quaternizations of tropane (**71a**, R = CH_3) and related compounds determined with the aid of NMR spectroscopy are summarized in Table 9. High stereoselectivities are observed in many of these quaternizations. As with the 3-azabicyclo[3.*n*.1]alkanols (Table 8) studied by House et al. (45,46), the configuration of the hydroxyl group at C-3 appears to have little effect on the stereochemistry of tropane quaternizations. Significantly, the conclusion that the preferred mode of *N*-oxidation is by axial attack fails to gain support from the similarities in stereochemistry seen for *N*-oxidations and *N*-alkylations of 2-substituted piperidines (Table 3).

---

[b] Most reactions were carried out in the refluxing solvent, which was ethanol unless noted otherwise.

[c] Percentages were estimated from data given in the reference. With one exception (73), NMR spectroscopy was used in all investigations. Unless noted otherwise, configurational assignments were based mainly on NMR correlations.

[d] In benzene.

[e] In acetonitrile.

[f] Similar stereochemistry was reported for the reaction with methyl bromide in methanol and in ether at room temperature.

[g] Configurational assignment based on optical rotation data.

[h] In methanol.

[i] A **77**:**78** ratio of 1.2 ± 0.2:1 was obtained in acetonitrile (49).

11. Quinolizidine and Related Compounds

Quinolizidines (**76**) can be regarded as a special class of piperidines in which rotation of the groups at nitrogen and C-2 is restricted. Unlike the other piperidines discussed so far, there are three rather than two reasonably likely conformations that a monosubstituted quinolizidine can be expected to assume during quaternization, and these lead either to the *trans* fused product (**77**) or one of the conformations of the *cis* fused product (**78a** or **78b**).

Quaternizations of quinolizidine and substituted quinolizidines with methyl iodide have been investigated by several groups. Although it appears that isolation and identification of major products were the major concerns in most of these studies, estimates of stereoselectivity can be made, and these are given in Table 10. Excepting the results reported for *t*-1-hydroxyquinolizidine and its esters, the data for monosubstituted quinolizidines suggest the following generalities (cf. ref. 56). Consider the base in the *trans* conformation: if the substituent is equatorial (*t*-1, *c*-2, *t*-3, or *c*-4), the major product is the *trans* fused salt (**77**); if the substituent is axial and on the same side of the ring as the lone pair on nitrogen (*c*-1 or *c*-3), the *cis* fused salt (**78**) is the major product; if the substituent is axial and *trans* to the lone pair (*t*-2 or *t*-4), both products are formed in near equal amounts. Interestingly, these generalities appear to apply as well to reactions of methyl iodide with heteroyohimbine alkaloids having the common skeleton **79** (59) and the reaction of methyl chloride with the 1,2-dehydroquinolizidine, yohimbol (**80**) (25). The unusual solvent effect reported for the quaternization of *t*-1-hydroxyquinolizidine seems curious, as does the different stereochemistry seen for quaternizations of the 1-methyl- and 1-phenyl-*c*-1-hydroxyquinolizidines (42b).

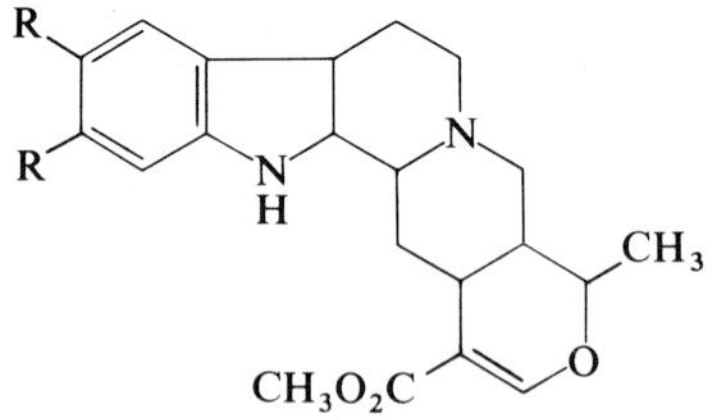

79
R = H or CH

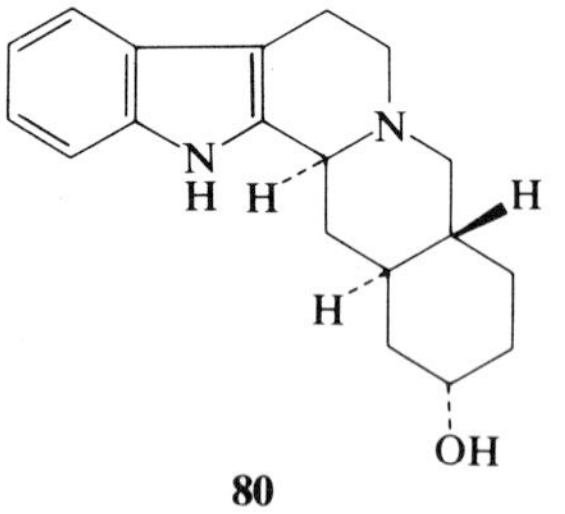

80

It should also be noted here that Fodor, Sallay, and Dutka (114) and Crow (73) examined the reaction of (−)-lupinine (**76**, R = *c*-1-CH_2OH) with ethyl iodoacetate, but the stereochemistry of the reaction is not clear.

12. Indolizidine and Related Compounds

Meyer and Sapianchiay (54) found that treatment of indolizidine (**81**) with methyl iodide in ethanol or ether, or methyl bromide in ether, gives a sharp-melting, crystalline product. The NMR spectra of these products showed that each consisted of nearly equal amounts of the *cis* and *trans* fused methosalts. The stereoselectivity seen in methylation of **81** is similar to that for monosubstituted *trans*-quinolizidines (**76**) with an equatorial substituent. Like **76**, indolizidine can be regarded as a piperidine that possesses 1,2-substituents with limited freedom of rotation.

Recently, Kotera, Hamada, and Mitsui (107) confirmed earlier reports (115–117) that two methiodides are formed from each of the related alkaloids caranine, pluviine, and lycorine (**82a–c**, respectively). They also found that β-dihydrocaranine (**83a**) gave two methiodides, but they could isolate only one product from the reaction of α-dihydrocaranine (**84a**), α-dihydropluviine (**84b**), or diacetyldihydrolycorine (**84c**) with methyl iodide. On the basis of conformational arguments supported by optical rotation and ultraviolet spectral data, they assigned the β-*N*-methyl (*cis*-quinolizidinium) configuration to the methiodides of **84a–c**. It should be noted that the Japanese workers failed to isolate a second methiodide of β-dihydropluviine (**83b**), and they ascribed this to experimental difficulties. Interestingly, they reported that **82c** and methyl iodide in methanol at 0°, and at room and reflux temperatures,

81 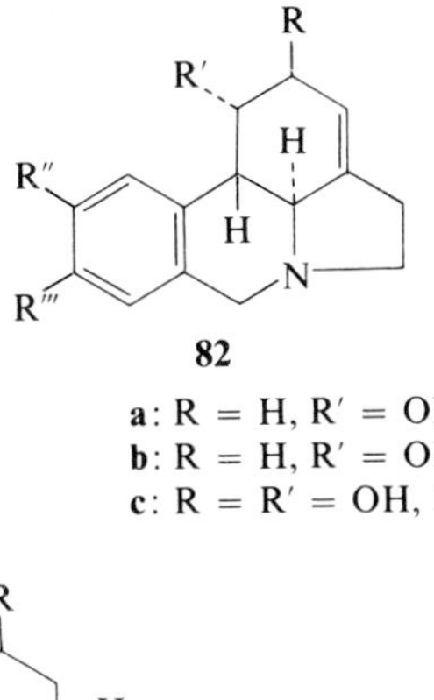 **82** 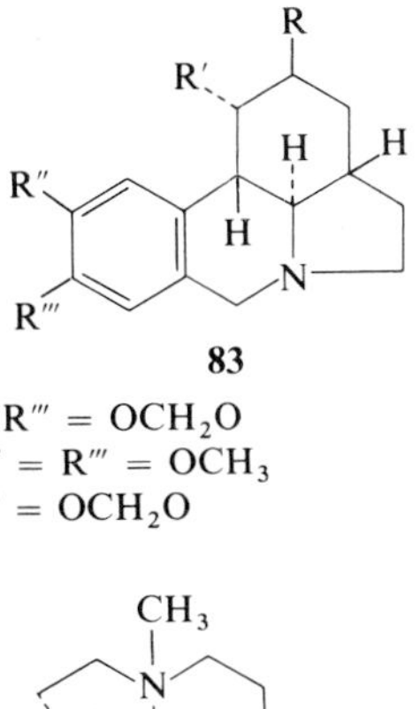 **83**

a: R = H, R′ = OH, R″R‴ = OCH_2O
b: R = H, R′ = OH, R″ = R‴ = OCH_3
c: R = R′ = OH, R″R‴ = OCH_2O

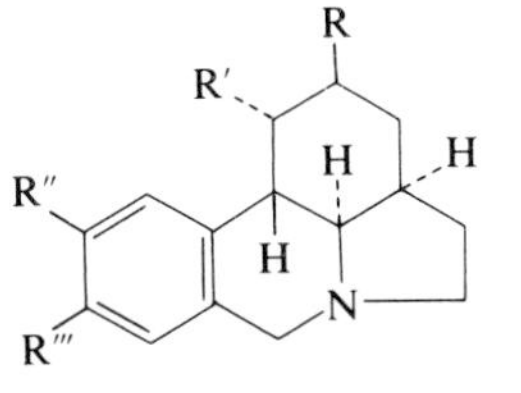

84a: R = H, R′ = OH, R″R‴ = OCH_2O
b: R = H, R′ = OH, R″ = R‴ = OCH_3
c: R = R′ = O_2CCH_3, R″R‴ = OCH_2O

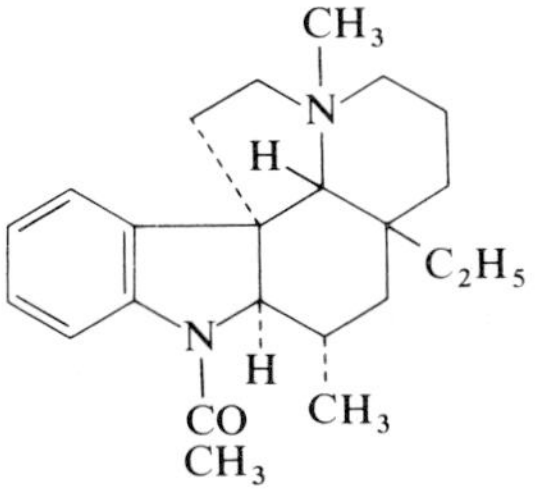

85

gave 5:1, 3:1, and 4:3 mixtures, respectively, of the α- and β-methiodides. As these analyses were based on fractional crystallization, they should only be regarded as approximations.

The structure of (±)-1-acetyl-16-methylaspidospermidine-4-methiodide (**85**) has been determined by X-ray crystallography (82), but the stereoselectivity of the reaction leading to its formation has not been studied.

B. Other Cyclic Amines with Six-membered Rings

1. 1,2,5,6-Tetrahydropyridines

Results obtained by Casy, Beckett, and Iorio (40) in their study of stereoselectivity in quaternizations of 1-alkyl-4-phenyl-5-methyl-1,2,5,6-tetrahydropyridines (**86**) are summarized in Table 11. With the exceptions of

TABLE 11
Stereoselectivity in Quaternizations of 1-Alkyl-4-phenyl-5-methyl-1,2,5,6-tetrahydropyridines[a]

ϕ H$_3$C N R R′X → ϕ H$_3$C N R R′ X^- + ϕ H$_3$C N R′ R X^-

86 **87** **88**

R	R′X	% **87**[b,c]	% **88**[b,c]
CH_3	C_2H_5I	70	30
CH_3	C_3H_5Br	56[d]	44[d]
CH_3	i-C_3H_7I	38	62
CH_3	ϕCH_2Cl	30[d]	70[d]
C_2H_5	CH_3I	90	10
C_3H_5	CH_3I	88[d]	12[d]
i-C_3H_7	CH_3I	82	18
ϕCH_2	CH_3I	Major product[d]	Trace[d]

[a] Data from ref. 40. The base was allowed to react with excess alkyl halide in ether or ether–acetone for one or two days at room temperature or 30 min at reflux temperature.

[b] Analysis by NMR spectroscopy. Percentages calculated from reported product ratio.

[c] Configurations were assigned to the salts by comparing their 1-methyl chemical shifts with those of related 1,3-dimethyl derivatives, by comparison of stereoselectivities (cf. Section III-F), and, for the 1-benzyl-1-methyl salts, by a partial equilibration experiment.

[d] Partial equilibration was not excluded but seems unlikely to have been significant.

quaternizations of the *N*-methyl base with isopropyl iodide and benzyl chloride, the major products were formed by axial attack, and reactions with methyl iodide were more stereoselective than alkylations of the *N*-methyl base. Comparison of the data in Table 11 with results obtained for similar reactions of 4-substituted 1-alkylpiperidines (Table 1) shows that a greater percentage of product is formed from the more stable amine conformer in reactions of methyl iodide and ethyl iodide with the 1-alkyltetrahydropyridines. On the other hand, reaction of isopropyl iodide or benzyl chloride with a 4-substituted 1-methylpiperidine gives a greater percentage of product corresponding to the more stable amine conformer than is obtained with the 1-methyltetrahydropyridine. (Also see ref. 128.)

2. *1,2,3,4-Tetrahydroquinolines*

In 1905, Scholtz and Pawlicki (4) reported the isolation of only one product from the quaternization of 1-ethyl-2-methyl-1,2,3,4-tetrahydroquinoline with benzyl iodide. This reaction has not been reinvestigated.

3. *1,2,3,4-Tetrahydroisoquinolines*

Stereoselectivity in reactions of 1,2-disubstituted 6,7-dimethoxy-1,2,3,4-tetrahydroisoquinolines (**89**) with alkyl iodides has been studied by Bernáth et al. (28,29), whose results are summarized in Table 12. With the exception of the reaction of methyl iodide and the *N*-benzyl-1-methyl base, the major products from methylation correspond to the more stable *trans* amine conformers. On the other hand, the major products from alkylations of the *N*-methyl bases correspond to the less stable *cis* amine conformers. Interestingly, the stereoselectivities (at least 95%) claimed for benzylations of the *N*-methyl bases are much greater than those observed for reactions of benzyl iodide with 1,2-dimethylpiperidine and 1-methyl-*trans*-decahydroquinoline. Also, increase in size of the C-1 alkyl group ($\phi \sim i\text{-}C_3H_7 > \text{DMBz} > CH_3$) favors formation of the quaternary salt corresponding to the more stable amine conformer; a similar structural variation in the 2-substituted piperidine series brings about the opposite result. (Also see refs. 131 and 133.)

4. *Tetrahydo-1,3-oxazines*

Gürne et al. (44) examined the 28 alkylations of 3,5-dialkyl-5-nitrotetrahydro-1,3-oxazines summarized in eq. 5. Previous studies (118,119) of dipole moments and NMR spectra of the bases had led to the conclusion that the ring exists in the chair form with the nitro group axial. It had also been inferred that *N*-alkyl groups less bulky than cyclohexyl or *t*-butyl assumed

TABLE 12
Stereoselectivity in Quaternizations of
1,2-Disubstituted 6,7-Dimethoxy-1,2,3,4-tetrahydroisoquinolines[a]

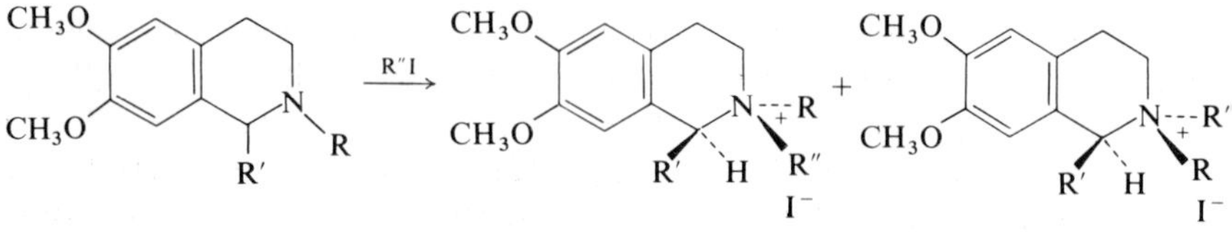

89 $\xrightarrow{R''I}$ **90** + **91**

R	R′	R″	% **90**[b]	% **91**[b]
CH_3	CH_3	C_2H_5	30	70
CH_3	CH_3	n-C_3H_7	25	75
CH_3	CH_3	n-C_4H_9	25	75
CH_3	CH_3	ϕCH_2	0	100
CH_3	i-C_3H_7	C_2H_5	40	60
CH_3	i-C_3H_7	ϕCH_2	5	95
CH_3	ϕ	C_2H_5	24	76
CH_3	ϕ	ϕCH_2	0	100
CH_3	DMBz[c]	C_2H_5	35	65
CH_3	DMBz[c]	ϕCH_2	0	100
C_2H_5	CH_3	CH_3	65	35
C_2H_5	i-C_3H_7	CH_3	83	17
C_2H_5	ϕ	CH_3	86	14
C_2H_5	DMBz[c]	CH_3	70	30
n-C_3H_7	CH_3	CH_3	65	35
n-C_4H_9	CH_3	CH_3	55	45
ϕCH_2	CH_3	CH_3	45	55
ϕCH_2	i-C_3H_7	CH_3	60	40
ϕCH_2	ϕ	CH_3	65	35
ϕCH_2	DMBz[c]	CH_3	60	40

[a] Data from refs. 28 and 29. Each base was treated with excess alkyl iodide in methanol, acetone, or ethanol at room or reflux temperature.

[b] Relative accuracy $\pm 5\%$. Stereochemical assignments to the *N*-benzyl-*N*-methyl salts were based on the results of equilibration experiments (Section III-E) and differences in degree of magnetic nonequivalence of the benzyl-methylene protons in the diastereomeric salts (Section III-A). Assignments to the other salts were based on the relative frequencies of their *N*-methyl resonances.

[c] DMBz = 2,4-dimethoxybenzyl.

the axial conformation. Yields of 75–90% were obtained from the reactions listed, but attempts to alkylate *N*-*t*-butyl and *N*-cyclohexyl bases were unsuccessful. Only one product was isolated from each alkylation; however, minor components in the reaction mixtures could have been lost through fractional crystallization. Different diastereomers were obtained when the order of attachment of the two *N*-alkyl groups was reversed, and comparison of the NMR spectra of the salts with those of the bases led the investigators to conclude that the stereochemistry of the reactions was as pictured, that is, alkylation occurred by equatorial attack.

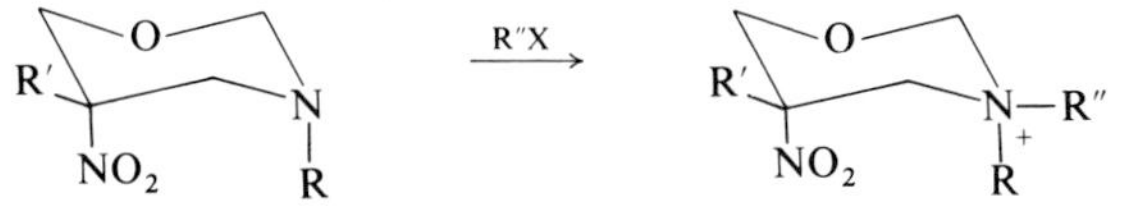

$$\xrightarrow{R''X} \qquad (5)$$

a: R = CH_3; R′ = CH_3; R″X = C_2H_5Br, C_2H_5I, n-C_3H_7Br, n-C_4H_9Br, ϕCH_2Br
b: R = CH_3; R′ = C_2H_5; R″X as in **a**
c: R = CH_3; R′ = n-C_3H_7; R″X = C_2H_5Br, C_2H_5I
d: R = CH_3; R′ = n-C_4H_9; R″X = C_2H_5I
e: R = CH_3; R′ = n-C_5H_{11}; R″X = C_2H_5I
f: R = C_2H_5; R′ = CH_3; R″X = CH_3Br, n-C_3H_7Br, n-C_4H_9Br, ϕCH_2Br, CH_3I
g: R = C_2H_5; R′ = C_2H_5; R″X as in **f**
h: R = C_2H_5; R′ = n-C_3H_7; R″X = CH_3Br, CH_3I
i: R = C_2H_5; R′ = n-C_4H_9; R″X = CH_3I
j: R = C_2H_5; R′ = n-C_5H_{11}; R″X = CH_3I

5. *Piperazines*

Imbach, Katritzky, and Kolinski (48) compared rates of methiodide formation for series of 1-alkyl-4-phenylpiperidines and 1-alkyl 4-t-butylpiperazines with those for model bi- and polycyclic amines, and concluded that the monocyclic amines, except the 1-*t*-butyl homologs, undergo methiodide formation predominantly by axial attack. Subsequently (33,37) their conclusion concerning the piperidines was shown to be incorrect. Note that the piperazine methiodides are equilibrated (eq. 6) by a combination of nitrogen inversion and ring inversion. As both of these processes certainly have lower energy barriers than methiodide formation, it is not possible to determine directly the stereochemistry of alkylation of a 1-alkylpiperazine.

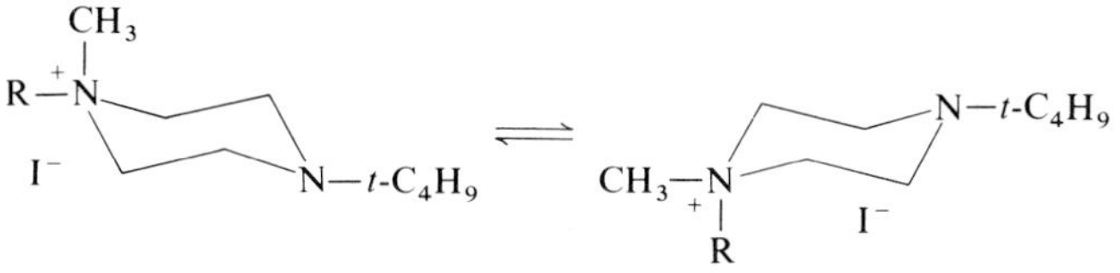

$$\rightleftharpoons \qquad (6)$$

C. Pyrrolidines

1. 2-Substituted Pyrrolidines

Stereoselectivity in quaternizations of 2-methylpyrrolidines (**92**) has been studied by McKenna et al. (18,26,27,37,38,52,53). Examination of their results, summarized in Table 13, shows that greatest stereoselectivity was seen in methylations and benzylations. Interestingly, the stereochemistry observed in these reactions parallels that for similar quaternizations of

TABLE 13
Stereoselectivity in Quaternizations of 2-Methylpyrrolidines[a]

92 —(R′I, $(CH_3)_2C{=}O$)→ **93** + **94**

R	R′	% **93**[b]	% **94**[b]	Method[c]	Ref.
CH_3	C_2H_5	<67	>33	A, F	18
CH_3	n-C_3H_7	~50	~50	—	18
CH_3	ϕCH_2[d]	67–75	25–33	E	37,52
C_2H_5	CH_3	83–95	5–17	A, F	18
n-C_3H_7	CH_3	67–83	17–33	A, F	18
ϕCH_2	CH_3[d]	82–88	12–18	E	37,52
ϕCD_2	ϕCH_2	75	25	A	38

[a] Unless noted otherwise, reactions were carried out at room temperature. Reactions of the *N*-ethyl base with *n*-propyl iodide and the *N*-*n*-propyl base with ethyl iodide were reported in ref. 18; mixtures were obtained, but product ratios were not given.

[b] Analysis by NMR spectroscopy. Calculated from product ratio reported in the reference.

[c] The method used for assigning configurations is one noted in the indicated part of Section III.

[d] Carried out at reflux temperature.

2-methylpiperidines (Table 3), with the five-membered ring bases giving greater percentages of product corresponding to the more stable *trans* amine conformer. (Also see ref. 129.)

2. Pyrrolizidines

Fodor, Uresch, Dutka, and Széll (91) found that 7-hydroxyheliotridane (**95a**) and ethyl bromoacetate gave an 80% yield of a salt with the same configuration as the lactone formed by intramolecular quaternization of the bromoacetate ester of **95a**. However, the lactone prepared in the same manner from retronecanol (**95b**) and ethyl bromoacetate had different configurations. This indicates that the preferred mode of quaternizations of these pyrrolizidines with ethyl bromoacetate leads to the *cis* fused salt. It should also be noted here that Meyer and Sapianchiay (54) assigned the *cis* fused configuration to the pyrrolizidine methiodide obtained by cyclization of 1-methyl-2-(3′-bromopropyl)pyrrolidine.

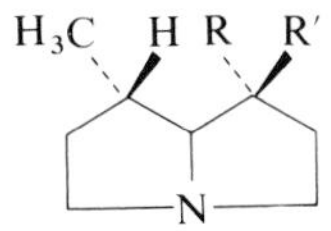

H_3C H R R′

N

95a: R = OH; R′ = H
b: R = H; R′ = OH

D. Aziridines

In Table 14 are summarized results of studies of stereoselectivity in quaternizations of 1,2,3-trialkylaziridines (**96**). *trans*-1,2-Dimethyl-3-isopropylaziridine shows greater stereoselectivities in deuteriomethylations than in other alkylations, and these reactions give product ratios smaller than the solvent-dependent conformational equilibrium constant of 3.8–4.2. High stereoselectivities are observed in quaternizations of the *cis*-2,3-dialkylaziridines and *trans*-1,2-dimethyl-3-*t*-butylaziridine, which can be expected to have much larger conformational equilibrium constants. In all reactions the major product is formed from the more stable amine conformer.

E. Acyclic Amines

Ötvös, Dutka, and Tüdős (120) quaternized *N*,*N*-dimethyl-*o*-toluidine with ^{14}C-enriched methyl iodide and then regenerated the base and methyl iodide by heating the salt in an aromatic solvent. They claimed that the specific activities of the base obtained from the reverse reactions in benzene and toluene were 77 and 72%, respectively, of that of the salt. Making the assumption that the more hindered methyl groups would be removed preferentially in the reverse reaction (cf. ref. 121), they interpreted their data as indicating that the preferred direction of attack by the methyl group during quaternization is at the side away from the *o*-methyl group. Note that if the specific activity measurements are correct, rotation about the *N*-aryl bond of the salt must be severely restricted (cf. ref. 122). (Also see ref. 132.)

Moynehan, Schofield, Jones, and Katritzky (55,56) and Meyer and Sapianchiay (54) have cyclized several 2-(ω-haloalkyl)-1-methyl-pyrrolidines

TABLE 14
Stereoselectivity in Quaternization of 1,2-Dialkyl-3-methylaziridines

$$\mathbf{96} \xrightarrow[\sim 30]{R''X} \mathbf{97} + \mathbf{98}$$

R	R′	R″X	Solvent	%97[a]	%98[a]	Method[b]	Ref.
CH_3	*cis*-CH_3	$CD_3OSO_2\phi$	CD_3OH	6	94	A	30
CH_3	*cis*-C_2H_5	$CD_3OSO_2\phi$	CD_3OH	5	95	A	30
CH_3	*cis-i*-C_3H_7	$CD_3OSO_2\phi$	CD_3OH	3	97	A	30
CH_3	*trans-i*-C_3H_7	CD_3I[c]	CD_3OH	71	29	G	31
CH_3	*trans-i*-C_3H_7	$CD_3OSO_2\phi$	CD_3OH	75	25	G	32
CH_3	*trans-i*-C_3H_7	$CD_3OSO_2\phi$	CH_3CN	73	27	G	23
CH_3	*trans-i*-C_3H_7	$CD_3OSO_2\phi$	ϕH	72	28	G	32
CH_3	*trans-i*-C_3H_7	C_2H_5OBs	CH_3OH	61	39	E	30
CH_3	*trans-i*-C_3H_7	C_2H_5OBs	CH_3CN	57	43	E	30
CH_3	*trans-i*-C_3H_7	*n*-$C_6H_{13}OBs$	CH_3OH	64	36	E	30
CH_3	*trans-i*-C_3H_7	$\phi CH_2OSO_2\phi$	CD_3OH	59	41	E	32
CH_3	*trans-i*-C_3H_7	$\phi CH_2OSO_2\phi$	ϕH	56	44	E	32
CH_3	*trans-t*-C_4H_9	$\phi CH_2OSO_2\phi$	CD_3OH	86	14	A, E	30
CH_3	*trans-t*-C_4H_9	$\phi CH_2OSO_2\phi$	$CD_3OH/\phi H$ (50:50)	92	8	A, E	30
CH_3	*trans-t*-C_4H_9	$\phi CH_2OSO_2\phi$	ϕH	93	7	A, E	30
C_2H_5	*trans-i*-C_3H_7	$CH_3OSO_2\phi$	CD_3OH	77	23	E	30
n-C_6H_{13}	*trans-i*-C_3H_7	$CH_3OSO_2\phi$	CD_3OH	76	24	E	30
ϕCH_2	*trans-i*-C_3H_7	$CH_3OSO_2\phi$	CD_3OH	72	28	A, E	30

[a] Analysis by NMR spectroscopy.

[b] The method used for assigning configurations is one noted in the indicated part of Section III.

[c] The 1:1 ratios of **97** and **98** reported (35) for this reaction and for that of *cis*-1,2-dimethyl-3-isopropylaziridine with methyl iodide in benzene were thermodynamically controlled product ratios.

and piperidines (**99**) capable of giving mixtures of diastereomeric salts, and their results are summarized in Table 15. Brown and McKenna (104) have also prepared mixtures of quaternary salts by cyclization of ω-tosyloxy tertiary amines. Although details of their work have not yet been described, they have noted that initial results indicate that the preferred course of intramolecular quaternization leads to the thermodynamically more stable salt. (Also see ref. 130.)

TABLE 15
Stereoselectivity in Intramolecular Quaternizations of 2-(ω-Haloalkyl)-1-methyl Cyclic Amines[a]

$$\mathbf{99} \longrightarrow \mathbf{100} \, (X^-) + \mathbf{101} \, (X^-)$$

m	n	R[b]	X	% **100**[c]	% **101**[c]	Ref.
3	3	H	Br	~100	~0	54
3	4	H	Br	~100	~0	54
4	3	H	Br	~100	~0	54
4	4	H	I	≫58[d]	≪42[d]	56
4	4	*cis*-3-CH_3	I	~100[e]	~0[e]	56
4	4	*trans*-3-CH_3	I	<15[e]	>85[e]	56
4	4	*cis*-5-CH_3	I	~100[e]	~0[e]	56
4	4	*trans*-5-CH_3	I	~50	~50	56

[a] Cyclizations to quinolizidinium iodides were carried out in chloroform at room temperature. Other cyclizations were carried out in absolute ethanol at or below room temperature.

[b] Orientation relative to the ω-haloalkyl group.

[c] Configurational assignments were based largely on NMR correlations. Analysis by NMR spectroscopy unless noted otherwise.

[d] Analysis by fractional crystallization.

[e] Apparently the analysis of the recrystallized product.

V. CONCLUSION

A. Summary

Studies of stereoselectivity in amine quaternizations have shown that, for many systems, selective synthesis of the members of a pair of diastereomeric *N*-alkylation products can be achieved by varying the order of attachment of the *N*-alkyl groups. For synthetic purposes, success in varying the degree of stereoselectivity by varying the solvent, leaving group, or temperature, or a combination of these variables, has been limited to reactions that show little stereoselectivity.

Although the degree of stereoselectivity is markedly dependent on the structure of the amine and the alkylating agent, certain patterns appear. Methylations and deuteriomethylations of homologous *N*-methyl and

N-alkyl bases are similar in both degree and manner of stereoselectivity. Alkylations of *N*-methyl amines give a greater proportion of product from the less stable amine conformer than do deuteriomethylations, and change of incoming group from ethyl or benzyl to alkoxycarbonylmethyl ($ROCH_2$) favors formation of the product from the less stable conformer. Quaternizations of 3-azabicyclo may be exceptions to these generalizations; however, if the transition states for these reactions have the piperidine ring in the twist-boat conformation, and this seems reasonable (51), the generalizations apply. With the possible exception of 1-methylpiperidines, another generalization that can be made is that the less stable amine conformer has the greater rate constant for quaternization. Reactions of 4-*t*-butyl-1-methylpiperidine with deuteriomethyl tosylate and ethyl tosylate give product ratios of 6.7:1 and 4.9:1, respectively, the major products corresponding to the more stable eq.-1-methyl conformer. Dipole moment measurements (123) and theoretical calculations (13), however, have been interpreted as indicating that the conformational equilibrium constants (eq.-1-methyl/ax.-1-methyl) of 4-phenyl-1-methylpiperidine and 1-methylpiperidine itself are between 2 and 4, that is, 7 to 15 times smaller than the corresponding constant for 1-methylcyclohexane. For those willing to accept that the conformational equilibrium constants of 1-methylpiperidines and the mode of quaternizations of 3-azabicyclononanes are not definitely established, the above generalizations are suggested as working hypotheses for predicting the stereochemistry for *N*-alkylations.

Limited data obtained in the 2-substituted piperidine and 6-substituted 1,2-dimethylpiperidine series (50) indicate that the stereochemical course of *N*-oxidation of an amine is intermediate between that for its reactions with methyl iodide and ethyl iodide. As a starting point for predicting the course of *N*-oxidations of other amines, it is suggested that comparable stereochemical results can be expected. Further, it is reasonable to expect that straightforward relationships between the stereochemistry of *N*-alkylations and the stereochemistry of other quaternizations, such as *N*-chlorinations (e.g., 124), will also be observed.

B. Related Problems

It seems appropriate to conclude this article by mentioning briefly two areas of research that are intimately related to study of the stereochemistry of amine quaternizations. The possibility of gaining information concerning the conformational equilibria of amines from stereoselectivity studies is being actively investigated. Two groups (32,47) have suggested that, in reactions of different amines with the same alkylating agent under identical conditions, the length of the new C—N bond in the transition state varies with the nucleophilicity of the amine; that is, the more nucleophilic the amine, the longer the new bond at the transition state. As the steric requirements of the incoming group obviously increase as the length of the new bond is decreased, it follows, at least as a first approximation, that the composition of diastereomeric salts produced will more closely reflect the conformational equilibrium constant of amines of greater nucleophilicity. The possibility of a relationship between the stereochemistry of *N*-alkylation of amines and *C*-alkylation of resonance-stabilized carbanions has also been investigated (125). The stereochemistry of the *C*-alkylations, in contrast to that of *N*-alkylations, was most easily rationalized on the basis of essentially trigonal hybridization of the nucleophilic atom in the transition state.

References

1. M. Scholtz, *Chem. Ber.*, **37**, 3627 (1904).
2. M. Scholtz and K. Bode, *Arch. Pharm.*, **242**, 568 (1904).
3. M. Scholtz, *Chem. Ber.*, **38**, 595 (1905).
4. M. Scholtz and P. Pawlicki, *Chem. Ber.*, **38**, 1289 (1905).
5. W. H. Mills, J. D. Parkin, and W. J. V. Ward, *J. Chem. Soc.*, **1927**, 2613.
6. M. Polonovski and M. Polonovski, *Bull. Soc. Chim. France (IV)*, **39**, 1147 (1926); **43**, 364 (1928).
7. M. Polonovski and M. Polonovski, *Bull. Soc. Chim. France (IV)*, **43**, 79 (1928).
8. G. Fodor, "The Tropane Alkaloids," in *The Alkaloids*, Vol. VI, R. H. F. Manske, Ed., Academic Press, New York, 1960, ch. 5; Vol. IX, 1967, ch. 7.
9. G. L. Closs, *J. Am. Chem. Soc.*, **81**, 5456 (1959).
10. A. T. Bottini and J. D. Roberts, *J. Am. Chem. Soc.*, **80**, 5203 (1958).
11. M. Saunders and F. Yamada, *J. Am. Chem. Soc.*, **85**, 1882 (1963).
12. E. L. Eliel, N. L. Allinger, S. J. Angyal, and G. A. Morrison, *Conformational Analysis*, Interscience, New York, 1965, ch. 7.
13. N. L. Allinger, J. A. Hirsch, and M. A. Miller, *Tetrahedron Lett.*, **1967**, 3729.
14. E. L. Eliel, *Stereochemistry of Carbon Compounds*, McGraw-Hill, New York, 1962, p. 436.
15. J. McKenna, J. M. McKenna, and A. Tulley, *J. Chem. Soc.*, **1965**, 5439.
16. D. Y. Curtin, *Rec. Chem. Progr.*, **15**, 111 (1954); ref. 14, pp. 149–156.
17. S. Winstein and N. J. Holness, *J. Am. Chem. Soc.*, **77**, 5562 (1955).
18. J. McKenna, J. M. McKenna, A. Tulley, and J. White, *J. Chem. Soc.*, **1965**, 1711.
19. K. Koczka and G. Bernáth, *Acta Chim. Acad. Sci. Hung.*, **51**, 393 (1967).
20. S. P. Findlay, *J. Am. Chem. Soc.*, **75**, 3204 (1953).
21. G. Fodor, *Acta Chim. Acad. Sci. Hung.*, **5**, 379 (1955).
22. K. Zeile and W. Schulz, *Chem. Ber.*, **88**, 1078 (1955).
23. A. T. Bottini, M. K. O'Rell, and C. C. Thut, unpublished.
24. G. Fodor and N. Mandava, unpublished.
25. M. F. Bartlett, B. Korzun, R. Sklar, A. F. Smith, and W. I. Taylor, *J. Org. Chem.*, **28**, 1445 (1963).
26. J. K. Becconsall and R. A. Y. Jones, *Tetrahedron Lett.*, **1962**, 1103.
27. J. K. Becconsall, R. A. Y. Jones, and J. McKenna, *J. Chem. Soc.*, **1965**, 1726.
28. G. Bernáth, J. Kóbor, K. Koczka, L. Radics, and M. Kajtár, *Tetrahedron Lett.*, **1968**, 225.

29. G. Bernáth, K. Koczka, J. Kóbor, L. Radics, and M. Kajtár, *Acta Chim. Acad. Sci. Hung.*, **55**, 331 (1968).
30. A. T. Bottini, B. F. Dowden, A. Bellettini, and L. C. Cooper, unpublished.
31. A. T. Bottini, B. F. Dowden, and L. Sousa, *J. Am. Chem. Soc.*, **87**, 3249 (1965).
32. A. T. Bottini, B. F. Dowden, and R. L. VanEtten, *J. Am. Chem. Soc.*, **87**, 3250 (1965).
33. A. T. Bottini and M. K. O'Rell, *Tetrahedron Lett.*, **1967**, 423.
34. A. T. Bottini and M. K. O'Rell, *Tetrahedron Lett.*, **1967**, 429.
35. A. T. Bottini and R. L. VanEtten, *J. Org. Chem.*, **30**, 575 (1965).
36. D. R. Brown, B. G. Hutley, J. McKenna, and J. M. McKenna, *Chem. Commun.*, **1966**, 719.
37. D. R. Brown, R. Lygo, J. McKenna, J. M. McKenna, and B. G. Hutley, *J. Chem. Soc., B*, **1967**, 1184.
38. D. R. Brown, J. McKenna, and J. M. McKenna, *J. Chem. Soc., B*, **1967**, 1195.
39. D. R. Brown, J. McKenna, J. M. McKenna, J. M. Stuart, and B. G. Hutley, *Chem. Commun.*, **1967**, 380.
40. A. F. Casy, A. H. Beckett, and M. A. Iorio, *Tetrahedron*, **22**, 2751 (1966).
41. H. Dorn, A. R. Katritzky, and M. R. Nesbit, *J. Chem. Soc., B.*, **1967**, 501.
42. (a) J. D. England, D. Temple, and J. Sam, *J. Med. Chem.*, **11**, 353 (1968); (b) D. Temple and J. Sam, *J. Heterocycl. Chem.*, **5**, 441 (1968).
43. G. Fodor, J. D. Medina, and N. Mandava, *Chem. Commun.*, **1968**, 581.
44. D. Gürne, T. Urbański, M. Witanowski, B. Karniewska, and L. Stefaniak, *Tetrahedron*, **20**, 1173 (1964).
45. H. O. House and C. G. Pitt, *J. Org. Chem.*, **31**, 1062 (1966).
46. H. O. House and B. A. Tefertiller, *J. Org. Chem.*, **31**, 1068 (1966).
47. H. O. House, B. A. Tefertiller, and C. G. Pitt, *J. Org. Chem.*, **31**, 1073 (1966).
48. J.-L. Imbach, A. R. Katritzky, and R. A. Kolinski, *J. Chem. Soc., B*, **1966**, 556.
49. C. D. Johnson, R. A. Y. Jones, A. R. Katritzky, C. R. Palmer, K. Schofield, and R. J. Wells, *J. Chem. Soc.*, **1965**, 6797.
50. Y. Kawazoe and M. Tsuda, *Chem. Pharm. Bull. (Tokyo)*, **15**, 1405 (1967).
51. R. Lygo, J. McKenna, and I. O. Sutherland, *Chem. Commun.*, **1965**, 356.
52. J. McKenna, J. M. McKenna, and J. White, *J. Chem. Soc.*, **1965**, 1733.
53. J. McKenna, J. White, and A. Tulley, *Tetrahedron Lett.*, **1962**, 1097.
54. W. L. Meyer and N. Sapianchiay, *J. Am. Chem. Soc.*, **86**, 3343 (1964).
55. T. M. Moynehan, K. Schofield, R. A. Y. Jones, and A. R. Katritzky, *Proc. Chem. Soc.*, **1961**, 218.
56. T. M. Moynehan, K. Schofield, R. A. Y. Jones, and A. R. Katritzky, *J. Chem. Soc.*, **1962**, 2637.
57. K. Schofield and R. J. Wells, *Chem. Ind. (London)*, **1963**, 572.
58. M. Shamma and J. B. Moss, *J. Am. Chem. Soc.*, **84**, 1739 (1962).
59. M. Shamma and J. M. Richey, *J. Am. Chem. Soc.*, **85**, 2507 (1963).
60a. C. C. Thut and A. T. Bottini, *J. Am. Chem. Soc.*, **90**, 4752 (1968).
60b. U. de la Camp, A. T. Bottini, J. Gal, C. C. Thut and A. G. Bellettini, manuscript in preparation.
61. N. Mandava and G. Fodor, *Can. J. Chem.*, **46**, 2761 (1968).
62. R. J. Bishop, G. Fodor, A. R. Katritzky, F. Soti, L. E. Sutton, and F. J. Swinbourne, *J. Chem. Soc., C.*, **1966**, 74.
63. A. F. Casy, *J. Med. Chem.*, **11**, 188 (1968).
64. A. F. Casy, A. H. Beckett, M. A. Iorio, and H. Z. Youssef, *Tetrahedron*, **21**, 3387 (1965).
65. C.-Y. Chen and R. J. W. Le Fèvre, *J. Chem. Soc.*, **1965**, 3473.
66. Y. Kawazoe, M. Tsuda, and M. Ohnishi, *Chem. Pharm. Bull. (Tokyo)*, **15**, 51 (1967).

67. M. Kotake, I. Kawasaki, T. Okamoto, S. Matsutani, S. Kusumoto, and T. Kaneko, *Bull. Chem. Soc. Japan*, **35**, 1335 (1962).
68. R. E. Lyle and C. R. Ellefson, *J. Am. Chem. Soc.*, **89**, 4563 (1967).
69. J. C. N. Ma and E. W. Warnhoff, *Can. J. Chem.*, **43**, 1849 (1965).
70. J. Trojánek, H. Komrsová, J. Pospísěk, and Z. Čekan, *Collect. Czech. Chem. Commun.*, **26**, 2921 (1961).
71. K. Koczka and G. Bernáth, *Acta Chim. Acad. Sci. Hung.*, **33**, 165 (1962).
72. G. Bernáth, J. A. Szabó, K. Koczka, and P. Vinkler, *Acta Chim. Acad. Sci. Hung.*, **51**, 339 (1967).
73. W. D. Crow, *Aust. J. Chem.*, **11**, 366 (1958).
74. K. L. Williamson, T. Howell, and T. A. Spencer, *J. Am. Chem. Soc.*, **88**, 325 (1966).
75. R. W. Horobin, J. McKenna, and J. M. McKenna, *Tetrahedron*, **Suppl.** 7, 35 (1966).
76. C. E. Johnson, Jr., and F. A. Bovey, *J. Chem. Phys.*, **29**, 1012 (1958).
77. H. M. McConnell, *J. Chem. Phys.*, **27**, 226 (1957).
78. J. D. M. Asher, J. M. Robertson, G. A. Sim, M. F. Bartlett, R. Sklar, and W. I. Taylor, *Proc. Chem. Soc.*, **1962**, 72.
79. C. C. Scott, G. A. Sim, and J. M. Robertson, *Proc. Chem. Soc.*, **1962**, 355.
80. C. H. MacGillavry and G. Fodor, *J. Chem. Soc.*, **1964**, 597.
81. P. L. Jackson and A. J. Smith, unpublished; cited in refs. 18 and 37.
82. O. Kennard, K. A. Kerr, D. G. Watson, J. K. Fawcett, and L. Riva di Sanseverino, *Chem. Commun.*, **1967**, 1286.
83a. G. Fodor and N. Mandava, 51st Annual Conference of the Chemical Institute of Canada, Vancouver, June, 1968, Abstracts, p. 56.
83b. N. Mandava, G. Fodor, D. Frehel, and M. J. Cooper, manuscript in preparation.
84. G. Fodor, K. Koczka, and J. Lestyán, *Magy. Kem. Foly.*, **59**, 242 (1953); *J. Chem. Soc.*, **1956**, 1411.
85. G. Fodor, J. Tóth, and I. Vincze, *Helv. Chim. Acta.*, **37**, 907 (1954).
86. G. Fodor, J. Tóth, and I. Vincze, *J. Chem. Soc.*, **1955**, 3504.
87. G. Fodor, I. Vincze, and J. Tóth, *Experientia*, **13**, 183 (1957); *J. Chem. Soc.*, **1961**, 3219.
88. Ö. Kovács, G. Fodor, and M. Halmos, *J. Chem. Soc.*, **1956**, 873.
89. W. A. Ayer and G. G. Iverach, *Can. J. Chem.*, **42**, 2514 (1964).
90. G. Bernáth and K. Koczka, *Chem. Ind. (London)*, **1960**, 1479; K. Koczka and G. Bernáth, *Acta Chim. Acad. Sci. Hung.*, **33**, 173 (1962).
91. G. Fodor, F. Uresch, F. Dutka, and T. Széll, *Collect. Czech. Chem. Commun.*, **29**, 274 (1964).
92. H. O. House, H. C. Müller, C. G. Pitt, and P. P. Wickham, *J. Org. Chem.*, **28**, 2407 (1963).
93. W. Hückel, *Theoretical Principles of Organic Chemistry*, Vol. 1, Elsevier, London, 1955, p. 517.
94. A. T. Bottini, R. L. VanEtten, and A. J. Davidson, *J. Am. Chem. Soc.*, **87**, 755 (1965).
95. M. Havel, J. Krupička, J. Sicher, and M. Svoboda, *Tetrahedron Lett.*, **1967**, 4009.
96. M. Havel, J. Krupička, M. Svoboda, J. Závada, and J. Sicher, *Collect. Czech. Chem. Commun.*, **33**, 1429 (1968).
97. É. A. Mistryukov and N. I. Aronova, *Izv. Akad. Nauk SSSR, Ser. Khim.*, **1967**, 783, 789.
98. S. Saeki, *Chem. Pharm. Bull. (Tokyo)*, **9**, 226 (1961).
99. M. Shamma and J. B. Moss, *J. Am. Chem. Soc.*, **83**, 5038 (1961).
100. M. Shamma and E. F. Walker, Jr., *Chem. Ind. (London)*, **1962**, 1866.
101. M. Shamma and E. F. Walker, Jr., *Experientia*, **19**, 460 (1963).
102. I. M. Skvortsov, I. V. Antipova, and A. A. Ponomarev, *Dokl. Akad. Nauk SSSR*, **178**, 1106 (1968).
103. A. R. Katritzky, *Rec. Chem. Progr.*, **23**, 223 (1962).

104. D. R. Brown and J. McKenna, unpublished, cited in ref. 37.
105. K. Koczka and G. Bernáth, *Chem. Ind. (London)*, **1958**, 1401.
106. R. Bognár and S. Szabó, *Tetrahedron Lett.*, **1964**, 2867.
107. K. Kotera, Y. Hamada, and R. Mitsui, *Tetrahedron*, **24**, 2463 (1968).
108. M. L. Stien, R. Ottinger, J. Reisse, and G. Chiurdoglu, *Tetrahedron Lett.*, **1968**, 1521.
109. F. R. Jensen and B. H. Beck, *J. Am. Chem. Soc.*, **90**, 3251 (1968).
110. A. Heusner and K. Zeile, *Chem. Ber.*, **91**, 2399 (1958).
111. G. Fodor, *Acta Chim. Acad. Sci., Hung.*, **5**, 421 (1955).
112. J. Meinwald and O. L. Chapman, *J. Am. Chem. Soc.*, **81**, 5800 (1959).
113. J. Meinwald, *J. Chem. Soc.*, **1953**, 712.
114. G. Fodor, I. Sallay, and F. Dutka, *Acta Phys. Chim.*, **2**, 78 (1957).
115. E. W. Warnhoff and W. C. Wildman, *J. Am. Chem. Soc.*, **79**, 2192 (1957).
116. H.-G. Boit, H. Ehmke, S. Uyeo, and H. Yajima, *Chem. Ber.*, **90**, 363 (1957); T. Kitagawa, S. Uyeo, and N. Yokoyama, *J. Chem. Soc.*, **1959**, 3741.
117. H. Kondo, H. Katsura, and S. Uyeo, *Chem. Ber.*, **71**, 1529 (1938).
118. D. Gürne and T. Urbański, *J. Chem. Soc.*, **1959**, 1912.
119. D. Gürne, L. Stefaniak, T. Urbański, and M. Witanowski, *Tetrahedron*, **20**, **Suppl. 1**, 211 (1964); Y. Allingham, R. C. Cookson, T. A. Crabb, and S. Vary, *ibid.*, **24**, 4625 (1968).
120. L. Ötvös, F. Dutka, and H. Tüdős, *Chem. Ind. (London)*, **1962**, 818.
121. B. G. Hutley, J. McKenna, and J. M. Stuart, *J. Chem. Soc., B*, **1967**, 1199.
122. A. Mannschreck and H. Muensch, *Tetrahedron Lett.*, **1968**, 3227.
123. R. J. Bishop, L. E. Sutton, D. Dineen, R. A. Y. Jones, A. R. Katritzky, and R. J. Wyatt, *J. Chem. Soc., B*, **1967**, 493.
124. E. E. van Tamelen, V. B. Haarstad, and R. L. Orvis, *Tetrahedron*, **24**, 687 (1968).
125. H. O. House, B. A. Tefertiller, and H. D. Olmstead, *J. Org. Chem.*, **33**, 935 (1968); H. O. House and T. M. Bare, *ibid.*, **33**, 943 (1968).

The following papers appeared too late for review.
126. R. Brettle, D. R. Brown, J. McKenna, and J. M. McKenna, *Chem. Commun.*, **1969**, 696. (These workers were apparently unaware of refs. 108 and 109.)
127. R. Brettle, D. R. Brown, J. McKenna, J. M. McKenna, and R. Mason. *Chem. Commun.*, **1969**, 339.
128. D. R. Brown and J. McKenna, *J. Chem. Soc., B*, **1969**, 570.
129. D. R. Brown, J. McKenna, and J. M. McKenna, *Chem. Commun.*, **1969**, 186.
130. D. R. Brown, J. McKenna, and J. M. McKenna, *J. Chem. Soc., B*, **1969**, 567.
131. J. Kóbor, G. Bernáth, L. Radics, and M. Kajtár, *Acta. Chim. Acad. Sci. Hung.*, **60**, 255 (1969).
132. J. McKenna, J. M. McKenna, and J. M. Stuart, *J. Chem. Soc., B*, **1969**, 698.
133. L. Radics, M. Kajtár, J. Kóbor, and G. Bernáth, *Acta. Chim. Acad. Sci. Hung.*, **60**, 381 (1969).

Nucleophilic Additions to Acetylenes*

SIDNEY I. MILLER AND RYUICHI TANAKA

Department of Chemistry, Illinois Institute of Technology, Chicago, Illinois

* Supported by the Public Health Service, grant GM 7021.

I. INTRODUCTION

A. Scope

An addition to an unsaturated bond constitutes one of the simplest, yet important, processes of organic chemistry. Certain aspects of this area, for example, electrophilic (**E**) additions to alkenes and alkynes (1), nucleophilic (**N**) attacks on alkenes (2), additions to carbonyl and substitutions in aromatic compounds have been treated. Nucleophilic additions to alkynes (**A**) have received far less attention, although excellent coverage of selected topics is available (3–9).

$$\mathbf{N} + \mathrm{RC{\equiv}CR'} + \mathbf{E} \longrightarrow \underset{\textit{anti}\text{ adduct}}{(\mathrm{R})(\mathbf{N})\mathrm{C{=}C}(\mathbf{E})(\mathrm{R'})} + \underset{\textit{syn}\text{ adduct}}{(\mathrm{R})(\mathbf{N})\mathrm{C{=}C}(\mathrm{R'})(\mathbf{E})} \tag{1}$$

In this review, we shall emphasize the kinetics, mechanisms, orientation, and stereoselectivity of such additions to acetylenes. Numerous useful results will be included in later tables and in Section IV, but detailed synthetic methods would have to be sought in our citations. Because the field is vast, we make no claim for completeness of coverage: our search for examples of eq. 1 extended approximately through 1968.

The preceding limits exclude large classes of reactions, for example, cycloadditions. Although the mechanisms of additions to alkynes of some carbenes, dipolarophiles (11), polyenes (12), transition metal complexes (13), etc., may have some nucleophilic character, such reactions are usually

$$\mathrm{RC{\equiv}CR} + {:}\mathrm{CCl_2} \longrightarrow \mathrm{R{-}C{=}C{-}R}\ (\text{cyclopropene},\ \mathrm{CCl_2}) \tag{2}$$

$$\mathrm{Ar}_i\mathrm{{-}C{\equiv}N^{+}{-}O^{-}} + \mathrm{H{-}C{\equiv}C{-}Ar}_j \longrightarrow \text{3-}\mathrm{Ar}_i\text{-5-}\mathrm{Ar}_j\text{-isoxazole} \tag{3}$$

regarded as no-mechanism, radical, redox, or nonclassifiable. Process 3, for example, does have characteristics of a polar process with the Hammett $\rho_i = 0.6$ for the "nucleophiles," and $\rho_j \simeq -0.3$ for some "electrophiles" (11). Among those excluded are the concerted cycloadditions which offer no *syn–anti* choice in eq. 1, when the cyclic product has fewer than eight atoms (14). Thus the directioselectivity or the *endo* versus *exo* selectivity of a

Diels-Alder reaction with an alkyne will not be considered. For much the same reasons, we shall have little to say about the many nucleophilic additions to dehydrobenzene and cycloalkynes; fortunately, these have been reviewed recently (15).

As is the case with many processes, for example, solvolysis and elimination, the stereochemical course of a nucleophilic addition to an alkyne may be simple, visibly complex, or even concealed as part of another process. In view of these possibilities we shall examine the stereoselectivity of elementary or single-step processes in some detail. When such steps are put together to make an overall scheme, the selectivity or lack of it should be more coherent.

For every nucleophile in a reaction there is an electrophile. Indeed, nucleophilic attack may lead, be simultaneous with, or follow electrophilic attack on a triple bond. This interrelation in addition mechanisms will often necessitate consideration of electrophilic additions (1).

Again, it has long been accepted that additions and eliminations are complementary: a reaction path traversed in one process is simply reversed for the other. For example, the reaction profiles of the Ad_N2 and ElcB processes are identical. Although most reaction conditions preclude the observation of both addition and elimination simultaneously, there are examples in which both are present (10).

$$I^- + HC{\equiv}CH + I_2 \rightleftharpoons \overset{H}{\underset{I}{>}}C{=}C\overset{I}{\underset{H}{<}} + I^- \tag{4}$$

More useful, is the mechanistic analogy that can be made between the two processes, for example, the acid-catalyzed addition of HX to an alkyne and the base-promoted elimination from an alkene, even though one process is not the microscopic reverse of the other. Clearly one theory might hold for both processes, and insights gained about one can be transferred to the other.

Concerning terminology, we use no new terms, although our usage or definitions here are worth mention. In this review *stereoselectivity* refers to a comparison of *syn* versus *anti* adducts in process 1 when both are present. *Stereospecificity* refers to additions that are exclusively *syn* or *anti* (14). Note that the actual structure of the product and the mode of addition need not have the same prefix. *Anti* addition leads to an *anti* adduct, which may, however, have *cis* in its name:

$$\begin{matrix} HC{\equiv}CCOOCH_3 \\ + \\ CH_3OH \end{matrix} \xrightarrow[\text{process}]{\textit{anti}} \overset{CH_3O}{\underset{H}{>}}C{=}C\overset{COOCH_3}{\underset{H}{<}} \tag{5}$$

methyl *cis*-β-methoxyacrylate

Another kind of orientation or the *directioselectivity* refers to alternate adducts, 1,2 or 2,1, that is R′NC = CER or R′EC = CNR (16). The term *regioselectivity* has also been used to distinguish these isomers (17).

B. Historical

Nucleophilic additions were carried out in the last century (18):

$$CH_3C{\equiv}CH + C_2H_5OH \xrightarrow{C_2H_5O^-} CH_3C(OC_2H_5){=}CH_2 \tag{6}$$

Some early reactions of synthetic importance were (19,20):

$$ADE^* + C_6H_5NHNH_2 \longrightarrow CH_3OOCC(NHNHC_6H_5){=}CHCOOCH_3 \tag{7}$$

$$\downarrow$$

(pyrazolone: ring N=C(COOCH$_3$)–CH$_2$–C(=O)–N(C$_6$H$_5$)–, i.e. C_6H_5–N, N=C–COOCH$_3$, C=O)

$$ADE^* + C_2H_5ONa \xrightarrow{C_2H_5OH} \begin{matrix} C_2H_5OOCCH{=}C(OC_2H_5)COOC_2H_5 \\ \textit{cis} \text{ and } \textit{trans} \\ + \\ C_2H_5OOCCH_2{-}C(OC_2H_5)_2COOC_2H_5 \end{matrix} \tag{8}$$

* ADE ≡ acetylenedicarboxylic ester or acid.

Although we have a citation from *1888* (18), it was probably not the first nucleophilic addition to an alkyne. Many of the subsequent examples have been collected in the surveys mentioned earlier.

We believe that the current intense interest in this field must, in large measure, have been triggered by the discoveries of Reppe's group on the utilization of acetylene, particularly in vinylations (21),

$$RQ{-}H + H{-}C{\equiv}C{-}H \longrightarrow H_2C{=}CHQR \tag{9}$$

The scientific and economic importance of this process has had a powerful impact on the field of acetylene chemistry (22).

The earliest report on kinetics concerns the addition of iodine to several substituted propiolic acids (23). Although the data are of uncertain quality, some of the rate constants are usable, provided that it is recognized that iodide ion is involved in the process, as in eq. 4 (10,24). The addition of eq. 10, dating back to 1928, is a well-behaved, second-order reaction (25), but recent work indicates that bisulfite additions to various alkynes are usually oxygen-promoted free radical processes (26). We refer to other rate studies later, but

$$CH_3C\equiv CCOO^- + HSO_3^- \longrightarrow \begin{matrix} H_3C & & H \\ & C{=}C & \\ ^-O_3S & & COO^- \end{matrix} \tag{10}$$

it is worth noting that the most coherent and sustained series, which spans some 20 years, is that of the Russian school (27–47).

Elsewhere we have cast Arthur Michael in the rather awkward role as the "father" of stereoselectivity (14). He was the first to recognize an *anti* preference in eliminations and additions, and he even obtained crude relative rates to support his *anti* rules (48). The ensuing spirited polemic in which he was engaged served to popularize his views (49,50).

Michael's work on stereoselectivity was largely based on dehydrohalogenations, decarboxylative dehalogenation, halogen addition, and hydrogen halide addition. Subsequently, other selectivities were recognized, for example, *syn* dihydroxylation by osmium tetroxide or potassium permanganate, *anti* dihydroxylation by perbenzoic acid, *syn* hydrogenation over catalysts, and *anti* reduction by dissolving sodium in alcohols (49,51,52). Without mechanistic distinctions, the juxtaposition of such results seem confusing; they turn up even in the 1940's (52). Once the importance of mechanism was recognized in the 1930's, for example, in the Walden inversion (53), in the elimination rates of *cis*- and *trans*-chlorohydrins (54), in preferred *anti* additions or eliminations of bromine (55), the Michael rule of favored *anti* processes could be applied rationally and selectively.

In the first of a related series of papers (56–60), Truce, et al. announced: "We have now observed that base-catalyzed additions of thiols to the acetylenic compounds . . . proceed in *trans* fashion" (56). After surveying the field, Miller independently concluded "that the examples cited collectively amount to a general *rule of nucleophilic addition* for alkynes analogous to the Michael rules of *trans* addition of halogens and halogen acids" (61,62).

In this same period a Bologna group recorded numerous examples of nucleophilic additions to alkynes and came to rather similar conclusions on the stereochemistry and mechanisms (63–67). The Truce and Montanari papers on thiols were important in several other ways: sequences of elimination-addition from haloalkynes were distinguished from addition-elimination; nucleophilic additions to alkynes, concealed in an overall process, were brought to light; and the different roles of base and nucleophile in elimination-addition steps were identified. The Miller papers provided kinetic evidence that a sterically stable carbanion, for example, could be formed in process 1 and resist isomerization, until attack by **E** occurred. Subsequent confirmations and violations of the *anti* rule are discussed presently.

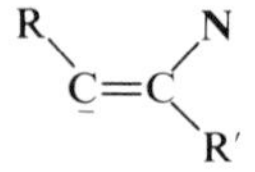

II. BASIC REQUIREMENTS FOR STEREOSELECTIVITY

We refer to several transition states or intermediates fairly frequently. The most important of these are given as **1–6**. The meaning of the labels on these species is clarified when they are used.

A. Bonding and Orbital Symmetry Requirements

When Ingold considered 1,2-eliminations, he often referred to the quantal preference for the *anti* over the *syn* process (53,68). The minimization of energy in this configuration apparently was seen in valence bond terms, and grew out of the analogy between the Walden inversion transition state and Heitler and London's linear model for H_3. Recently the valence bond models for *syn* versus *anti* transition states have been recalled and qualitatively described (14). This theory does, in fact, favor the 1,2-*anti* transition state in which nonbonding interactions are minimized.

Another rationale for the *anti* preference was suggested after the discovery that acetylene had a planar *trans*-bent first excited state and a planar *cis*-bent second excited state (69). Subsequent MO calculations appeared to confirm this preference (70), but this theoretical line of argument has not been continued and has not been especially convincing for ethylene (71).

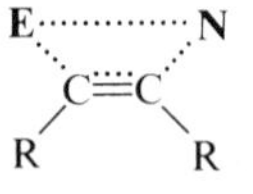

1. *syn*, $Ad_{EN}2$ 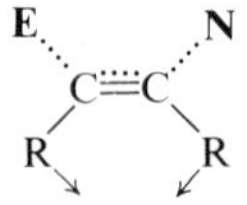 **2.** *syn*, $Ad_{EN}3$ 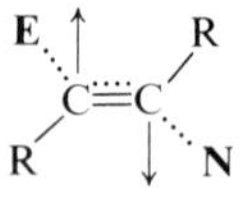 **3.** *anti*, $Ad_{EN}3$ 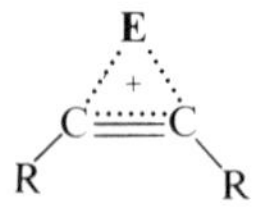 **4a.** Ad_E2

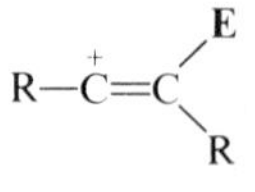

4b. Ad_E2 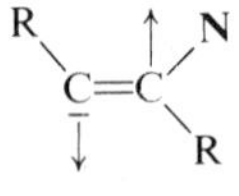 **5a.** *anti*, Ad_N2 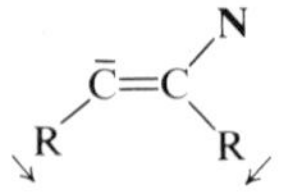 **5b.** *syn*, Ad_N2 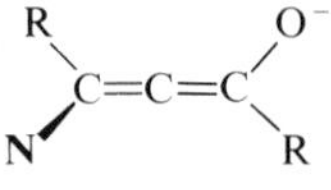 **6**

In 1965 Woodward and Hoffmann made a substantial breakthrough in the theory of the stereochemistry of molecular reactions (72–76). The requirement that orbital symmetry be conserved in the bonding orbitals of reactants, transition state, and products is not limited to such "no mechanism" reactions. Fukui soon demonstrated that the same stereospecificity rules carried over to "some mechanism" or polar and radical reactions (77–82). Moreover, his frontier electron theory calculations reinforced the qualitative arguments. Together, these contributions constitute a fundamental basis for predicting how the bonding factor influences the stereochemical course of reaction. A review on applications of the theory, including heterolytic processes, has appeared (14).

Even a brief outline of Fukui's frontier electron approach to additions should be useful. Suppose (**N** – **E**),($\mathbf{N}^-$ + $\mathbf{E}^+$), or (**N·** + **E·**) are to be added in *concerted* fashion to the ends of an acyclic polyene or polyyne $(C{=}C)_k$ (78). Effectively, one σ electron pair is to be fed into the π system. Now, the symmetry of the lowest vacant πMO (LVMO) of the polyene will be symmetric (*s*) or antisymmetric (*a*), depending on the π system. It will be recalled that the LVMO's alternate in symmetry as *n* increases, ethylene (*a*), butadiene (*s*), hexatriene (*a*), etc. (14). As indicated in Figure 1, the choice

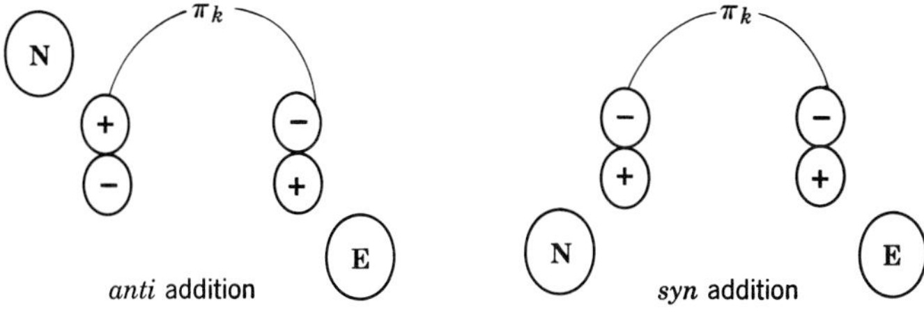

Figure 1. Frontier electron description of *anti* addition to an antisymmetric LVMO and *syn* addition to a symmetric LVMO.

TABLE 1
Predicted Stereochemistry for Concerted Additions,
$\mathbf{N} + (\pi)_k + \mathbf{E} \rightarrow \mathbf{N}{-}(\pi)_{k-1}{-}\mathbf{E}$

π Chain $(C{=}C)_k$	π electrons, $2k$	Stereoselectivity: Ground state	Stereoselectivity: Excited state
1	2	*anti*	*syn*
2	4	*syn*	*anti*
3	6	*anti*	*syn*

between *syn* or *anti* concerted addition is determined by the arrangement of maximum overlap (minimum energy). In principle, such concerted additions should be stereospecific: 1,2-*anti*; 1,4-*syn*; 1,6-*anti*; etc. (Table 1). In the excited states all of these specificities would be reversed (14).

As illustrations, we cite two typical additions that proceed with high stereoselectivity (83,84):

$$(C_6H_5)_2PLi + C_6H_5C{\equiv}CH + R_2NH \xrightarrow{\text{THF}} \underset{H}{\overset{(C_6H_5)_2P}{}}\!\!>C{=}C<\!\!\underset{H}{\overset{C_6H_5}{}} \tag{11}$$

$$ArSO_2C{\equiv}CH + CH_3O^- \xrightarrow{CH_3OH} \underset{H}{\overset{ArSO_2}{}}\!\!>C{=}C<\!\!\underset{H}{\overset{OCH_3}{}} \tag{12}$$

(In later sections we list many more examples of *anti*-stereospecific or *anti*-stereoselective additions.) On the other hand, we know of no authentic 1,4 or 1,6 nucleophilic additions to polyynes that are stereoselective. We find it interesting, however, that the electrophilic additions to a 1,3-diyne have been formulated as giving *syn* products, although complete proofs of structure were lacking (85).

$$C_6H_5{-}C{\equiv}C{-}C{\equiv}C{-}C_6H_5 \xrightarrow{-20^\circ} \begin{cases} \xrightarrow[34\%]{N_2O_4} \underset{O_2N}{\overset{H_5C_6}{}}\!\!>C{=}C{=}C{=}C<\!\!\underset{NO_2}{\overset{C_6H_5}{}} \\ \xrightarrow[3\%]{Br_2} \underset{Br}{\overset{H_5C_6}{}}\!\!>C{=}C{=}C{=}C<\!\!\underset{Br}{\overset{C_6H_5}{}} \end{cases} \tag{13}$$

In the course of this review, therefore, almost all of our discussion refers to 1,2-additions unless otherwise specified.

A number of additions, which at first glance may *appear* to be *syn*, are best regarded as successive *anti* 1,2-additions (86,87):

$$(RC{\equiv}C)_2C{=}O + R'NH_2 \longrightarrow \text{2,6-R}_2\text{-1-R}'\text{-4-pyridone} \tag{14}$$

$$HC{\equiv}C{-}C{\equiv}CH + C_6H_5AsH_2 \xrightarrow{n\text{-}C_4H_9Li} \text{1-}C_6H_5\text{-arsole} \tag{15}$$

The conservation of orbital symmetry in five- or six-centered transition states may also be treated. In Figure 2 we have pictured the concerted *syn* additions of carboxylic acids and dimeric hydrogen halides to an alkyne and a hydrogen acid to an α-ethynylcarbonyl compound. If the transition states are fairly represented by the figure, then each involves six electrons in the change or a pseudoaromatic cycle. Such transition states are electronically allowed (80) and geometrically feasible.

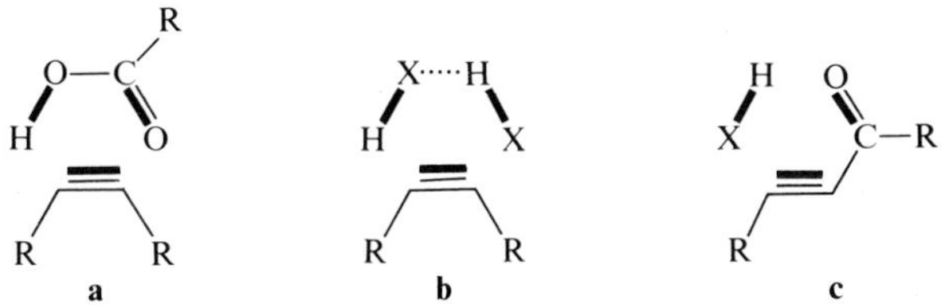

Figure 2. Six-center additions to RC≡CR: (*a*) carboxylic acid, *syn*; (*b*) dimeric $(HX)_2$, *syn*; (*c*) HX to an acetylenic carbonyl compound. The bonds to be broken have been thickened.

Unfortunately the transition states pictured in Figure 2 are difficult to validate. 1,2-*Anti* addition is usual, even when the rate expression could be appropriate; for example, rate $= k[\mathrm{A}]\,[\mathrm{HX}]^2$ for **b** or rate $= k[\mathrm{A}]\,[\mathrm{HX}]$ for **a** or **c**. Nevertheless, these transition states provide a satisfactory rationale for some *syn* additions. The sufficient condition of allowedness is met, if such transition states are postulated.

Another group of reactions, including certain dipolarophile additions (e.g., eq. 3), could involve pseudoaromatic transition states (88,89). Some of

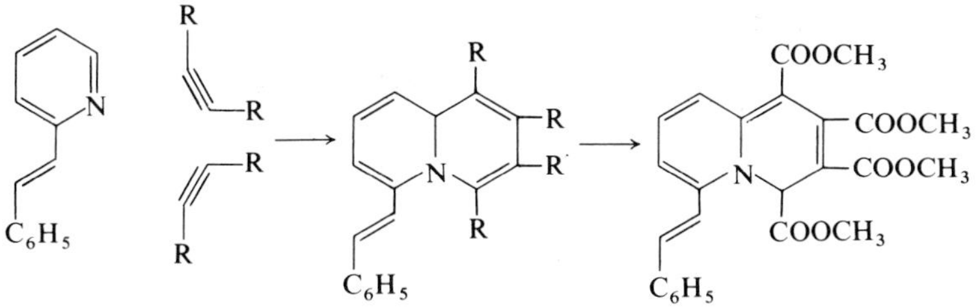

(16)

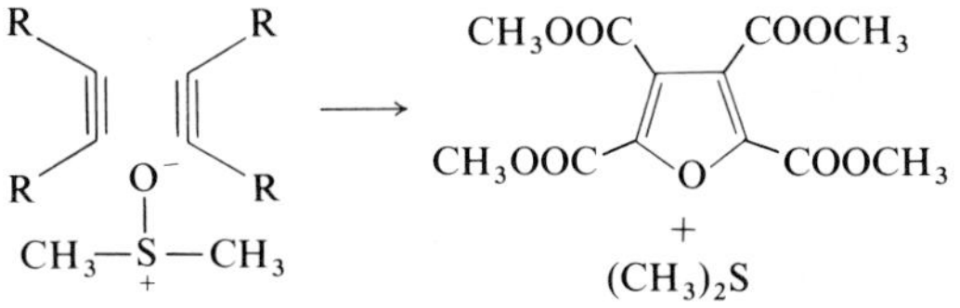

(17)

these reactions probably proceed stepwise, since in some examples proton donors trap the first carbanion (90):

$$C_5H_5N + ADE \xrightarrow{CH_3OH} \text{[1-(H}_5\text{C}_2\text{OOC)C}=\text{CHCOOC}_2\text{H}_5\text{-2-H-2-OCH}_3\text{-1,2-dihydropyridine]} \tag{18}$$

For this reason processes 17 and 18 probably involve two *syn* additions for completion, in violation of the orbital symmetry rules. Note, however, that the acetylenes used generally have at least one activating ester function (e.g., R = $COOCH_3$) and that isomerization of the carbanion via a linear delocalized structure could be facile (see Section III-B).

An unsymmetrical reagent (e.g., HX) with a lone pair on X need not follow the rules of Table 1. Consider 1,2-addition and the conservation of orbital symmetry. The relevant correlation diagram is given in Figure 3 and involves the net conversion of the occupied $\pi_{C\equiv C}$, p_z, and σ_{H-X} orbitals into σ_{C-X}, σ_{C-H}, and p_x orbitals of the product (14). Although such a process is

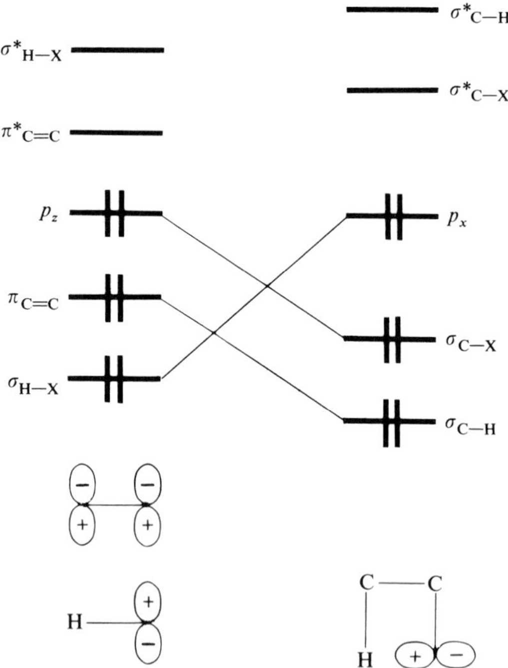

Figure 3. Orbital correlation diagram for the concerted 1,2-additions of HX to C=C. Reactants are on the left, products on the right. The *p* orbitals are sketched.

allowed, we know of no certain examples of this bimolecular *syn* addition. However, cases of the reverse or unimolecular elimination in the gas phase are known (91). Other factors, for example, geometry and conformation, bear on this allowed path and are discussed shortly.

Stereoselectivity should also depend on energy factors. If **E—N** is added 1,2 to an acetylene, either **E** or **N** may make the first contact. As a guide to the probable shape of the intermediate, we can use Hückel MO levels of Figure 4 for allyl or cyclopropenyl as models (92). Depending on the number

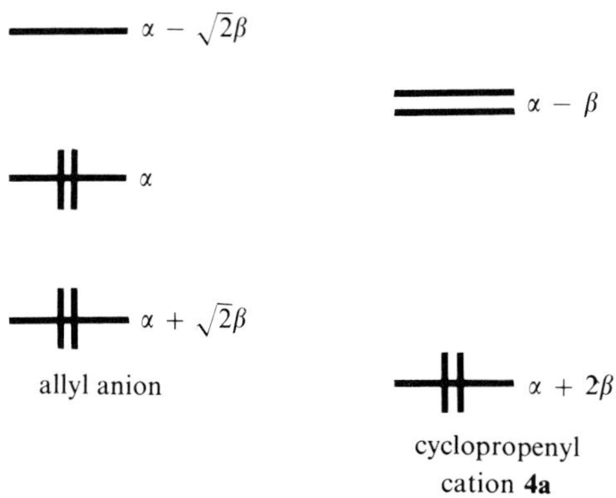

Figure 4. Hückel levels for a three-center π system.

of bonding electrons, the three-center system will be linear or cyclic. The adduct of E^+ and acetylene should be cyclic, for **4a** is favored energetically over the acyclic form **4b** by 1.17β. On the other hand, the adduct of N^- and acetylene should be acyclic, for **5a** is favored over the cyclic form by 0.83β. If, for the present, we assume that *anti* carbanion is formed and that it reacts before it isomerizes, the overall addition will be *anti*.

The preceding theory does not literally forbid the "wrong" addition; it simply says that this should be of higher energy. There may, in fact, be valid reasons for violations of the rules. In certain molecules the usual sequence of π levels may be scrambled; or there may be steric, electrostatic, stability, or delocalization factors. All of these may be aligned with, or opposed to, the orbital symmetry (energy) predictions. As a working procedure, we take the orbital symmetry conservation rules as setting the standard or norm for stereospecific additions. The other contributions, whether large or small, will be regarded as possible perturbations of the norm (14).

Perhaps the most striking group of exceptions to the *anti* rule is found in the nucleophilic additions to benzyne (15). The *syn* specificity is general and may stem from steric, stability, delocalization factors, or all three.

(19)

B. Geometry and Conformation

In the older literature, there are some intuitive speculations on the role of steric and symmetry factors in additions to alkynes (49,50). We are not, however, aware of any modern discussion of the possible conformational factors involved.

In Figure 5 we give a scale diagram for the possible additions of the hydrogen halides or halogens to acetylene, according to the transition state model **1**. The only bonding concept implicit in the model is that the *p* orbitals change to sp^2 orbitals; that is, they move *outward* on the side of the entering **E—N** molecule. The transition state geometry is "half-way" between acetylene and ethylene, that is, 1.27 Å for the C—C distance and the incipient bonds make angles of 105° with the C—C bond; to be specific we have used C—X bond distances appropriate to alkene carbon, but the transition-state distances are probably somewhat larger. In Table 2 we compare the distance of closest approach for HX or X_2 in the scale model with the reported bond

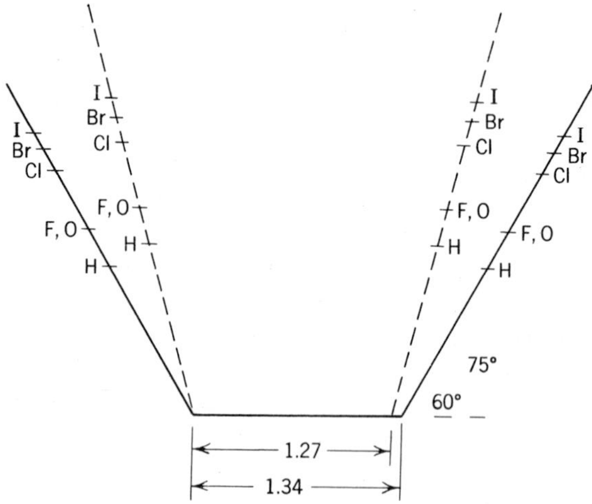

Figure 5. Scale diagram for the possible additions of HX or X_2 to an alkyne. The alkene bonds are solid, the transition state bonds are dashed. The triple and double bond distances of 1.20 and 1.34 A give the scale.

lengths (93). Only if these distances are within 10% of one another, have we rated them as compatible (+) and the model as dimensionally feasible.

One of the consequences of this geometric model seems to be that most **E—N** are mismatched to the triple bond. For the **E—N** that do fit (e.g., Cl_2, ClBr, ClI, Br_2) *syn* additions are in violation of the orbital symmetry rules. On the other hand, none of the hydrogen acids can reasonably "reach" to allow concerted bond-making, even though such additions are allowed. Even if the rigid dimensional requirements of the model are dropped, considerable distortion of normal bond dimensions must take place before *any* four-center transition state of type **1** can be formed. On purely geometric grounds, therefore, we believe addition via **1** should be of high energy. Transition states **2–5** do not have this particular problem.

TABLE 2
E—N Distances (93) for a Model Transition State of Additions to Acetylenes: r_{C-C} = 1.27 Å, ∠CCE = CCN = 105°

E—N	*r*, Å	*r* (transition state model), Å	Compatibility
HF	0.92	1.92	−
HCl	1.27	2.11	−
HBr	1.41	2.2	−
HI	1.61	2.3	−
FF	1.42	1.94	−
ClCl	1.99	2.17	+
BrBr	2.28	2.25	+
II	2.67	2.32	−
ClF	1.63	2.1	−
ClBr	2.14	2.2	+
ClI	2.25	2.3	+
BrF	1.76	2.2	−

In 1,2-additions to alkenes or elimination from alkanes there is a conformational preference for the *anti* process, because nonbonding repulsions have been minimized (53,68,94). In acetylene additions (or alkene eliminations), this is also true. Transition states **2** and **3**, and those that led to **5a** and **5b**, are to be compared. The arrows on the nonbonding groups indicate their motion from the acetylene: there is obviously less repulsion in the *anti* forms **3** and **5a**. This "*cis*" effect, which makes most eclipsed forms higher energy than corresponding staggered forms, is aligned with the orbital symmetry factor to favor 1,2-*anti* addition.

It is conceivable, of course, that steric interference between the incoming groups and the β substituents in transition states **3**, **4b**, and **5a** could be large enough to overcome the normal conformational effect just discussed. In an interesting opposition of a steric factor against the *anti* preference Truce et al. added mesitylenethiol to mesitylacetylene (95):

$$2,4,6\text{-}(CH_3)_3C_6H_2S^- + 2,4,6\text{-}(CH_3)_3C_6H_2C{\equiv}CH \xrightarrow{C_2H_5OH} 2,4,6\text{-}(CH_3)_3C_6H_2CH{=}CHSC_6H_2(CH_3)_3\text{-}2,4,6 \quad (cis) \tag{20}$$

Other examples of the type

$$R{-}C{\equiv}C{-}H + R'O^- \xrightarrow{ROH} \underset{R}{\overset{H}{}}\!\!>C{=}C<\!\!\underset{OR}{\overset{H}{}} \tag{21}$$

in which R = Ar, ArS, COOH, COOR″, etc., and R′ = alkyl or aryl, are available (Tables 8–10) (61,96). Note that the less stable isomer may be formed in each case. In the face of steric effects *anti* selectivity is maintained.

C. Electrostatic Factor

Benson et al. have developed an electrostatic model of four-center reactions that was first suggested and used by Maccoll (91,97,98). These workers believe that many molecular four-center reactions have polar or even partially ionic transition states. Briefly, a geometry for the transition state is assumed, each atom is assigned a partial charge, and electrostatic relations are used to estimate the transition state. What is most significant is that calculated activation energies for hydrogen halide addition to alkenes agree well with observed values (97). For the purpose at hand we are interested in a comparison of *syn* versus *anti* additions according to a coulombic model. If a carbanion is formed in the slow step (Ad_N2) as in eq. 22, clearly

$$-C{\equiv}C- + N^- \xrightarrow{\text{slow}} \begin{cases} {}^{\delta-}N\cdots C{\equiv}C^{\delta-} \ (\text{trans}) \xrightarrow{E^+} anti \\ {}^{\delta-}N\cdots C{\equiv}C_{\delta-} \ (\text{cis}) \xrightarrow{E^+} syn \end{cases} \tag{22}$$

the larger separation of charges will be favored in the *anti* form. If a termolecular process ($Ad_{NE}3$) is involved as in eq. 23, a coulombic calculation

$$-C\equiv C- + E^{+} + N^{-} \longrightarrow \begin{cases} N^{\delta-}\cdots C\equiv C\cdots E^{\delta+} & \textit{anti} \\ \overset{\delta+}{C}\equiv\overset{\delta-}{C}\ (N^{\delta-},\ E^{\delta+}\ \text{same side}) & \textit{syn} \end{cases} \tag{23}$$

shows that the *syn* transition state would be favored in the gas phase. In any medium, but especially in a high dielectric solvent, the energy advantage of *syn* over *anti* would be reduced; further, strong solvation would favor the more polar *anti* form and could possibly reverse the overall preference. Unfortunately, such calculations could not be made in any reliable way at this time. We shall see later, in connection with amine additions, that the coulombic factor may enter in the way we have indicated, that is, more *anti* product in the more polar solvent (Section IV-B-1). Unfavorable coulombic factors, however, have usually been insufficient to overcome stereoelectronic factors, for example, in substitution (S_N2) or elimination (E_N2) (14).

In Figure 5 we have constrained the transition states to take up plausible positions for incipient bonds. In his four-center model, Benson places the atoms at reasonable distances that are *very much like reactants and very little like products*. Despite this aspect of the model, which seems a weakness to us, the model does work (97). Several reactions (e.g., 21) have already been given in which *anti*-1,2 addition proceeds, despite an unfavorable electrostatic effect. Although Truce et al. thought they had found some exceptions in the

$$RC\equiv CCOO^{-} + p\text{-}C_7H_7S^{-} \xrightarrow{C_2H_5OH} (p\text{-}C_7H_7S)(H)C{=}C(H)(COO^{-}) \underset{p\text{-}C_7H_7S^{-}}{\rightleftharpoons} (p\text{-}C_7H_7S)(H)C{=}C(COO^{-})(H) \tag{24}$$

additions to propiolate anions ($R = H$ or C_6H_5) (99,100), it was later demonstrated that the *anti* adduct did indeed form first (101,102). In eq. 24 the *cis* to *trans* isomerization, catalyzed by *p*-toluenethiolate or promoted in some other way, had confused the issue. As a result of such work, it does not appear that substituents can normally exert a large enough electrostatic effect to contravene *anti* addition.

III. RATES AND MECHANISM

A. Rate Laws

Nucleophilic additions to alkynes (**A**) include both simple and complex rate laws. In general, we require an expression of the type

$$-d[\mathrm{A}]/dt = [\mathbf{A}]\sum_i k_i[\mathbf{N}][\mathbf{E}]^n \tag{25}$$

in which $n = 0$ or 1. Equation 25 allows neither for a first-order rate law, which seems highly improbable indeed, nor for fairly complex rate laws formulated for special mechanisms.

All in all, however, relatively few kinetic studies have been made. Tables 3 and 4 list most of the available data for simple second-order or third-order nucleophilic additions. We shall have occasion to refer to these tables in considering substituent effects, medium effects, etc.

An example of mixed second- and third-order kinetic dependence is found in the addition in methanol:

$$CH_3COOH + ADE \xrightarrow[CH_3OH]{Q^+I^-} (CH_3OOC)(H)C{=}C(I)(COOCH_3) \tag{26}$$

The iodide is in the form of n-butylquinolinium iodide (Q^+I^-) (38).

$$-d[\mathrm{I}^-]/dt = (k_2 + k_3[CH_3COOH])[\mathbf{A}][\mathrm{I}^-] \tag{27}$$

Another kind of competition, i.e., between chlorine versus hydrogen chloride addition to acetylene dicarboxylic ester, results in a similar law (30):

$$ADE \xrightarrow[H^+]{LiCl} (CH_3OOC)(H)C{=}C(Cl)(COOCH_3)$$
$$ADE \xrightarrow{Cl_2} (CH_3OOC)(Cl)C{=}C(Cl)(COOCH_3) \tag{28}$$

$$-d[\mathbf{A}]/dt = (k_2 + k_3[Cl_2])[\mathbf{A}][LiCl] \tag{29}$$

Occasionally, a fourth-order rate law crops up. For example, the addition of mercuric chloride in the presence of lithium chloride follows the law (47),

$$-d[HgCl_2]/dt = k_4[\mathbf{A}][HgCl_2][LiCl]^2 \tag{30}$$

TABLE 3

Second-Order Kinetics in the Nucleophilic Additions (Ad_N2) to Alkynes[a]

Alkyne	Nucleophile	Solvent	Temp., °C	k, M^{-1} sec^{-1}	E_{act}, kcal/mole	$\Delta S^{\ddagger}$, eu	Ref.
$C_6H_5C{\equiv}CCOOC_2H_5$	$C_6H_5S^-$	C_2H_5OH	0	2.18×10^{-2}	12.5	−12	142*
4-$CH_3OC_6H_4C{\equiv}CCOOC_2H_5$	4-$CH_3C_6H_4S^-$	C_2H_5OH	0	7.2×10^{-3}	14.2	−14	142*
$C_6H_5C{\equiv}CH$	CH_3O^-	CH_3OH	108.5	7.6×10^{-4}	28.5	−8	62
$CH_3OOCC{\equiv}CCOOCH_3$	I^-	$CH_3OH(HOAc)$	40	9.9×10^{-4}	12.3	−35	38*
	I^-	$CH_3OH(Cl_3CCOOH)$	40	1.9×10^{-3}			38*
	I^-	$C_2H_5OH(HOAc)$	40	7.6×10^{-4}			37*
	I^-	*n*-$C_3H_7OH(HOAc)$	40	7.6×10^{-4}			37*
	LiI	HOAc	35	2.8×10^{-2}	14	−23	34*
	LiBr	HOAc	40	2.5×10^{-4}	18.2	−18.9	35*
	LiBr	62% HOAc	30	3.3×10^{-5}			44*
	LiCl	HOAc	30	5.7×10^{-6}			35
	Cl^-	DMF (HOAc)	40	8.8×10^{-4}	15.9	24	39
	HBr	C_6H_6	40	1.07×10^{-2}	3.3		27*
	SCN^-	$CH_3OH(HOAc)$	35	4.25×10^{-2}			37*
	Aziridine	Dioxane	37	5.74			153*
	Aziridine	CH_3CN	37	5.67			153*
	Piperidine	CH_3CN	37	445			153*
	$C_6H_{11}NH_2$	CH_3CN	37	17.7			153*
$C_6H_5C{\equiv}C{-}CN$	SCN^-	$CH_3OH(HOAc)$	35	$5 \times 10^{-}2$			33*
$HC{\equiv}CCOOCH_3$	SCN^-	$CH_3OH(HOAc)$	35	9.7×10^{-2}			33*
	Aziridine	C_6H_6	37	3.2×10^{-3}			153*
	Aziridine	CH_3CN	37	2.4×10^{-2}			153*
$HC{\equiv}CCN$	$(CH_3)_3SnH$	*n*-C_3H_7CN	20	1.13×10^{-4}			312*
$HC{\equiv}CCOOCH_3$	$(CH_3)_3SnH$	*n*-C_3H_7CN	20	1.25×10^{-5}			312*
$NCC{\equiv}CCN$	$(CH_3)_3SnH$	*n*-C_3H_7CN	20	$>1.5 \times 10^{-1}$			312*

[a] Not all of the data will be given here. The availability of data closely related to one or more entries is indicated by an asterisk on the reference number.

TABLE 4

Third-Order Kinetics in the Nucleophilic Additions ($Ad_{NE}3$) to Alkynes[a]

Alkyne	Nucleophile	Electrophile	Solvent	Temp., °C	k, M^{-2} sec^{-1}	E_{act} kcal/mole	$\Delta S^{\ddagger}$, eu	Ref.
$HC{\equiv}CCOOH$	I^-	H^+	H_2O	60	6.4×10^{-5}			43
$HC{\equiv}CCOOCH_3$	I^-	HOAc	$CHCl_3$	40	1.0×10^{-4}			40
$HC{\equiv}CCOOCH_3$	HBr	HBr	HOAc	40	4.0×10^{-3}	21.2		41*
$HOOCC{\equiv}CCOOH$	Br^-	H^+	H_2O	40	1.23×10^{-4}	16.1		42*
	I^-	H^+	H_2O	40	4.5×10^{-3}	15.4		43
$CH_3OOCC{\equiv}CCOOCH_3$	$(LiCl)^2$	$(HOAc)^0$	CH_3OH	50	1.7×10^{-3}			45,46*
	HN_3	HN_3	90% HOAc	25	3.05×10^{-3}			32
	I^-	HOAc	CH_3NO_2	40	4.4×10^{-3}	12.5	−31	36*
	I^-	HOAc	CH_3CN	40	6.6×10^{-3}	17.1	−16	36*
	I^-	HOAc	i-C_3H_7OH	40	6.5×10^{-4}			37*
	I^-	HOAc	$CHCl_3$	40	1.92×10^{-2}	12.6		40
	I^-	Cl_3CCOOH	$CHCl_3$	40	4.4×10^{-2}	13.2		40
	I^-	HOAc	DMF	40	1.5×10^{-3}	15.2	−24.5	39
	Br^-	HOAc	DMF	40	1.5×10^{-3}	10.3	−40.6	39
	SCN^-	HOAc	DMF	40	307×10^{-3}	8.2	−45.5	39

[a] Not all of the data will be given here. The availability of data closely related to one or more entries is indicated by an asterisk on the reference number.

It is suggested that lithium chloride dimer or $LiCl_2^-$ may be the chloride donor, in which case the reaction could be considered one of third order. Another example was given by Otsuka and Murahashi for the vinylation of carbazole in hydrocarbon solvents (103):

$$\text{rate} = k[C_2H_2][\text{carbazole}][KOH]^2 \tag{31}$$

Complex or nonintegral rate laws have been reported for a few additions to alkynes. Furukawa et al. reported the form

$$v = k[\mathbf{A}]^{1.5}[NaOC_2H_5][C_2H_5OH]^{1.4} \tag{32}$$

for the addition to acetylene (**A**); eq. 32 was later revised to give another fairly complex expression (104),

$$v = \frac{k([HC{\equiv}CH] + [HC{\equiv}C^-])[ROH][NaOR]}{[ROH] + K[HC{\equiv}CH]} \tag{33}$$

Then Fahey and Lee reported the form

$$-d[\mathbf{A}]/dt = [\mathbf{A}](k[HCl]^{2.6} + k[Q^+Cl^-][HCl]^{1.4}) \tag{34}$$

in which $\mathbf{A}$ = hexyne and Q^+Cl^- = tetramethylammonium chloride (105). This has since been modified and simplified by the inclusion of medium dependence and hydrogen chloride association in acetic acid (106).

Because of its industrial importance, additions to acetylene have received much attention. Of necessity, the rate studies had to be carried out in an autoclave and often involved two or three phases (solid, gas, solution) and forcing conditions. It is not surprising, therefore, that the results in the literature are sometimes inconsistent. Wassermann and Bedrintzeva suggested that the addition of ethanol to acetylene had an activation energy of 36 kcal/mole and was first order in each reactant in the presence of potassium hydroxide (107). [Their mechanism has been questioned (62).] Rigamonti and Bernardi report E_{act} = 16–18 kcal/mole for similar additions, but it is not clear how they obtained these figures (108).

We have remarked previously on the fourth-order rate law (eq. 31) for a vinylation (103). Apparently this is a special case, for Otsuka et al. proposed the rather formidable expression for vinylation (109–112),

$$\text{rate} = \frac{K_1[ROM][A][ROH]}{1 + K_2[A] + K_3[ROH] + K_4[A][ROH]} \tag{35}$$

Equation 35 is perhaps the most complex of several expressions mentioned in the Japanese literature [e.g., eq. 33 (104,113)]. It is a Michaelis-Menton type of expression in which K_1–K_4 may consist of rate and/or equilibrium constants, and several complexes (ROM·A, ROM·ROH, ROM·ROH·A) are postulated. Such intermediates were proposed to accommodate varied observations, for example, on changing medium, reactant alcohol (ROH), and alkali alkoxide (ROM). We are inclined to believe, however, that such systems are probably conventional in a kinetic sense; we would ascribe the complexity to environmental effects; that is, in dilute solution a rate law such as eq. 25 would be followed. This would have to be modified, of course, for medium effects, changing base strength (114), as well as complex formation under the conditions thus far studied. Equation 35 could be useful empirically to fit acetylene uptake under specified conditions.

B. General Mechanisms

Process 1 is simple enough, but there are complications: nucleophilic additions to alkynes may often be concealed; what may seem to be a simple addition may be something else, or the visible product may be the result of equilibrium rather than kinetic control. By way of background to the addition mechanisms we catalog a few of the recognized complications.

Apparent displacements on a haloalkene (115) or halobenzene (15) may involve an alkyne (96). True nucleophilic displacement usually goes with

$$C_6Cl_5CH{=}CHCl \xrightarrow[\quad]{C_2H_5O^-} \!\!\!\!\!\!\!/\!\!/ \;\; C_6Cl_5CH{=}CHOC_2H_5$$

$$C_6Cl_5CH{=}CHCl \xrightarrow{C_2H_5O^-} C_6Cl_5C{\equiv}CH \xrightarrow{C_2H_5O^-,\,C_2H_5OH} C_6Cl_5CH{=}CHOC_2H_5 \qquad (36)$$

retention, that is, *cis*-reactant → *cis*-product and *trans*-reactant → *trans*-product (2,116). On the other hand, the elimination–addition sequence (eq. 36) usually proceeds from *cis* or *trans* reactant to *anti* adduct (96,117).

Apparent nucleophilic displacements on a haloalkyne may involve addition–elimination, particularly when a source of protons is available:

$$HC{\equiv}CX \xrightarrow[RQH]{p\text{-}C_7H_7S^-} \underset{Y}{\overset{p\text{-}C_7H_7S}{}}\!\!\!>C{=}C<\!\!\!\underset{H}{\overset{X}{}} \xrightarrow{base}$$

$$p\text{-}C_7H_7SC{\equiv}CH \xrightarrow[RQH]{p\text{-}C_7H_7S^-} cis\text{-}p\text{-}C_7H_7SCH{=}CHSC_7H_7\text{-}p \qquad (37)$$

$$(Y = F, X = H;\ Y = H, X = Cl, Br)$$

When X = F in eq. 37, the first adduct can be isolated (118); when X = Br in eq. 37, the last product can be isolated (119); when X = Cl in eq. 37, both intermediates and the product can be isolated (57,58,117).

As an alternative to nucleophilic substitution at an alkyne carbon, Viehe has suggested a mechanism involving the Fritsch–Buttenberg–Wiechell (FBW) rearrangement (16):

$$t\text{-}C_4H_9C{\equiv}CCl + C_6H_5S^- \longrightarrow (t\text{-}C_4H_9C)(C_6H_5S)C{=}CHCl \xrightarrow{LiNR_2} t\text{-}C_4H_9C{\equiv}CSC_6H_5 \quad (38)$$

Depending on the system, the adduct may be isolable. It is in those systems in which the adduct rearranges that we have a concealed nucleophilic addition to alkynes.

Base-catalyzed acetylene–allene rearrangements, discovered by Favorsky, may occur when propargylic hydrogen is available. Either initial addition or initial rearrangement is possible (18,120–123):

$$CH_3CH_2C{\equiv}CH \xrightarrow{C_2H_5O^-} CH_3CH_2C(OC_2H_5){=}CH_2 \longrightarrow CH_3CH{=}C{=}CH_2 \quad (39)$$

$$(RC{\equiv}C)_2CH_2 \xrightarrow[C_2H_5OH]{C_2H_5O^-} RC{\equiv}CCH{=}C{=}CHR \xrightarrow[C_2H_5OH]{C_2H_5O^-} R(C{\equiv}C)_2CH_2R \quad (40)$$

$$C_6H_5SO_2C{\equiv}CCH_3 \overset{N}{\rightleftharpoons} C_6H_5SO_2CH{=}C{=}CH_2 \overset{N}{\rightleftharpoons} C_6H_5SO_2CH_2C{\equiv}CH \quad (41)$$

Incidentally, nucleophilic substitutions of the 1,3 type (S_N2') which lead to allenes

$$R_2C(Cl)\text{—}C{\equiv}CH + (CH_3)_3N \longrightarrow \begin{cases} R_2C(N^+(CH_3)_3)C{\equiv}CH,\ Cl^- \\ R_2C{=}C{=}CHN(CH_3)_3{}^+,\ Cl^- \end{cases} \quad (42)$$

can also be regarded as versions of process 1, but such "additions" are probably best discussed with allenes and will not be treated here.

Concerning equilibrium versus kinetic control, we have already remarked in connection with eq. 24 that this has sometimes led to erroneous conclusions on the stereoselectivity of addition. Either heat, light, or basic, acidic, or radical catalysis could be involved. The following are representative examples among many that are available:

$$HC{\equiv}CCl \xrightarrow{C_6H_5S^-} (C_6H_5S)(H)C{=}C(Cl)(H) \underset{}{\overset{heat}{\rightleftharpoons}} (C_6H_5S)(H)C{=}C(H)(Cl)$$

$$HC{\equiv}CCl \xrightarrow{C_2H_5O^-} HC{\equiv}CSC_6H_5 \xrightarrow{C_6H_5S^-} (C_6H_5S)(H)C{=}C(SC_6H_5)(H) \overset{h\nu}{\rightleftharpoons} (C_6H_5S)(H)C{=}C(H)(SC_6H_5) \quad (43)$$

(from refs. 65, 66)

$$\underset{\text{H}}{\overset{\text{C}_6\text{Cl}_5}{}}\text{C=C}\underset{\text{H}}{\overset{\text{OC}_2\text{H}_5}{}} \xrightleftharpoons{\text{C}_2\text{H}_5\text{O}^-} \underset{\text{H}}{\overset{\text{C}_6\text{Cl}_5}{}}\text{C=C}\underset{\text{OC}_2\text{H}_5}{\overset{\text{H}}{}} \tag{44}$$

(from ref. 96)

$$\text{HC}\equiv\text{CCOOCH}_3 + \text{C}_3\text{H}_7\text{NH}_2 \longrightarrow \underset{\text{H}}{\overset{\text{C}_3\text{H}_7\text{NH}}{}}\text{C=C}\underset{\text{COOCH}_3}{\overset{\text{H}}{}} \xrightleftharpoons{\text{H}^+} \underset{\text{H}}{\overset{\text{C}_3\text{H}_7\text{—NH}}{}}\text{C=C}\underset{\text{H}}{\overset{\text{COOCH}_3}{}} \tag{45}$$

(from ref. 124)

$$\text{HC}\equiv\text{COR} \begin{cases} \xrightarrow{\text{R'S}\cdot} \text{R'SCH=CHOR} \\ \xrightarrow{\text{R'S}^-} \text{CH}_2\text{=C(OR)SR} \end{cases} \tag{46}$$

(from ref. 125)

$$\underset{\text{H}}{\overset{\text{C}_6\text{H}_5\text{SO}_2}{}}\text{C=C}\underset{\text{CH}_3}{\overset{\text{SC}_6\text{H}_5}{}} \underset{\text{or } h\nu}{\overset{\text{CH}_3\text{O}^-}{\rightleftharpoons}} \underset{\text{H}}{\overset{\text{C}_6\text{H}_5\text{SO}_2}{}}\text{C=C}\underset{\text{SC}_6\text{H}_5}{\overset{\text{CH}_3}{}} \tag{47}$$

(from refs. 126, 127)

$$(\text{C}_2\text{H}_5)_2\text{NCH}_2\text{C}\equiv\text{C—CH=CH—N(C}_2\text{H}_5)_2 \ (cis) \xrightleftharpoons{\text{distill}} (\text{C}_2\text{H}_5)_2\text{NCH}_2\text{C}\equiv\text{C—CH=CH—N(C}_2\text{H}_5)_2 \ (trans) \tag{48}$$

(from ref. 127)

Still another ambiguity in mechanism may be traced to the rate law. Because of the equilibrium,

$$\text{I}_2 + \text{I}^- \rightleftharpoons \text{I}_3^- \tag{49}$$

the published rate law (24),

$$\text{rate} = k[\text{I}_3^-][\text{C}_6\text{H}_5\text{C}\equiv\text{CCOO}^-] \tag{50}$$

could equally be,

$$\text{rate} = k\text{K}[\text{I}^-][\text{I}_2]\text{C}_6\text{H}_5\text{C}\equiv\text{CCOO}^-] \tag{51}$$

Because we can use the notion of microscopic reversibility for the analogous addition of iodine to acetylene and the reverse elimination in the presence of iodide (eq. 4), we are inclined to write the transition state **7**,

$$\left(\underset{\text{H}}{\overset{\text{I}}{}}\text{C}\equiv\text{C}\underset{\text{I}\cdots\text{I}}{\overset{\text{H}}{}}\right)^- \qquad \underset{\text{H}}{\overset{\text{H}\cdots\text{X}}{}}\text{C}\equiv\text{C}\underset{\text{H}\cdots\text{X}}{\overset{\text{H}}{}}$$

7 **8**

and assume that the third order rate law is correct (10).

Another example of mechanistic ambivalence is found in the hydrogen chloride addition to acetylene dicarboxylic ester (ADE) (45,46).

$$\text{rate} = k[\text{ADE}][\text{LiCl}]^2 \tag{52}$$

Dvorko and Shilov take into account the equilibrium 53, but consider that the attacking nucleophile is probably $LiCl_2^-$ or Li_2Cl_2.

$$Li_2Cl_2 \rightleftharpoons 2\,LiCl \rightleftharpoons LiCl_2^- + Li^+ \tag{53}$$

Since third-order rate constants are reported, we list them with other third-order constants in Table 4. Note too that the electrophile to be added in this case is a proton, which does not appear in eq. 52.

Laws of the type 54 or 55 could be equivocal with respect to mechanism.

$$\text{rate} = k[\mathbf{A}][\text{HX}]^2 \tag{54}$$

$$\text{rate} = k[\mathbf{A}][\text{HOAc}] \tag{55}$$

Although it is certainly possible to have a transition state **8**, the equilibriums (56) might apply, depending on the medium and on the acid. In aqueous

$$H^+ + HX_2^- \rightleftharpoons H_2X_2 \rightleftharpoons 2HX \rightleftharpoons 2H^+ + 2X^- \tag{56}$$

solution, for example, the rate law for strong acids could be of the form of eq. 57. Even a rate law of the simple form of eq. 55, may be the end result

$$\text{rate} = k[\mathbf{A}][X^-][H^+] \tag{57}$$

of suitable corrections for acetic acid dimerization, as was the case in chloroform (38). Finally, the addition of trifluoroacetic acid to phenylacetylene appears to be second order in the dimer (128)!

$$\text{rate} = k[C_6H_5C{\equiv}CH][(CF_3COOH)_2]^2 \tag{58}$$

Let us return to the authentic examples of eq. 1 for which the transition states and intermediates have been pictured in **1**–**6**. To these can now be attached the free-energy profiles in Figure 6, which cover many additions.

Even though it may seem like jargon in an isolated paper, Ingold's terminology for mechanism is both useful and appropriate here (53,106).

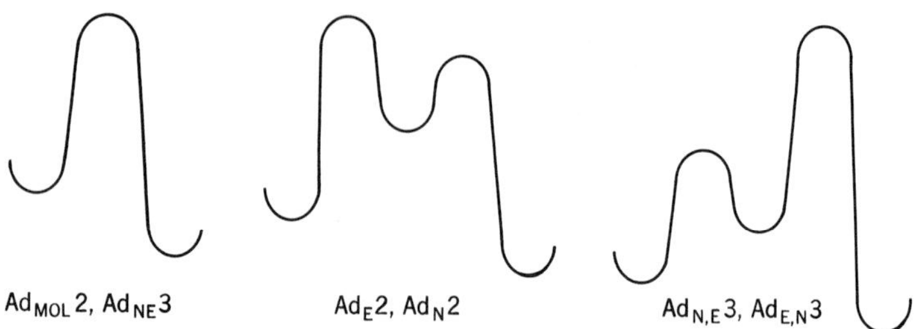

Figure 6. Common free energy profiles for addition reactions. The abscissa is the reaction coordinate.

Our modifications are minor and will be spelled out.

Ad_N2	addition, nucleophilic, bimolecular
Ad_E2	addition, electrophilic, bimolecular
$Ad_{EN}2$	addition, molecular, bimolecular
Ad 3	addition, termolecular

The use of $Ad_{N,E}3$, $Ad_{E,N}3$, Ad 3 should be restricted to attacks in which the nucleophile leads, the nucleophile follows, and the sequence is uncertain, respectively. It will be recognized, particularly with the aid of Figure 6, that each addition is the reverse of an elimination (e.g., E1, E1cB, E2).

Except for the $Ad_{EN}2$ category which can only be *syn*, none of the preceding mechanisms imply anything about stereochemistry. Here we use $Ad_{EN}2$ rather than $Ad_{MOL}2$ to distinguish the polar from the nonpolar molecular additions. In this way we omit a process such as eq. 2 or a Diels-Alder reaction, but include the *syn* additions of polar species such as the hydrogen halides, carboxylic acids, and interhalogens (28,29,129). The

$$\text{ADE} \xrightarrow[\text{CH}_3\text{COOH}]{\text{HBr}} \textit{cis-}\text{ and }\textit{trans-}\text{CH}_3\text{OOCCH}{=}\text{CBrCOOCH}_3 \tag{59}$$

$$\text{C}_6\text{H}_5\text{C}{\equiv}\text{CCH}_3 \xrightarrow[\text{CH}_3\text{COOH}]{\text{HCl}}$$

$$\textit{cis-}\text{ and }\textit{trans-}\text{C}_6\text{H}_5\text{CCl}{=}\text{CHCH}_3 + \textit{cis-}\text{and }\textit{trans-}\text{C}_6\text{H}_5\text{C(OOCCH}_3){=}\text{CHCH}_3 \tag{60}$$

chief difficulty with *syn* additions of this type is that there is no firm evidence for the $Ad_{EN}2$ process. Indeed, the context of the work usually suggests an alternative. Process 60 is kinetically of third order and appears to involve ion pairs (129)—we shall return to these presently. Process 59 may have a free radical component, associated with the *syn* adduct; the presence of ionic bromide increases the proportions of *anti* adduct (29). Although there is evidence for the $Ad_{EN}2$ process (and the reverse unimolecular elimination) in the gas phase (91,97,98), it is difficult to validate such a process in solution.

In both the Ad_N2 and Ad_E2 mechanisms an intermediate is formed in the slow step (eq. 61). Initially, the intermediate will be taken as a single species; later, the question of ion pairs will be considered. One or more counter-species may compete for this intermediate in fast steps. The simplest free energy profile is given in Figure 6, in which only a single step completes the addition.

$$\mathrm{RC{\equiv}CR'} + \mathrm{E^+} \xrightarrow{\text{slow}} \begin{cases} \text{cyclic cation (R)C{=}C(R') bridged by } \overset{+}{\mathrm{E}} \xrightarrow{\mathrm{N^-}} \text{(R)(E)C{=}C(N)(R')} \\ \mathrm{R{-}\overset{+}{C}{=}C(E)(R')} \xrightarrow{\mathrm{N^-}} \text{(R)(E)C{=}C(R')(N)} \text{ or (via } \mathrm{N^-}\text{) } \text{(R)(E)C{=}C(N)(R')} \end{cases} \quad (61)$$

Just enough of the Ad_E2 process is described here to supplement the picture of Ad_N processes (1). As we have seen, simple Hückel theory suggests that the cyclic intermediate of scheme 61 should be preferred to the open cation, because of its lower energy. Substituents (R), however, could de-localize the positive charge on the β carbon so that an open carbonium ion is always a viable alternative. From the structure of the cyclic cation it would appear that *anti* addition is almost inevitable, but, beginning with the open cation, the stability of the product should be a major influence in determining the overall stereoselectivity. In the addition of hydrogen chloride to 3-hexyne, for example, the entry of chloride ion yields >99% *anti* adduct (eq. 34) (106). On the other hand, the addition of trifluoroacetic acid to 3-hexyne gives almost equal amounts of *syn* and *anti* adducts (130). These stereochemical and kinetic results do not permit us to assign a definite mechanism.

The Ad_N2 scheme is as follows:

$$\mathrm{RC{\equiv}CR'} + \mathrm{N^-} \xrightarrow[1]{\text{slow}} \text{(R)}\bar{\mathrm{C}}{=}\text{C(N)(R')} \xrightarrow[2]{\mathrm{E^+}} \text{(R)(E)C{=}C(N)(R')}$$
$$\big\downarrow 3 \quad \big\uparrow -3 \qquad\qquad (62)$$
$$\text{(R)}\bar{\mathrm{C}}{=}\text{C(N)(R')} \xrightarrow[4]{\mathrm{E^+}} \text{(E)(R)C{=}C(N)(R')}$$

(Obviously, another electrophile, say E'^+, could compete for the carbanions in scheme 62, although reports on such competition are unknown to us.) In a few cases the kinetic evidence for the Ad_N2 process is reasonably complete.

Russian workers have shown that reaction 63 was first order in chloride ion, first order in alkyne, and zero order in acetic acid (31,39). Likewise, reaction

$$CH_3OOCC{\equiv}CCOOCH_3 \xrightarrow[\text{slow}]{Cl^-} CH_3OOCCCl{=}\bar{C}COOCH_3 \xrightarrow[\text{fast}]{CH_3COOH} \underset{Cl}{\overset{CH_3OOC}{}}C{=}C\underset{COOCH_3}{\overset{H}{}} \tag{63}$$

64 was shown to be first order in phenylacetylene, first order in methoxide

$$C_6H_5C{\equiv}CH \xrightarrow[\text{slow}]{CH_3O^-} C_6H_5\bar{C}{=}CHOCH_3 \xrightarrow[\text{fast}]{CH_3OH} \underset{H_5C_6}{\overset{H}{}}C{=}C\underset{OCH_3}{\overset{H}{}} \tag{64}$$

ion, and zero order in methanol (62). Further examples of second-order kinetics which fit this mechanism are given in Table 3.

The first intermediate in the Ad_N2 process is a vinyl anion. In a few instances such a species has been detected and trapped (131):

$$(C_6H_5)_3P \xrightarrow{ADE} \underset{CH_3OOC}{\overset{(C_6H_5)_3P^+}{}}C{=}\bar{C}\overset{COOCH_3}{} \xrightarrow{CO_2} \underset{CH_3OOC}{\overset{(C_6H_5)_3P^+}{}}C{=}C\underset{COO^-}{\overset{COOCH_3}{}} \tag{65}$$

Several lines of argument have been used to rationalize its stereostability. The configurational stability of some organometallics, for example, vinyl Grignard or vinyllithium reagents, which have sometimes been cited in this context, is not necessarily relevant unless the free carbanions are involved (132,133). The analogy between vinyl anions and the stable isoelectronic oximes or azo compounds has been made (133). Kinetic arguments have been used to estimate lower limits on the barriers of *trans*-$ClCH = CCl^-$ and *trans*-$BrCH = CBr^-$ as ca. 30 kcal/mole (133). Very crude calculations indicate that the barriers to isomerization of vinyl and alkylvinyl anions are even higher, ca. 40 kcal/mole (132).

In any given system it is perhaps not so much a question of how stereostable the vinyl anion is but how fast it reacts with the waiting electrophile(s). All available data indicate that base-catalyzed deuterium for hydrogen exchange is faster than isomerization in the alkenes $HCX = CXH$ (133), $Ar'ArC = CHX$ (134), and $C_6H_5CH = CHC_6H_5$ (135). Of necessity, proton addition to a vinyl anion would then be faster than isomerization of the anion. This appears to be confirmed more directly for amine adducts (see Section IV-B).

Orbital considerations have suggested that the open *anti* anion should be formed. Theoretical arguments and experimental observations suggest that these anions should be stereostable. All of this provides a satisfying rationale for the *anti* course of 1,2-additions.

Certain substituents (e.g., carbonyl and aryl) could conceivably delocalize the charge on the carbanion, either partially or completely, and render the linear carbanion energetically attractive. In this case the attacking nucleophile need not attack the anion stereospecifically, although a more stable product could create an *anti* bias. However, we are inclined to believe that the bent anions are probably of lower energy than the delocalized anions, chiefly because the stable isoelectronic nitrogen analogs are also bent.

Up to this point we have considered attack by an external electrophile. With amines as nucleophiles, particularly to activated alkynes, we have competition between internal (k_i) and external (k_e) proton delivery (eq. 66). In the absence of external protons, whether from solvent or reactant amine, a *syn* adduct is produced. Only when external protons become available does *anti* product begin to appear. This indicates that if the intermediate zwitterion does in fact exist in three configurations their interconversions must be faster than the proton transfers from the solvent. Huisgen et al. use a single intermediate, the allenic form (136). In any case the zwitterions appear to survive both in an aprotic solvent, in which they form a small amount of *anti* adduct, and in a fairly acidic protic solvent, in which they form a substantial amount of *syn* product.

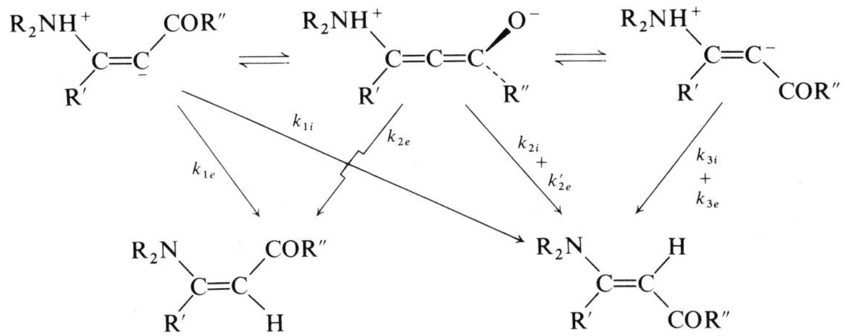

(66)

Both concerted and stepwise versions of the free energy profiles of the Ad 3 process are given in Figure 6. According to the orbital symmetry conservation rules, the concerted addition of **N** and **E** should be 1,2-*anti*, 1,4-*syn*, etc., stereospecific, as given in Table 1. Since the probability of a termolecular collision is low, we conceive of most third-order reactions as two-step processes, in which the second step is rate determining. Whether the nucleophile or electrophile leads the attack, the reaction profile (Fig. 6) has the same form. As for the steochemistry, the Ad_N2 and $Ad_{N,E}3$ and the Ad_E2 and $Ad_{E,N}3$ processes should be similar.

In order to examine the Ad 3 mechanism in detail, consider the $Ad_{N,E}3$ sequence:

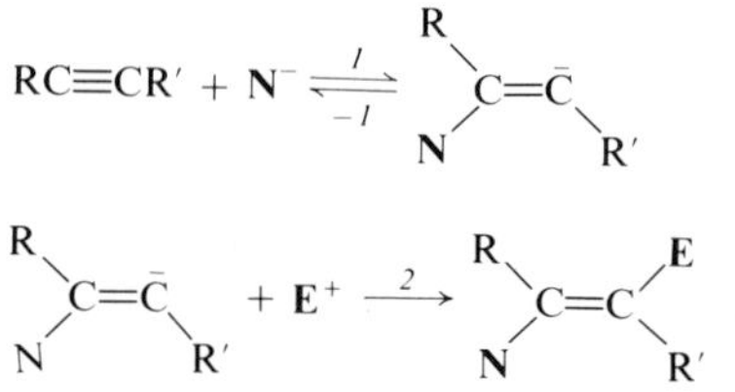

$$\mathrm{RC{\equiv}CR'} + \mathbf{N}^- \underset{-1}{\overset{1}{\rightleftharpoons}} \mathrm{(R)(\mathbf{N})C{=}\bar{C}R'}$$
$$\mathrm{(R)(\mathbf{N})C{=}\bar{C}R'} + \mathbf{E}^+ \xrightarrow{2} \mathrm{(R)(\mathbf{N})C{=}C(\mathbf{E})R'} \tag{67}$$

In the most general case we have two successive second-order reactions for which an integrated rate law cannot be obtained, except in special cases. Next in complexity is a sequence in which the steady-state condition applies to the vinyl anion of scheme 67. Then

$$k_{\mathrm{obs}} = \frac{k_1 k_2[\mathrm{RC{\equiv}CR'}][\mathbf{N}^-][\mathbf{E}^+]}{k_{-1} + k_2[\mathbf{E}^+]} \tag{68}$$

Further reduction comes about if $k_{-1} \gg k_2[\mathbf{E}^+]$ or vice versa. Finally telescoping the two steps of eq. 67 into one yields the termolecular process. Clearly, whether the nucleophile leads or follows the electrophile will only be obvious from some of the rate laws that apply to Ad 3 processes. For the others we shall need independent evidence to characterize the addition.

We know of no example of a rate law of the type 68 or its $Ad_{E,N}3$ counterpart. (The Michaelis-Menton expression for the *rate* in eq. 35 is only superficially similar for a *rate constant* in eq. 68.) There is, however, the simpler example

$$k_{\mathrm{obs}} = k[\mathrm{I}^-][\mathrm{I}_2][\mathbf{A}] \tag{69}$$

for $\mathbf{A} \equiv$ acetylene or phenylpropiolic acid, which we have classed as a probable $Ad_{E,N}3$ process.

Whether an electrophile or a nucleophile leads the attack on an alkyne in the Ad 3 process can sometimes be deduced from a group of related reactions; for example,

$$\mathrm{rate} = k[\mathrm{Cl}^-][\mathrm{CH_3OOCC{\equiv}CCOOCH_3}] \tag{70}$$

and

$$\mathrm{rate} = k[\mathbf{N}][\mathrm{CH_3OOCC{\equiv}CCOOCH_3}][\mathrm{HOAc}] \tag{71}$$

in which **N** may be Br^-, I^-, SCN^-, N_3^- (31,32,36,37,39,42). That the reactivity in this series follows the order of nucleophilic power, of itself indicates that the nucleophile is involved in the transition state. Coupled with the fact that the rate law for the slowest nucleophile, chloride, is second order, this reactivity series does indicate an $Ad_{N,E}3$ process. If the rate of protonation of the anion in the series may be regarded as approximately constant, rate law 70 would hold for the slow nucleophiles and a rate law such as 71 would hold for the fast nucleophiles (42).

In general, structure-reactivity work on Ad 3 reactions could reveal dominant sensitivity to either electrophilic or nucleophilic power of the attacking species independent of their order of entry. One must, however, be cautious about interpreting such observations as evidence for any given version of the Ad 3 process.

C. Structure and Reactivity

Systematic structure-reactivity work on process 1 is scarce. Often as not one must piece together the results of different workers to obtain a general view (one should not, of course, compare reactivities derived from different mechanisms). Here we shall consider structural variations in the electrophile (alkyne) and nucleophile in turn.

Electron-withdrawing substituents can change the reactivity of a triple bond by many orders of magnitude. At one extreme there are the forcing conditions, e.g., 150–180°, of the Reppe vinylation of acetylene; at the other extreme, are the almost "spontaneous" additions, for example, amination of benzyne (15). The reactions of a highly activated alkyne are interesting. The upper path of 72 is rapid at room temperature (137); the lower process goes

$$CF_3C{\equiv}CCF_3 \begin{cases} \xrightarrow[\text{ca. } 25^\circ]{C_2H_5OH,\ C_2H_5O^-} CF_3C(OC_2H_5){=}CHCF_3 \\ \xrightarrow[\text{HOAC, } 60^\circ]{\text{NaOAc, } Ac_2O} CF_3C(OAc){=}CHCF_3 \end{cases} \tag{72}$$

fairly readily (138), despite the low nucleophilic power of acetate. The reactivity of some alkynes may in fact become a nuisance during their preparation, because of addition. The formation of vinyl ethers has been noted, e.g., with $C_6Cl_5C{\equiv}CH$ in eq. 36 (96), p-$O_2NC_6H_4C{\equiv}CH$ (139,140) and m-$F_3CC_6H_4C{\equiv}CH$ (141).

On the basis of their experience with thiolate additions to terminal alkynes ($RC{\equiv}CH$) of the type in eqs. 20, 24, 37, and 43, Truce and Heine suggest an order of decreasing reactivity for R: COAr > COOR > COO^- > C_6H_5 > Cl > ArS > alkyl (99). From several of the Russian papers we can put together an order for alkyne susceptibility to halide attack:

$$CH_3OOCC{\equiv}CCOOCH_3 \gg HOOCC{\equiv}CCOOH > HC{\equiv}CCOOH > HC{\equiv}CCH_2OH,$$

$$^-OOCC{\equiv}CCOO^-, \quad RC{\equiv}CCOO^-, \quad C_6H_5C{\equiv}CCH_2N(C_2H_5)_2 \quad (33,34,42,43,46)$$

There is only one Hammett study of the effect of substituents in alkynes: the second-order process (Ad_N2) of eq. 73 gives the correlation of eq. 74 at 0° (142). The ρ value is moderately large and positive, indicating that electron-withdrawing groups in the alkyne facilitate addition.

$$ArC{\equiv}CCOOC_2H_5 + p\text{-}O_2NC_6H_4S^- \xrightarrow{C_2H_5OH} Ar\,(p\text{-}O_2NC_6H_4S)C{=}CHCOOC_2H_5 \tag{73}$$

$$\log k(M^{-1}\ \sec^{-1}) = 2.58\sigma + 0.00175 \tag{74}$$

Our information with respect to the reactivity of nucleophiles is somewhat more quantitative. Although the reaction conditions for a given substrate may vary somewhat, we can use the data of Table 3 as well as the data given in many of the Russian papers to set up the order: $k(SCN^-) > k(I^-) > k(Br^-) \gg k(Cl^-)$ (31,32,36,37,39,42). This sequence is consistent with the generalization, a "soft" electrophile, e.g., an alkyne, reacts faster with "soft" nucleophiles than with "hard" ones (143).

Several groups have investigated the nucleophilicity of alkoxides toward acetylene, but the observations are often inconsistent and the causes of the observed orders are not yet clear. Judging by the acidity of the alcohols, i.e. $pK(3^{ary}) > pK(2^{ary}) > pK(1^{ary})$ (144,145), one might expect the rates for alkoxide attack on acetylene to be $k(3^{ary}) > k(2^{ary}) > k(1^{ary})$. Furukawa et al. reported the vinylation sequence $k(CH_3OH) > k(t\text{-}C_4H_9OH) > k(n\text{-}C_4H_9(C_2H_5)CHCH_2OH) > k(i\text{-}C_4H_9OH) > k(CH_2 = CHCH_2OH)$, in which the expected structure-reactivity order is not followed (113). However, if relatively acidic compounds were added to isobutyl alcohol, its vinylation rate was retarded (113). The findings of Otsuka et al. were similar (109–111). Later, Holly found a rather different order for the butoxides, in the solvent ethyleneglycol dimethyl ether, i.e., $k(3^{ary}){:}k(2^{ary}){:}k(1^{ary})$ = 1:2.4:5.4, and concluded that F-strain was the dominant factor (146).

In view of several correlations of the Taft *polar* substituent parameter (σ^*) in reactions involving the RO-function (147), we are inclined to regard bulk steric effects of R as unimportant. Certainly, there is ready access to the π electrons of most alkynes. We believe that these "substituent effects" arise largely from medium effects and discussion of them will be postponed (Section III-D).

Ban et al. reported a Hammett $\rho \simeq -0.55$ for the vinylation of phenols

$$ArO^- + HC{\equiv}CH \longrightarrow ArOCH{=}CH_2 \tag{75}$$

at 160° in isobutyl alcohol (112). Likewise, Krishnamurthy and Miller give eq. 76 for process 77 (142) at 0°.

$$\log k\ (M^{-1}\ \sec^{-1}) = -0.92\sigma + 0.00323 \tag{76}$$

$$C_6H_5C{\equiv}CCOOC_2H_5 + ArS^- \xrightarrow{C_2H_5OH} C_6H_5(ArS)C{=}CHCOOC_2H_5 \tag{77}$$

The ρ values in these cases are negative, indicating that electron-withdrawing groups in the phenolate or thiolate decrease the rate of addition.

In Ad 3 reactions, we must take into account the power of the electrophile, but pertinent data are scarce. Dvorko et al. reported that the Ad 3 addition of iodide and a proton to acetylene dicarboxylic ester is about twice as fast in trichloroacetic acid as in acetic acid (38,40).

One can summarize all of these data on reactivity as follows: base strength appears to be important *within* a given group or family, for example, phenols or thiols; nucleophilic power increases as the size of the element in a given series increases, for example, $Cl^- < Br^- < I^-$ or $RO^- < RS^-$. In terms of "hard" and "soft" concepts, the softest nucleophiles react most rapidly with the soft alkyne. These reactivity orders, for example, $SCN^- > I^-$, are reasonable for proton solvents in which they have been obtained (148,149).

D. Medium Effects

Most of the available data on environmental effects on process 1 were usually incidental or peripheral to another line of research. After some preliminary observations, we shall take up the few systems which have been studied.

Consider the tentative generalization: the optimum system for Ad_N2 reaction involves a soft nucleophile, a soft electrophile and a soft solvent. The attack of thiolate on an alkyne in dimethylformamide comprises such a system (119,182). By contrast, "hard" hydroxide ion in the "hard" solvent water does not add to unactivated alkynes. Another pertinent factor which could make for retardation is the strong solvation of small electronegative species in proton solvents (149).

In Figure 7 we give schematic representations of the predicted effects of solvation on process 1 (319). Of necessity, the basis of the models used by us derives from studies of the solvation of ions, molecules, and transition states

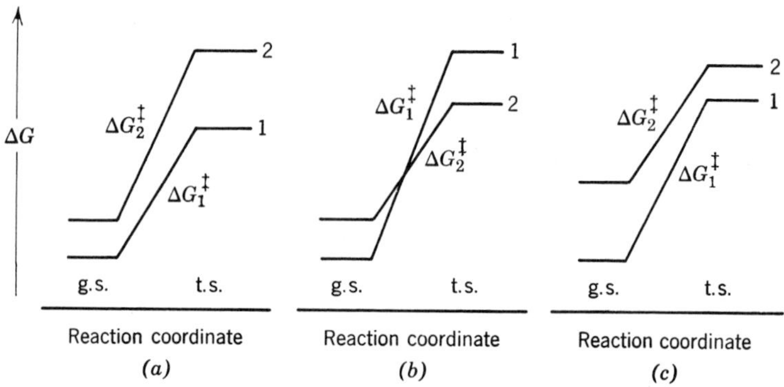

Figure 7. Models of Solvation. Free-energy profiles of a reaction in two solvents in which solvation of ground states (g.s.) and transition states (t.s.) are compared. Ethanol could be solvent 1 and dimethylformamide could be solvent 2. (*a*) t.s. solvation predominates and reaction in solvent 1 is favored; (*b*) both g.s. and t.s. solvation favor reaction in solvent 2; (*c*) g.s. solvation predominates and reaction in solvent 2 is favored.

(149,319–321). We can use protic and polarizable aprotic solvents, for example, ethanol and dimethylformamide as typical extremes of behavior. Pattern *a* of Figure 7 presumably holds for molecule–molecule reactions, e.g., amines adding to alkynes. Pattern *c* of Figure 7 holds in the case of strong solvation by a protic solvent, for example, in an ion–molecule reaction. In pattern *b* of Figure 7, which again could apply to an ion–molecule reaction, energy terms deriving from strong solvation of the ion in the proton solvent and of the polarizable transition state in the aprotic solvent make for a faster reaction in the aprotic solvent. In the Ad 3 process, we would expect similar solvation patterns.

Recently Dvorko and Travchook have found highly individual patterns for additions to ADE over the whole range of methanol–DMF solutions (252). As DMF is added to methanol, the following trends in the rates of Ad_N2 and $Ad_{N,E}3$ reactions occur: decrease with KCNS and HCNS; increase with chloride salt and HCl; initial decrease then increase with KI and HI. The monotonically increasing or decreasing rates are certainly consistent with patterns *a* or *b* of Figure 7. On the other hand, a minimum (or maximum) in the rate in a "mixed" solvent indicates that specific solvent–solvent and solvent-sorting interactions must be present.

Different workers have noted varying orders of reactivity of alkoxides toward alkynes. It seems pertinent to consider the equilibriums 78 and 79 as well as the medium in which the process occurs. The polarity of the alkoxide environment can change drastically, as indicated by trends in the dielectric constants (150,151):

$$ROH + R'O^- \rightleftharpoons R'OH + RO^- \tag{78}$$

$$RO^-M^+ + R'O^- \rightleftharpoons R'O^-M^+ + RO^- \tag{79}$$

	ε(25°)		ε(25°)
CH_3OH	32.63	1-octanol	9.8
C_2H_5OH	24.30	2-octanol	7.7
n-C_3H_7OH	20.1	3-octanol	6.8
n-C_4H_9OH	17	4-octanol	5.0
sec-C_4H_9OH	16	4-methyl-heptan-4-ol	2.9
t-C_4H_9OH	11		

The availability of alkoxide for vinylation is governed by the relative acidity (eq. 78) and by the relative coordinating power of the metal ion (eq. 79). Moreover, the strongest base among the butoxide ions (e.g., *t*-butoxide) is likely to be most strongly coordinated; for example, the high rates of methoxide additions (113) are unexpected on the basis of base strength. These can perhaps be ascribed both to the higher polarity of methanol and to the greater availability of methoxide via eq. 79, when another alcohol is the solvent (111).

Otsuka et al. have made some interesting observations on the vinylation of alcohols (103,109–112). In the presence of alkali alkoxide (M^+OR^-), the relative rates for isobutyl alcohol are: Li^+ 0.1, Na^+ 0.76, K^+ 1.00, Rb^+ 0.83. This order is consistent with the fact that the smaller ions coordinate or form aggregates with the alkoxide and reduce its effective concentration (111,150). The use of benzyltrimethylammonium hydroxide appears to inhibit the addition altogether (111); in part this may be ascribed to the conversion by the hydroxide ion of the alcohol to form water, which retards vinylation (see below).

Otsuka et al. could find no clear dependence on the dielectric constant in the vinylation of ethoxyethanol at 160° (111):

	ε	$k \times 10^5\ M^{-1}\ \text{sec}^{-1}$
t-C_4H_9OH	11.4	8.6
$C_6H_5CH_3$	2.38	7.0
Dioxane	2.28	3.2

Since the alkali alkoxide would be highly aggregated in any of these solvents, it would be difficult to decide what is being compared. The addition of water also retarded vinylation (111,146). Since water can cut down the concentration of alkoxide by forming hydroxide and reduce the reactivity of the alkoxide by hydrogen bonding (148), the reduction in the rate is plausible.

As mentioned earlier (Section III-A), we are inclined to regard the rate laws of Otsuka et al. as reflecting complex environmental effects rather than unusual mechanisms. For example, a maximum was found in the rate of vinylation of ethoxyethanol as its concentration was increased in the solvent *t*-butyl alcohol. In vinylations, the kinetic order in an alcohol was zero, or unity, or varied between these limits (111). These and the other observations were fitted to the Michaelis-Menton expression. Contrary to findings in simpler systems, Otsuka et al. concluded that free alkoxide (as opposed to complexed forms) cannot be the active nucleophile in vinylation (111). For this reason, we are inclined to question his mechanism and his rate laws.

In the Ad_N2 and $Ad_{N,E}3$ additions of acids to acetylene dicarboxylic ester (ADE), Russian workers found that proton solvents are "slowest." The addition of hydrogen iodide follows the order: *i*-C_3H_7OH, 1; $HCON(CH_3)_2$, 2; CH_3NO_2, 7; CH_3CN, 10; $CHCl_3$, 30 (36). Indeed, Dvorko and Mironova point out that if the protonic solvent is strongly hydrogen bonding, the rate of halide attack is reduced and the kinetic dependence on acid (acetic) may become zero (36). (We might point out here that the high rate in chloroform may be due in part to the fact that dimeric acetic acid is a stronger acid than the monomer (152).)

The addition of hydrogen bromide to ADE in a mixed solvent follows the order: pure acetic acid, 4; 33% dioxane in acetic acid, 7; 66% dioxane in acetic acid, 11 (41). Effectively, the nucleophile, bromide, becomes more available as the dioxane solvates the proton (41):

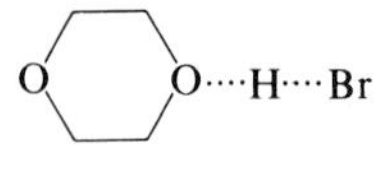

O⟨ ⟩O····H····Br

On the other hand, the addition of water to acetic acid retards the attack of bromide ion, presumably because of greater solvation (hydrogen bonding) of the bromide by water than by acetic acid (44).

In Ad_N2 reactions of amines to alkynes, a zwitterion is probably formed in the rate-determining step. Here the activation energy for charge separation should be reduced with increasing solvent polarity. Plots of log k versus Brownstein's solvent parameter in Figure 8 (153) are in accord with this hypothesis. The slopes of the lines in Figure 8 also indicate that the additions to ADE are somewhat less sensitive than those to methyl propiolate. One might suppose, therefore, the charge separation in the transition state is somewhat less in the more reactive ester (ADE) (153).

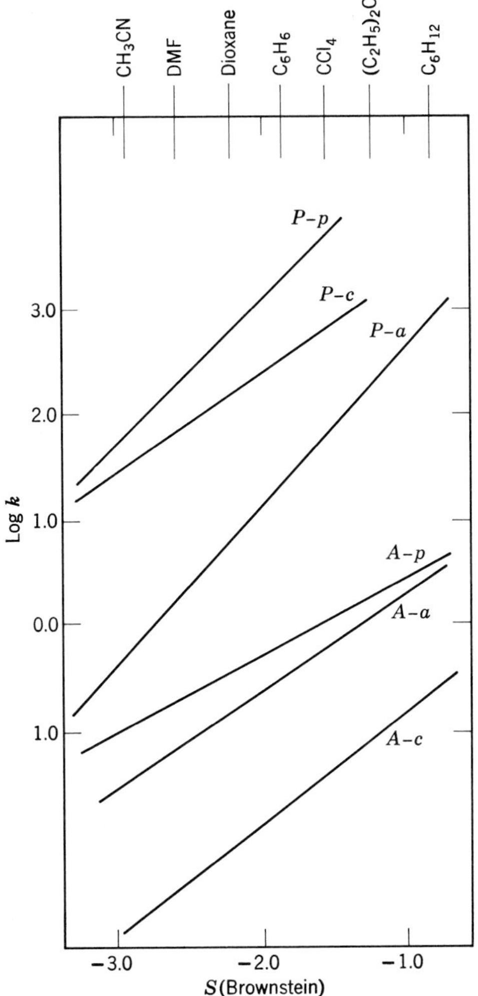

Figure 8. Adapted from Ref. *153*. Solvent dependence of the rates of addition at 37° in M^{-1} sec^{-1}. The systems studied were: P-p, methyl propiolate piperidine; P-c, methyl propiolate-cyclohexylamine; P-a, methyl propiolate-aziridine; A-p, dimethyl acetylenedicarboxylate-piperidine; A-a, dimethyl acetylene dicarboxylate-aziridine; A-c, dimethyl acetylenedicarboxylate, cyclohexylamine.

E. **Directioselectivity** (16)

In contrast to the rigid stereoelectronic requirements on *syn* versus *anti* additions, the choice between 1,2- and 2,1-orientation seems to depend both on the entering nucleophile and the substituents on the alkyne. On the other hand, the direction of intramolecular nucleophilic additions which lead to five- or six-membered rings is not strongly influenced by the properties of the participating centers.

For terminal alkynes without hetero substituents the direction usually is given by the Markovnikov rule and is unexceptional (see also eq. 39):

$$CH_3C{\equiv}CH + C_2H_5O^- \xrightarrow{C_2H_5OH} CH_3C(OC_2H_5){=}CH_2 \qquad (80)$$

(from ref. 8)

$$RC{\equiv}CH + ArS^- \longrightarrow \begin{cases} ArSCR{=}CH_2 \text{ (major product)} \\ \xrightarrow{\text{radical ?}} ArSCH{=}CHR \end{cases} \qquad (81)$$

(from refs. 8, 60)

$$C_6H_5Se^- + n\text{-}C_4H_9{-}C{\equiv}CH \xrightarrow[120^\circ]{C_2H_5OH} \begin{cases} n\text{-}C_4H_9(C_6H_5Se)C{=}CH_2 \text{ (major product)} \\ \xrightarrow{\text{radical ?}} n\text{-}C_4H_9CH{=}CHSeC_6H_5 \end{cases} \qquad (82)$$

(from ref. 154)

Additions of mercaptans to *t*-butylacetylene in the presence of base appear to be anti-Markovnikov (8). Since oxygen was not excluded in these reactions,

$$(CH_3)_3C{-}C{\equiv}CH + RSH \xrightarrow[(O_2)]{KOH} \begin{cases} (CH_3)_3C{-}CH{=}CHSR \text{ (ca. 100\%)} \\ (CH_3)_3C(R{-}S)C{=}CH_2 \text{ (ca. 0\%)} \end{cases} \qquad (83)$$

it is possible that the reactions went via free radicals (8). Although a steric factor was considered too (8), there is little precedent for large steric effects in eq. 1 (see eqs. 20 and 21); indeed, the steric requirements of the nucleophile do not seem to be unusual in 83.

Activated acetylenes also give straightforward products; examples of both inductive and conjugative substituents are known.

$$HC{\equiv}CCH(OC_2H_5)_2 + C_2H_5O^- \xrightarrow{C_2H_5OH} C_2H_5OCH{=}CHCH(OC_2H_5)_2 \qquad (84)$$

(from ref. 8)

$$C_6H_5{-}\overset{O}{\overset{\|}{C}}{-}C{\equiv}CH + S_2O_3^{=} \longrightarrow C_6H_5COCH{=}CHS_2O_3^- \qquad (85)$$

(from ref. 155)

$$C_6H_5SO_2C{\equiv}CCH_3 + C_6H_5SO_2^- \longrightarrow (C_6H_5SO_2)(H)C{=}C(SO_2C_6H_5)(CH_3) \qquad (86)$$

(from ref. 123)

$$(i\text{-}C_3H_7O)_2P(O)C{\equiv}CH + C_2H_5O^- \longrightarrow (i\text{-}C_3H_7O)_2P(O)CH{=}CHOC_2H_5 \quad (87)$$

(from ref. 156)

$$(C_6H_5)_3P^+C{\equiv}CC_6H_5 + C_6H_5S^- \longrightarrow (C_6H_5)_3P^+CH{=}C(C_6H_5)SC_6H_5 \quad (88)$$

(from ref. 157)

$$HC{\equiv}C{-}CN + (CH_3)_3PbH \begin{cases} \xrightarrow{\text{polar}} H_2C{=}C(CN)Pb(CH_3)_3 \quad 94\% \\ \xrightarrow{\text{radical}} (CH_3)_3PbCH{=}CHCN \quad 6\% \end{cases} \quad (89)$$

(from ref. 158)

Additions to divinylacetylene and to eneynes have been reviewed (159,286). The interpretation is simplified if nucleophilic attacks are kept distinct from electrophilic attacks. Although ground-state electron density is best represented by **9**, it is probably better to examine the transition states, for which intermediates **10–13** are models. Since alkynes are more electro-

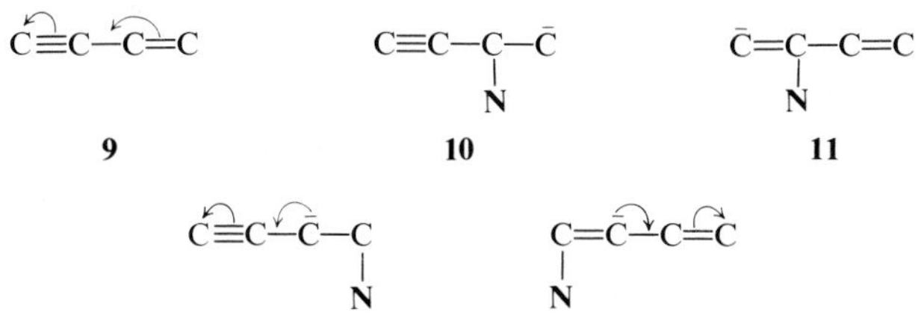

$$C{\equiv}C{-}C{=}C \quad (\mathbf{9}) \qquad C{\equiv}C{-}C(N){-}\bar{C} \quad (\mathbf{10}) \qquad \bar{C}{=}C(N){-}C{=}C \quad (\mathbf{11})$$

$$C{\equiv}C{-}\bar{C}{-}C(N) \quad (\mathbf{12}) \qquad C(N){=}\bar{C}{-}C{=}C \quad (\mathbf{13})$$

negative than alkenes, we would rate **10** as the least probable; moreover it lacks conjugation. As for **11–13**, examples of all three have been given, chiefly by Russian workers.

$$R'C{\equiv}C{-}CH{=}CH_2 \xrightarrow[\text{2. } H_2O]{\text{1. LiN}} R'CH{=}C{=}CHCH_2N \quad (90)$$

($R' = H$, alkyl; **N** = NR_2^-, PR_2^-, R^-, OR^-)

(from refs. 5, 8, 159)

In scheme 91 we take up a specific example. The product distribution for this complex system is also included (160,161):

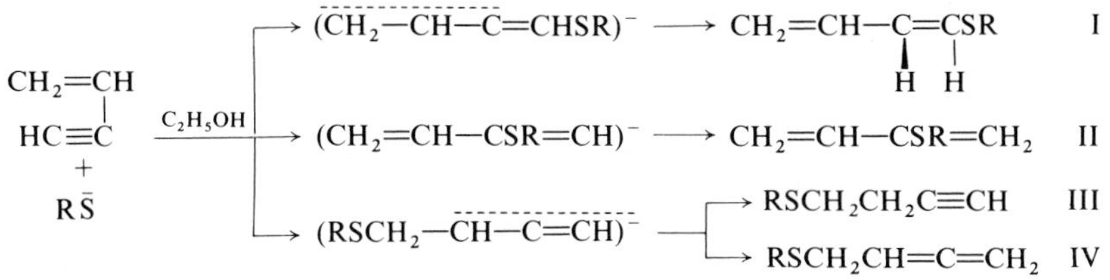

$$CH_2{=}CH{-}C{\equiv}CH + R\bar{S} \xrightarrow{C_2H_5OH} \begin{cases} (CH_2{-}CH{-}C{=}CHSR)^- \longrightarrow CH_2{=}CH{-}CH{=}CHSR \quad \text{I} \\ (CH_2{=}CH{-}CSR{=}CH)^- \longrightarrow CH_2{=}CH{-}CSR{=}CH_2 \quad \text{II} \\ (RSCH_2{-}CH{-}C{=}CH)^- \longrightarrow \begin{cases} RSCH_2CH_2C{\equiv}CH \quad \text{III} \\ RSCH_2CH{=}C{=}CH_2 \quad \text{IV} \end{cases} \end{cases}$$

(91)

R	I	II	III	IV
CH_3	50%	12	7	25
C_2H_5	58	10	5	27
n-C_3H_7	58	7	5	30
i-C_3H_7	64	3	6	27
n-C_4H_9	64	5	4	27
t-C_4H_9	69	0	6	25

Although the overall trends are plausible, the distribution is not easy to explain in detail. The importance of charge delocalization in **12** and **13** may account for their predominance over **11** in eqs. 90 and 91, but the factors which favor opposite modes of terminal attack in 90 and 91 are obscure. Indeed, we would not have predicted any yield of II. Then, the apparent steric effect on the yield of II is also puzzling. As between the alternate modes of terminal attack, there is some precedent favoring the alkyne rather than the alkene end (4,201), and the ratio of (I)/((III) + (IV)) is not unreasonable. In fact, the final distribution of I, III, and IV is similar to that one might expect from their equilibration (121). However, all of these rationalizations are weakened by the observations that the allene (IV) becomes the exclusive product in aprotic solvents such as dioxane or tetrahydrofuran (160).

Since the direction of addition to hetero-substituted alkynes may seem confusing, we give the theoretical background. Arens pointed out that the nature of the polarization in an ethynyl ether and in an ethynyl thioether are

$$\overset{\delta-}{-C}\equiv\overset{\delta+}{C} \longrightarrow \bar{O}R \qquad\qquad \overset{\delta+}{-C}\equiv C \longrightarrow \overset{\delta-}{S}R$$

different (125). As a *donor*, oxygen can conjugate with the triple bond, while as an *acceptor*, sulfur can conjugate with the triple bond, because it is able to accommodate more than eight valence electrons. For this reason additions of the following type can be found:

$$HC{\equiv}COC_2H_5 + C_2H_5O^- \xrightarrow{C_2H_5OH} H_2C{=}C(OC_2H_5)_2 \tag{92}$$

(from ref. 125)

$$HC{\equiv}COC_2H_5 + C_2H_5S^- \xrightarrow[/]{C_2H_5OH} H_2C{=}C(OC_2H_5)SC_2H_5 \tag{46a}$$

(from ref. 125)

$$HC{\equiv}CSC_2H_5 + C_2H_5O^- \longrightarrow C_2H_5OCH{=}CHSC_2H_5 \tag{93}$$

(from ref. 125)

$$p\text{-}C_7H_7SC{\equiv}CH + p\text{-}C_7H_7S^- \longrightarrow cis\text{-}p\text{-}C_7H_7SCH{=}CHSC_7H_7\text{-}p \tag{37a}$$

(from ref. 65)

$$HC{\equiv}COC_2H_5 + HNR_2 \longrightarrow H_2C{=}C(OC_2H_5)(NR_2) \tag{94}$$

(from ref. 125)

$$HC{\equiv}CSC_2H_5 + RNH_2 \longrightarrow RNHCH{=}CHSC_2H_5 \tag{95}$$

(from ref. 125)

It is interesting that the directive power of the hetero substituent is maintained even in conjugated systems (8,314):

$$HC{\equiv}C{-}C{\equiv}CH \xrightarrow[60\text{–}110^\circ]{KOH,\ ROH} HC{\equiv}C{-}CH{=}CHOR \longrightarrow$$

$$(HC{\equiv}C{-}CH_2CH(OR)_2) \longrightarrow CH_3C{\equiv}CCH(OR)_2 \tag{96}$$

$$HC{\equiv}C{-}C{\equiv}CH \xrightarrow{RS^-Na^+} (HC{\equiv}C)(H)C{=}C(SR)(H) \longrightarrow RSCH{=}C(H){-}C(H){=}C(SR)(H) \tag{97}$$

With cooperative substituents on the alkyne there is one orientation,

$$RSC{\equiv}COC_2H_5 \xrightarrow{R'S^-} RSCH{=}C(OC_2H_5)SR' \tag{98}$$

(from ref. 162)

$$R'C{\equiv}COC_2H_5 \xrightarrow{LiNR_2} (R'CLi{=}C(OC_2H_5)NR_2) \longrightarrow R'C{\equiv}CNR_2 \tag{99}$$

$(R{=}C_6H_5)$

(from refs. 163, 290)

On the other hand, antagonistic substituents on the alkyne retard the rate of addition, and with mercaptans, may even facilitate radical paths (8,125):

$$C_2H_5OC{\equiv}CCH_3 + C_2H_5SH \longrightarrow C_2H_5OCH{=}C(CH_3)SC_2H_5 \tag{100}$$

$$ROOC{-}C{\equiv}C{-}SC_2H_5 + C_2H_5SH \longrightarrow ROOC{-}C(SC_2H_5){=}CHSC_2H_5 \tag{101}$$

Additions to other first row heteroalkynes appear to be similar to those of the ethers.

$$HC{\equiv}CF \xrightarrow{NaNH_2} CH_3CN \tag{102}$$

(from ref. 164)

$$HC{\equiv}CN(C_2H_5)_2 \xrightarrow{H_2O} CH_3CON(C_2H_5)_2 \tag{103}$$

(from ref. 165)

The results with heavier element heteroalkynes probably parallel those of the thioethers, unless there are opposing substituent effects. In processes 37 and 38 the predicted orientation is obtained; in eq. 104 it may be followed if, as Viehe supposes (16), the FBW elimination-rearrangement sequence of eq. 38 is followed. Halogen abstraction along the lower path of eq. 104 is another possibility for which there is precedent (16). Incidentally, the formation in eq. 104 of a 1,2-adduct, for R = n-C_4H_9 and a 1,1-adduct for R = *t*. C_4H_9 was rationalized by invoking steric hindrance (166). By proposing the alternate routes in eq. 104, for which there is precedent, we have a more consistent (but not more certain) mechanistic picture.

$$\mathrm{RC{\equiv}C{-}Br} + \mathrm{(C_2H_5)_2NH} \longrightarrow \begin{cases} \mathrm{((C_2H_5)_2NC(R){=}CHBr)} \longrightarrow \mathrm{(C_2H_5)_2NCR{=}CHN(C_2H_5)_2} \\ \quad\downarrow \text{FBW} \qquad\qquad\qquad\qquad\qquad \downarrow \text{FBW} \\ \mathrm{(RC{\equiv}CN(C_2H_5)_2)} \longrightarrow \mathrm{RCH{=}C(N(C_2H_5)_2)_2} \\ \quad\uparrow \\ \mathrm{(RC{\equiv}C^- \quad {}^+BrHN(C_2H_5)_2)} \end{cases} \qquad (104)$$

(from ref. 166)

In eq. 38 and along the upper branch of eq. 104 the effects of substituents are aligned. But in eqs. 73 and 105, these are opposed. The directing power of the phenyl overwhelms that of the chloro group in 105, while the power of carbethoxy overcomes that of the phenyl group in eq. 73 or 77 (142).

$$\underset{(X\,=\,\mathrm{Cl,\,Br})}{\mathrm{C_6H_5{\equiv}CX}} \xrightarrow[\mathrm{C_2H_5OH}]{\mathrm{C_2H_5O^-}\ 100^\circ} \mathrm{C_6H_5CH{=}CXOC_2H_5} \qquad (105)$$

(from ref. 167)

Directioselectivity in cyclization of alkynes has been mentioned but not really explained (86). Certainly, the substituents R and R′ will direct according to their electronic properties. What seems more critical to us, however, is

[Scheme (106): R′-substituted aminoalkyne (NH, alkyne carbons labelled β and α, terminal R) cyclizes —favored→ five-membered ring with exocyclic =C(H)R at N; —not favored→ six-membered ring with ring C(H)=C(R)–N.] (106)

the fact that the nucleophilic site, N, is closer to the β-carbon atom. Of course, at elevated temperatures the promotion of higher amplitudes in the bending modes would tend to reduce the importance of the spatial factor. According to available precedent, in the five-membered ring there may be also a rate advantage of forming an *exo* double bond as compared with the internal double bond in the six-membered ring (168,169). Nevertheless, equilibrium data presumably favor the six-membered ring: "Ease of ring formation is not synonymous with ring stability" (168).

In eq. 14 both substituents favor the six-membered ring and this is formed, but we have also found Michael addition followed by frequent *anti*-Michael addition with diphenylethynyl ketone (86). In eq. 106a note that the preference for the five-membered cycle as well as the electron-accepting character of sulfur cooperate to form product against the possible effect of ene–yne conjugation (125):

$$\text{(cyclic } C{\equiv}C{-}S{-}C_2H_5,\ O^- Li^+) \longrightarrow \text{(dihydrofuran)}{-}CH_2SC_2H_5 \cdot \qquad (106a)$$

Even the carbonyl group cannot direct *all* attack to the conventional β-carbon

$$(C_6H_5C{\equiv}C)_2C{=}O \xrightarrow[130^\circ]{ArNH_2} \text{2,6-}(C_6H_5)_2\text{-}N\text{-Ar-4-pyridone} + \text{2-}C_6H_5\text{-5-}(=CHC_6H_5)\text{-}N\text{-Ar-pyrrol-3-one}$$

$$(C_6H_5C{\equiv}C)_2C{=}O \xrightarrow[C_2H_5OH,\ 100^\circ]{H_2S} \text{2,6-}(C_6H_5)_2\text{-4-thiopyrone}$$

$$(C_6H_5C{\equiv}C)_2C{=}O \xrightarrow[CH_3OH,\ 0\text{–}10^\circ]{(H_2N)_2CS \text{ or } (C_6H_5NH)_2CS} \text{2-}C_6H_5\text{-5-}(=CHC_6H_5)\text{-thiophen-3-one} \qquad (107)$$

and give a six-membered ring (86,280). Note too that conjugative stabilization of the anion intermediate by the carbonyl is improbable because of ring strain. In Section IV we shall cite other examples which appear to confirm the preference shown in eq. 106.

IV. SELECTIVITY AND CHANGING NUCLEOPHILES IN SYNTHESES

In this last section we survey additions to alkynes with emphasis on the range of nucleophiles and the variety of reactions. In a few instances (e.g., amine additions) their unique character or special interest requires elaboration of the more general theoretical discussions given in previous sections. Because of the availability of reviews, our own discussion can be limited and selective (3–9).

A. Formation of Carbon–Carbon Bonds

1. Organometallic Nucleophiles

With organometallic reagents it is sometimes difficult to distinguish among Ad_N, Ad_E, and Ad_{MOL} processes. Trialkylaluminums add *syn* to alkynes (170). It is probable that here the electrophilic center leads the attack giving an intermediate similar to **4a** (14). For this reason we have largely bypassed this interesting research area.

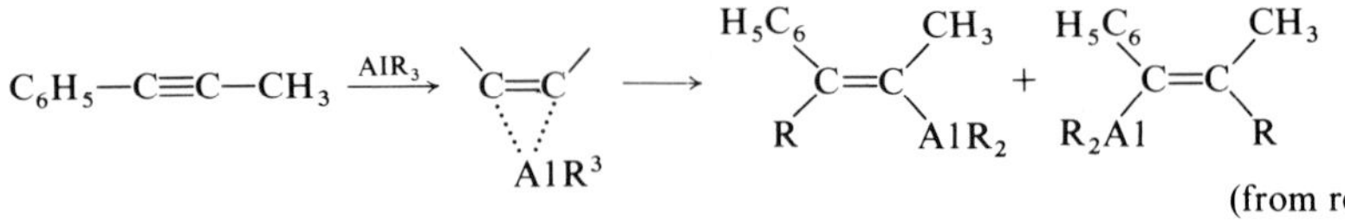

$$C_6H_5{-}C{\equiv}C{-}CH_3 \xrightarrow{AlR_3} \underset{\dot{A}lR^3}{>C{=}C<} \longrightarrow \begin{matrix} H_5C_6 & & CH_3 \\ & C{=}C & \\ R & & AlR_2 \end{matrix} + \begin{matrix} H_5C_6 & & CH_3 \\ & C{=}C & \\ R_2Al & & R \end{matrix} \quad (108)$$

(from ref. 170)

Normally, Grignard reagents do not add to triple bonds (171). At relatively high temperatures the following reaction does occur (172):

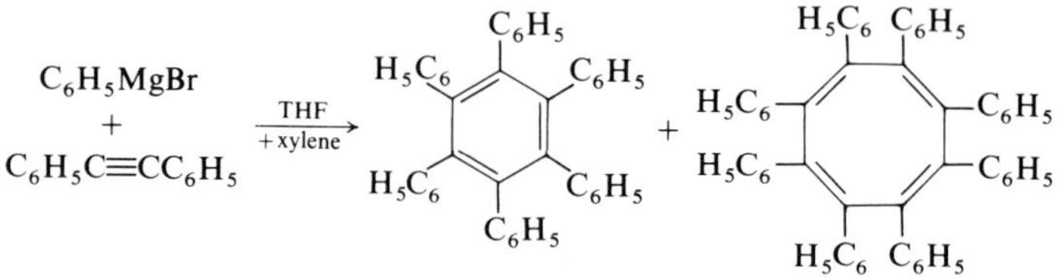

$$C_6H_5MgBr + C_6H_5C{\equiv}CC_6H_5 \xrightarrow[+\text{xylene}]{\text{THF}} (C_6H_5)_6C_6 \text{ (hexaphenylbenzene)} + (C_6H_5)_8C_8 \text{ (octaphenylcyclooctatetraene)} \quad (109)$$

Likewise, organolithium additions become possible for certain activated alkynes:

$$CH_3(C{\equiv}C)_4CH_3 + CH_3Li \xrightarrow[2.\ H_2O]{1.\ \text{ether}} CH_3(C{\equiv}C)_3CH{=}C(CH_3)_2 \quad (110)$$

(from ref. 173)

$$C_6H_5C{\equiv}CC_6H_5 + C_6H_5Li \xrightarrow[2.\ CO_2]{1.\ \text{ether}} (C_6H_5)_2C{=}C(C_6H_5)COOH \quad (111)$$

(from ref. 174)

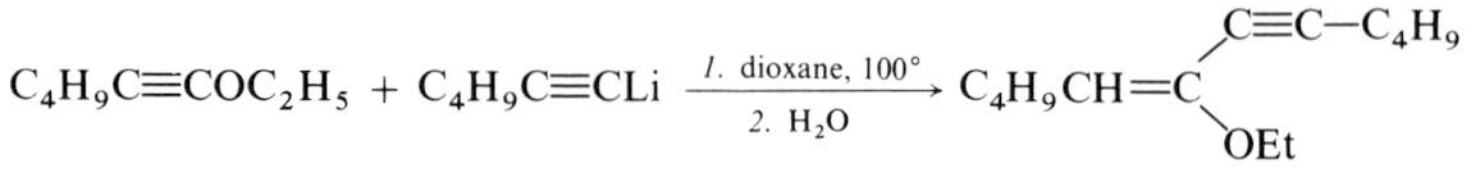

$$C_4H_9C{\equiv}COC_2H_5 + C_4H_9C{\equiv}CLi \xrightarrow[2.\ H_2O]{1.\ \text{dioxane, }100^\circ} C_4H_9CH{=}C(C{\equiv}C{-}C_4H_9)(OEt) \quad (112)$$

(from ref. 175)

Somewhat unexpectedly, the reaction apparently analogous to eq. 111, also involves addition and subsequent ring metallation (171):

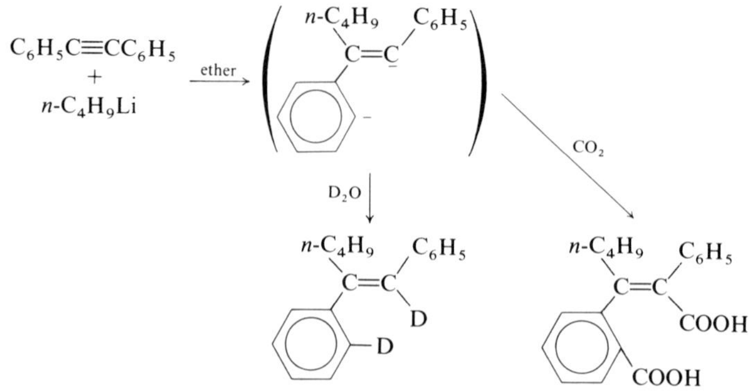

$$C_6H_5C{\equiv}CC_6H_5 + n\text{-}C_4H_9Li \xrightarrow{\text{ether}} \left((n\text{-}C_4H_9)(o\text{-}C_6H_4^-)C{=}C^-(C_6H_5) \right)$$

$$\xrightarrow{D_2O} (n\text{-}C_4H_9)(o\text{-}DC_6H_4)C{=}C(C_6H_5)D \qquad \xrightarrow{CO_2} (n\text{-}C_4H_9)(o\text{-}HOOCC_6H_4)C{=}C(C_6H_5)COOH \quad (113)$$

Two additions with α-phenyl isopropylpotassium have also been reported (176).

$$C_6H_5\bar{C}(CH_3)_2K^+ \xrightarrow{C_6H_5C\equiv CC_6H_5} C_6H_5CK{=}C(C_6H_5)C(CH_3)_2C_6H_5$$

$$C_6H_5\bar{C}(CH_3)_2K^+ \xrightarrow{C_6H_5C\equiv CCH_3} C_6H_5CK{=}C(CH_3)C(CH_3)_2C_6H_5 \qquad (114)$$

The choice of forming four-, five-, and six-membered rings by intramolecular addition of organometallic compounds to alkynes has been explored. It is not really surprising that closure to a four-membered ring does not occur in eq. 115 (177). On the other hand, five-membered rings do form and may be favored over six-membered rings (177,178).

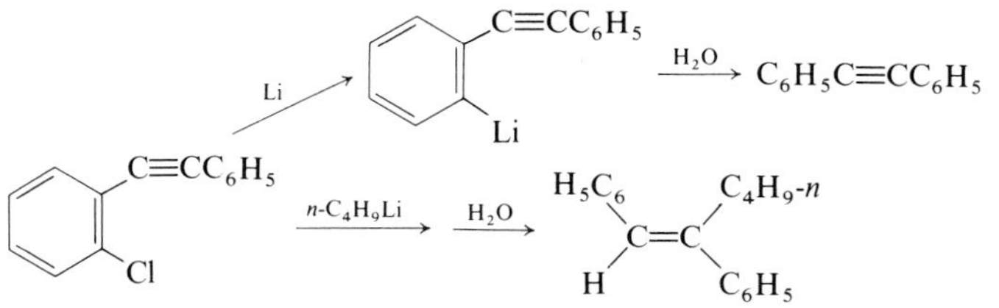

(115)

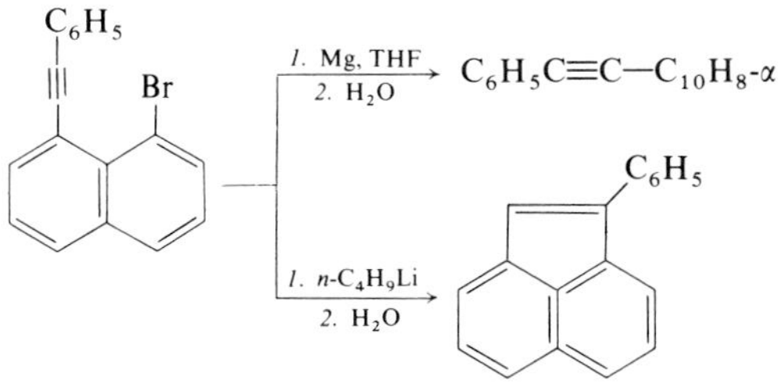

(116)

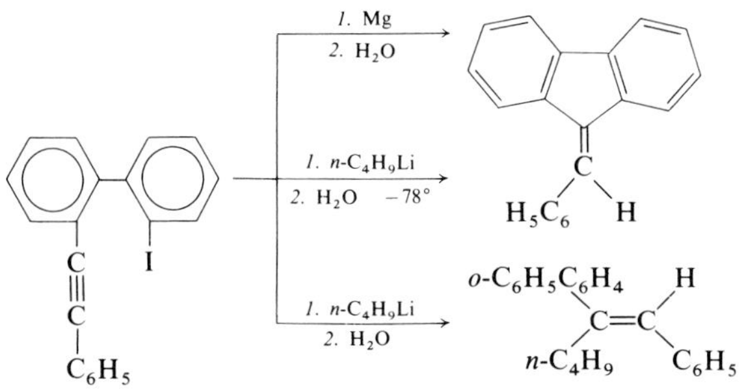

(117)

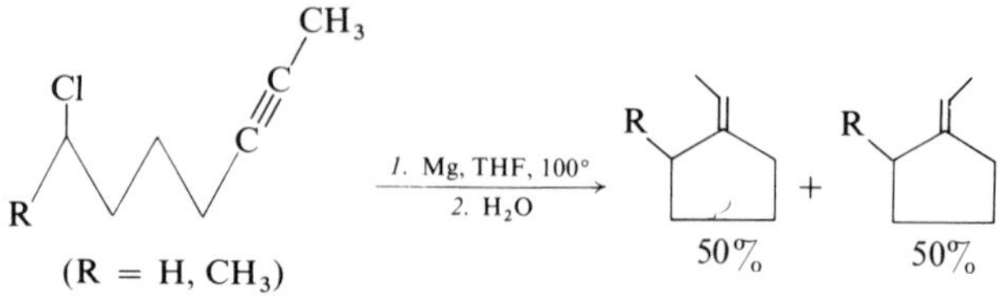

$$\text{R–CH(Cl)–(CH}_2)_4\text{–C}\equiv\text{C–CH}_3 \xrightarrow[2.\ H_2O]{1.\ Mg,\ THF,\ 100^\circ} \text{(50\%)} + \text{(50\%)} \qquad (118)$$

(R = H, CH_3)

2. Carbanions

Additions of carbon acids to alkynes in the presence of base generally requires some sort of activation in the total system. Seefelder's finding that malonic esters or cyanoacetic esters add to acetylene in the presence of zinc or cadmium salts may be taken as evidence of electrophilic catalysis (R = alkyl, aryl) (179).

$$RCH(COOC_2H_5)_2 + HC{\equiv}CH \xrightarrow[\text{alkali}]{Zn^{2+}\text{ or }Cd^{2+}} (HC{\equiv}C)(R)C(COOC_2H_5)_2 \qquad (119)$$

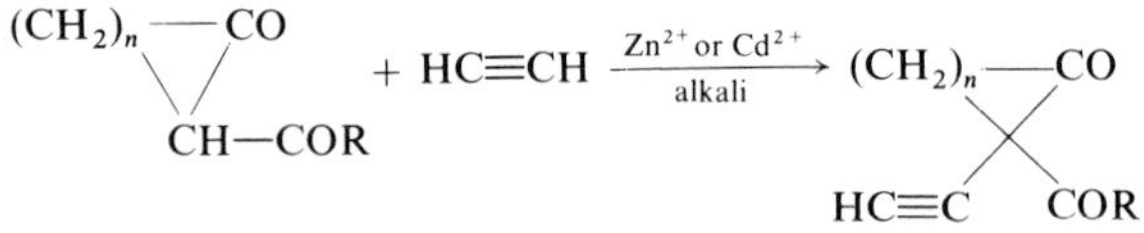

$$(CH_2)_n\text{—CO—CH—COR (ring)} + HC{\equiv}CH \xrightarrow[\text{alkali}]{Zn^{2+}\text{ or }Cd^{2+}} (CH_2)_n\text{—CO—C(HC{\equiv}C)(COR) (ring)} \qquad (120)$$

In eq. 121a the chloro substituents may be mildly activating.

$$ClC{\equiv}CCl + Na^{+\,-}C(C_2H_5)(COOC_2H_5)_2 \longrightarrow ClC{\equiv}CC(C_2H_5)(COOC_2H_5)_2 + (C_2H_5OOC)C(C_2H_5)CCl{=}CHC(C_2H_5)(COOC_2H_5)_2 \qquad (121a)$$

(from ref. 202)

A powerful nucleophile could also be effective (180):

$$C_6H_5C{\equiv}CC_6H_5 \xrightarrow[\text{NaH, }26^\circ]{\text{DMSO}} C_6H_5CH{=}C(C_6H_5)CH_2SOCH_3 \qquad (121)$$

88% *anti* adduct

$$C_6H_5(C{\equiv}C)_2C_6H_5 \xrightarrow[\text{NaH, }36^\circ]{\text{DMSO}} C_6H_5C{\equiv}C{-}CH{=}C(C_6H_5)CH_2SOCH_3 \qquad (122)$$

84% *anti* adduct

(Isomers in eqs. 121 and 122 arise from postequilibration of the *anti* products.) The electron-withdrawing ethoxide group appears to make the base-catalyzed tetramerization (eq. 123) possible (181). Since four *syn* additions are involved, there is a real question, however, as to whether nucleophilic processes are involved.

$$4\,HC{\equiv}COC_2H_5 \xrightarrow[NH_3,\,-33^\circ]{NaNH_2} \quad \begin{matrix} C_2H_5O & & OC_2H_5 \\ & \text{(cyclooctatetraene ring)} & \\ C_2H_5O & & OC_2H_5 \end{matrix} \tag{123}$$

The ease of carbanion addition may depend very much on the structure of the carbon acid when the degree of activation of the triple bond is small. In ethanolic ethoxide, 2-nitropropane adds to *p*-nitrophenylacetylene, whereas ethyl cyanoacetate and diethyl malonate do not add (180).

$$p\text{-}O_2NC_6H_4C{\equiv}CH + CH_3CH_2NO_2 \xrightarrow[C_2H_5OH]{C_2H_5O^-} \begin{matrix} p\text{-}O_2NH_4C_6 & & CNO_2(CH_3)_2 \\ & C{=}C & \\ H & & H \end{matrix} \tag{124}$$

But the use of an appropriate solvent, e.g., dimethylsulfoxide (DMSO) may facilitate the vinylations (204). The *anti* adduct was usually formed as

$$C_6H_5(R)CHCN + HC{\equiv}CR' \xrightarrow[(C_2H_5)_3\overset{+}{N}CH_2C_6H_5,\ Cl^-]{DMSO} \begin{matrix} H_5C_6 & & CN & & \\ & C & & & R' \\ R & & C{=}C & & \\ & & H & & H \end{matrix} \tag{125}$$

$(R = \text{alkyl},\ C_6H_5CH_2,\ C_2H_5NCH_2CH_2;\ R' = H,\ C_6H_5)$

indicated in eq. 125, although some *syn* and 1,1-alkene were formed when $R = C_6H_5CH_2$. Under similar conditions diphenylacetonitrile could not be made to add to the indicated alkynes, and diphenylacetylene and 1-hexyne did not react with the indicated phenylacetonitriles. Although there was no comment on the role of the quaternary salt (204), it may simply provide base of sufficient strength to react with the phenylacetonitrile but not with the terminal alkyne.

The possibility of forming a cycle also seems to provide driving force for addition in eq. 126 (183,184). In this group are a number of propargylic rearrangements, an area which has been reviewed recently (185):

$$\text{R—C}\equiv\text{C—C(R')=C(R')—CH}_2\text{X} \xrightarrow[\text{C}_2\text{H}_5\text{O}^-]{\text{CH}_2(\text{COOC}_2\text{H}_5)_2} \text{R—C}\equiv\text{C—C(R')=C(R')—CH}_2\text{CH(COOC}_2\text{H}_5)_2 \xrightarrow[\text{2. H}^+]{\text{1. C}_2\text{H}_5\text{O}^-} \text{cyclopentene, } (\text{C}_2\text{H}_5\text{OOC})_2, =\text{CHR}, \text{R'}, \text{R'} \tag{126}$$

(X = Cl, Br.;
R = H, CH_3, $C_4H_9(C{\equiv}C)_2$,
$C_4H_9(C{\equiv}C)_3$)

$$\text{X—C}_6\text{H}_4\text{—OCH}_2\text{C}\equiv\text{CR} \xrightarrow{\text{C}_6\text{H}_5\text{N(C}_2\text{H}_5)_2} \text{X—(4-R-2H-chromene)} \tag{127}$$

The effect of *meta* substituents (X) is such that electron donating substituents are associated with higher product yields. This suggests that the reaction probably involves nucleophilic attack by aryl carbon on the alkyne (electrophile) (186). In the following sequence, alternatives for the critical step are indicated (187):

$$(\text{C}_6\text{H}_5\text{C}\equiv\text{CCH}_2)_2\text{X} \xrightarrow[t\text{-C}_4\text{H}_9\text{OH}]{t\text{-C}_4\text{H}_9\text{O}^-\text{K}^+} \text{[intermediates, or]} \longrightarrow \text{4-C}_6\text{H}_5\text{-naphtho-fused ring with X} \tag{128}$$

($X{=}CH_2, S, N{-}CH_3, O$)

Michael additions of carbon acids (e.g., fluorene-9-carboxylic ester, triphenylmethane, malonic ester, α-cyano ester, β-keto esters, β-ketones) to activated triple bonds, e.g., acetylenic esters, have been known for a long time (6). Some recent examples follow:

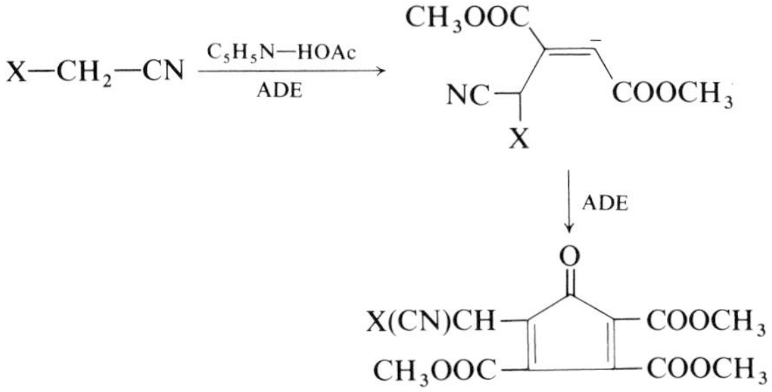

$$X{-}CH_2{-}CN \xrightarrow[ADE]{C_5H_5N{-}HOAc} \text{NC(X)CH–C(COOCH}_3\text{)=CH}^-\text{–COOCH}_3 \xrightarrow{ADE} \text{cyclopentadienone: X(CN)CH, CH}_3\text{OOC, COOCH}_3\text{, COOCH}_3 \quad (129)$$

$(X = CN, COOC_2H_5)$

(from ref. 188)

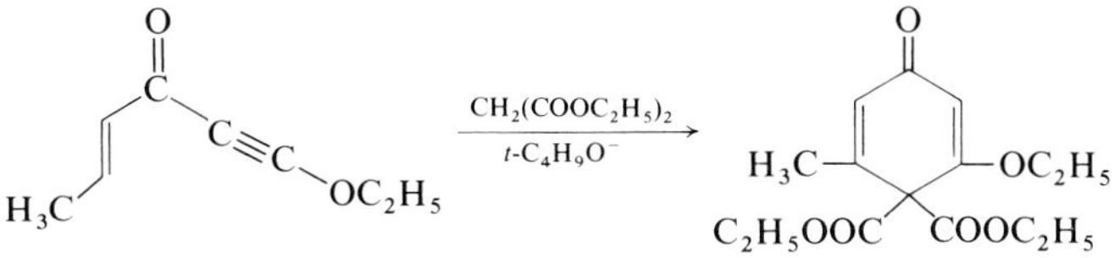

$$H_3C{-}CH{=}CH{-}C(=O){-}C{\equiv}C{-}OC_2H_5 \xrightarrow[t\text{-}C_4H_9O^-]{CH_2(COOC_2H_5)_2} \text{cyclohexadienone: H}_3\text{C, OC}_2\text{H}_5\text{, C(COOC}_2\text{H}_5)_2 \quad (130)$$

$$(n\text{-}C_4H_9C{\equiv}C)_2C{=}O + CH_2(COOC_2H_5)_2 \xrightarrow{base} \text{2-pyrone: } n\text{-}H_9C_4\text{, C}_2\text{H}_5\text{OOC, C}{\equiv}\text{CC}_4\text{H}_9\text{-}n \quad (131)$$

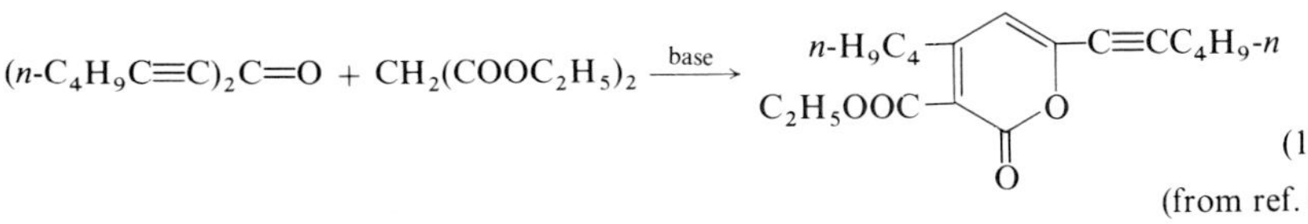

(from ref. 86)

In reaction 130 nucleophilic addition to the triple bond takes precedence over addition to the double bond (189). It is interesting that dimethylsulfoxonium methylide gives a stable ylid in eq. 132, but gives none isolable with several other alkynes, for example, ethyl propiolate, ADE, and benzoylacetylene (190).

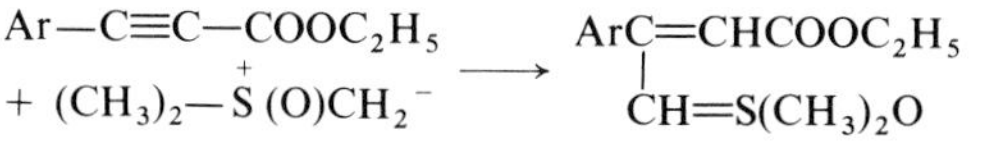

$$\begin{array}{l} Ar{-}C{\equiv}C{-}COOC_2H_5 \\ + (CH_3)_2{-}\overset{+}{S}(O)CH_2^- \end{array} \longrightarrow \begin{array}{l} ArC{=}CHCOOC_2H_5 \\ \quad | \\ \ CH{=}S(CH_3)_2O \end{array} \quad (132)$$

Eneamines are an interesting class of reagents whose additions to alkynes involve either the carbon α or β to the nitrogen (16,191,192).

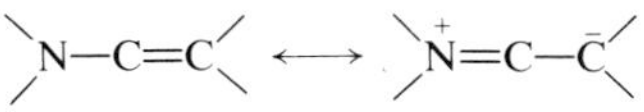

$$>N{-}C{=}C< \longleftrightarrow >\overset{+}{N}{=}C{-}\overset{-}{C}<$$

Only the β-carbon additions appear to involve the eneamines as nucleophiles. Representative examples will illustrate their utility in syntheses.

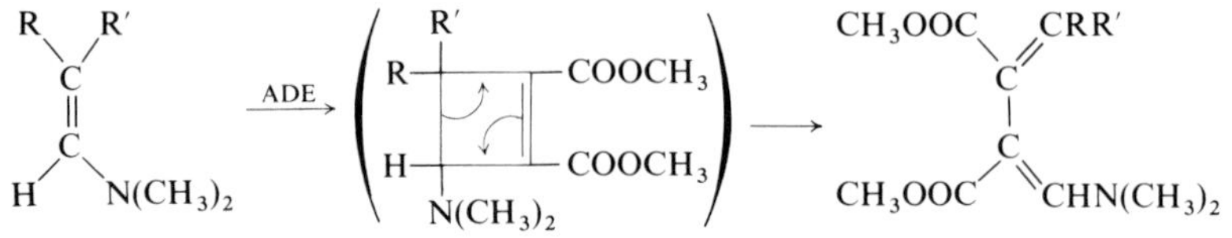

(133)

(from ref. 192)

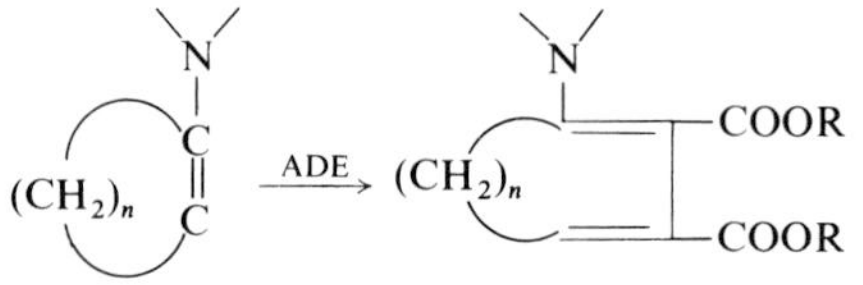

(134)

(n = 3, 5, 6, 10)
(from ref. 192)

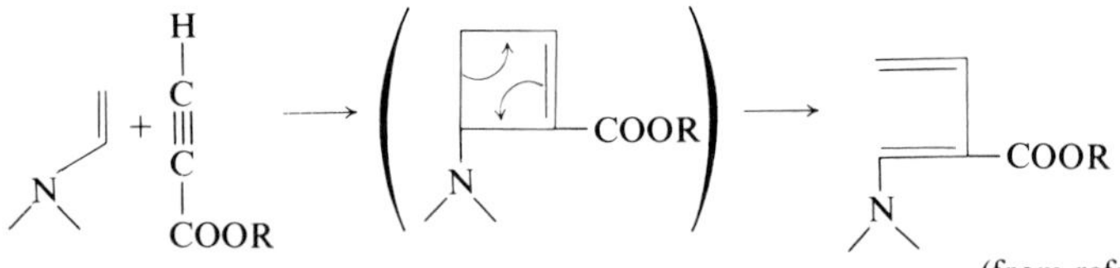

(135)

(from refs. 192, 193)

In some reactions the first adduct, which presumably is a zwitterion and precedes the cyclobutene, can be trapped (194):

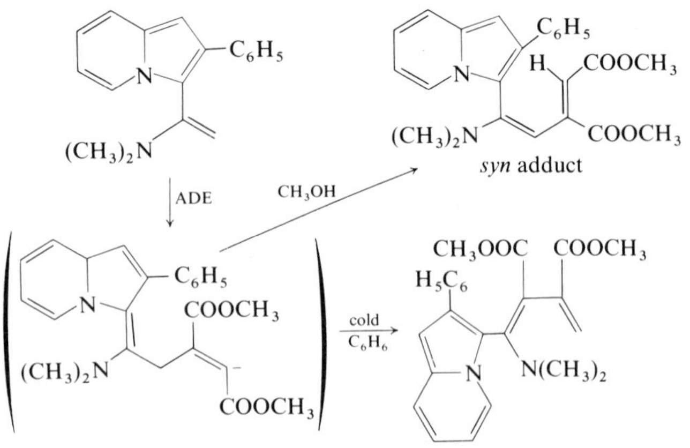

syn adduct

(136)

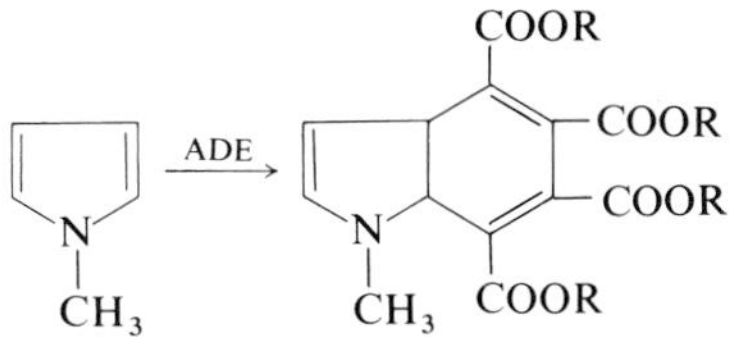

(137)

(from ref. 195)

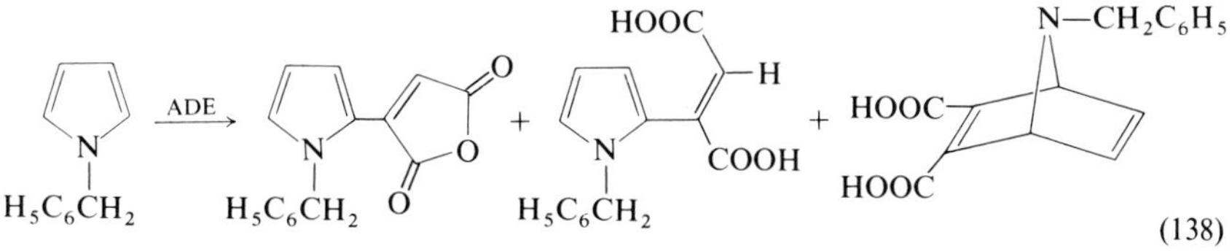

(138)

(from ref. 191)

The presence of a cyclobutene in at least some of these reactions has been demonstrated by its isolation (196). However, its subsequent fate may vary from that given in eqs. 137 or 138 as in eq. 139 (192).

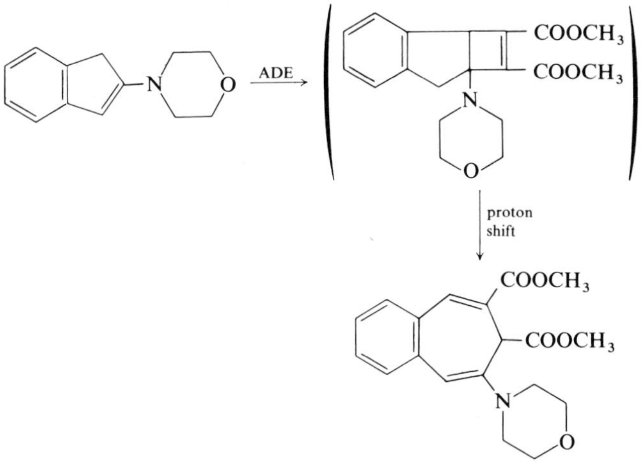

(139)

B. Formation of Carbon–Nitrogen Bonds

Following Reppe, the usual conditions for adding primary and secondary amines to simple alkyl acetylenes involve elevated temperatures, pressure, and a catalyst, for example, zinc or cadmium acetate and cuprous chloride (5,191,197).

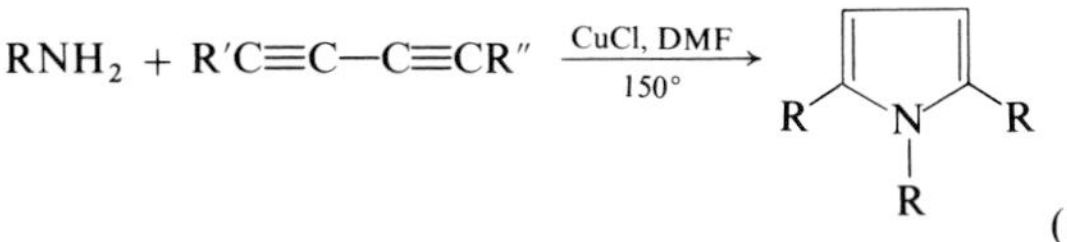

$$RNH_2 + R'C{\equiv}C{-}C{\equiv}CR'' \xrightarrow[150°]{CuCl,\ DMF}$$ (140)

(R=H, Ar, Alk)
(from ref. 198)

Such conditions seem more complex than in process 1, and the nucleophilic component, if any, is not obvious. By contrast nucleophilic additions to activated alkynes can be quite facile.

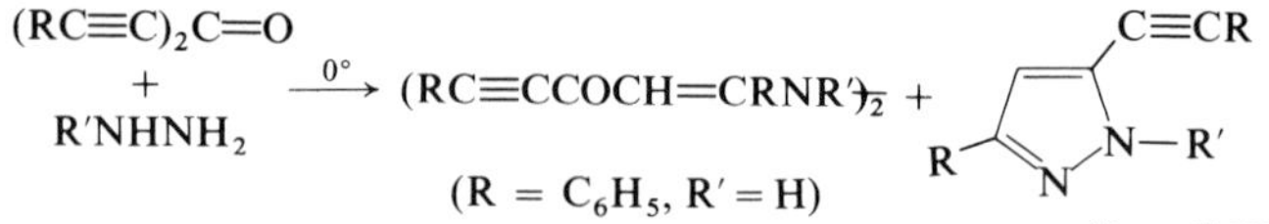

$$(RC{\equiv}C)_2C{=}O + R'NHNH_2 \xrightarrow{0^\circ} (RC{\equiv}CCOCH{=}CRNR'\!\!)_2 + \text{pyrazole (3-R, 5-C}{\equiv}\text{CR, N-R')} \qquad (141)$$

$(R = C_6H_5, R' = H)$

$(R = C_6H_5, n\text{-}C_4H_9)$
$(R' = H, Ar)$
(from ref. 86)

$$RC{\equiv}CC_6H_5 \xrightarrow{NaNHNH_2} C_6H_5CH_2{-}CR{=}N{-}NH_2 \longrightarrow (C_6H_5CH_2{-}CR{=}N\!\!)_2 \qquad (142)$$

$(R = \text{alkyl})$
(from ref. 199)

1. *Primary and Secondary Nitrogen*

Unlike the alkylalkynes (8), the examples we give here generally involve activated alkynes. The exclusive *syn* adduct in 143 and the small amount of

$$F_3CC{\equiv}CH \xrightarrow{(CH_3)_2NH} trans\text{-}CF_3CH{=}CHN(CH_3)_2 \qquad (143)$$

(from ref. 200)

$$F_3CC{\equiv}CCF_3 \xrightarrow{(CH_3)_2NH} (F_3C)(H)C{=}C(N(CH_3)_2)(CF_3) + (F_3C)(H)C{=}C(CF_3)(N(CH_3)_2) \qquad (144)$$

6 parts; 1 part

(from ref. 200)

syn adduct in 144 were attributed to postisomerization, presumably by nucleophilic catalysis (eq. 44). In eqs. 145 and 146 the eneyne product was often a *cis–trans* mixture (127). As mentioned earlier (eq. 48), the *anti*- product of eq. 146 isomerizes on distillation (127).

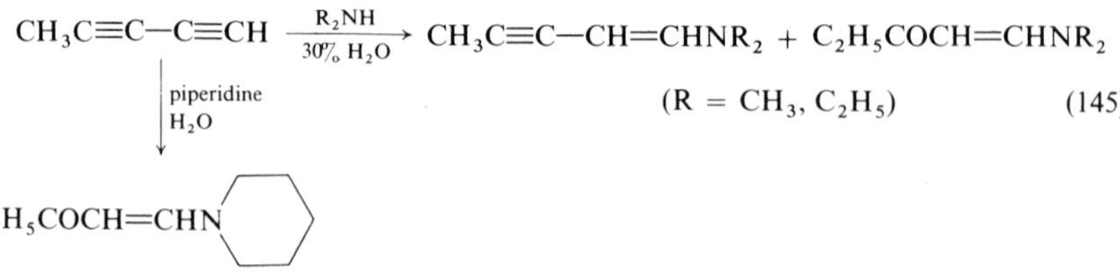

$$CH_3C{\equiv}C{-}C{\equiv}CH \xrightarrow[30\%\ H_2O]{R_2NH} CH_3C{\equiv}C{-}CH{=}CHNR_2 + C_2H_5COCH{=}CHNR_2 \qquad (145)$$

$(R = CH_3, C_2H_5)$

$$CH_3C{\equiv}C{-}C{\equiv}CH \xrightarrow{\text{piperidine}, H_2O} C_2H_5COCH{=}CHN(CH_2)_5$$

$$(C_2H_5)_2NCH_2{-}C{\equiv}C{-}C{\equiv}CH \xrightarrow[50\%\ H_2O]{(C_2H_5)_2NH} cis\text{-}(C_2H_5)_2NCH_2C{\equiv}C{-}CH{=}CHN(C_2H_5)_2 \qquad (146)$$

The presence of displaceable substituents on the triple bond introduces familiar mechanistic complications (see eqs. 38, 99, 102, 104, 105).

$$HC{\equiv}COC_2H_5 + H_2NCH_2CH_2OH \longrightarrow \text{(2-methyl-2-oxazoline, } H_3C\text{–)} \quad (147)$$

(from ref. 125)

$$HC{\equiv}CCl + C_6H_5NH_2 \longrightarrow \text{indole (N–H)} \quad (148)$$

(from ref. 202)

$$ClC{\equiv}CCl + (C_2H_5)_2NH \longrightarrow [(C_2H_5)_2N]_2C{=}CHCl \quad (149)$$

(from ref. 202)

$$(C_2H_5OOC)_2C(C_2H_5)C{\equiv}CCl \xrightarrow{(C_2H_5)_2NH} (C_2H_5OOC)_2C(C_2H_5){-}C(N(C_2H_5)_2){=}CHN(C_2H_5)_2 \quad (150)$$

(from ref. 202)

$$C_6H_5C{\equiv}COCH_3 \xrightarrow{NH_2^-} C_6H_5CH_2CN \quad (151)$$

(from ref. 203)

Clearly, addition to an alkyne must occur at some stage in the mechanism of such reactions.

Numerous additions to activated alkynes have been reported (3,6,8). We can give only a few recent examples here. In eqs. 151a and 152 the type of heterocycle obtained appears to depend on the reaction conditions

$$\text{R–C}_6\text{H}_3(\text{COC}_6\text{H}_5)(\text{NH}_2) \xrightarrow[\text{reflux, } CH_3OH]{\text{ADE}} \text{R-substituted 4-}C_6H_5\text{-quinoline-2,3-(COOCH}_3)_2 \quad (151a)$$

$(R = H, Cl, Br, NO_2)$

(from ref. 229)

$$C_6H_5COCH(R')NHR \xrightarrow{\text{ADE}} H_5C_6C(=O){-}CH(R'){-}N(R){-}C(COOCH_3){=}CH(COOCH_3) \longrightarrow \text{pyrrole: 2-}H_5C_6\text{, 3-R}'\text{, N-R, 4,5-(COOCH}_3)_2 \quad (152)$$

$(R = H, Ar;\ R' = H, CH_3)$

(from ref. 205)

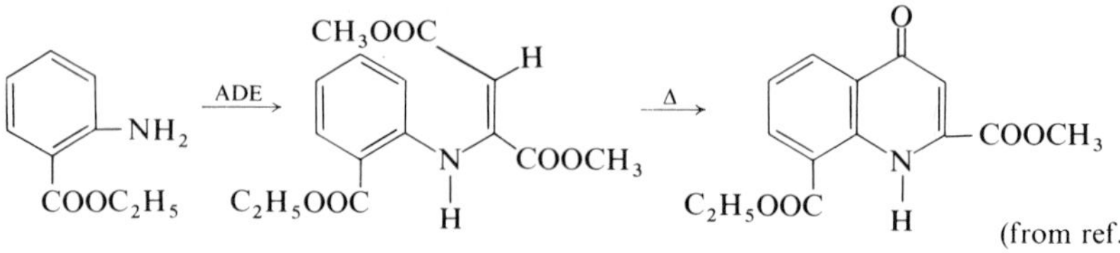

(153)

(from ref. 205)

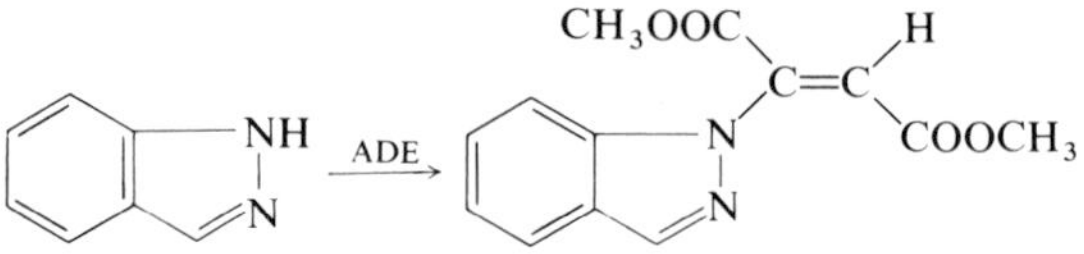

(154)

(from ref. 206)

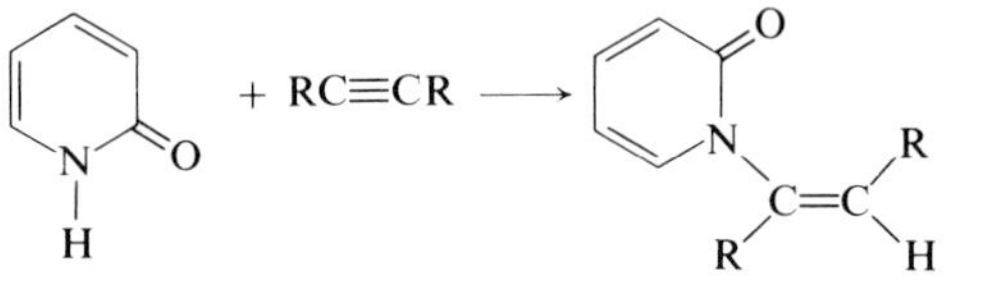

(155)

($R = CF_3, COOCH_3$)

(from ref. 207)

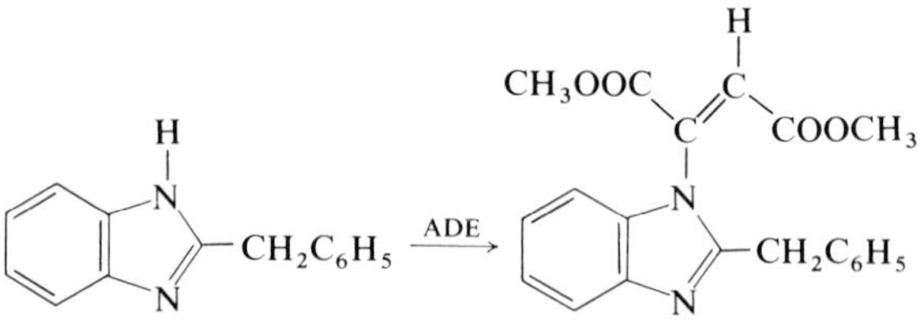

(156)

(from ref. 208)

We turn now to the fascinating and controversial area of the stereoselectivity of additions to activated alkynes. An account of the stereochemical course of amine additions becomes greatly simplified, if one is adequately armed with hindsight. Any chronological approach is roughly equivalent to working through a complex puzzle in which pieces are supplied at random over an extended period. Therefore we summarize the key observations first for a process of the general type of eq. 157, in which the activating R'' group may be $COOR'''$, COR''', $CONR_2'''$, CN, SO_2R'''. The question of

$$R_2NH + R'C{\equiv}CR'' \longrightarrow R_2NC(R'){=}CHR'' \qquad (157)$$

stereoselectivity in eq. 157 hinges on kinetic product control. Although several research groups had recognized this as a problem (3,209,210), only recently has it been discovered that trace catalysts such as acid, on the surfaces of the glass reaction vessels or from the atmosphere, or impurities in the reagents could affect the *anti* to *syn* product ratio (124). Product differences arising from kinetic versus equilibrium control are given in Table 5.

TABLE 5
Per Cent of *Syn* Adduct in Amine Additions to Acetylene Carboxylic Esters (0.5 *M*, 0 ~ 25°), According to Kinetic and Equilibrium Control (124)[a]

Amine	Solvent	Kinetic control, %	Equilibrium control, %
$HC\equiv CCOOCH_3$			
n-$C_4H_9NH_2$	C_6H_6	70	20
i-$C_3H_7NH_2$	CCl_4	70	20
$C_6H_{11}NH_2$	CCl_4	75	14
$C_6H_{11}NH_2$	C_6H_6	75	12
$C_6H_{11}NH_2$	C_6H_{12}	70	—
t-$C_4H_9NH_2$	C_6H_6	70	20
Piperidine	C_6H_6	90	100
$CH_3OOCC\equiv CCOOCH_3$			
$C_6H_{11}NH_2$	C_6H_6	70	100
$C_6H_{11}NH_2$[b]	C_6H_6	60	100
$C_6H_{11}NH_2$[c]	C_6H_6	80	100
$C_6H_5NH_2$	C_6H_6	80	100
Piperidine	C_6H_6	90	100

[a] Some variation from these figures may be found, if the reaction conditions are changed (136).
[b] The ester was added dropwise to cyclohexylamine (1:1) at 15°.
[c] The amine was added dropwise to the ester (1:1) at 15°.

Of the several mechanisms for acid-catalyzed isomerization, addition elimination and elimination addition (eqs. 36, 37) have been discussed. For the case at hand eqs. 158 and 159 do not seem to be important, since amine

$$\begin{matrix} RNH & & COOCH_3 \\ & C{=}C & \\ H & & H \end{matrix} + RNH_2 \underset{}{\overset{H^+}{\rightleftharpoons}} (RNH)_2CH{-}CH_2COOCH_3 \overset{H^+}{\rightleftharpoons} \begin{matrix} RNH & & H \\ & C{=}C & \\ H & & COOCH_3 \end{matrix} \tag{158}$$

$$\begin{matrix} RNH & & COOCH_3 \\ & C{=}C & \\ H & & H \end{matrix} \overset{H^+}{\rightleftharpoons} RNH_3^+ + HC\equiv CCOOCH_3 \overset{H^+}{\rightleftharpoons} \begin{matrix} RNH & & H \\ & C{=}C & \\ H & & COOCH_3 \end{matrix} \tag{159}$$

exchange between the adducts and other amines is not observed (211). On the other hand, acid-catalyzed eneamine isomerization (eq. 45) is facile (211).

In the absence of acid catalysis, thermal activation might become a factor in the isomerization process, but it does seem less important.

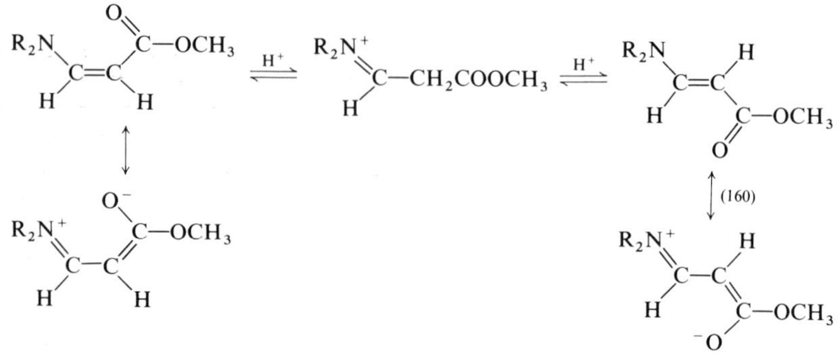

$$\begin{array}{ccccc}
(R_2N)HC{=}CH{-}C({=}O){-}OCH_3 & \overset{H^+}{\rightleftharpoons} & R_2N^+{=}CH{-}CH_2COOCH_3 & \overset{H^+}{\rightleftharpoons} & (R_2N)HC{=}CH{-}C({=}O){-}OCH_3 \\
\updownarrow & & & & \updownarrow \\
R_2N^+{=}CH{-}CH{=}C(O^-){-}OCH_3 & & & & R_2N^+{=}CH{-}CH{=}C(O^-){-}OCH_3
\end{array} \qquad (160)$$

There is an interesting consequence of eq. 160, namely that aziridine adducts should be relatively stable. This, of course, follows from the fact that the immonium structure,

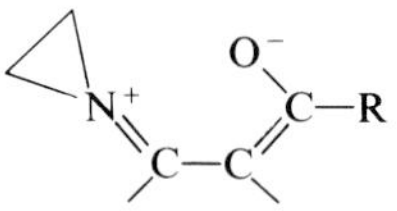

$$(CH_2)_2N^+{=}C{-}C{=}C(O^-){-}R$$

would be of high energy (3,209). Aziridine adducts do in fact conform nicely to this prediction, being among the most resistant to isomerization.

In order that the limit of equilibrium product control be defined, we cite typical data (211):

Syn adducts		*Anti* adducts	
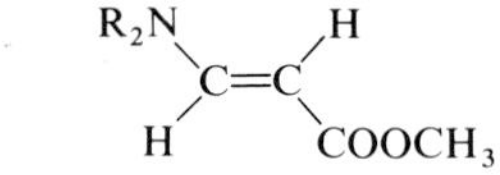 $R_2N(H)C{=}C(H)COOCH_3$	more stable	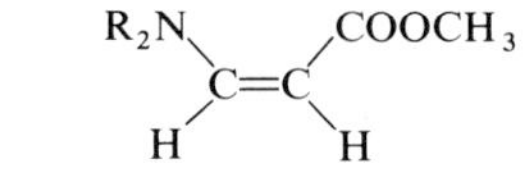 $R_2N(H)C{=}C(H)COOCH_3$	less stable
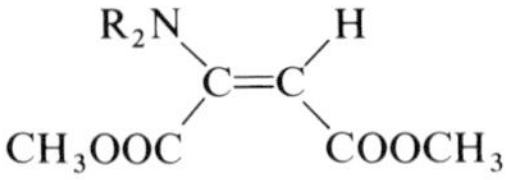 $R_2N(CH_3OOC)C{=}C(H)COOCH_3$	more stable	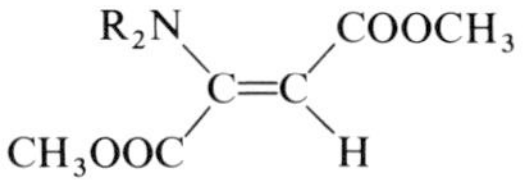 $R_2N(CH_3OOC)C{=}C(H)COOCH_3$	less stable
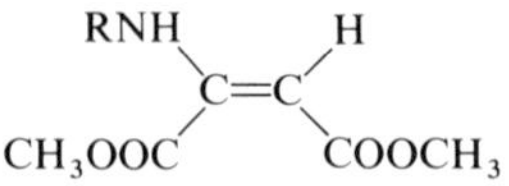 $RNH(CH_3OOC)C{=}C(H)COOCH_3$	less stable	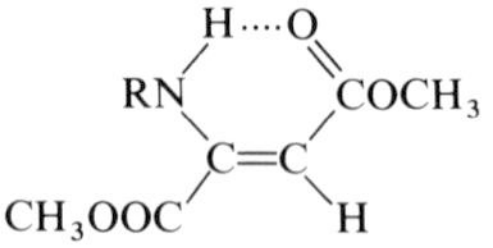 $RN{-}H{\cdots}O{=}C(OCH_3)$, $(CH_3OOC)C{=}C(H)$	more stable
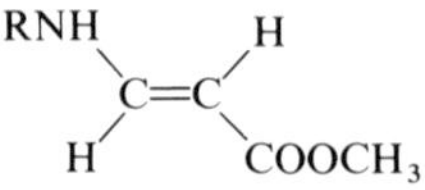 $RNH(H)C{=}C(H)COOCH_3$	less stable	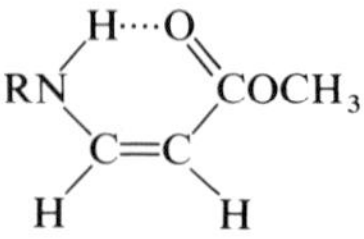 $RN{-}H{\cdots}O{=}C(OCH_3)$, $(H)C{=}C(H)$	more stable

These relative stabilities seem to carry over to related systems, e.g., (209).

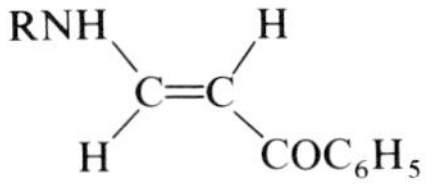

RNH(H)C=C(H)COC_6H_5 — less stable; RN—H····O=C(C_6H_5)—C(H)=C(H) — more stable

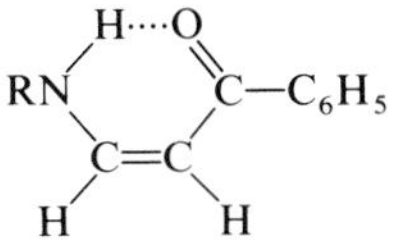

Note that the preceding stabilities indicate that the steric (bulk) requirements of dialkylamino (R_2N—) exceed those carboalkoxy (CH_3OOC—), but that the possibility for intramolecular hydrogen bonding overcomes an unfavorable steric factor. In what appears to be a more qualitative study, McMullen and Stirling find that the proportions of *syn* adduct $RCH{=}CHNH(CH_2C_6H_5)$ at equilibrium falls when R changes (209): $C_6H_5CO > C_6H_5N(CH_3)CO > CH_3OOC > C_6H_5CH_2SO > C_6H_5SO_2 > C_6H_5CH_2SO_2$.

Superimposed on these factors is the fact that the equilibrium ratio could change with solvent. It has been suggested, for example, that inter- versus intramolecular hydrogen bonding could become a factor. With respect to the adducts just discussed, the *syn* adduct is favored by a change in solvent from chloroform to DMSO (209). The equilibrium factor has some

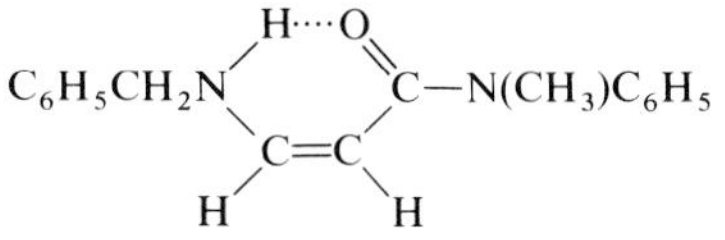

$C_6H_5CH_2N$—H····O=C($N(CH_3)C_6H_5$)—C(H)=C(H) — *anti* adduct; $O{=}S(CH_3)_2$····H—$N(CH_2C_6H_5)$—C(H)=C(H)$CON(CH_3)C_6H_5$ — *syn* adduct

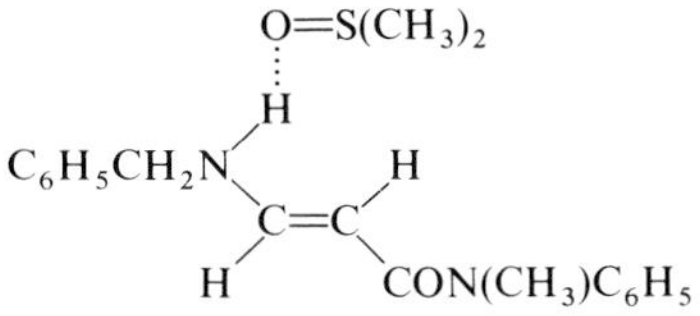

interesting consequences. Iwanami determined that the following process was *anti*-stereospecific (212):

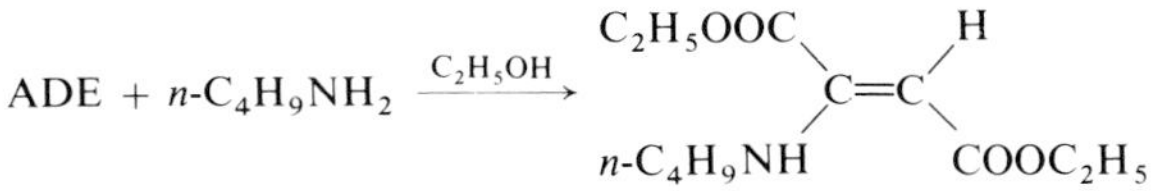

$$\text{ADE} + n\text{-}C_4H_9NH_2 \xrightarrow{C_2H_5OH} (C_2H_5OOC)(n\text{-}C_4H_9NH)C{=}C(H)(COOC_2H_5) \quad (161)$$

In view of the availability of external protons and the greater stability of the *anti* adduct in this case it is not surprising that the *anti* addition rule was confirmed. Winterfeldt and Preuss determined that the following process was often *syn*-stereospecific (213):

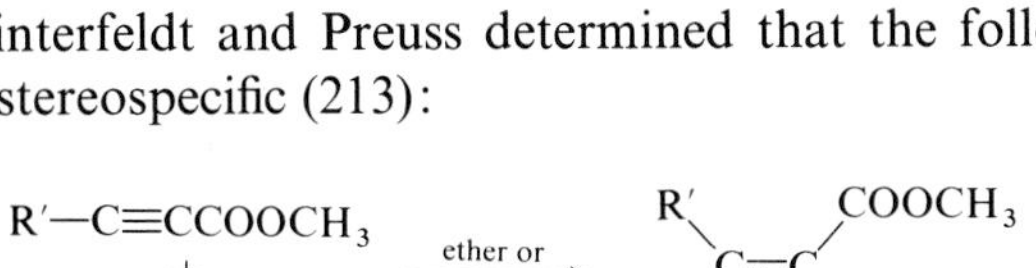

$$R'{-}C{\equiv}CCOOCH_3 + R_2NH \xrightarrow[t\text{-}C_4H_9OH]{\text{ether or}} (R')(R_2N)C{=}C(COOCH_3)(H) \quad (162)$$

In view of the availability of internal protons and the greater stability of the *syn* adduct in this case it is not surprising that the *anti* rule was "violated."

The second major factor appears to depend greatly on the availability of protons in the solvent (see Section III-D) (214). In **14** and **15** we distinguish

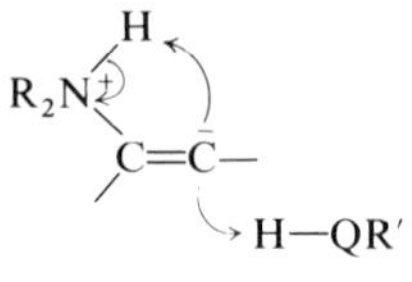

14

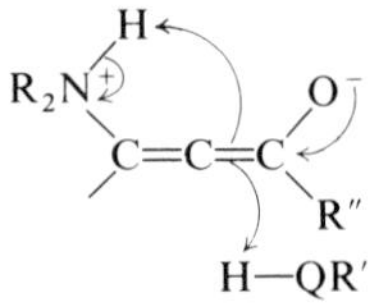

15

between so-called "internal" and "external" delivery of the proton, alternatives which lead to *syn* or *anti* adducts, respectively. That this competition is important is illustrated in eq. 163 and more fully in Tables 6 and 7.

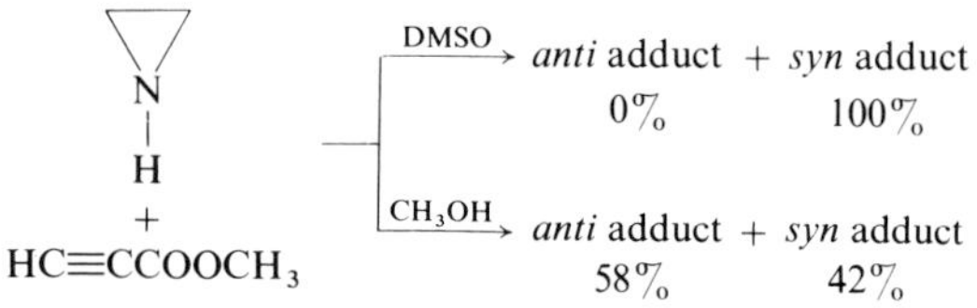

$$\text{aziridine (N–H)} + HC{\equiv}CCOOCH_3 \begin{cases} \xrightarrow{\text{DMSO}} \textit{anti}\text{ adduct } (0\%) + \textit{syn}\text{ adduct } (100\%) \\ \xrightarrow{CH_3OH} \textit{anti}\text{ adduct } (58\%) + \textit{syn}\text{ adduct } (42\%) \end{cases} \tag{163}$$

In proton solvents or in high concentrations of amine, external delivery becomes important. Thus the proportions of *anti* adducts increase with the acidity of the proton medium (Table 7). Likewise, the proportions of *anti* adduct are greater when the alkyne is added slowly to an excess of the amine, rather than when the amine is added to the alkyne. In aprotic solvents, both primary and secondary amines and ADE give *syn* adducts predominantly, and the transfer appears to be intramolecular. This was fully confirmed in an elegant kinetic study (Table 3) in which the concentrations of amine were reduced until exclusive *syn* addition was demonstrated (153).

With this background we can see why there are discrepancies in the selectivities reported by different research groups. First, one should be careful about deducing product structures from reports in which the problem of equilibrium versus kinetic control either was not recognized or was not dealt with (63,153). Second, it is unwise to assign a structure for an amine adduct on the basis of the *anti* rule, as has sometimes been done. Third, one should be cautious about drawing mechanistic conclusions from systems in which the more stable isomer has been formed, whether or not the *anti* rule may have been violated. On this basis, apparent discrepancies in the product proportions, e.g., in aziridine additions (136,210), or apparent violations of the *anti* rule (124,136,210–217), e.g., the pyrazole addition to ADE or methyl propiolate (136), may be regarded as explained.

TABLE 6
Per Cent of *Syn* Adduct in the Kinetically Controlled Addition of Primary and Secondary Amines to Dimethylacetylene Dicarboxylate (ADE) at 0° (136)

Amine	Dioxane	Benzene	Methanol
$(n\text{-}C_4H_9)_2NH$	95	88	30
Aziridine	95	87	34
$(C_2H_5)_2NH$	93	89	38
$(CH_3)_2NH$	90	83	31
Piperidine	85	81	38
$C_6H_5NHCH_3$	85	78	38
$C_6H_5CH_2NH_2$	87	66	35
$C_6H_{11}NH_2$	82	66	30
$n\text{-}C_4H_9NH_2$	79	61	40
Pyrazole (90°)	75		27
Imidazole (50°)	35		19

TABLE 7
Additions of Aziridine to Two Alkynes in Several Solvents. The Per Cent of *Syn* Adduct is Given (136)[a]

Solvent	$HC{\equiv}CCOOCH_3$ (at 25°)	$CH_3OOCC{\equiv}CCOOCH_3$ (at 0°)
$HCON(CH_3)_2$	97	97
$(CH_3)_2SO$	96	95[c]
Dioxane	95	95
$(C_2H_5)_2O$	92[b]	
CH_3CN	92	
C_6H_6	86	87
CCl_4	85	
C_6H_{12}	85	83
$n\text{-}C_4H_9OH$	66	
$n\text{-}C_3H_7OH$	58	52
C_2H_5OH	53	
CH_3OH	47	34[d]
Aziridine	67	
C_2H_5SH	24	

[a] Equal volumes of 4*M* solutions were mixed and analyzed by NMR at 35°.
[b] Ref. 213 gives 88%.
[c] Value from ref. 214.
[d] Ref. 214 gives 33%.

2. *Cyclizations Involving Nitrogen Compounds*

Intramolecular attack appears to be particularly favorable for closure (279).

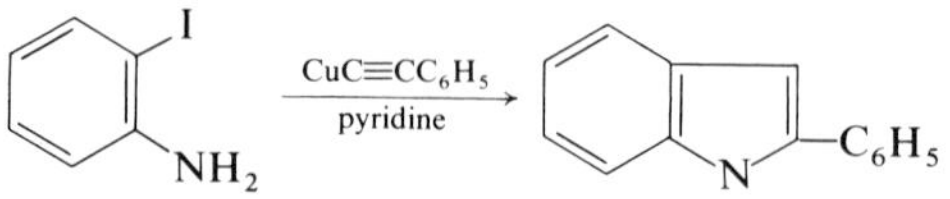

$$o\text{-}IC_6H_4NH_2 \xrightarrow[\text{pyridine}]{CuC{\equiv}CC_6H_5} \text{2-}C_6H_5\text{-indole} \quad (164)$$

It turns out, however, that the reaction indicated is far from simple, since *o*-aminophenylphenylacetylene does not cyclize under the reaction conditions (including copper and cuprous iodide). One must presume that the cyclization takes place in the copper complex or that the amine addition precedes coupling.

If one or the other reagent is polyfunctional in eq. 1, the overall process

16 (Z = NH, O) **17** (Z = NH) **18** (Z = NH, O, S) **19** 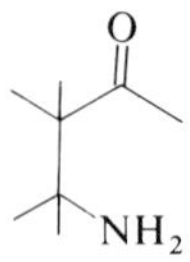 **20**

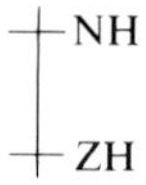

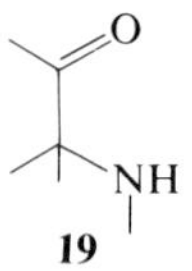

can lead to a cyclic compound. In **16–20** we can give some of the nitrogen nucleophiles which have been used in this way. (Nucleophile centers involving sulfur will be treated later.) Polyfunctional nucleophiles have been mentioned, of course, previously (e.g., eqs. 133–141, 147, 148, 152, and 153) without special comment. Because of their great importance in nitrogen chemistry we single them out here. Some of the older literature has been reviewed (3,6) or simply cited (86).

Hydrazine additions are of interest because they may lead to pyrazoles [e.g., eqs. 141 and 165 (19,86,230,316)]:

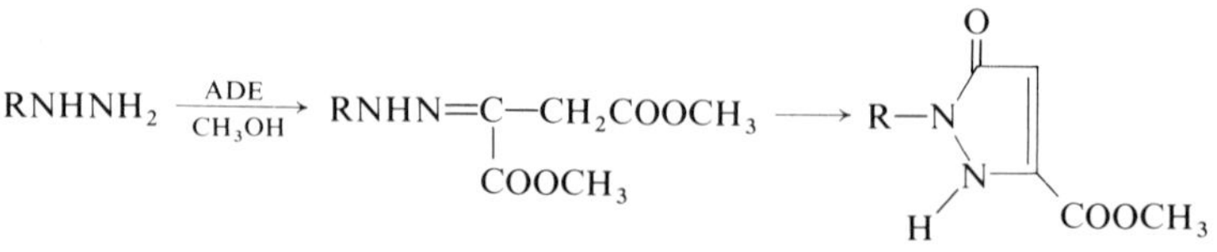

$$RNHNH_2 \xrightarrow[CH_3OH]{ADE} RNHN{=}C(COOCH_3){-}CH_2COOCH_3 \longrightarrow \text{R-N(pyrazolone, N-H, COOCH}_3) \quad (165)$$

Hydroxylamine can produce isoxazoles.

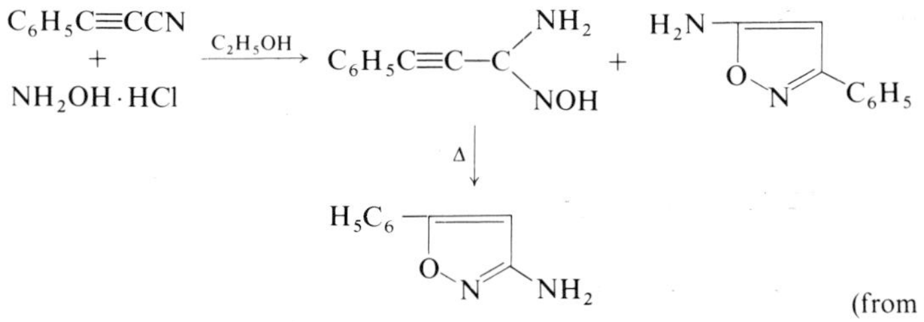

$$C_6H_5C{\equiv}CCN + NH_2OH \cdot HCl \xrightarrow{C_2H_5OH} C_6H_5C{\equiv}C{-}C(NH_2){=}NOH + \text{3-phenyl-5-aminoisoxazole } (H_2N, C_6H_5)$$

$$\xrightarrow{\Delta} \text{5-phenyl-3-aminoisoxazole } (H_5C_6, NH_2) \quad (166)$$

(from ref. 247)

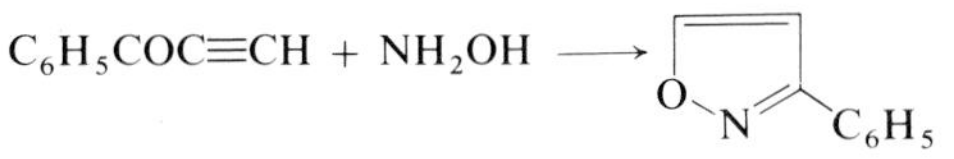

$$C_6H_5COC{\equiv}CH + NH_2OH \longrightarrow \text{3-phenylisoxazole } (C_6H_5) \quad (166a)$$

(from ref. 236)

Unlike the molecular additions of many covalent azides, eq. 167 appears to represent a genuine nucleophilic process.

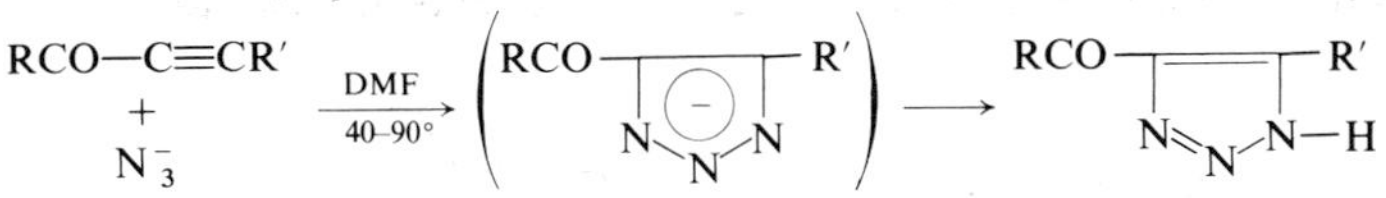

$$RCO{-}C{\equiv}CR' + N_3^- \xrightarrow[40\text{–}90°]{DMF} \left(\text{triazolide anion } (RCO, R')\right) \longrightarrow \text{1,2,3-triazole } (RCO, R', N{-}H) \quad (167)$$

($R' = H, C_6H_5$)

(from ref. 233)

Other diamine reactions are straightforward.

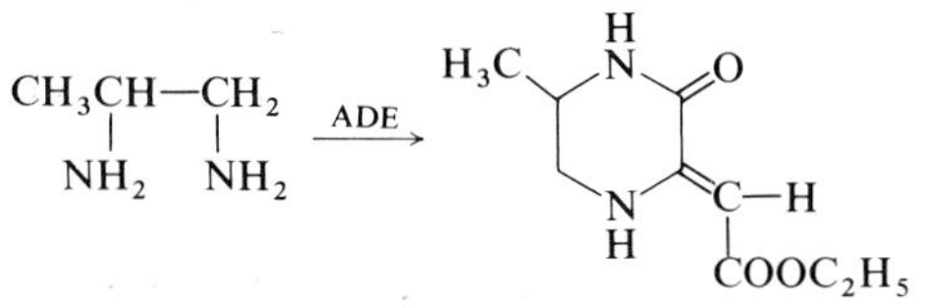

$$CH_3CH(NH_2){-}CH_2NH_2 \xrightarrow{ADE} \text{piperazinone } (H_3C, O, {=}C(H)COOC_2H_5) \quad (168)$$

(from ref. 231)

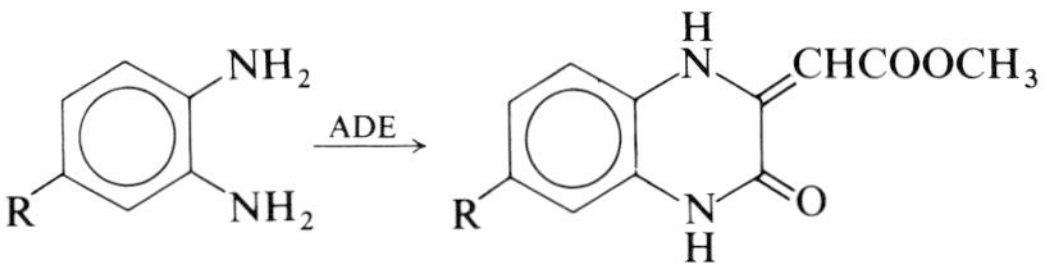

$$\text{R-}o\text{-phenylenediamine} \xrightarrow{ADE} \text{quinoxalinone } ({=}CHCOOCH_3, {=}O, R) \quad (169)$$

($R = H, NO_2$)

(from ref. 232)

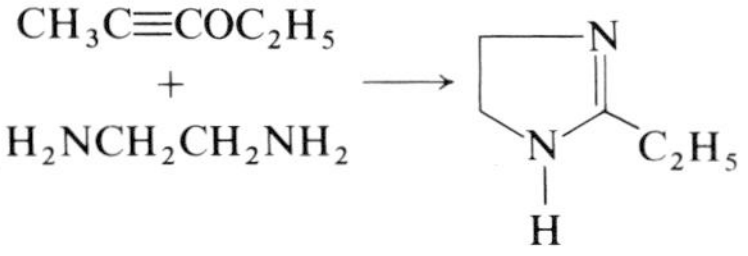

$$CH_3C{\equiv}COC_2H_5 + H_2NCH_2CH_2NH_2 \longrightarrow \text{2-ethyl-2-imidazoline } (C_2H_5, N{-}H) \quad (170)$$

(from ref. 125)

Syntheses with nucleophiles **17** produce various cycles. In eq. 171 we have competition

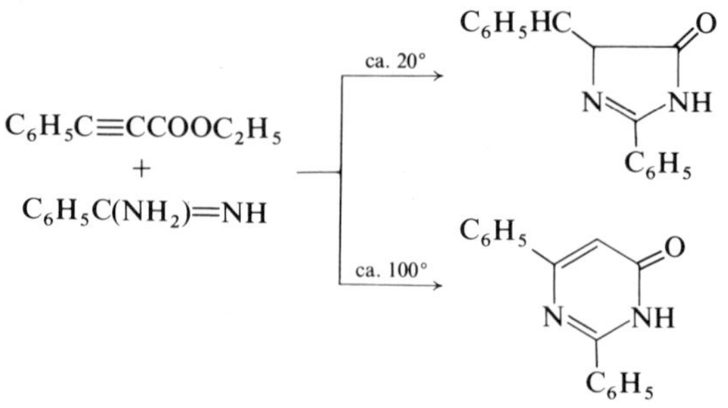

$$C_6H_5C{\equiv}CCOOC_2H_5 + C_6H_5C(NH_2){=}NH \xrightarrow{\text{ca. }20^\circ} \text{(imidazolinone: } C_6H_5HC{=}, \text{ C=O, NH, N, } C_6H_5) \qquad \xrightarrow{\text{ca. }100^\circ} \text{(pyrimidinone: } C_6H_5, \text{ C=O, NH, N, } C_6H_5) \tag{171}$$

(from ref. 234)

between closures to the five- and six-membered rings. Sasaki et al. have found that the semicarbazide and substituted guanidines add readily to ADE, but hippuric acid esters, α-amino acids, and amides fail to add (235).

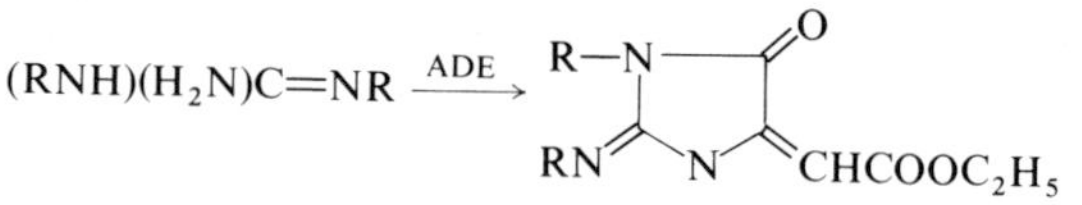

$$(RNH)(H_2N)C{=}NR \xrightarrow{\text{ADE}} \text{(ring: R–N, C=O, RN=C, N, =CHCOOC}_2\text{H}_5) \tag{172}$$

$(R = H, C_6H_5)$

$$H_2NCONHNH_2 \xrightarrow{\text{ADE}} H_2NCONH{=}C(COOC_2H_5)(CH_2COOC_2H_5) \tag{173}$$

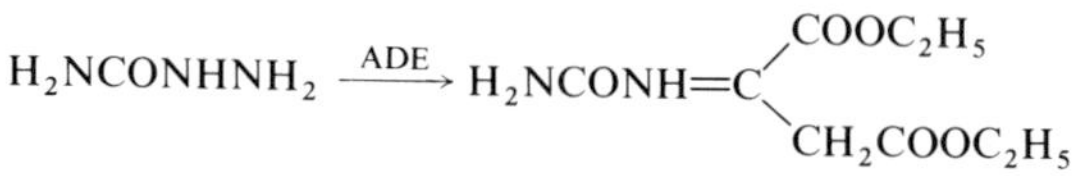

Note the open chain product in eq. 173. As we shall see in connection with

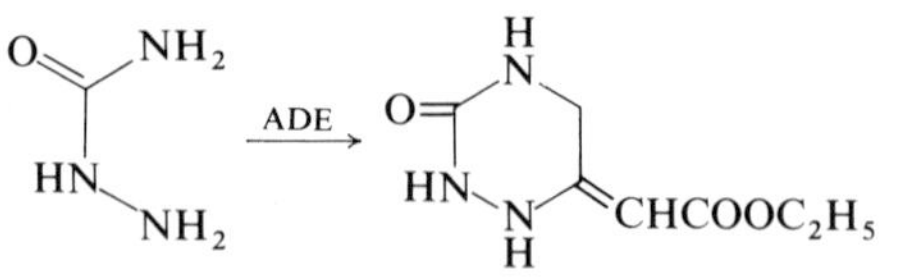

$$O{=}C(NH_2)NHNH_2 \xrightarrow{\text{ADE}} \text{(ring: O=C, HN, N–H, N–H, =CHCOOC}_2\text{H}_5) \tag{174}$$

the thio analogs (eq. 254), the structures of the products of eq. 174 may have to be revised (230).

Syntheses of pyrroles with nucleophiles, **19**, have been given in eq. 152. The general procedure has been discussed (90); further examples are appended.

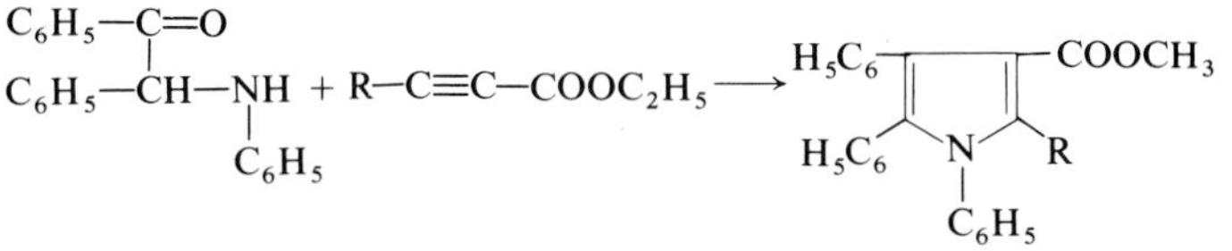

C_6H_5—C=O
C_6H_5—CH—NH(C_6H_5) + R—C≡C—$COOC_2H_5$ ⟶ [pyrrole: H_5C_6, H_5C_6, $COOCH_3$, R, N—C_6H_5] (175)

(R = H, $COOC_2H_5$)

(from ref. 246)

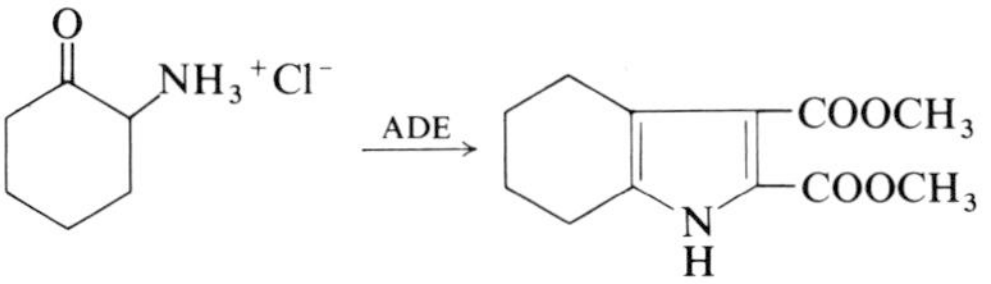

[2-oxocyclohexyl]$NH_3^+Cl^-$ —ADE→ [tetrahydroindole: $COOCH_3$, $COOCH_3$, N—H] (176)

(from ref. 237)

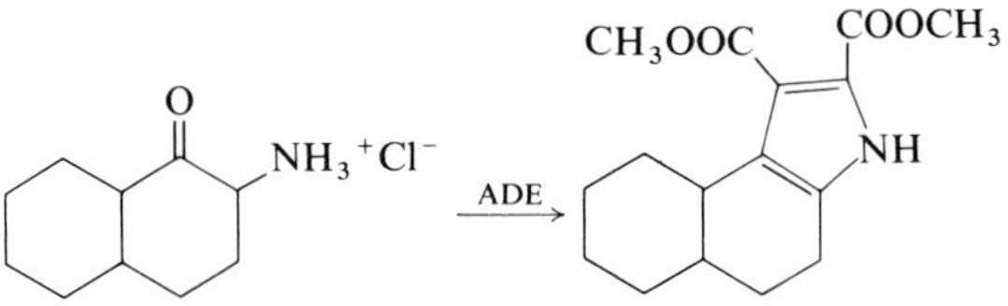

[oxodecalinyl]$NH_3^+Cl^-$ —ADE→ [CH_3OOC, $COOCH_3$, NH] (177)

(from ref. 237)

If another carbon is inserted as in **20**, six-membered cycles may be obtained as in eq. 151a (229). The first adducts are often isolable. Variations at the carbonyl function can lead to seven- or five-membered rings (eqs. 179, 180).

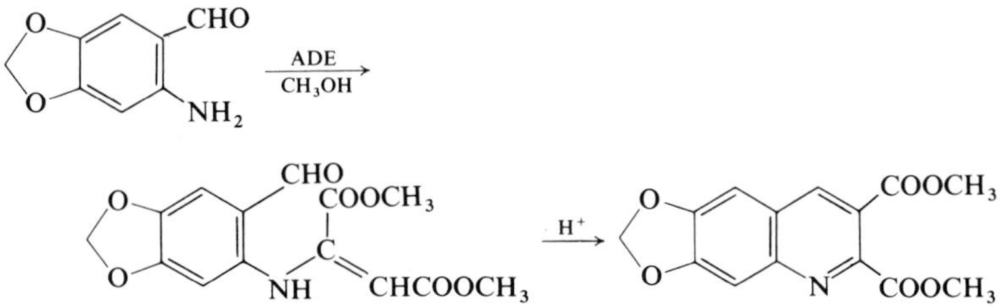

[methylenedioxy: CHO, NH_2] —ADE, CH_3OH→

[methylenedioxy: CHO, NH—C($COOCH_3$)=$CHCOOCH_3$] —H^+→ [quinoline: $COOCH_3$, $COOCH_3$, N] (178)

(from ref. 90)

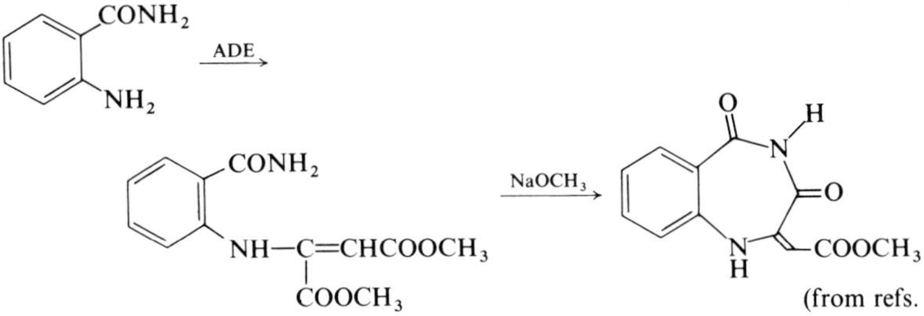

[benzene: $CONH_2$, NH_2] —ADE→

[benzene: $CONH_2$, NH—C($COOCH_3$)=$CHCOOCH_3$] —$NaOCH_3$→ [O, N—H, =O, N—H, $COOCH_3$] (179)

(from refs. 232, 239)

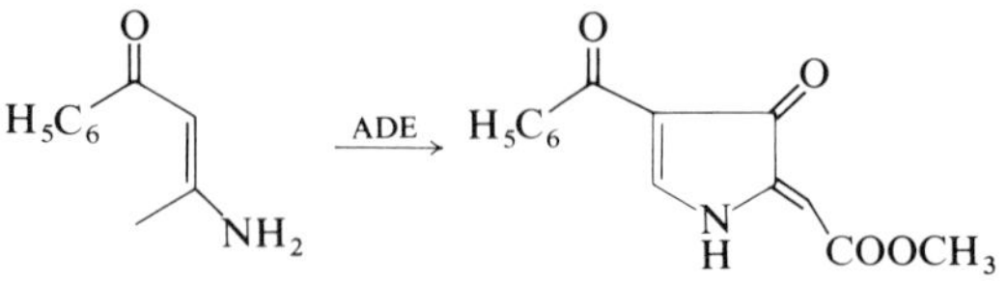

$$H_5C_6C(=O)CH=C(CH_3)NH_2 \xrightarrow{ADE} \text{(pyrrolinone, } H_5C_6CO\text{-substituted, =CHCOOCH}_3\text{)} \quad (180)$$

(from ref. 90)

3. *Tertiary Nitrogen Compounds* (*3,9*)

Simple zwitterion adducts of tertiary nitrogen and alkynes have not been isolated.

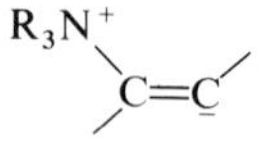

$$R_3N^+\text{—}C{=}\bar{C}$$

21

The 1,3-dipole in eq. 181 may have been formed but was not the isolable

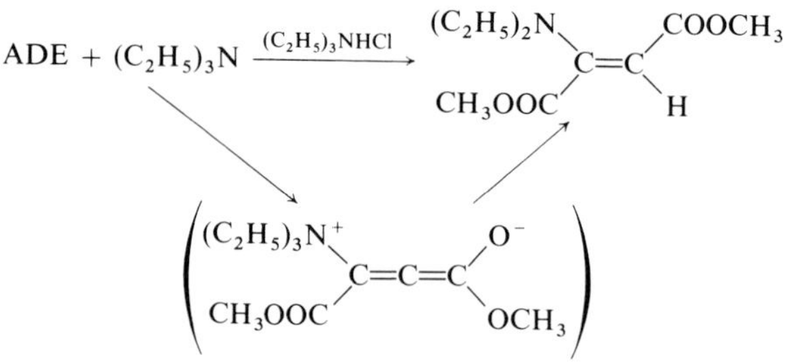

$$ADE + (C_2H_5)_3N \xrightarrow{(C_2H_5)_3NHCl} (C_2H_5)_2N(CH_3OOC)C{=}C(COOCH_3)H$$

$$\left((C_2H_5)_3N^+(CH_3OOC)C{=}C{=}C(O^-)OCH_3\right) \quad (181)$$

product (218). Nevertheless, this kind of zwitterion (1,3-dipole) appears to be the key intermediate in a variety of reactions in which special structural features of the alkyne and amine facilitate further reaction. The most important characteristic here seems to be that the alkyne have an activating substituent.

Evidence for the intermediate **21** comes from one-step trapping experiments with proton donors (see eq. 18) as well as with other reagents:

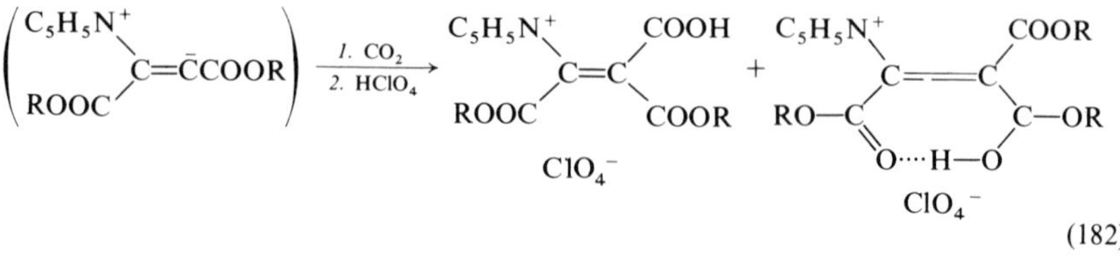

$$\left(C_5H_5N^+(ROOC)C{=}\bar{C}COOR\right) \xrightarrow[2.\ HClO_4]{1.\ CO_2} C_5H_5N^+(ROOC)C{=}C(COOH)COOR\ ClO_4^- + C_5H_5N^+(RO\text{—}C{=}O\cdots H\text{—}O\text{—}C\text{—}OR)C{=}C(COOR)\ ClO_4^- \quad (182)$$

(from ref. 219)

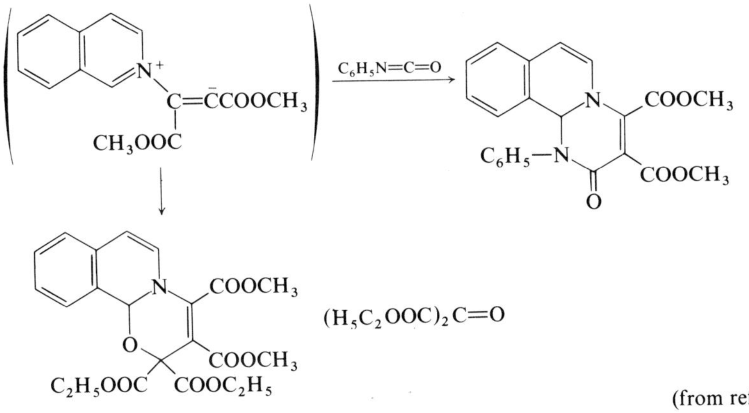

(183)

(from ref. 220)

Often as not the trapping is intramolecular.

Apart from the relatively simple processes of 181–183, the zwitterion may enter into several reaction sequences, depending on the participating reactants. The number and variety of products that can be prepared just from ADE, for example, are truly remarkable (9).

The reactions of substituted pyridines with ADE appear to depend on the basicity of the pyridine (221). It has been suggested that 2-chloro, 2-bromo, or 2-cyano-pyridine do not give the usual adduct **22** because they are insufficiently nucleophilic. 3-Cyanopyridine gives a low yield of the expected quinolizine **23** and 4-cyanopyridine gives the indolizine **24**, possibly

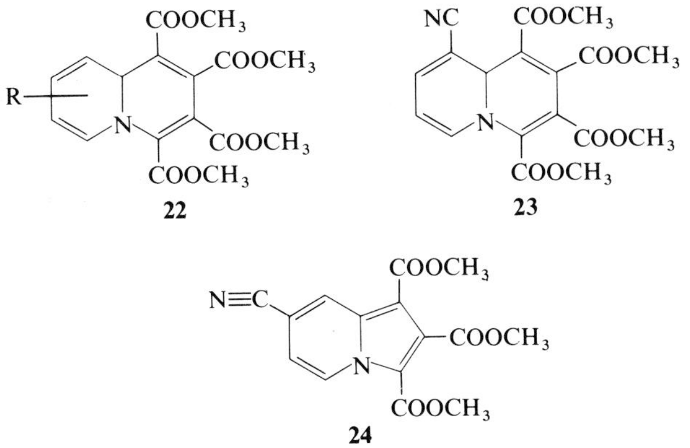

22 **23**

24

as outlined in Acheson (9). 2-Methoxypyridine yields still other products, e.g., **25** and **26**. Moreover, polymerization of ADE, initiated by the zwitterion, can also be in competition with cyclization. The electronegative substituents may retard ring closure sufficiently so that the polymer (tar) formation can occur.

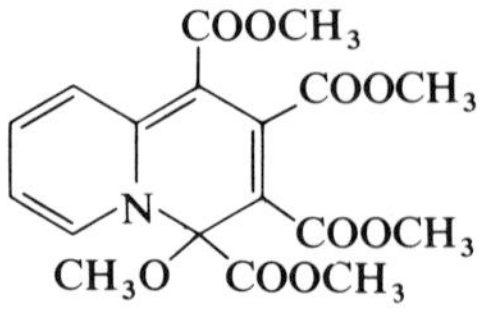

In the following example one of the 1 : 2 adducts is typical of the reaction of pyridines or quinolines with ADE, whereas the other is obviously also dependent on the α-methyl in quinoline (222). Indeed, α-picoline also gives an azepine (222). Recognizing the role of the zwitterion and the fact that it

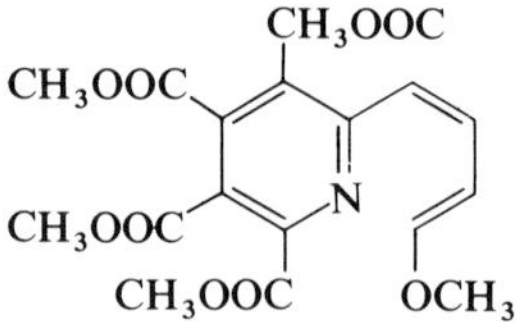

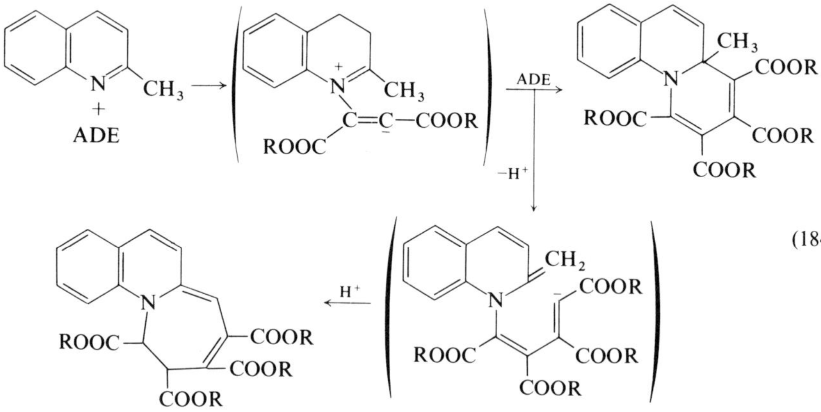

(184)

may generate daughter zwitterions, we can rationalize and occasionally anticipate further observations.

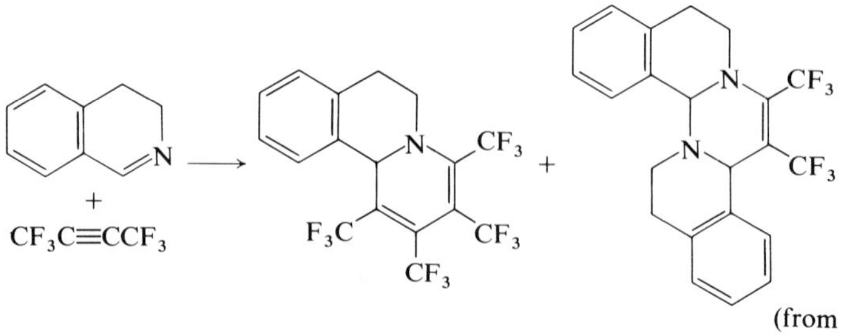

(185)

(from ref. 223)

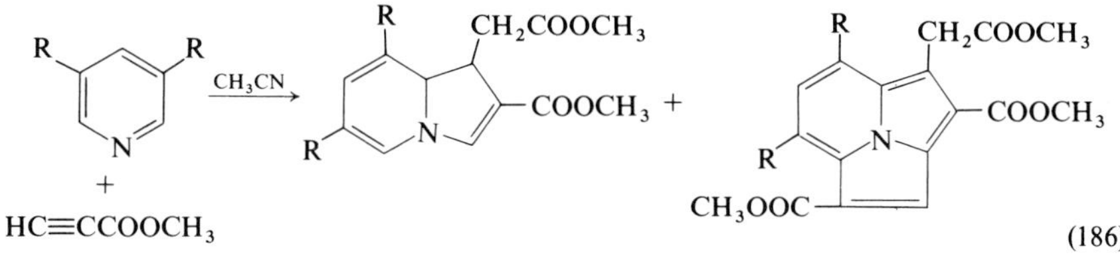

CH_3CN ; $HC{\equiv}CCOOCH_3$; CH_2COOCH_3 ; $COOCH_3$; CH_3OOC

(186)

(from ref. 224)

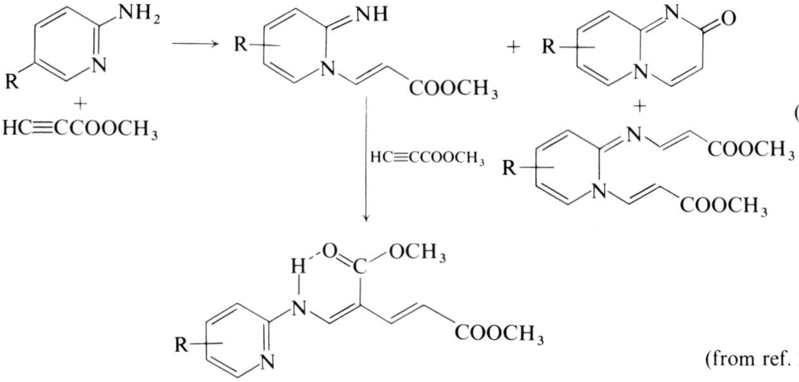

NH_2 ; NH ; $HC{\equiv}CCOOCH_3$; $COOCH_3$

(187)

(from ref. 225)

H_3C ; CH_3 ; 3(ADE)

H_3COOC ; $COOCH_3$; CH_3 or

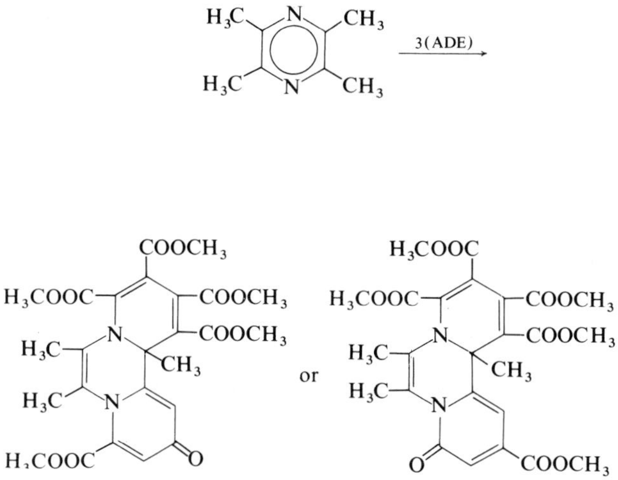

(188)

(from ref. 226)

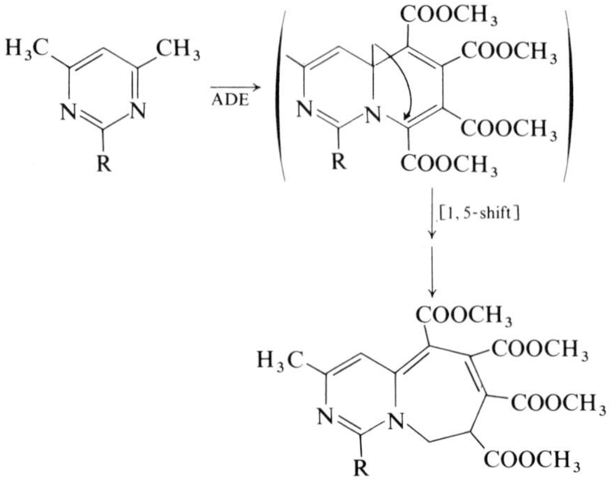

(189)

$(R = H, CH_3)$
(from ref. 226)

Our proposal of the methyl migration in eq. 189 is speculative.

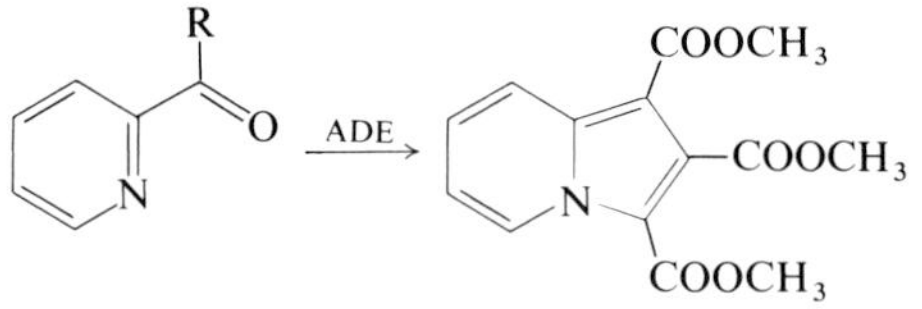

(190)

$(R = CH_3, C_6H_5, C_6H_5CH_2)$
(from ref. 240)

The reaction of Schiff bases, which may be regarded as open-chain analogs of pyridines, depend largely on the nature of the *N*-substituents (227,228):

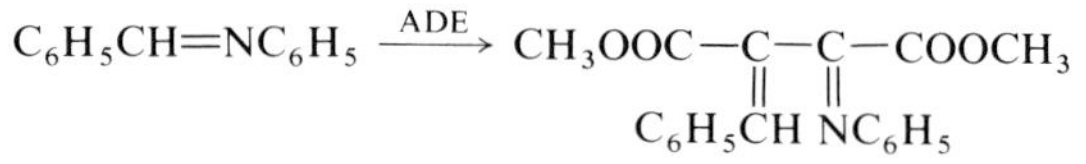

$$C_6H_5CH{=}NC_6H_5 \xrightarrow{ADE} CH_3OOC{-}C(=CHC_6H_5){-}C(=NC_6H_5){-}COOCH_3 \quad (191)$$

$$C_6H_5CH{=}NCH_3 \xrightarrow{ADE} \quad (192)$$

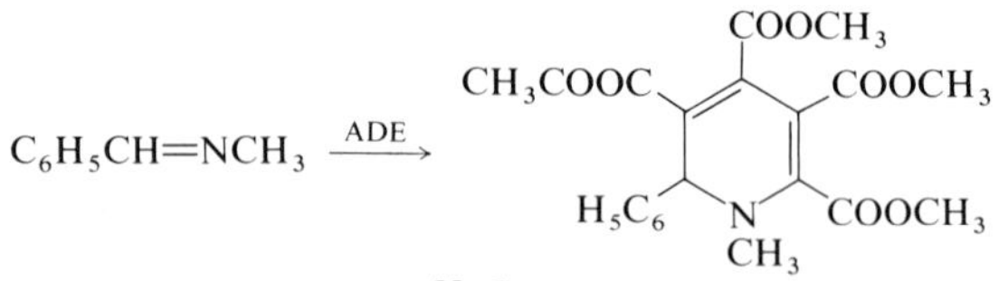

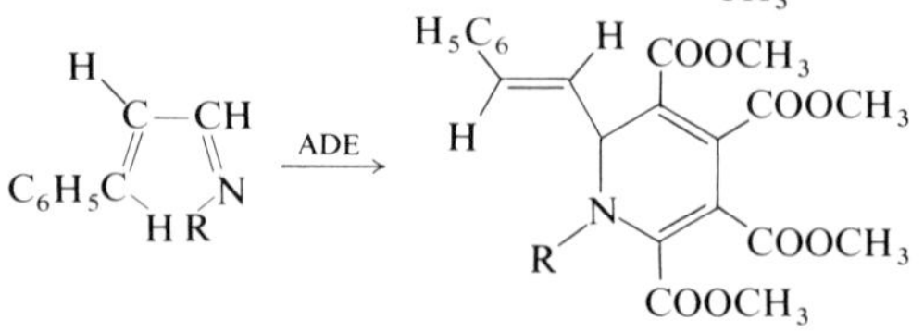

(193)

The reactions of ketoximes are related to these Schiff bases, for the initial role of the hydroxyl group is minor in eq. 194 (241). Other workers have

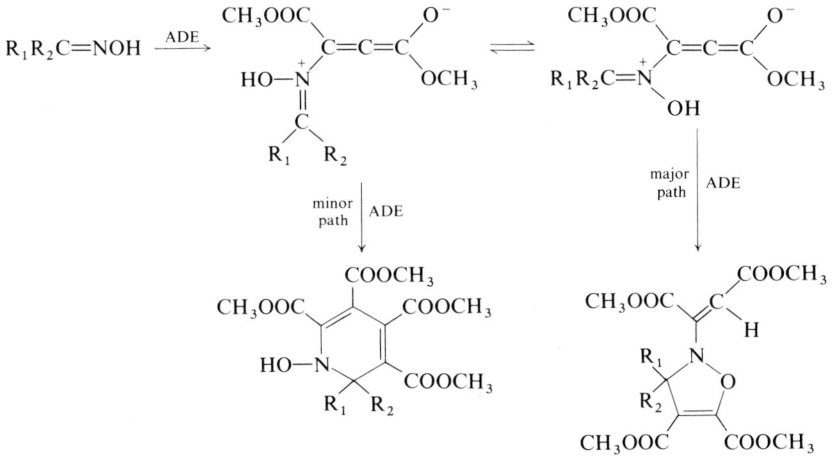

$$R_1R_2C{=}NOH \xrightarrow{ADE} (HO{-}\overset{+}{N}{=}CR_1R_2)(CH_3OOC)C{=}C{=}C(O^-)(OCH_3) \rightleftharpoons (R_1R_2C{=}\overset{+}{N}(OH))(CH_3OOC)C{=}C{=}C(O^-)(OCH_3)$$

minor path, ADE: 1-hydroxy-2,2-R_1R_2 dihydropyridine bearing CH_3OOC and three $COOCH_3$ groups

major path, ADE: isoxazoline (N-substituted by $CH_3OOC{-}C{=}CH{-}COOCH_3$; ring carbons bearing R_1, R_2, CH_3OOC, $COOCH_3$)

(194)

obtained a ring closure from formaldoxime (242) corresponding to the minor path of eq. 194.

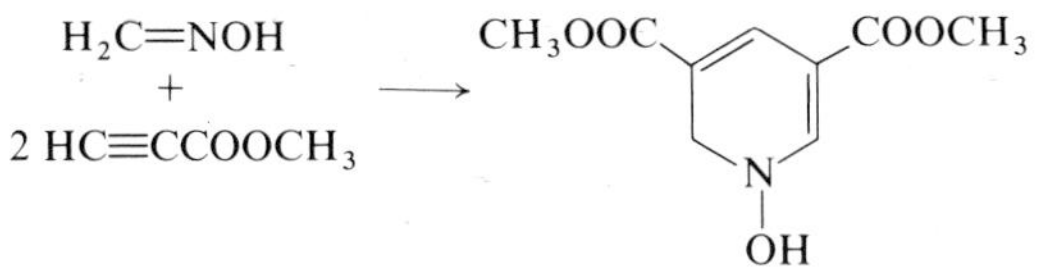

$$H_2C{=}NOH + 2\ HC{\equiv}CCOOCH_3 \longrightarrow \text{3,5-bis}(COOCH_3)\text{-1-hydroxy-dihydropyridine (N–OH)} \qquad (195)$$

C. Phosphorus Nucleophiles

Because few of the additions of phosphorus compounds follow the straightforward course of eq. 1, we give only a few in Table 8. The others show a variety of mechanisms often related to those found with amines.

With primary and secondary phosphines, the additions may be slow or proceed by radical mechanisms, especially for addition to those alkynes with little or no activation. Diphenylphosphine, for example, does not appear to react with phenylacetylene (83). A typical radical process is given in eq. 196 (243).

$$HP(CH_2CH_2CN)_2 + CH_3(CH_2)_nC{\equiv}CH \xrightarrow{ABN} CH_3(CH_2)_nCH{=}CHP(CH_2CH_2CN)_2 \qquad (196)$$

$(n = 4, 5)$

TABLE 8
Stereoselective Nucleophilic Additions of Group V Nucleophiles to Alkynes ($R—C\equiv C—R'$) to give $R—C(N)=CH—R'$[a]

R, R′	N	Conditions	*Anti* adduct, %[b]	Ref.
H, $COOC_2H_5$	$(CH_3)_2NH$	C_2H_5OH	0[c]	210*
H, $COOC_2H_5$	n-$C_3H_7NH_2$	C_2H_5OH	61[c]	210*
H, $C_6H_5N(CH_3)CO$	$(C_6H_5CH_2)_2NH$	CH_3OH	0[c]	209*
H, p-$C_7H_7SO_2$	2,6-Dimethylpiperidine	C_2H_5OH	0[c]	210*
H, $COOCH_3$	$(C_2H_5)_2NH$	t-C_4H_9OH	0[c]	213*
NC, CN	Piperidine	$(C_2H_5)_2O$	0[c]	216*
H, t-$C_4H_9SO_2CH_2$	$CH_3O(CH_2)_3NH_2$	$CDCl_3$	80[c]	209*
H_5C_2OOC, $COOC_2H_5$	$C_6F_5NH_2$	Na, THF	100	308
CH_3OOC, $COOCH_3$	o-$C_6H_4(NH_2)_2$	CH_3OH	100	232*
t-$C_4H_9SO_2$, $SO_2C_4H_9$-t	$C_6H_5NH_2$	Neat	100	315*
H, $PO(OC_3H_7$-$i)$	$(C_2H_5)_2NH$	Neat	?	156
C_6H_5, $P(C_6H_5)_3^+$	Piperidine	Neat	?	157*
n-C_4H_9, $COC\equiv CC_4H_9$-n	p-$(CH_3)_2NC_6H_4NH_2$	CH_3OH	?	86*
n-C_4H_9, $COC\equiv CC_4H_9$-n	N_2H_4	C_2H_5OH	?	86*
F_3C, CF_3	$(C_6H_5)_2PH$	Neat	?[c]	200
C_6H_5, $COOC_2H_5$	$(C_6H_5)_3P$	H_2O, HCl	100	244*
H, CF_3	$(CH_3)_2AsH$	Neat	?[c]	200*
C_6H_5, C_6H_5	$(C_6H_5)_2As^-Li^+$	THF, R_2NH	62[c]	313*

[a] This table is intended to be representative rather than exhaustive. Additional examples and leading references are indicated by an asterisk on the citation. Further examples are surveyed in refs. 3 and 5.

[b] This is the fraction of the total product that is *anti*.

[c] This may not be the kinetically controlled yield.

The overall stereochemistry of metal phosphide addition to alkynes depends critically on familiar factors; for example, the configurational stability of the first intermediate and the availability of protons (83). There is also evidence that the lithium phosphide intermediates may be "more

$$C_6H_5C\equiv CC_6H_5 + (C_6H_5)_2P^-Na^+ \xrightarrow{R_2NH} \begin{matrix} H_5C_6 & & P(C_6H_5)_2 \\ & C=C & \\ H & & C_6H_5 \end{matrix} \qquad (197)$$

covalent" or that the ions are highly aggregated. In eq. 197 the probable intermediate is a carbanion, which can deprotonate an amine readily.

$$C_6H_5C{\equiv}CH + (C_6H_5)_2PLi \xrightarrow{THF} \begin{cases} \xrightarrow{\text{no amine}} & (C_6H_5)_2P(H)C{=}C(H)C_6H_5 \text{ (trans)} \\ \xrightarrow[70\%]{R_2NH \text{ or } RNH_2} & (C_6H_5)_2P(H)C{=}C(H)C_6H_5 \text{ (cis)} \end{cases} \quad (198)$$

$$C_6H_{13}C{\equiv}CH + (C_6H_5)_2PLi \longrightarrow (C_6H_5)_2P(H)C{=}C(H)C_6H_{13} \text{ (trans)} \quad (199)$$

$$C_6H_5C{\equiv}CC_6H_5 + (C_6H_5)_2PLi \xrightarrow{THF} \begin{cases} \xrightarrow{R_3N} & \text{Syrup} \\ \xrightarrow{(C_2H_5)_2NH} & H_5C_6(H)C{=}C(C_6H_5)P(C_6H_5)_2 \quad 90\% \\ \xrightarrow{n\text{-}C_4H_9NH_2} & H_5C_6(H)C{=}C(P(C_6H_5)_2)C_6H_5 \xrightarrow{80^\circ} \text{(the } 90\% \text{ product)} \end{cases} \quad (200)$$

In eqs. 198–200 the first intermediate may be formed in polar aggregates, possibly **27** or **28**. Note too the wrong directiospecificity in eq. 199.

$$\begin{array}{c} \diagdown C{\equiv}C \diagup \\ Li \cdots\quad\cdots PR_2 \\ R_2P\cdots Li \\ \textbf{27} \end{array} \qquad \begin{array}{c} \diagdown C{\equiv}C \diagup \\ H \cdots\quad\cdots PR_2 \\ R_2'N\cdots Li \\ \textbf{28} \end{array}$$

In any case the amine finds it more difficult to get to the anionic site before isomerization occurs. Further, the secondary amine (e.g., of eq. 200) appears to be less effective than the primary amine as a proton donor in capturing any *anti* intermediates (83).

When a proton is available, tertiary phosphines add to activated alkynes in the expected manner, but the first products may not survive (200,244,245):

$$(C_6H_5)_3P + C_6H_5C{\equiv}CCOOH \xrightarrow{H^+} \left((C_6H_5)_3P^+(H_5C_6)C{=}C(COOH)H \right) \xrightarrow{H_2O} (C_6H_5)_3PO + trans\text{-}C_6H_5CH{=}CHCOOH \quad (201)$$

$$(C_6H_5)_3P + \text{ADE} \xrightarrow{\text{HBr}} \left[\begin{array}{c}(C_6H_5)_3P^+ \quad COOH \\ C{=}C \\ HOOC \quad H\end{array}\right] Br^- \longrightarrow \textit{trans-}(C_6H_5)_3P^+CH{=}CHCOOH, Br^- + CO_2 \quad (202)$$

$$HC{\equiv}CC_6H_5 \xrightarrow{(C_6H_5)_3P} (C_6H_5)_3P^+CH{=}\bar{C}C_6H_5 \xrightarrow{H_2O} [(C_6H_5)_3P^+CH{=}CHC_6H_5, OH^-] \longrightarrow (C_6H_5)_2P(=O){-}HC(C_6H_5)(CH_2C_6H_5) \quad (203)$$

$$C_6H_5C{\equiv}CP(C_4H_9\text{-}n)_3^+ Br^- + (C_6H_5)_3P \xrightarrow{H^+} (C_6H_5)_3P(H_5C_6)C{=}CH(C_4H_9\text{-}n)_3P^+Br^- \quad (204)$$

(from ref. 238)

When protons are not available, the zwitterion formed from tertiary phosphines and alkynes may attack another molecule (eq. 65) or rearrange internally. The alternatives are illustrated in eqs. 205 (249) and 206 (250).

$$C_6H_4 \text{ (benzyne)} \xrightarrow{(C_6H_5)_3P} o\text{-}^-C_6H_4P(C_6H_5)_3^+ \begin{cases} \xrightarrow{CH_3I} o\text{-}CH_3C_6H_4P(C_6H_5)_3^+I^- \\ \xrightarrow{-C_6H_6} \text{5-phenyldibenzophosphole } (C_{12}H_8PC_6H_5) \end{cases} \quad (205)$$

$$C_6H_4 \text{ (benzyne)} \xrightarrow{(C_6H_5)_2PCH_3} \left(o\text{-}^-C_6H_4P^+(C_6H_5)_2CH_2{-}H \right) \longrightarrow (C_6H_5)_3P{=}CH_2 \quad (206)$$

Although the course of the triphenylphosphine additions to ADE was a controversial subject a few years ago, we can now give the probable reaction paths and products:

$(C_6H_5)_3P$ + ADE $\xrightarrow{-50°\ \text{ether}}$ $[(C_6H_5)_3P^+\text{–}C(COOCH_3)\text{=}C^-\text{–}COOCH_3]$

$\xrightarrow{(C_6H_5)_3P}$ $(C_6H_5)_3P\text{=}C(COOCH_3)\text{–}C(COOCH_3)\text{=}P(C_6H_5)_3$ **30**

$\xrightarrow{CO_2}$ $(C_6H_5)_3P^+\text{–}C(COOCH_3)\text{=}C(COOCH_3)\text{–}COO^-$ **29** $\not\rightarrow$ **33** (1,4-diphosphorin: H_5C_6, C_6H_5, C_6H_5 on each P; CH_3OOC, $COOCH_3$)

29 $\xrightarrow{-CO_2;\ 50°,\ CH_3OH\ \text{or}\ CH_3I}$ **31** ($(C_6H_5)_3P$, $COOCH_3$, $COOCH_3$, CH_3OOC, CH_3OOC, $P(C_6H_5)_3$)

$\longrightarrow$ **32** ($(C_6H_5)_3P^+$, $COOCH_3$, CH_3OOC, $COOCH_3$, $COOCH_3$) $\xrightarrow{?}$ (phosphole: C_6H_5, H_5C_6, C_6H_5, P; CH_3OOC, $COOCH_3$, CH_3OOC, $COOCH_3$)

32 $\longrightarrow$ CH_3OOC, C_6H_5, $(C_6H_5)_2P$, $COOCH_3$, CH_3OOC, $COOCH_3$ $\xleftarrow{?}$ (phosphole)

(207)

We begin with a postulated 1:1 adduct. In the presence of excess phosphine or carbon dioxide this can be trapped to give **29** or **30** (251,131). The 1:2 zwitterion was detected at $-50°$ in the 1950's, but its decomposition route was worked out some years later (131,251,253,254). The reactions that we indicate as *not* occurring were postulated in the early work. Likewise, eq. 208 shows an early claim (255) and its correction (256).

$$NC{-}C{\equiv}C{-}CN + (C_6H_5)_3P \xrightarrow[CH_3CN]{0^\circ} \text{(phosphole, not formed)} ;\quad (C_6H_5)_3P{=}C(CN){-}C(CN){=}C(CN){-}C(CN){=}P(C_6H_5)_3 \tag{208}$$

For this reason the phosphole structure proposed in eq. 209 (257) cannot be regarded as securely established.

$$\text{1,2,5-}(C_6H_5)_3\text{-phosphole} \xrightarrow{ADE} \text{spiro-P}(C_6H_5)\text{ with } C_6H_5, COOCH_3 \text{ substituents} \tag{209}$$

Trialkyl phosphites can add even to slightly activated alkynes. In an aprotic medium, there are three patterns beginning with the initial zwitterion (258,259):

$$RC{\equiv}CH + (C_2H_5O)_3P \longrightarrow \left(RC^-{=}CH{-}P^+(OC_2H_5)_3 \longrightarrow \ldots\right) \longrightarrow (RHC{=}CH)P(O)(OC_2H_5)_2 + C_2H_4 \tag{210}$$

The activated alkynes are somewhat more versatile:

$$R'C{\equiv}CCOOH + (RO)_3P \longrightarrow \left((RO)_2P^+(OR)C(R'){=}C^-{-}C(O)OH\right) \longrightarrow \left((RO)_2PO(R')C{=}C{=}C(OR)(OH)\right) \longrightarrow (RO)_2PO(R')C{=}CHCOOR \tag{211}$$

$$(CH_3O)_3P + \text{ADE} \longrightarrow (CH_3O)_2\overset{+}{P}(-O-CH_3)-C(COOCH_3)=\bar{C}-COOCH_3 \longrightarrow$$

$$(CH_3O)_2P(CH_3OOC)C=C(CH_3)COOCH_3 \longrightarrow CH_3OOC-\underset{(CH_3O)_2P=O}{\overset{CH_3}{C}}-\underset{P(OCH_3)_2=O}{\overset{CH_3}{C}}-COOCH_3 \qquad (212)$$

Other phosphite additions have been studied by Russian workers (260–262). For certain alkynes (e.g., ethyl phenylpropiolate) the diadduct becomes the major product. Unfortunately the geometry of the monoadducts was not determined.

$$HC{\equiv}CCOOC_2H_5 + {}^-OP(OC_2H_5)_2 \xrightarrow[C_2H_5OH,\ 40^\circ]{C_2H_5O^-} (C_2H_5O)_2\underset{\underset{O}{\|}}{P}CH{=}CHCOOC_2H_5 + ((C_2H_5O)_2\underset{\underset{O}{\|}}{P})_2CHCH_2COOC_2H_5 \qquad (213)$$

$$\text{ADE} + {}^-OP(OR)_2 \longrightarrow CH_3COOCCH{=}CH(COOCH_3)\overset{\overset{O}{\|}}{P}(OR)_2 + CH_3OOC\underset{(RO)_2P=O}{CH}-\underset{P(OR)_2=O}{CH}COOCH_3 \qquad (214)$$

$$H_5C{\equiv}CCOOC_2H_5 + (C_2H_5O)_2P(S)H \longrightarrow C_6H_5C(P(S)(OC_2H_5)_2){=}CHCOOC_2H_5 + C_6H_5\underset{(C_2H_5O)_2P=S}{CH}-\underset{P(OC_2H_5)_2=S}{CH}COOC_2H_5 \qquad (215)$$

Secondary phosphine oxides in the presence of base also add to alkynes (248,263).

$$(C_6H_5)_2P(O)H \xrightarrow{\text{ADE}} \begin{array}{l} (C_6H_5)_2P(O)CHCOOC_2H_5 \\ \qquad\qquad\quad | \\ (C_6H_5)_2P(O)CHCOOC_2H_5 \end{array} \qquad (216)$$

One can make a rough analogy between phosphorane and imine additions to triple bonds. In both cases there are two sites which can attack the alkyne (eqs. 191, 192). Either carbon or phosphorus may appear to lead the attack; synthetically the distinction may be irrelevant if a transient cyclobutene lies on the reaction path (253,266). The difference in the fate of the

$$(C_6H_5)_3P^+—\bar{C}RR' + \text{ADE} \longrightarrow \left(\begin{array}{c} R'RC—C—COOCH_3 \\ (C_6H_5)_3P^+ \;\; \underline{C}COOCH_3 \end{array} \right) \longrightarrow \begin{array}{c} RR'C{=}CCOOCH_3 \\ (C_6H_5)_3P{=}CCOOCH_3 \end{array} \qquad (217)$$

$$\downarrow \qquad\qquad \nearrow\!\!\swarrow$$

$$\left(\begin{array}{c} R'R—\square—COOR \\ (C_6H_5)_3—P—\square—COOR \end{array} \right) \not\longrightarrow \begin{array}{c} R—C——C—COOR \\ (C_6H_5)_3P \qquad C—H \\ ROOC \end{array}$$

(from refs. 264, 265)

$$C_6H_5COCH{=}P(C_6H_5)_3 + \text{ADE} \xrightarrow[\text{reflux}]{CH_3OH} \begin{array}{c} O \qquad COOCH_3 \\ H_5C_6 \qquad\qquad H \\ (C_6H_5)_3P \qquad COOCH_3 \end{array}$$

$$\downarrow \text{dry ether}$$

$$\begin{array}{c} C_6H_5CO—CH—CCOOCH_3 \\ (C_6H_5)_3\,P^+ \;\; ^-CCOOCH_3 \end{array} \longrightarrow \begin{array}{c} O \qquad COOCH_3 \\ H_5C_6 \qquad\qquad P(C_6H_5)_3 \\ COOCH_3 \end{array} \qquad (218)$$

$$\swarrow\!\!\!/\!\!/ \qquad\qquad \searrow\!\!\!/\!\!/$$

$$\begin{array}{c} CH_3OOC—\;O\;—C_6H_5 \\ CH_3OOC—\;P\;—C_6H_5 \\ H_5C_6 \quad C_6H_5 \end{array} \qquad \begin{array}{c} O \qquad COOCH_3 \\ H_5C_6 \qquad\qquad P(C_6H_5)_3 \\ CH_3OOC \end{array}$$

(from ref. 254)

betaine between eqs. 217 and 218 seems to depend on the greater acidity of its mobile proton in eq. 218.

Some details of the mechanism of ylid reactions with benzyne have been established by labeling experiments (267). This may also be directly applicable to the phenyl transfer from **32** in eq. 207, in which case the phosphole may be an intermediate in the process.

$$(C_6H_5)_2P{=}CHC_6H_5 \; (C_6H_4^*) + C_6H_6 \longrightarrow \ldots \qquad (219)$$

Phosphinimines and phosphazines are similar to the phosphoranes. Reaction 220 is facilitated by electron donating substituents (R) (265).

$$(C_6H_5)_3P{=}NC_6H_4R\text{-}p \xrightarrow{\text{ADE}} (C_6H_5)_3P{=}C(COOCH_3){-}C(COOCH_3){=}NC_6H_4R\text{-}p \qquad (220)$$

$$(C_6H_5)_3P{=}N{-}N{=}CHCOOR \xrightarrow{\text{ADE}} (C_6H_5)_3P{=}N{-}N{=}C(COOCH_3){-}C(COOCH_3){=}C(H)COOR \qquad (221)$$

$$(C_6H_5)_3P{=}N{-}N{=}C_{13}H_8 \xrightarrow{\text{ADE}} (C_6H_5)_3P{=}C(COOCH_3){-}C(COOCH_3){=}N{-}N{=}C_{13}H_8 \qquad (222)$$

D. Arsenic Nucleophiles

In general, arsines give simple adducts. In process 223 Cullen found the order of nucleophilic reactivity (200),

$$k((CH_3)_2AsH) > k(C_6H_5(CH_3)AsH) > k((C_6H_5)_2AsH)$$

$$RR'AsH + R''C{\equiv}CR''' \longrightarrow RR'As{-}C(R''){=}CH{-}R''' \qquad (223)$$

The simple arsole synthesis of eq. 15 has been elaborated (87) (e.g., eq. 224).

$$C_6H_5AsH_2 + RC{\equiv}C{-}C{\equiv}CR \xrightarrow[C_6H_6]{n\text{-}C_4H_9Li} \text{2,5-}R_2\text{-1-}(C_6H_5)\text{arsole} \tag{224}$$

($R = C_6H_5$, p-C_7H_7, p-C_6H_4Cl, CH_3, α-naphthyl)

There is also a close analogy between alkali metal arsenide and phosphide additions (83,314)

$$C_6H_5C{\equiv}CC_6H_5 + LiAs(C_6H_5)_2 \xrightarrow{THF} \begin{cases} \xrightarrow{(C_2H_5)_2NH} (H_5C_6)(H)C{=}C(C_6H_5)(As(C_6H_5)_2) \\ \rightleftharpoons \\ \xrightarrow{n\text{-}C_4H_9NH_2} (H_5C_6)(H)C{=}C(As(C_6H_5)_2)(C_6H_5) \end{cases} \tag{225}$$

Further examples are given in Table 8.

In eq. 226, triphenylarsine follows a path completely different from triphenylphosphine (eq. 207) (253):

$$(C_6H_5)_3As \xrightarrow[H_2O]{ADE} (C_6H_5)_3As{=}C(COOCH_3){-}C({=}O)COOCH_3 + CH_3OOCCH{=}HCCOOCH_3 \tag{226}$$

E. Oxygen Nucleophiles

Previous discussion has touched fairly frequently on the additions of oxygen nucleophiles. We have cited appropriate examples in connection with vinylation (eqs. 9, 75), rate laws (eqs. 32–35, 64), addition mechanisms (eqs. 36, 64), reactivity (eqs. 72, 75), stereoselectivity (eq. 12), directioselectivity (eqs. 21, 80, 84, 92, 94), competing reactions and complications of the addition process (eqs. 36, 39, 43, 105). A recent fairly conventional example is given by Shostakovskii et al. (268):

$$(R_2S)_2C(CH_3)CH_2OH + HC{\equiv}CH \xrightarrow[KOH]{90\text{–}140°} (R_2S)_2C(CH_3)CH_2OCH{=}CH_2 \tag{227}$$

Other examples are given in Table 9; some have been collected elsewhere (8).

TABLE 9
Stereoselective Additions of Group VI Nucleophiles (**N**—H) to Alkynes ($R—C\equiv C—R'$) to give $R—C(\mathbf{N})=CH—R'$[a]

R, R′	N	Conditions	*Anti* adduct, %[b]	Ref.
H, $COOCH_3$	CH_3O^-	$t\text{-}C_4H_9OH$, base	32[c]	213*
H, $COOC_2H_5$	$i\text{-}C_3H_7O^-$	$t\text{-}C_4H_9OH$, base	0[c]	213*
CH_3OOC, $COOCH_3$	$C_6H_5O^-$	$(C_2H_5)_2O$, base	53[c]	213*
$p\text{-}C_7H_7SO_2$, CH_3	$C_2H_5O^-$	C_2H_5OH	100	84*
NC, CN	$t\text{-}C_4H_9O^-$	$t\text{-}C_4H_9OH$	18[c]	216*
$t\text{-}C_4H_9SO_2$, $SO_2C_4H_9\text{-}t$	$C_6H_5O^-$	C_6H_5OH	100	315*
H, $PO(OC_3H_7\text{-}i)_2$	$C_2H_5O^-$	C_2H_5OH	?[c]	156
CH_3, CH_3	$p\text{-}C_7H_7S^-$	C_2H_5OH	100	60*
H, COOH	$o\text{-}C_7H_7S^-$	C_2H_5OH	100	66*
H, $C(CH_3)=CH_2$	$C_2H_5S^-$	CH_3OH	100	310*
$HC\equiv C(CH_2)_4$, SCH_3	$C_2H_5S^-$	CH_3OH	?[c]	314*
H, $CH=CH_2$	CH_3S^-	C_2H_5OH	100	161*
$t\text{-}C_4H_9SO_2$, $SO_2C_4H_9\text{-}t$	$(CH_3)_3CSH$	Neat	100	315*
H, $PO(OC_3H_7\text{-}i)_2$	$C_2H_5S^-$	C_2H_5SH	?[c]	156
C_6H_5, $OC—C\equiv CC_6H_5$	$p\text{-}C_7H_7S^-$	CH_3OH	?	86*
CH_3, $SO_2C_6H_5$	$C_6H_5SO_2^-$	CH_3OH	100	123
H, COOH	$p\text{-}O_2NC_6H_4Se^-$	C_2H_5OH	100	289*
C_6H_5, $COOC_2H_5$	$C_6H_5Se^-$	C_2H_5OH	?	288*

[a] This table is intended to be representative rather than exhaustive. Additional examples and leading references are indicated by an asterisk on the citation. Further examples are surveyed in refs. 3, 8, and 159.

[b] This is the fraction of the total product that is *anti*.

[c] This may not be the kinetically controlled yield.

In the presence of alkali methanol adds to certain 1,3-dynes (8,269):

$$HC\equiv C—C\equiv CH \xrightarrow[CH_3OH]{KOH} HC\equiv C—CH=CHOCH_3 \quad (228)$$

$$CH_3C\equiv C—C\equiv CH \xrightarrow[CH_3OH-C_6H_6]{KOH,\ 125^\circ} CH_3C\equiv C—CH=CHOCH_3 \quad (229)$$

$$(CH_3)_2C(OR)—C\equiv C—C\equiv CH \xrightarrow[CH_3OH]{KOH} (CH_3)_2C(OR)C\equiv C—CH=CHOCH_3 \quad (230)$$

On the other hand, terminal alkynes, which can become nonterminal do so (18,121) and thereby resist nucleophilic addition:

$$RCH_2C\equiv C—C\equiv CH \overset{base}{\rightleftharpoons} RC\equiv C—C\equiv C—CH_3 \quad (231)$$

$$(R=CH_3,\ C_6H_5CH_2)$$

This retardation has been ascribed to the electron-donating alkyl substituents (4). As the degree of unsaturation is increased, reactivity also increases. According to Bohlmann, the time in minutes for 25% reaction in the addition of sodium methoxide to $CH_3(C{\equiv}C)_nCH_3$ at 65° changes drastically from $t = \infty$ ($n = 1$) and $t > 5000$ ($n = 2$), to $t = 260$ ($n = 3$) or $t = 10$ ($n = 4$) (270). The two opposing effects of structure are evident in the observation that the reactivity of the compound $t\text{-}C_4H_9(C{\equiv}C)_5C_4H_9\text{-}t$ falls between $CH_3(C{\equiv}C)_3CH_3$ and $CH_3(C{\equiv}C)_4CH_3$ (8,274).

Additions of hydroxy compounds to activated alkynes have been known for a long time (20). Recently, Winterfeldt reported on thermal additions to these alkynes and has proposed that these violate the *anti* rule (3). Further, his data on the additions of methanol, 2-propanol, and phenol to ADE, methyl propiolate, and dicyanoacetylene appear to support this view (see Table 9).

In our opinion the "neutral" thermal additions are probably acid-catalyzed (201) and *not* molecular. Only in a molecular addition such as eq. 232 could the alcohol be termed a nucleophile. Judging by previous

$$-C{\equiv}C- + RO{-}H \longrightarrow \overset{-}{C}{=}C(OR) \quad H^+ \tag{232}$$

$$R'C{\equiv}CCOOCH_3 + ROH \rightleftharpoons \left[RO^- \; R'C{\equiv}C{-}C({=}\overset{+}{O}H)OCH_3 \longleftrightarrow RO^- \; R'{-}\overset{+}{C}{=}C{=}C(OH)OCH_3 \right] \longrightarrow (RO)(R')C{=}C{=}C(OH)(OCH_3) \longrightarrow (RO)(R')C{=}CHCOOCH_3$$

$$R'C{\equiv}CCOOCH_3 + ROH \xrightarrow{?} \left[R{-}O{\cdots}H{\cdots}O{=}C(OCH_3){-}C{\equiv}C{-}R' \right] \longrightarrow (RO)(R')C{=}C{=}C(OH)(OCH_3) \tag{233}$$

experience with oxygen nucleophiles, we believe that this Ad_N2 process might occur under forcing conditions or with activated alkynes (309). Taking into account the mechanistic questions and the fact that *syn* adducts of ADE and methyl propiolate are more stable than *anti* adducts (213,216), we believe that more quantitative studies are in order on the matter of stereoselection. In the thermal reactions the problem may change from one of the stereoselectivity of elementary reactions to questions about the lifetimes of labile intermediates. The role of "internal" and "external" proton donors to analogs of **14** and **15** should be examined, as noted previously for amines (Sections III-D, and IV-B) (153).

Winterfeldt has also studied base-catalyzed additions to methyl propiolate, ADE, and dicyanoacetylene, and found substantial to exclusive production of *syn* adducts (213,216). As a measure of his skepticism of the *anti* rule, Winterfeldt cites the example which appears to be kinetically controlled (3):

$$\begin{array}{c} \text{ROOC—C}\equiv\text{C—C(H)=C(H)—COOR} \\ + \\ CH_3OH + CH_3O^-Na^+ \end{array} \begin{array}{l} \xrightarrow{\sim 50\%} \text{(H)(ROOC)C=C(OCH}_3\text{)—C(H)=C(H)(COOR)} \\ \qquad + \\ \xrightarrow[\sim 50\%]{} \text{(H)(ROOC)C=C(OCH}_3\text{)—C(H)=C(H)(COOR)} \end{array} \qquad (234)$$

Indeed, it appears that *syn* adducts may be thermodynamically favored, e.g., eq. 235. For this reason the stereoselectivity of these reactions require quantitative studies.

$$\begin{array}{c} HC\equiv CCOOCH_3 \\ + \\ CH_3OH \end{array} \begin{array}{l} \longrightarrow \text{(CH}_3\text{O)(H)C=C(H)(COOCH}_3\text{)} \\ \qquad \upharpoonleft\downharpoonright \\ \longrightarrow \text{(H)(CH}_3\text{O)C=C(H)(COOCH}_3\text{)} \end{array} \qquad (235)$$

A novel oxygen nucleophile may be found in dimethylsulfoxide (89):

$$\begin{array}{c} (CH_3)_2SO \\ + \\ RC\equiv CCOOCH_3 \end{array} \xrightarrow{\text{heat}} \left(\begin{array}{c} CH_3OOC\overset{-}{C}\;\; S^+(CH_3)_2 \\ \| \quad\;\; | \\ R\text{—C—O} \end{array} \right) \longrightarrow \text{(R)(}^-\text{O)C=C(SCH}_3\text{)(COOCH}_3\text{)} \qquad (236)$$

$(R = H, COOCH_3)$

Intramolecular addition of alcohol is a useful method of ring closure (271). The formation of dioxanes is straightforward (272):

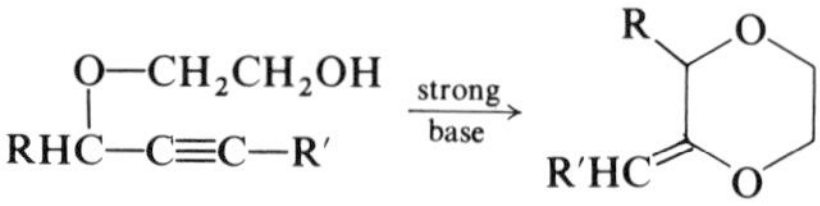

$$\mathrm{RHC(O{-}CH_2CH_2OH){-}C{\equiv}C{-}R'} \xrightarrow[\text{base}]{\text{strong}} \text{2-R-3-(R'HC=)-1,4-dioxane} \quad (237)$$

$$\mathrm{R'R{-}C(NR''{-}CH_2CH_2OH){-}C{\equiv}C{-}R'''} \xrightarrow{\text{base}} \text{oxazolidine (R, N-R'', }CH{=}CH_2\text{)} + \text{morpholine (R', R, R'''HC=, N-R'')} + \text{oxazepine (R', R, N-R'', R''')} \quad (238)$$

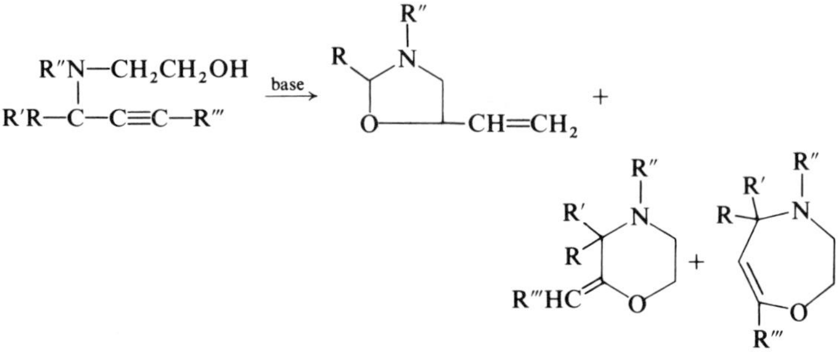

The five-membered ring in 238 is formed when R′ = H and when strong base is used; this has been ascribed to the preliminary acetylene–allene rearrangement. Otherwise the morpholine may be formed. When R‴ = aryl, chloro or methylthio, the morpholine is formed; but when R‴ = methyl or ethyl the third product oxazepine is formed. The inductive effect appears to determine the directioselectivity, that is, to form a six- or a seven-membered cycle (271).

An oxygen analog of the α-aminoethyl ketone cyclizations (see **19** and **20**) follows a similar course (90):

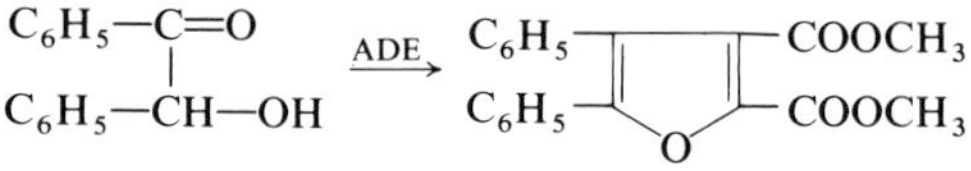

$$\mathrm{C_6H_5{-}C{=}O,\ C_6H_5{-}CH{-}OH} \xrightarrow{\text{ADE}} \text{2,5-}(C_6H_5)_2\text{-furan-3,4-}(COOCH_3)_2 \quad (239)$$

Related closures may take place within these same molecules:

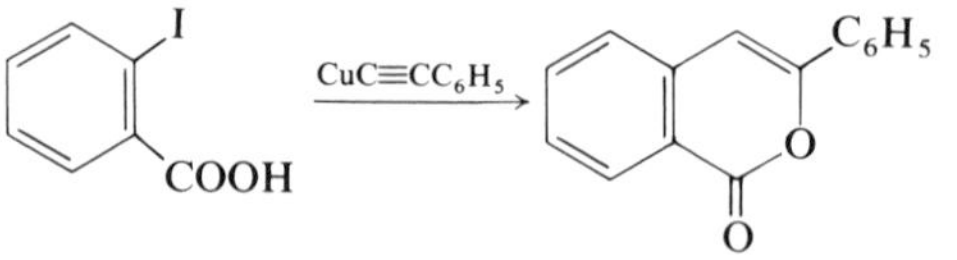

$$\text{o-iodobenzoic acid (I, COOH)} \xrightarrow{CuC{\equiv}CC_6H_5} \text{3-}C_6H_5\text{-isocoumarin} \quad (240)$$

(from ref. 279)

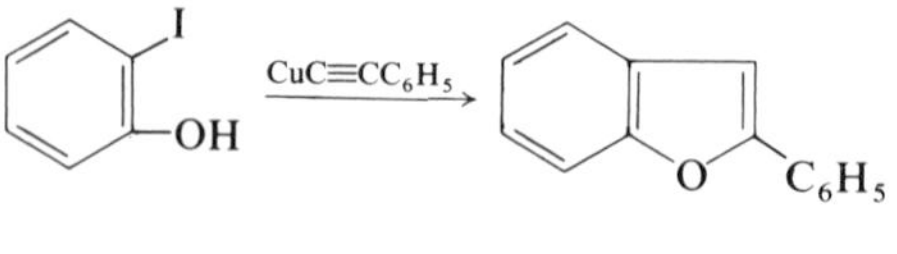

$$\text{o-iodophenol (I, OH)} \xrightarrow{CuC{\equiv}CC_6H_5} \text{2-}C_6H_5\text{-benzofuran} \quad (241)$$

(from ref. 279)

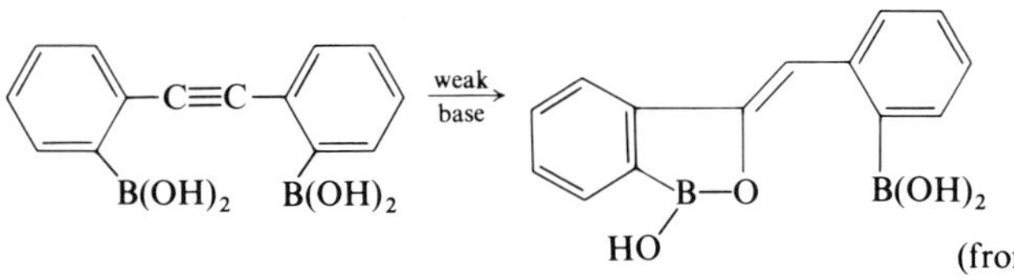

(242)

(from ref. 273)

F. Sulfur Nucleophiles

Sulfur compounds have often been referred to previously (e.g., eqs. 20, 37, 38, 43, 45, 73, 93, 95). The cyclic thioamide and the more usual thiolate give stereospecific additions. Other examples are given in Table 9.

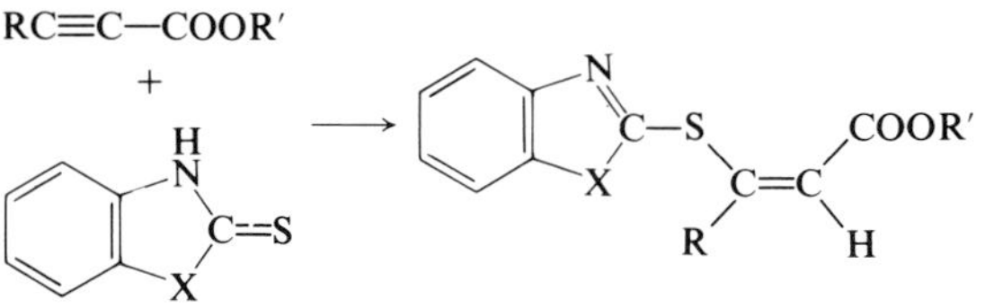

(243)

(X = NH, S)

(from ref. 281)

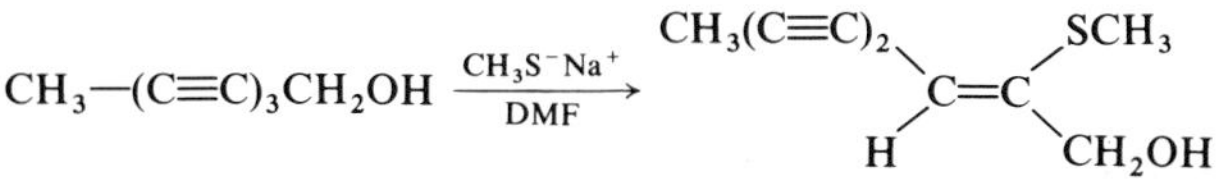

$$CH_3-(C\equiv C)_3CH_2OH \xrightarrow[DMF]{CH_3S^-Na^+} CH_3(C\equiv C)_2(H)C=C(SCH_3)(CH_2OH) \quad (244)$$

(from ref. 277)

Since thiols add to triple bonds, both in the presence and absence of base, the problem of competing processes must be recognized. As indicated in the discussion on directioselectivity (eqs. 81, 83) radical processes normally are associated with low stereoselectivity and "wrong" directioselectivity (8,60,125). An interesting example of competition is found in eq. 245 (8), but

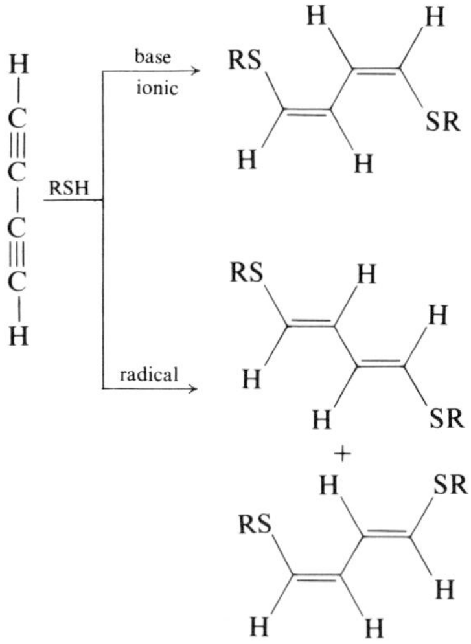

(245)

other radical or radical and ionic processes have been given (275,276)

$$CH_3C{\equiv}CH \xrightarrow{C_6H_5SH} CH_2{=}C(CH_3)SC_6H_5 + C_6H_5SCH{=}CHCH_3 \qquad (246)$$

The addition of nucleophiles other than thiolates (e.g., sulfinates, thiosulfates, and sulfites) have been mentioned in eqs. 85, 86, and 10. One must remember, however, that like thiolate, sulfinate or sulfite may undergo radical reactions (26,281).

Bifunctional sulfur species may lead to cyclic compounds. A good route to α,α'-disubstituted thiophenes is given in eq. 247 (278). The elements of

$$R{-}C{\equiv}C{-}C{\equiv}C{-}R' + H_2S \xrightarrow[\text{base}]{20\text{–}80^\circ} \text{R-(thiophene-2,5-diyl)-R}' \qquad (247)$$

(R = H, CH_3, C_6H_5, COOH, etc.)

hydrogen sulfide appear to add to diethynylketone in two ways (see eq. 107); the addition from thiourea goes under mild conditions (86), while the direct addition proceeds under relatively forcing conditions (280). It would appear that cyclization to the five- rather than the six-membered ring is preferred. Another such example is based on the disulfide (278):

$$R{-}(C{\equiv}C)_2{-}CH_3 \xrightarrow{Na_2S_2} RC{\equiv}C{-}C(H){=}C(CH_3){-}S{-}SH \qquad (249)$$

$\not\rightarrow$ (R–(1,2-dithiin)–CH_3, with S–S in the ring)

$\rightarrow$ R–CH= (1,2-dithiole ring bearing CH_3, with S–S in the ring)

(R = H, CH_3, OCH_3, $CH_3(C{\equiv}C)_2{-}$)

The additions of bifunctional nucleophiles involving sulfur give both open chain and cyclic products (282,283). It will be noted that technically only the first step falls within the scope of process 1. According to two studies

(90,284), a general scheme may be formulated as follows:

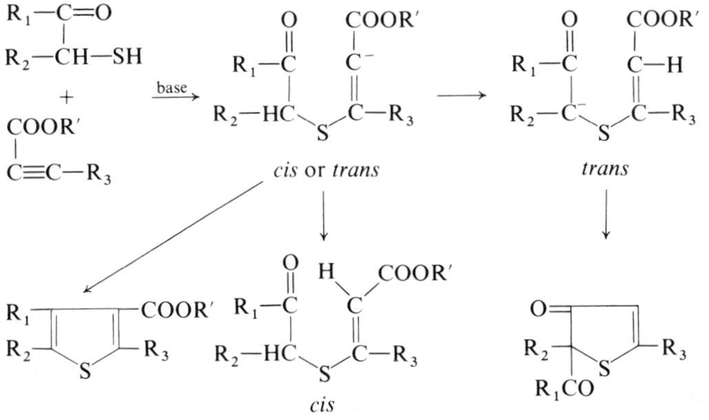

$$R_1{-}CO{-}CH(R_2){-}SH + R'OOC{-}C{\equiv}C{-}R_3 \xrightarrow{\text{base}} R_1CO{-}CH(R_2){-}S{-}C(R_3){=}C^{-}{-}COOR' \; (\textit{cis}\text{ or }\textit{trans}) \longrightarrow R_1CO{-}C^{-}(R_2){-}S{-}C(R_3){=}CH{-}COOR' \; (\textit{trans}) \tag{250}$$

cis or *trans* → thiophene (R_1, R_2, COOR′, R_3 substituted) and $R_1CO{-}CH(R_2){-}S{-}C(R_3){=}CH{-}COOR'$ (*cis*); *trans* → $O{=}$ dihydrothiophene with R_2, R_1CO, R_3

Scheme 250 could serve to include an even broader area, if NH or O were used in place of S, but the choices that are actually realized do differ—see eqs. 175–177, 239. It was supposed that the first open chain adducts under the mildest conditions should be *anti*, whenever mobile protons were available (90), but *syn* adducts are often found and may even be the exclusive products in the aprotic solvents (284). As with amines, stereoselectivity is

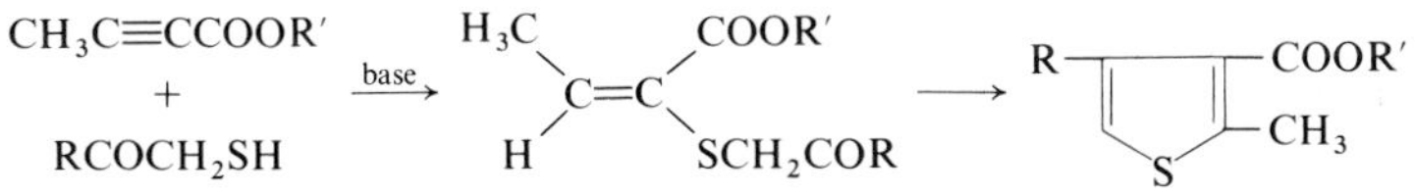

$$CH_3C{\equiv}CCOOR' + RCOCH_2SH \xrightarrow{\text{base}} (H_3C)(H)C{=}C(COOR')(SCH_2COR) \longrightarrow \text{2-R-3-COOR}'\text{-4-}CH_3\text{-thiophene} \tag{251}$$

probably determined by proton availability, *cis–trans* conversion rates of the anion, and *cis–trans* stabilities. Quantitative work is not yet available, however.

When sulfur and nitrogen centers are both present in the nucleophile, it is usually the former that leads the attack. In process 252 no amine attack was

$$(C_6H_5C{\equiv}C)_2C{=}O + 2{-}H_2NC_6H_4SH \longrightarrow (C_6H_5(2{-}H_2NC_6H_4S)C{=}CH)_2CO \tag{252}$$

observed (86). Although both centers react in eq. 253, the nature of the products usually indicates the greater nucleophilicity of the thiolate. Incidentally reactions 253 and 254 correct earlier versions of these additions. Equation 255 is to be compared with eq. 179 in which the ortho-aminobenzamide proceeds under similar conditions to yield an azepine.

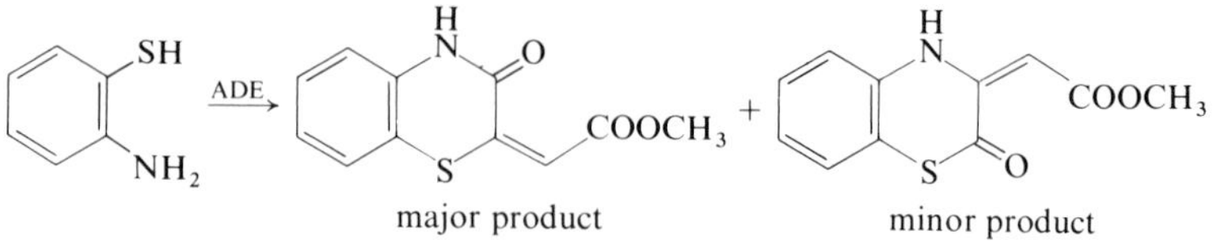

(253)

major product

minor product

(from ref. 285)

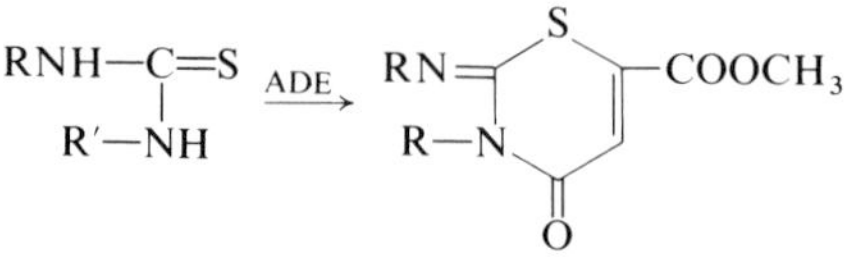

$RNH{-}C{=}S$ / $R'{-}NH$ $\xrightarrow{ADE}$ (254)

(from ref. 230)

($R = CH_3, CH_2{=}CH, C_6H_5$; $R = NH_2, CH_3$)

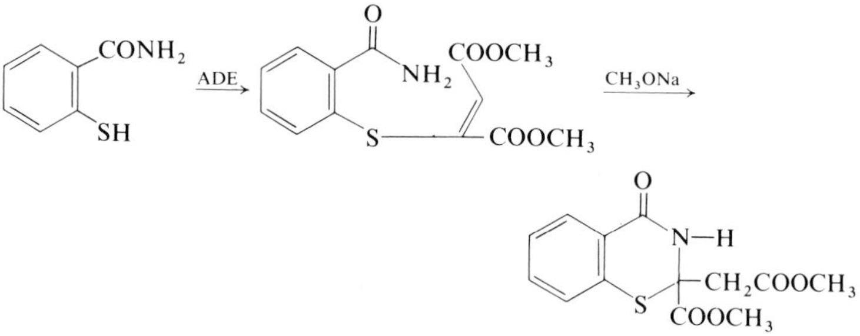

$\xrightarrow{ADE}$ $\xrightarrow{CH_3ONa}$ (255)

(from ref. 239)

When oxygen and sulfur are both present in the nucleophile, the latter leads the attack (284). Cyclization of the product in 256 leads to a benzoxathiole rather than the thiophene of eq. 250.

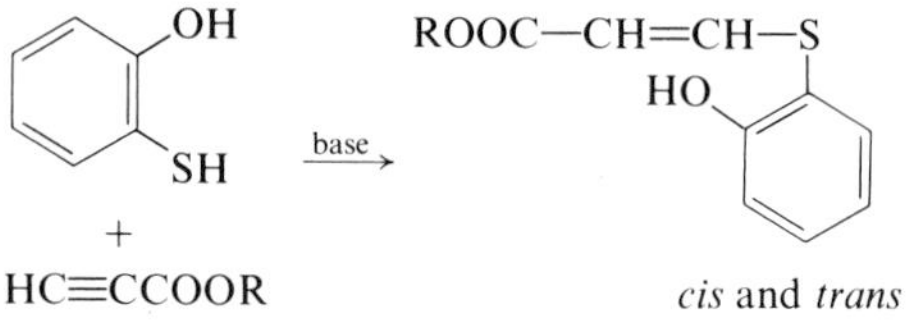

$HC{\equiv}CCOOR$ $\xrightarrow{base}$ $ROOC{-}CH{=}CH{-}S$ (256)

cis and *trans*

G. Selenium Nucleophiles

A number of examples involving divalent selenium in process 1 have been reported (e.g., eq. 82). These appear to be parallel to the results with analogous thiols (154,287–289):

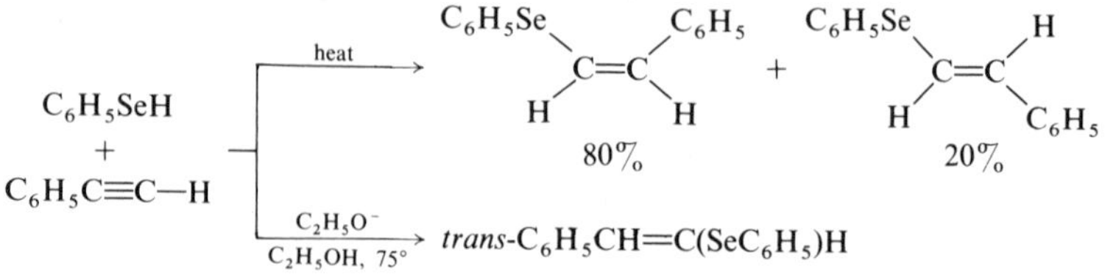

$$C_6H_5SeH + C_6H_5C{\equiv}C{-}H \xrightarrow{heat} \text{(Z)-}C_6H_5Se(H)C{=}C(H)C_6H_5\ (80\%) + \text{(E)-}C_6H_5Se(H)C{=}C(H)C_6H_5\ (20\%) \tag{257}$$

$$C_6H_5SeH + C_6H_5C{\equiv}C{-}H \xrightarrow[C_2H_5OH,\ 75^\circ]{C_2H_5O^-} trans\text{-}C_6H_5CH{=}C(SeC_6H_5)H$$

Additions of selenophenols or hydrogen selenide to activated alkynes have also been reported (288,289). It seems generally agreed that these base-catalyzed additions proceed chiefly in the *anti* sense. Even the *syn* adduct in

$$\text{H}_3\text{C-(2-pyridyl)-C}\equiv\text{CH} + \text{C}_6\text{H}_5\text{Se}^- \longrightarrow \text{H}_3\text{C-(2-pyridyl)-C(H)=C(H)SeC}_6\text{H}_5 \qquad (258)$$

(from refs. 154, 287)

eq. 257 is rationalized by a heterolytic cyclic process, which is given by the Russian workers (287) as shown in

$$\text{C}_6\text{H}_5\text{-C}\equiv\text{C-H} \cdots \text{H-Se(C}_6\text{H}_5\text{)}\cdots\text{H-Se-C}_6\text{H}_5$$

H. Miscellaneous Additions

In this section we comment cursorily on several reagents which are not usually classed with nucleophiles. These include several reductants, metal hydrides, as well as organometallic reagents. Here we shall do little more than indicate the connections of our subject with that of the areas of reduction and organometallic chemistry, see Table 10. As for the additions of halide, azide, and acetate, these have been essentially described in part III and Tables 1 and 2 concerned with rates and mechanisms.

Nucleophilic reduction of alkynes may be initiated by a solvated electron, an electron generated electrochemically, or by various metal hydrides. It has been suggested that a dissolving metal, usually from periodic groups I or II gives up electrons to ammonia, amines, ethers, and hydroxylic solvents (291,292). Alternatively, electrolysis in a medium of amine and salt produces

$$\text{D-C}\equiv\text{C-D} \xrightarrow{\text{Na/C}_2\text{H}_5\text{OH}} \text{(H)(D)C=C(D)(H)} \qquad (259)$$

(from ref. 296)

$$\text{cyclic alkyne} \xrightarrow{\text{K/NH}_3} \text{cyclic alkene} \qquad (260)$$

(from ref. 293)

$$n\text{-C}_3\text{H}_7\text{C}\equiv\text{C(CH}_2)_4\text{C}\equiv\text{CH} \xrightarrow[2.\ \text{NH}_4\text{Cl}]{1.\ \text{Na/NH}_3} \text{(H)(}n\text{-H}_7\text{C}_3\text{)C=C(H)(CH}_2)_4\text{C}\equiv\text{CH} \qquad (261)$$

(from ref. 294)

electrons (295). In the presence of proton donors electrons from both these sources initiate chiefly *anti* reduction of the alkyne (51,52,292–295). A

$$n\text{-}C_2H_5{-}C{\equiv}C{-}C_4H_9\text{-}n \xrightarrow[\text{electrolysis}]{CH_3NH_2,\ LiCl} \text{octenes} \tag{262}$$

(33% *trans*-2-; 4% *cis*-2-; 37% *trans*-3-; 24% *trans*-4-) (from ref. 295)

reasonable mechanism is given in scheme 263. Incidentally, direct transfer of the electron from the metal to the alkyne is not precluded.

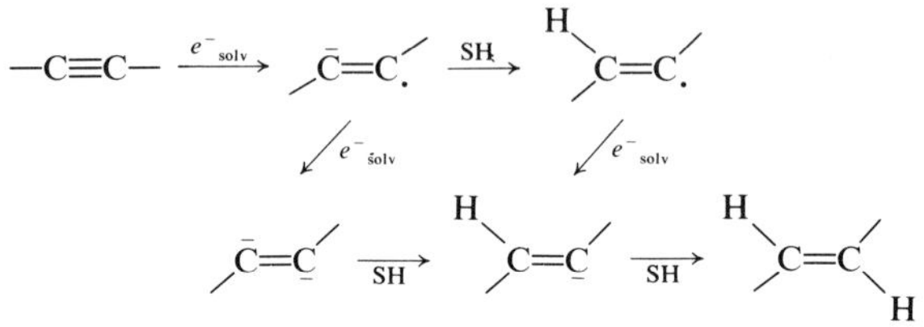

$$-C{\equiv}C- \xrightarrow{e^-_{solv}} \bar{C}{=}\dot{C} \xrightarrow{SH} (H)C{=}\dot{C}$$
$$\downarrow e^-_{solv} \qquad\qquad \downarrow e^-_{solv} \tag{263}$$
$$\bar{C}{=}\bar{C} \xrightarrow{SH} (H)C{=}\bar{C} \xrightarrow{SH} (H)C{=}C(H)$$

Under these reducing conditions terminal alkynes may be converted to metal acetylides which do not accept the electron readily (eq. 261). If a proton source more acidic than acetylene is present (e.g., eq. 259), the reduction proceeds normally.

These reductions may produce a number of isomeric compounds. Base-catalyzed isomerization of an alkyne is a typical complication (122). In the present context there is the case of cyclodecyne for which relative rates have

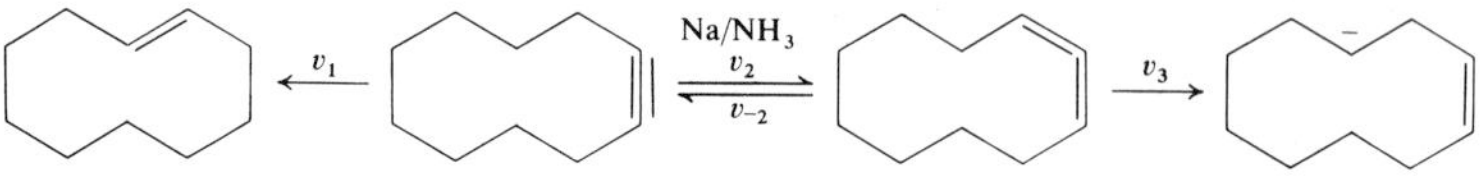

$$\textit{trans}\text{-cyclodecene} \xleftarrow{v_1} \text{cyclodecyne} \underset{v_{-2}}{\overset{Na/NH_3,\ v_2}{\rightleftharpoons}} \text{allene} \xrightarrow{v_3} \textit{cis}\text{-cyclodecene} \tag{264}$$

$$(v_2 > v_3 \gg v_1;\ v_2 \simeq v_{-2})$$

been deduced (293). Likewise, Benkeser and Tincher attribute the variety of products in eq. 262 to pre- or postisomerization rather than to the lack of *anti* selectivity in the reduction step (295).

Although lithium aluminum hydride does not reduce simple alkynes readily (297), highly stereoselective reductions are feasible at ca. 110–130° in an autoclave (317). On treatment with lithium aluminum hydride and then water, 2-pentyne yields >98% *trans*-pentene in tetrahydrofuran and approximately equal amounts of the *cis*- and *trans*-pentenes in toluene. In ether solvents it is proposed that the *anti* adduct,

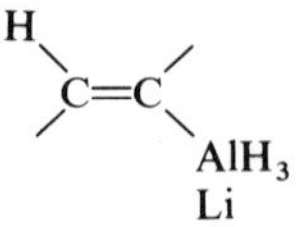

$$(H)C{=}C(AlH_3Li)$$

reacts with the water to give the alkene (317). Substituted prop-2-ynyl alcohols and certain polyunsaturated alcohols are conveniently reduced (298,299,318). Hydride is apparently transferred in the slow step within an aluminate complex (298). The final product is an allene.

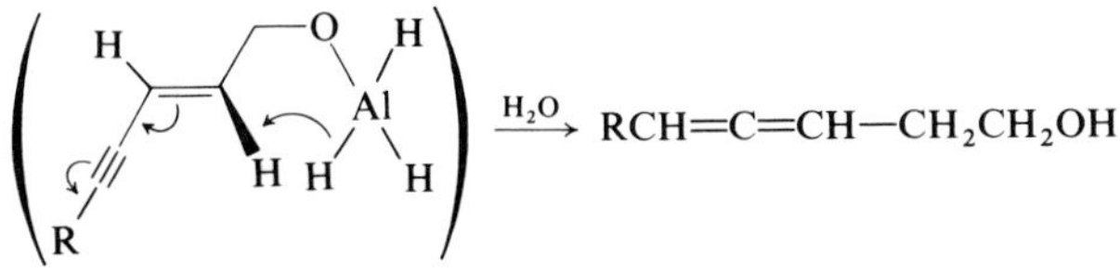

$$\left(\text{aluminate complex: R–C}\equiv\text{C–CH=C(H)–CH}_2\text{–O–AlH}_2\text{–H}\right) \xrightarrow{H_2O} RCH{=}C{=}CH{-}CH_2CH_2OH \qquad (265)$$

(from ref. 299)

The attacks of other metal hydrides may begin with the electrophilic site (eqs. 266, 267), may be initiated by radicals (eq. 268) or take place within the metal coordination sphere (13).

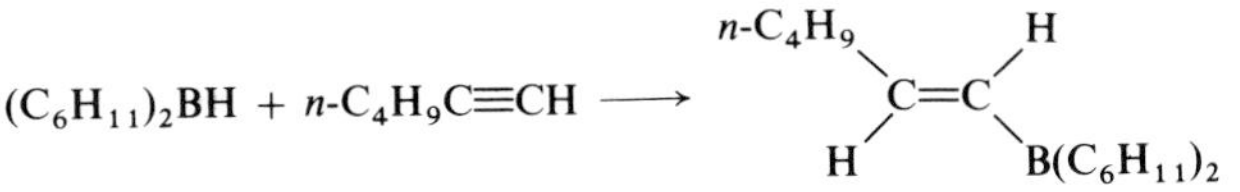

$$(C_6H_{11})_2BH + n\text{-}C_4H_9C{\equiv}CH \longrightarrow (n\text{-}C_4H_9)(H)C{=}C(H)(B(C_6H_{11})_2) \qquad (266)$$

(from ref. 300)

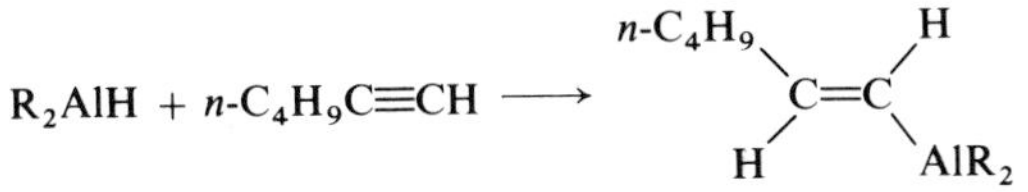

$$R_2AlH + n\text{-}C_4H_9C{\equiv}CH \longrightarrow (n\text{-}C_4H_9)(H)C{=}C(H)(AlR_2) \qquad (267)$$

(from ref. 301)

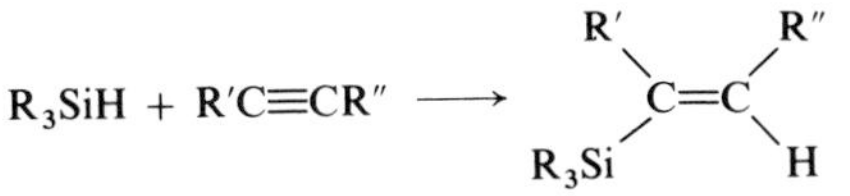

$$R_3SiH + R'C{\equiv}CR'' \longrightarrow (R')(R_3Si)C{=}C(R'')(H) \qquad (268)$$

(from refs. 302, 303)

This leaves a few hydrides whose reactions appear to be genuinely nucleophilic (see also eq. 89)

$$(CH_3)_2AsH + HC{\equiv}CCF_3 \longrightarrow (CH_3)_2AsCH{=}CHCF_3 \qquad (269)$$

(from ref. 304)

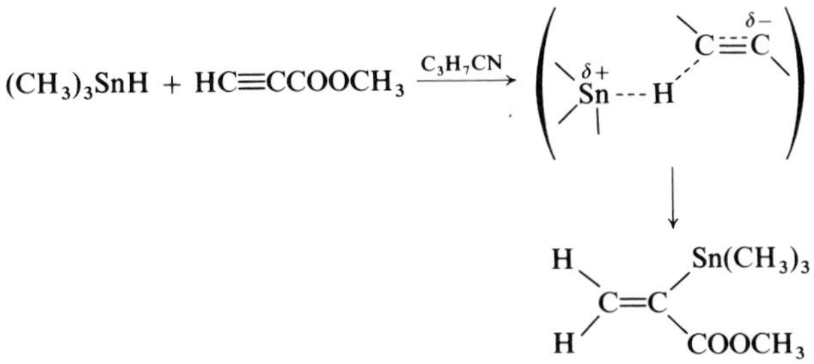

$$(CH_3)_3SnH + HC{\equiv}CCOOCH_3 \xrightarrow{C_3H_7CN} \left(\overset{\delta+}{Sn}\text{---}H\text{---}C{\equiv}\overset{\delta-}{C}\right) \longrightarrow (H)(H)C{=}C(Sn(CH_3)_3)(COOCH_3) \qquad (270)$$

(from ref. 311)

A minor concurrent process in eq. 270 is radical in character and different in directioselectivity; it yields a *cis–trans* mixture (311).

The additions of certain metal complex anions to alkynes seem to conform to the Ad_N process:

$$F_3C{-}C{\equiv}CH \xrightarrow[THF]{Fe(CO)_2cp} F_3C{-}CH{=}CH{-}Fe(CO)_2cp \quad (271)$$

cis

(cp = cyclopentadienyl) (from ref. 305)

$$C_6F_5{-}C{\equiv}CH + \text{anion complexes} \xrightarrow{THF} \begin{cases} HC_6F_4CH{=}CH(Fe(CO)_2cp) \\ C_6F_5CH{=}CH(Mn(CO)_4P(C_6H_5)_3) \end{cases} \quad (272)$$

(from ref. 306)

TABLE 10
Miscellaneous Stereoselective Nucleophilic (**N**) Additions to Alkynes $R{-}C{\equiv}C{-}R'$ to give $R{-}CN = C(\mathbf{E}){-}R'$[a]

R, R′	N(E)	Conditions	*Anti*[b] adduct, %	Ref.
H, C_6H_5CO	$(CF_3)_3C^-(H^+)$	$(C_2H_5)_3N$	100	307
H, $PO(OC_3H_7\text{-}i)_2$	$(C_2H_5OOC)_2CH^-(H^+)$	C_2H_5OH	?[c]	156
C_6H_5, $P(C_6H_5)_3{}^+$	*N*-Morpholino-1-cyclohexene	?	?	157
H, CN	$H^-(Sn(CH_3)_3{}^+)$	Neat	100	311*
H, $SC_4H_9\text{-}n$	$H^-(Sn(C_2H_5)_3{}^+)$	Neat	0[c]	311*
NC, CN	$H^-(Sn(CH_3)_3{}^+)$	Neat	100	311*
H, $COOCH_3$	$D^-(Sn(C_2H_5)_3{}^+)$	Neat	93	312*
H, CF_3	$Re(CO)_5{}^-(H^+)$	THF	100	305*
H, C_6F_5	$Mn(CO)_5{}^-(H^+)$	THF	?	306*
C_2H_5, C_2H_5	$H^-(LiAlH_3{}^+)$	THF, 125°	~100	317*
C_6H_5, CH_3	$H^-(LiAlH_3{}^+)$	THF, 66°	>99, 8	317*

[a] This table is intended to be representative rather than exhaustive. Additional examples and leading references are indicated by an asterisk on the citation.

[b] This is the fraction of the total product that is *anti*.

[c] This may not be the kinetically controlled yield.

References

1. P. B. D. De La Mare and R. Bolton, *Electrophilic Additions to Unsaturated Systems*, Elsevier, Amsterdam, 1966.
2. S. Patai and Z. Rappoport in *The Chemistry of Alkenes*, S. Patai, Ed., Interscience, London, ch. 8.
3. E. Winterfeldt, *Angew. Chem. Intern. Ed.*, **6**, 423 (1967).

4. F. Bohlmann, *Angew. Chem.*, **69**, 82 (1957).
5. I. A. Chekulaeva and L. V. Kondrat'eva, *Usp. Khim.*, **34**, 1583 (1965); *Russian Chem. Rev.*, **34**, 669 (1965).
6. A. W. Johnson, *The Chemistry of Acetylene Compounds*, Vol. II, Edward Arnold, London, 1950.
7. R. A. Raphael, *Acetylenic Compounds in Organic Synthesis*, Academic Press, New York, 1955.
8. M. F. Shostakovskii, A. V. Bogdanova and G. I. Plotnikova, *Usp. Khim.*, **33**, 129 (1964); *Russian Chem. Rev.*, **33**, 66 (1964); M. F. Shostakovskii, E. P. Gracheva, and N. K. Kul'bovskaya, *Usp. Khim.*, **30**, 493 (1961).
9. R. M. Acheson, *Advan. Heterocyclic Chem.*, **1**, 125 (1963).
10. S. I. Miller and R. M. Noyes, *J. Am. Chem. Soc.*, **74**, 3403 (1952).
11. A. Dondoni, *Tetrahedron Letters*, 2397 (1967).
12. J. Hamer, *1,4-Cycloaddition Reactions*, Academic Press, New York, 1967.
13. J. P. Candlin, K. A. Taylor, and D. T. Thompson, *Reactions of Transition-Metal Complexes*, Elsevier, New York, 1968, pp. 238–250.
14. S. I. Miller, *Advan. Phys. Org. Chem.*, **6**, 185 (1968).
15. R. W. Hoffmann, *Dehydrobenzene and Cycloalkynes*, Academic Press, New York, 1967.
16. H. G. Viehe, *Angew. Chem. Intern. Ed.*, **6**, 767 (1967).
17. A. Hassner, *J. Org. Chem.*, **33**, 2684 (1968).
18. A. Favorsky, *J. Prakt. Chem.*, **37**, 382, 532 (1888).
19. E. Bucherer, *Chem. Ber.*, **22**, 2929 (1889).
20. A. Michael and J. E. Bucher, *Chem. Ber.*, **29**, 1792 (1896).
21. W. Reppe, *Ann.*, **601**, 81 (1956).
22. S. A. Miller, *Acetylene*, Academic Press, New York, 1965, pp. 23ff.
23. T. C. James and J. J. Sudborough, *J. Chem. Soc.*, **91**, 1042 (1907).
24. E. A. Moelwyn-Hughes and A. R. Legard, *J. Chem. Soc.*, **1933**, 424.
25. E. Hagglund and A. Ringborn, *Z. Anorg. Allgem. Chem.*, **169**, 96 (1928).
26. E. E. Gilbert, *Sulfonation and Related Reactions*, Interscience, London, 1965, pp. 151–152.
27. I. V. Smirnov-Zamkov and G. A. Piskovitina, *Ukr. Khim. Zh.*, **30**, 1076 (1964); *Chem. Abstr.*, **62**, 5160 (1965).
28. I. V. Smirnov-Zamkov and E. A. Shilov, *Dokl. Akad. Nauk SSSR*, **67**, 671 (1949); *Chem. Abstr.*, **43**, 8824 (1949).
29. E. A. Shilov and I. V. Smirnov-Zamkov, *Izv. Akad. Nauk SSSR, Ser. Khim.*, **1951**, 32; *Chem. Abstr.*, **45**, 7952 (1951).
30. E. A. Shilov and A. E. Shilov, *Ukr. Khim. Zh.*, **20**, 279 (1954); *Chem. Abstr.*, **50**, 185 (1956).
31. E. A. Shilov and A. E. Shilov, *Dokl. Akad. Nauk SSSR*, **91**, 873 (1953); *Chem. Abstr.*, **48**, 10557 (1954).
32. V. G. Ostroverkhov and E. A. Shilov, *Ukr. Khim. Zh.*, **23**, 615 (1957); *Chem. Abstr.*, **52**, 7828 (1958).
33. V. G. Ostroverkhov, *Ukr. Khim. Zh.*, **23**, 474 (1957); *Chem. Abstr.*, **52**, 6196 (1958).
34. V. G. Ostroverkhov and E. A. Shilov, *Ukr. Khim. Zh.*, **22**, 743 (1956); *Chem. Abstr.*, **51**, 7813 (1957).
35. V. G. Ostroverkhov and E. A. Shilov, *Ukr. Khim. Zh.*, **22**, 590 (1956); *Chem. Abstr.*, **51**, 6515 (1957).
36. G. F. Dvorko and D. F. Mironova, *Ukr. Khim. Zh.*, **32**, 362 (1966); *Chem. Abstr.*, **65**, 7014 (1966).
37. G. F. Dvorko and D. F. Mironova, *Ukr. Khim. Zh.*, **31**, 1289 (1965); *Chem. Abstr.*, **65**, 7014 (1966).

38. G. F. Dvorko, T. F. Karpenko, D. F. Mironova, and E. A. Shilov, *Ukr. Khim. Zh.*, **31**, 1177 (1965); *Chem. Abstr.*, **65**, 7014 (1966).
39. G. F. Dvorko and D. F. Mironova, *Ukr. Khim. Zh.*, **31**, 195 (1965).
40. G. F. Dvorko and T. F. Karpenko, *Ukr. Khim. Zh.*, **31**, 75 (1965); *Chem. Abstr.*, **62**, 14465 (1965).
41. G. F. Dvorko, *Ukr. Khim. Zh.*, **29**, 1188 (1963); *Chem. Abstr.*, **60**, 6713 (1964).
42. G. F. Dvorko and E. A. Shilov, *Ukr. Khim. Zh.*, **29**, 165 (1963); *Chem. Abstr.*, **59**, 6222 (1963).
43. G. F. Dvorko and E. A. Shilov, *Ukr. Khim. Zh.*, **29**, 73 (1963); *Chem. Abstr.*, **59**, 6222 (1963).
44. G. F. Dvorko and E. A. Shilov, *Ukr. Khim. Zh.*, **28**, 1073 (1962); *Chem. Abstr.*, **59**, 6222 (1963).
45. G. F. Dvorko and E. A. Shilov, *Ukr. Khim. Zh.*, **28**, 626 (1962); *Chem. Abstr.*, **58**, 12385 (1963).
46. G. F. Dvorko, *Dopov. Akad. Nauk. Ukr. RSR.*, **1959**, 498; *Chem. Abstr.*, **54**, 2162 (1960).
47. G. F. Dvorko and E. A. Shilov, *Ukr. Khim. Zh.*, **28**, 833 (1962); *Chem. Abstr.*, **59**, 1449 (1963).
48. A. Michael, *J. Prakt. Chem.*, **52**, 305 (1895).
49. P. F. Frankland, *J. Chem. Soc.*, **1912**, 654.
50. P. Pfeiffer, *Z. Phys. Chem.*, **48**, 40 (1904).
51. S. Goldschmidt, in *Stereochemie*, Vol. 1, K. Freudenberg, Ed., Franz Deuticke, Leipzig, 1933, p. 556ff.
52. C. S. Marvel, in *Organic Chemistry*, Vol. I, H. Gilman, Ed. Wiley, New York, 1943, pp. 460–462.
53. C. K. Ingold, *Structure and Mechanism in Organic Chemistry*, Cornell University Press, Ithaca, 1953, ch. 7, 8.
54. P. D. Bartlett, *J. Am. Chem. Soc.*, **57**, 224 (1935).
55. S. Winstein, D. Pressman, and W. G. Young, *J. Am. Chem. Soc.*, **61**, 1645 (1939).
56. W. E. Truce, J. A. Simms, and M. M. Boudakian, *J. Am. Chem. Soc.*, **78**, 695 (1956).
57. W. E. Truce, M. M. Boudakian, R. F. Heine, and R. J. McManimie, *J. Am. Chem. Soc.*, **78**, 2743 (1956).
58. W. E. Truce and M. M. Boudakian, *J. Am. Chem. Soc.*, **78**, 2748 (1956).
59. W. E. Truce and M. M. Boudakian, *J. Am. Chem. Soc.*, **78**, 2752 (1956).
60. W. E. Truce and J. A. Simms, *J. Am. Chem. Soc.*, **78**, 2756 (1956).
61. S. I. Miller, *J. Am. Chem. Soc.*, **78**, 6091 (1956).
62. S. I. Miller and G. Shkapenko, *J. Am. Chem. Soc.*, **77**, 5038 (1955).
63. F. Montanari, *Boll. Sci. Fac. Chim. Ind. Bologna*, **16**, 31 (1958).
64. L. Maioli and G. Modena, *Boll. Sci. Fac. Chem. Ind. Bologna*, **16**, 86 (1958).
65. F. Montanari, *Gazz. Chim. Ital.*, **86**, 736, 747 (1956).
66. F. Montanari and A. Negrini, *Gazz. Chim. Ital.*, **87**, 1073 (1957).
67. F. Montanari and A. Negrini, *Gazz. Chim. Ital.*, **87**, 1061 (1957).
68. M. L. Dhar, E. D. Hughes, C. K. Ingold, A. M. Mandour, G. A. Maw, and L. I. Woolf, *J. Chem. Soc.*, **1948**, 2118.
69. C. K. Ingold, *J. Chem. Soc.*, **1954**, 2991.
70. L. Burnelle, *Tetrahedron*, **20**, 2403 (1964).
71. L. Burnelle, *Tetrahedron*, **21**, 49 (1965).
72. R. B. Woodward and R. Hoffmann, *J. Am. Chem. Soc.*, **87**, 395 (1965).
73. R. Hoffmann and R. B. Woodward, *J. Am. Chem. Soc.*, **87**, 2046 (1965).
74. R. B. Woodward and R. Hoffmann, *J. Am. Chem. Soc.*, **87**, 2511 (1965).
75. R. Hoffmann and R. B. Woodward, *J. Am. Chem. Soc.*, **87**, 4388, 4389 (1965).

76. R. B. Woodward, in *Aromaticity*, Special Publication No. 21, The Chemical Society, London, 1967.
77. K. Fukui, *Tetrahedron Letters*, **1965**, 2009.
78. K. Fukui, *Tetrahedron Letters*, **1965**, 2427.
79. K. Fukui and H. Fujimoto, *Tetrahedron Letters*, **1965**, 4303.
80. K. Fukui and H. Fujimoto, *Bull. Chem. Soc. Japan*, **39**, 2116 (1966); **40**, 2018 (1967).
81. K. Fukui and H. Fujimoto, *Tetrahedron Letters*, **1966**, 251.
82. K. Fukui, *Bull. Chem. Soc. Japan*, **39**, 498 (1966).
83. A. M. Aguiar and T. G. Archibald, *Tetrahedron Letters*, **1966** 5471, 5541.
84. L. Di Nunno, G. Modena, and G. Scorrano, *J. Chem. Soc.*, (*B*), **1966**, 1186.
85. H. H. Schlubach and W. Rott, *Ann.*, **594**, 59 (1955); H. H. Schlubach and E. W. Trautschold, *ibid.*, **594**, 67 (1955).
86. T. Metler, A. Uchida, and S. I. Miller, *Tetrahedron*, **24**, 4285 (1968).
87. G. Märkl and H. Hauptmann, *Tetrahedron Letters*, **1968**, 3257.
88. R. M. Acheson and R. S. Feinberg, *J. Chem. Soc. (C)* **1968**, 351.
89. E. Winterfeldt, *Chem. Ber.*, **98**, 1581 (1965).
90. J. B. Hendrickson, R. Rees, and J. F. Templeton, *J. Am. Chem. Soc.*, **86**, 107 (1964).
91. A. Maccoll and P. J. Thomas, *Progr. Reaction Kinetics*, **4**, 119 (1967).
92. C. A. Coulson and A. Streitwieser, *Dictionary of Π-Electron Calculations*, W. H. Freeman, San Francisco, 1965.
93. L. E. Sutton, Ed., *Tables of Interatomic Distances and Configurations in Molecules and Ions*, The Chemical Society, London, 1958.
94. D. V. Banthorpe, *Elimination Reactions*, Elsevier, Amsterdam, 1963.
95. W. E. Truce, H. G. Klein, and R. B. Kruse, *J. Am. Chem. Soc.*, **83**, 4636 (1961).
96. G. Huett and S. I. Miller, *J. Am. Chem. Soc.*, **83**, 408 (1961).
97. S. W. Benson and G. R. Haugen, *J. Am. Chem. Soc.*, **87**, 4036 (1965).
98. S. W. Benson and A. N. Bose, *J. Chem. Phys.*, **39**, 3463 (1963).
99. W. E. Truce and R. F. Heine, *J. Am. Chem. Soc.*, **79**, 1770, 5311 (1957).
100. W. E. Truce and D. L. Goldhamer, *J. Am. Chem. Soc.*, **81**, 5795, 5798 (1959).
101. F. Montanari, *Tetrahedron Letters*, **1960**, 18.
102. W. E. Truce, W. Bannister, B. Groten, H. Klein, R. Kruse, A. Levy, and E. Roberts, *J. Am. Chem. Soc.*, **82**, 3799 (1960).
103. S. Otsuka and S. Murahashi, *Kogyo Kagaku Zasshi*, **59**, 511 (1956).
104. T. Tsuruta, H. Komatsubara, and J. Furukawa, *Bull. Chem. Soc. Japan*, **28**, 552 (1955).
105. R. C. Fahey and D-J. Lee, *J. Am. Chem. Soc.*, **89**, 2780 (1967).
106. R. C. Fahey and D-J. Lee, *J. Am. Chem. Soc.*, **90**, 2124 (1968).
107. E. S. Wassermann and A. B. Bedrintzeva, *Compt. Rend. Acad. Sci. U.S.S.R.*, **33**, 34 (1941); *Chem. Abstr.*, **37**, 6539 (1943).
108. R. Rigamonti and L. Bernardi, *Chim. Ind.*, **34**, 561 (1952).
109. S. Otsuka, Y. Matsui, and S. Murahashi, *Nippon Kagaku Zasshi*, **77**, 766 (1956).
110. S. Otsuka, Y. Matsui, and S. Murahashi, *Nippon Kagaku Zasshi*, **80**, 1153 (1959).
111. S. Otsuka, Y. Matsui, and S. Murahashi, *Nippon Kagaku Zasshi*, **80**, 1157 (1959).
112. M. Ban, T. Yamamoto, and S. Otsuka, *Nippon Kagaku Zasshi*, **77**, 176 (1956).
113. J. Furukawa, T. Ando, and M. Yokoyama, *Bull. Inst. Chem. Res., Kyoto Univ.*, **31**, 220 (1953).
114. C. H. Rochester, *Quart. Rev.*, **20**, 511 (1966).
115. S. D. Ross, W. A. Leach, and I. Kuntz, *J. Am. Chem. Soc.*, **74**, 2908 (1952).
116. S. I. Miller and P. K. Yonan, *J. Am. Chem. Soc.*, **79**, 5931 (1957).
117. W. E. Truce, in *Organic Sulfur Compounds*, Vol. I, N. Kharasch, Ed., Pergamon Press, London, 1961, ch. 12.

118. V. Concialini, G. Modena, and F. Taddei, *Boll. Sci. Fac. Chim. Ind., Bologna*, **21**, 207 (1963).
119. G. R. Ziegler, C. A. Welch, C. E. Orzech, S. Kikkawa, and S. I. Miller, *J. Am. Chem. Soc.*, **85**, 1648 (1963).
120. A. Favorsky, *J. Prakt. Chem.*, **44**, 208 (1891).
121. I. M. Mathai, H. Taniguchi, and S. I. Miller, *J. Am. Chem. Soc.*, **89**, 115 (1967).
122. H. Taniguchi, I. M. Mathai, and S. I. Miller, *Tetrahedron*, **22**, 867 (1966).
123. C. J. M. Stirling, *J. Chem. Soc.*, **1964**, 5856.
124. K. Herbig, R. Huisgen, and H. Huber, *Chem. Ber.*, **99**, 2546 (1966).
125. J. F. Arens, *Advan. Org. Chem.*, **2**, 117 (1960); W. Drenth, in *The Chemistry of Organic Sulfur Compounds*, Vol. II, N. Kharasch and C. Y. Meyers, Eds., Pergamon Press, Oxford, 1966, ch. 7.
126. K. Bowden, E. A. Braude, and E. R. H. Jones, *J. Chem. Soc.*, **1946**, 945.
127. B. P. Gusev, E. A. Él'perina and V. F. Kucherov, *Izv. Akad. Nauk SSSR Ser. Khim.*, **1967**, 2345; *ibid.*, **1966**, 1803.
128. A. G. Evans, E. D. Owen, and B. D. Phillips, *J. Chem. Soc.*, **1964**, 5021.
129. R. C. Fahey and D-J. Lee, *J. Am. Chem. Soc.*, **88**, 5555 (1966).
130. P. E. Peterson and J. E. Duddey, *J. Am. Chem. Soc.*, **88**, 4990 (1966).
131. M. A. Shaw, J. C. Tebby, J. Ronayne, and D. H. Williams, *J. Chem. Soc. (C)*, **1967**, 944.
132. G. W. Koeppl, D. S. Sagatys, G. S. Krishnamurthy, and S. I. Miller, *J. Am. Chem. Soc.*, **89**, 3396 (1967).
133. S. I. Miller and W. G. Lee, *J. Am. Chem. Soc.*, **81**, 6313 (1959).
134. J. G. Pritchard and A. A. Bothner-By, *J. Phys. Chem.*, **64**, 1271 (1960).
135. D. H. Hunter and D. J. Cram, *J. Am. Chem. Soc.*, **86**, 5478 (1968).
136. R. Huisgen, B. Giese, and H. Huber, *Tetrahedron Letters*, **1967**, 1883.
137. A. L. Henne and M. Nager, *J. Am. Chem. Soc.*, **74**, 650 (1952).
138. A. L. Henne, J. V. Schmitz, and W. G. Finnegan, *J. Am. Chem. Soc.*, **72**, 4195 (1950).
139. S. J. Cristol, A. Begoon, W. P. Norris and P. S. Ramey, *J. Am. Chem. Soc.*, **76**, 4559 (1954).
140. G. Marchese, G. Modena, and F. Naso, *Tetrahedron*, **24**, 663 (1968).
141. I. J. Solomon and R. Filler, *J. Am. Chem. Soc.*, **85**, 3492 (1963).
142. G. S. Krishnamurthy and S. I. Miller, *J. Am. Chem. Soc.*, **83**, 3961 (1961).
143. R. G. Pearson, *J. Chem. Educ.*, **45**, 581, 643 (1968).
144. G. B. Barlin and D. D. Perrin, *Quart. Rev.*, **20**, 75 (1966).
145. J. Hine and M. Hine, *J. Am. Chem. Soc.*, **74**, 5267 (1952).
146. E. D. Holly, *J. Org. Chem.*, **24**, 1752 (1959).
147. P. R. Wells, *Chem. Rev.*, **63**, 171 (1963).
148. R. Alexander, E. C. F. Ko, A. J. Parker, and T. J. Broxton, *J. Am. Chem. Soc.*, **90**, 5049 (1968).
149. A. J. Parker, *Advan. Phys. Org. Chem.*, **5**, 173 (1967); *Advan. Org. Chem.*, **5**, 1 (1965); *Quart. Rev.*, **16**, 163 (1962).
150. F. Franks and D. J. G. Ives, *Quart. Rev.*, **20**, 1 (1966).
151. A. L. McClellan, *Tables of Experimental Dipole Moments*, W. H. Freeman, San Francisco, 1963.
152. M. M. Davis and H. B. Hetzer, *J. Natl. Bur. Std.*, **60**, 569 (1958).
153. B. Giese and R. Huisgen, *Tetrahedron Letters*, **1967**, 1889.
154. E. G. Kataev and V. N. Petrov, *Zh. Obshch. Khim.*, **32**, 3699 (1962); *Chem. Abst.*, **59**, 481 (1963).
155. H. Distler, *Angew. Chem. Intern. Ed.*, **6**, 544 (1967).
156. B. C. Saunders and P. Simpson, *J. Chem. Soc.*, **1963**, 3351.
157. H. Hoffmann and H. Förster, *Tetrahedron Letters*, **1964**, 983.
158. A. J. Leusink and G. J. M. Van Der Kerk, *Rec. Trav. Chim.*, **84**, 1617 (1965).

159. A. A. Petrov, *Usp. Khim., Russian Chem. Rev.*, **29**, 489 (1960).
160. E. N. Prilezhaeva, V. N. Petrov, and G. S. Vasil'ev, *Izv. Akad. Nauk SSSR, Ser. Khim.*, **1967**, 1879.
161. V. N. Petrov, G. M. Andrianova, and E. N. Prilezhaeva, *Izv. Akad. Nauk SSSR, Ser. Khim.*, **1966**, 2180.
162. J. R. Nooi and J. F. Arens, *Rec. Trav. Chim.*, **80**, 244 (1961).
163. P. P. Montijn, E. Harryvan, and L. Brandsma, *Rec. Trav. Chim.*, **83**, 1211 (1964).
164. H. G. Viehe and E. Franchimont, *Chem. Ber.*, **95**, 319 (1962).
165. W. K. Kwok, W. G. Lee, and S. I. Miller, *J. Am. Chem. Soc.*, **91**, 468 (1969).
166. V. Wolf and H. Piater, *Ann.*, **696**, 90 (1966); V. Wolf and W. Block, *ibid.*, **637**, 119 (1960).
167. J. U. Nef, *Ann.*, **308**, 264 (1899).
168. E. L. Eliel, *The Stereochemistry of Carbon Compounds*, McGraw-Hill, New York, 1962, p. 198.
169. M. Hanack, *Conformation Theory*, Academic Press, New York, 1965, p. 79.
170. J. J. Eisch and W. C. Kaska, *J. Am. Chem. Soc.*, **88**, 2213 (1966).
171. J. E. Mulvaney, Z. G. Gardlund, S. L. Gardlund, and D. J. Newton, *J. Am. Chem. Soc.*, **88**, 476 (1966); J. E. Mulvaney and L. J. Carr, *J. Org. Chem.*, **33**, 3286 (1968).
172. M. Tsutsui, *Chem. Ind. (London)*, **1962**, 780.
173. F. Bohlmann, E. Inhoffen, and J. Politt, *Ann.*, **604**, 207 (1957).
174. J. J. Eisch and W. C. Kaska, *J. Am. Chem. Soc.*, **84**, 1501 (1962).
175. J. G. A. Kooyman, H. P. G. Hendriks, P. P. Montijn, L. Brandsma, and J. F. Arens, *Rec. Trav. Chem.*, **87**, 69 (1968).
176. K. Ziegler and H. Dislich, *Chem. Ber.*, **90**, 1107 (1957).
177. S. A. Kandil and R. E. Dessy, *J. Am. Chem. Soc.*, **88**, 3027 (1966).
178. H. G. Richey, Jr. and A. M. Rothman, *Tetrahedron Letters*, **1968**, 1457.
179. M. Seefelder, *Ann.*, **652**, 107 (1962).
180. I. Iwai and J. Ide, *Chem. Pharm. Bull. (Tokyo)*, **13**, 663 (1965).
181. J. F. H. Braams, H. J. T. Bos, and J. F. Arens, *Rec. Trav. Chim.*, **86**, 193 (1968).
182. A. K. Kuriakose and S. I. Miller, *Tetrahedron Letters*, **1962**, 905.
183. M. V. Mavrov, A. R. Derzhinskii, and V. F. Kucherov, *Izv. Akad. Nauk. SSSR Ser. Khim.*, **1965**, 1460.
184. M. V. Mavrov and V. F. Kucherov, *Izv. Akad. Nauk. SSSR Ser. Khim.*, **1967**, 1559.
185. I. Iwai, in "Mechanisms of Molecular Migrations," Vol. 2, B. S. Thyagarajan, Ed., Wiley—Interscience, New York, 1969.
186. I. Iwai and J. Ide, *Chem. Pharm. Bull. (Tokyo)*, **10**, 926 (1962).
187. I. Iwai and J. Ide, *Chem. Pharm. Bull. (Tokyo)*, **12**, 1094 (1964).
188. E. LeGoff and R. B. LaCount, *J. Org. Chem.*, **29**, 423 (1964).
189. G. Stork and M. Tomasz, *J. Am. Chem. Soc.*, **86**, 471 (1964).
190. J. Ide and Y. Kushida, *Tetrahedron Letters*, **1966**, 1787.
191. C. W. Kruse and R. F. Kleinschmidt, *J. Am. Chem. Soc.*, **83**, 216 (1961).
192. K. C. Brannock, R. D. Burpitt, V. W. Goodlett, and J. G. Thweatt, *J. Org. Chem.*, **28**, 1964 (1963); *ibid.*, **29**, 818 (1964).
193. C. F. Huebner, L. Dorfman, M. M. Robinson, E. Donoghue, W. G. Pierson, and P. Strachan, *J. Org. Chem.*, **28**, 3134 (1963).
194. W. K. Gibson and D. Leaver, *J. Chem. Soc. (C)*, **1966**, 324.
195. L. Mandell and W. E. Blanchard, *J. Am. Chem. Soc.*, **79**, 6189 (1957).
196. G. A. Berchtold and G. F. Uhlig, *J. Org. Chem.*, **28**, 1459 (1963).
197. K. C. Brannock, R. D. Burpitt, and J. G. Thweatt, *J. Org. Chem.*, **28**, 1462 (1963).
198. K. E. Schulte, J. Reisch, and H. Walker, *Chem. Ber.*, **98**, 98 (1965).
199. T. Kaufmann and J. Sobel, *Chem. Ber.*, **99**, 1843 (1966).

200. W. R. Cullen, D. S. Dowson, and G. E. Styan, *Can. J. Chem.*, **43**, 3392 (1965).
201. S. I. Miller, *J. Org. Chem.*, **21**, 247 (1956).
202. E. Ott, G. Dittus and H. Weissenburger, *Chem. Ber.*, **76**, 84 (1943); E. Ott and G. Dittus, *ibid.*, **76**, 80 (1943).
203. T. L. Jacobs and W. R. Scott, *J. Am. Chem. Soc.*, **75**, 5497 (1953).
204. M. Mokosza, *Tetrahedron Letters*, **1966**, 5489.
205. S. K. Khetan, J. G. Hiriyakkanavar, and M. V. George, *Tetrahedron*, **24**, 1567 (1968).
206. R. M. Acheson and M. W. Foxton, *J. Chem. Soc. (C)*, **1968**, 389.
207. R. M. Acheson and P. A. Tasker, *J. Chem. Soc. (C)*, **1967**, 1542.
208. R. M. Acheson and W. R. Tully, *J. Chem. Soc. (C)*, **1968**, 1623.
209. C. H. McMullen and C. J. M. Stirling, *J. Chem. Soc. (B)*, **1966**, 1217.
210. W. E. Truce and D. G. Brady, *J. Am. Chem. Soc.*, **88**, 3543 (1966).
211. R. Huisgen, K. Herbig, A. Siegl, and H. Huber, *Chem. Ber.*, **99**, 2526 (1966); private communication.
212. Y. Iwanami, *Nippon Kagaku Zasshi*, **83**, 593 (1962).
213. E. Winterfeldt and H. Preuss, *Chem. Ber.*, **99**, 450 (1966).
214. J. H. Dolfini, *J. Org. Chem.*, **30**, 1298 (1965).
215. H. Reimlinger and C. H. Moussebois, *Chem. Ber.*, **98**, 1805 (1965).
216. E. Winterfeldt, W. Krohn, and H. Preuss, *Chem. Ber.*, **99**, 2572 (1966).
217. E. Winterfeldt, *Chem. Ber.*, **97**, 1952 (1964).
218. R. J. Alaimo and D. G. Farnum, *Can. J. Chem.*, **43**, 700 (1965).
219. R. M. Acheson and A. O. Plunkett, *J. Chem. Soc.*, **1964**, 2676.
220. R. Huisgen, M. Morikawa, K. Herbig, and E. Brunn, *Chem. Ber.*, **100**, 1094 (1967).
221. R. M. Acheson and D. A. Robinson, *J. Chem. Soc. (C)*, **1968**, 1629.
222. R. M. Acheson, J. M. F. Gagan, and D. R. Harrison, *J. Chem. Soc., (C)*, **1968**, 362.
223. M. Morikawa and R. Huisgen, *Chem. Ber.*, **100**, 1616 (1967).
224. R. M. Acheson and D. A. Robinson, *Chem. Commun.*, **1967**, 175; *J. Chem. Soc. (C)*, **1968**, 1633.
225. J. G. Wilson and W. Bottomley, *J. Heterocyclic Chem.*, **4**, 360 (1967).
226. R. M. Acheson, M. W. Foxton, and J. K. Stubbs, *J. Chem. Soc. (C)*, **1968**, 926.
227. H. R. Snyder, H. Cohen, and W. J. Tapp, *J. Am. Chem. Soc.*, **61**, 3561 (1939).
228. J. M. F. Gagan, *J. Chem. Soc. (C)*, **1966**, 1121.
229. E. C. Taylor and N. D. Heindel, *J. Org. Chem.*, **32**, 1666 (1967).
230. J. W. Lown and J. C. N. Ma, *Can. J. Chem.*, **45**, 939, 953 (1967).
231. Y. Iwanami, Y. Kenjo, K. Nishibe, M. Kajiura, and S. Isoyama, *Bull. Chem. Soc. Japan*, **37**, 1740 (1964).
232. N. D. Heindel, V. B. Fish, and T. F. Lemke, *J. Org. Chem.*, **33**, 3997 (1968).
233. A. N. Nesmeyanov and M. I. Rybinskaya, *Dokl. Akad. Nauk SSSR*, **158**, 408 (1964).
234. S. Ruhemann and H. E. Stapleton, *J. Chem. Soc.*, **77**, 239 (1900).
235. H. Sasaki, H. Sakata, and Y. Iwanami, *Nippon Kagaku Zasshi*, **85**, 704 (1964).
236. K. Bowden, E. A. Braude, and E. R. H. Jones, *J. Chem. Soc.*, **1946**, 953.
237. U. K. Pandit and H. O. Huisman, *Rec. Trav. Chim.*, **85**, 311 (1966).
238. J. Dickstein and S. I. Miller, unpublished data.
239. N. D. Heindel and V. B. Fish, *J. Org. Chem.*, **32**, 2678 (1967).
240. E. Winterfeldt, *Chem. Ber.*, **98**, 3537 (1965).
241. E. Winterfeldt and W. Krohn, *Angew. Chem. Intern. Ed.*, **6**, 709 (1967).
242. M. Ochiai, M. Obayashi, and K. Morita, *Tetrahedron*, **23**, 2641 (1967).
243. M. M. Rauhut, *J. Org. Chem.*, **26**, 5138 (1961).
244. H. Hoffmann and H. J. Diehr, *Chem. Ber.*, **98**, 363 (1965).
245. D. W. Allen and J. C. Tebby, *Tetrahedron*, **23**, 2795 (1967).

246. D. S. James and P. E. Fanta, *J. Org. Chem.*, **27**, 3346 (1962).
247. L. Lopez and J. Barrans, *Compt. Rend. Ser. C.*, **263**, 557 (1966).
248. L. D. Quin and H. G. Anderson, *J. Org. Chem.*, **31**, 1206 (1966).
249. L. Horner and H. Matsuda, *Angew. Chem. Intern. Ed.*, **3**, 231 (1964).
250. D. Seyferth and J. M. Burlitch, *J. Org. Chem.*, **28**, 2463 (1963).
251. M. A. Shaw, J. C. Tebby, R. S. Ward, and D. H. Williams, *J. Chem. Soc. (C)*, **1967**, 2442.
252. G. F. Dvorko and T. P. Travchook, *Org. Reactivity*, **5**, 995 (1968).
253. J. B. Hendrickson, R. E. Spenger, and J. J. Sims, *Tetrahedron*, **19**, 707 (1963).
254. J. B. Hendrickson, C. Hall, R. Rees, and J. F. Templeton, *J. Org. Chem.*, **30**, 3312 (1965).
255. G. S. Reddy and C. D. Weis, *J. Org. Chem.*, **28**, 1822 (1963).
256. M. A. Shaw, J. C. Tebby, R. S. Ward, and D. H. Williams, *J. Chem. Soc. (C)*, **1968**, 1609.
257. A. N. Hughes and S. Uaboonkul, *Tetrahedron*, **24**, 3437 (1968).
258. K. M. Kirillova, V. A. Kukhtin, and T. M. Sudakova, *Dokl. Akad. Nauk SSSR*, **149**, 316 (1963).
259. C. E. Griffin and T. D. Michell, *J. Org. Chem.*, **30**, 2829 (1965).
260. A. J. Kirby and S. G. Warren, *The Organic Chemistry of Phosphorus*, Elsevier, Amsterdam, 1967, pp. 45ff.
261. A. N. Pudovik and D. Kh. Yarmukhametova, *Izv. Akad. Nauk SSSR, Ser. Khim.*, **1954**, 636.
262. A. N. Pudovik and R. G. Kuzovleva, *Zh. Obshch. Khim.*, **35**, 354 (1965).
263. G. M. Campbell and I. D. R. Stevens, *Chem. Commun.*, **1966**, 505.
264. H. J. Bestmann and O. Rothe, *Angew. Chem.*, **76**, 569 (1964).
265. G. W. Brown, R. C. Cookson, and I. D. R. Stevens, *Tetrahedron Letters*, **1964**, 1263.
266. S. T. D. Gough and S. Trippett, *Proc. Chem. Soc.*, **1961**, 302.
267. E. Zbiral, *Tetrahedron Letters*, **1964**, 3963.
268. M. F. Shostakovskii, A. S. Atabin, A. I. Mikhalova, N. P. Vasil'ev, and L. P. Dmitrieva, *Izv. Akad. Nauk SSSR, Ser. Khim.*, **1967**, 1380.
269. B. P. Gusev and V. F. Kucherov, *Izv. Akad. Nauk SSSR, Ser. Khim.*, **1964**, 1318; *ibid.*, **1962**, 1062, 1067.
270. F. Bohlmann, H. Sinn, J. Politt, and E. Inhoffen, *Chem. Ber.*, **89**, 1281 (1956).
271. R. D. Dillard and N. R. Easton, *J. Org. Chem.*, **31**, 122 (1966).
272. W. J. Croxall and N. D. Dawson, U.S. Patent, 3,021,341 (Feb. 13, 1962).
273. R. L. Letsinger and L. R. Nazy, *J. Am. Chem. Soc.*, **81**, 3013 (1959).
274. F. Bohlmann and H-G. Viehe, *Chem. Ber.*, **88**, 1017 (1955).
275. B. S. Kupin and A. A. Petrov, *Zh. Org. Khim.*, **3**, 975 (1967), *Chem. Abstr.*, **67**, 99786 (1967).
276. W. H. Mueller and K. Griesbaum, *J. Org. Chem.*, **32**, 856 (1967).
277. F. Bohlmann and H. Hummel, *Chem. Ber.*, **101**, 2506 (1968).
278. F. Bohlmann and E. Bresinsky, *Chem. Ber.*, **100**, 107 (1967).
279. C. E. Castro and R. D. Stephens, *J. Org. Chem.*, **28**, 2163 (1963).
280. F. Gaudemar-Bardone, *Ann. Chim.* (13) **3**, 52 (1958).
281. E. I. Grinblat and I. Ya. Postovskiî, *Zh. Obshch. Khim.*, **31**, 394 (1961).
282. H. Fiesselmann and P. Schipprak, *Chem. Ber.*, **89**, 1897 (1956).
283. H. Fiesselmann and W. Böhm, *Chem. Ber.*, **89**, 1902 (1956).
284. F. Bohlmann and E. Bresinsky, *Chem. Ber.*, **97**, 2109 (1964).
285. S. M. Kalbag, M. D. Nair, P. Rajagopalan, and C. N. Talaty, *Tetrahedron*, **23**, 1911 (1967).
286. S. A. Vartanyan, *Usp. Khim.*, **33**, 517 (1964); *Russian Chem. Rev.*, **33**, 243 (1964).
287. L. M. Kataeva, I. V. Anonimova, L. K. Yuldasheva, and E. G. Kataev, *Zh. Obshch. Khim.*, **32**, 3965 (1962); *Chem. Abstr.*, **59**, 407 (1963).
288. J. Gosselck and E. Wolters, *Z. Naturforsch.*, **176**, 131 (1962).
289. L. Chierichi and F. Montanari, *Gazz. Chim. Ital.*, **86**, 1269 (1956).

290. M. E. Kuene and P. J. Sheeran, *J. Org. Chem.*, **33**, 4406 (1968).
291. E. J. Hart, "Solvated Electron," in *Advances in Chemistry Series 50*, American Chemical Society Publications, Washington, 1965.
292. M. Smith, in *Reduction*, R. L. Augustine, Ed., Marcel Dekker, New York, 1968, ch. 2.
293. J. Sicher, M. Svoboda, and J. Závada, *Coll. Czech. Commun.*, **30**, 413, 421 (1965).
294. N. A. Dobson and R. A. Raphael, *J. Chem. Soc.*, **1955**, 3558.
295. R. A. Benkeser and C. A. Tincher, *J. Org. Chem.*, **33**, 2727 (1968).
296. B. S. Rabinovitch and T. S. Looney, *J. Am. Chem. Soc.*, **75**, 2652 (1953).
297. M. N. Rerick, in *Reduction*, R. L. Augustine, Ed., Marcel Dekker, New York, 1968, ch. 1.
298. M. V. Mavrov and V. F. Kucherov, *Usp. Khim.*, **36**, 533 (1967); *Russian Chem. Rev.*, **36**, 233 (1967).
299. F. Bohlmann, R. Enkelmann, and W. Plettner, *Chem. Ber.*, **97**, 2118 (1964).
300. G. Zweifel, H. Arzoumanian, and C. C. Whitney, *J. Am. Chem. Soc.*, **89**, 3652 (1967).
301. G. Zweifel, J. T. Snow, and C. C. Whitney, *J. Am. Chem. Soc.*, **90**, 7139 (1968).
302. A. G. Brook, K. H. Pannell, and D. G. Anderson, *J. Am. Chem. Soc.*, **90**, 4374 (1968).
303. J. A. Shikhiev, G. F. Askerov, S. V. Garaeva, and M. I. Aliev, *Zh. Obsch. Khim.*, **37**, 317 (1967).
304. W. R. Cullen and W. Leeder, *Inorg. Chem.*, **5**, 1004 (1966).
305. R. J. Goodfellow, M. Green, N. Mayne, A. J. Rest, and F. G. A. Stone, *J. Chem. Soc. (A)*, **1968**, 177.
306. M. I. Bruce, D. A. Harbourne, F. Waugh, and F. G. A. Stone, *J. Chem. Soc. (A)*, **1968**, 895.
307. S. T. Kocharyan, E. M. Rokhlin, and I. L. Knunyants, *Izv. Akad. Nauk SSSR, Ser. Khim.*, **1966**, 1849.
308. G. M. Brooke and R. J. D. Rutherford, *J. Chem. Soc. (C)*, **1967**, 1189.
309. J. Ficini and C. Barbara, *Tetrahedron Letters*, **1966**, 6425.
310. T. L. Jacobs and A. Mihailovski, *Tetrahedron Letters*, **1967**, 2607.
311. A. J. Leusink, H. A. Budding, and J. W. Marsman, *J. Organometal. Chem.*, **9**, 285 (1967).
312. A. J. Leusink, H. A. Budding, and W. Drenth, *J. Organometal. Chem.*, **9**, 295 (1967).
313. A. M. Aguiar, T. G. Archibald, and L. A. Kapcak, *Tetrahedron Letters*, **1967**, 4447.
314. A. A. Petrov and M. P. Forost, *Zh. Organ. Khim.*, **2**, 1178, 1358 (1966).
315. H. J. Backer and J. Strating, *Rec. Trav. Chim.*, **73**, 565, 709 (1954).
316. R. Fusco, in *Pyrazoles, Pyrazolines, Pyrazolidines, Indazoles, and Condensed Rings*, R. H. Wiley, Ed., Interscience, New York, 1967, Part 1.
317. E. F. Magoon and L. H. Slaugh, *Tetrahedron*, **23**, 4509 (1967).
318. B. B. Molloy and K. L. Hauser, *Chem. Commun.*, **1968**, 1017.
319. Y. Kondo, Y. Honjo, and N. Tokura, *Bull. Chem. Soc. Japan*, **41**, 987 (1968).
320. R. F. Rodewald, K. Mahendran, J. L. Bear, and R. Fuchs, *J. Am. Chem. Soc.*, **90**, 6698, (1968).
321. P. Haberfield, L. Clayman, and J. S. Cooper, *J. Am. Chem. Soc.*, **91**, 787 (1969).

Selectivity in Addition Reactions of Allenes

MARJORIE C. CASERIO

University of California, Irvine, California

The nature of the cumulated double bonds of allenic compounds creates several possible modes of addition of an attacking reagent. An unsymmetrical reagent XY clearly has the option of placing either X or Y at the central carbon with Y or X at the terminal carbon to give products of structure $-\overset{|}{C}{=}C\,(X)\underset{|}{\overset{|}{C}}{-}Y$ or $-\overset{|}{C}{=}C\,(Y)\underset{|}{\overset{|}{C}}{-}X$. Further options are available if the two π-bonds happen to be differently substituted. The reactivities of the two bonds are likely to be different, and the reagent may therefore be expected to show some selectivity between them, as well as between the two carbons of a given bond. Also, when the two substituents at one terminus are different, two directions of approach of the attacking reagent with respect to these substituents are possible, giving rise to *cis* and *trans* isomers. In total there are *eight* possible products of addition of XY to an unsymmetrically substituted allene of the type ab C=C=C cd.

$$\text{ab C=C=C cd} \xrightarrow{XY} \text{ab C=C(X)–C(Y) cd} \quad (\textit{cis} \text{ and } \textit{trans}); \quad \text{ab C=C(Y)–C(X) cd} \quad (\textit{cis} \text{ and } \textit{trans})$$

$$\text{ab C=C=C cd} \xrightarrow{XY} \text{ab C(Y)–C(X)=C cd} \quad (\textit{cis} \text{ and } \textit{trans}); \quad \text{ab C(X)–C(Y)=C cd} \quad (\textit{cis} \text{ and } \textit{trans})$$

The purpose of the present chapter is to review the available experimental data on the orientation and stereochemistry of addition reactions of allenes, and to try insofar as possible to assess the factors that determine the observed orientations (1). We have restricted the discussion to include electrophilic, nucleophilic, and free-radical addition reactions of allenes. Cycloaddition reactions, hydrogenation, oxidation, and π-complex formation are not specifically considered in this chapter.

The orientation of addition of an unsymmetrical reagent to an unsymmetrical double bond is generally considered indicative of the structure of the reaction intermediates involved. For the sake of argument, we will consider the addition of an electrophilic reagent $X^{\delta+}—Y^{\delta-}$ to an allene. Attachment of the electrophile X^+ to a terminal carbon indicates that intermediate vinylic cations **1** are probably formed. Alternatively, attachment of X^+ at the central carbon indicates that allylic cations **2** may be formed. Furthermore, selectivity between the terminal and central carbons can be taken as a measure of the relative stabilities of vinylic and allylic ions **1** and **2**. However, it is important to note that the allylic ion formed as the result of central attack is probably nonplanar and does not resemble a resonance-stabilized allylic cation **3**. The terminal 2*p* orbital of **2** is orthogonal to the π-bonding orbitals, and electron delocalization is therefore at a minimum. The stability of **2** is probably close to that of an alkyl cation, and to achieve allylic resonance stabilization, the ion must rotate 90° about the σ bond.

Alternative forms of nonplanar ions **2** are bridged ions **4** in which the locus of initial attack has lost some of its meaning. Bridged ions of this type are quite often invoked to explain the frequent occurrence of stereospecific *trans* addition of halogens, sulfenyl halides, and mercuric salts to alkenes, but a precise description of the bonding in bridged ions of this type remains

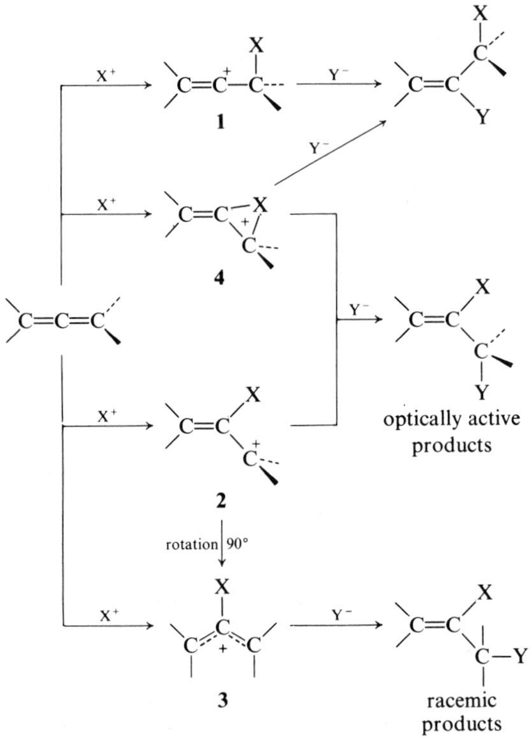

obscure. Thus a clear distinction between the various formalisms of bridged ions as π-complexes **4a**, symmetrical σ-complexes **4b**, or unsymmetrical σ-complexes with appreciable carbonium ion character **4c**, is not always possible, as will be noted in subsequent discussion.

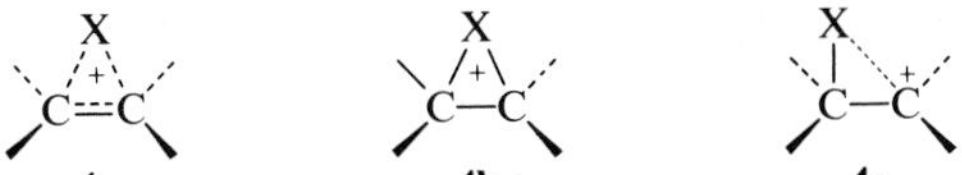

The question of interest as regards central attack by electrophilic, nucleophilic, or radical reagents is whether, in the course of reaction, resonance-stabilized allylic intermediates **3** do indeed intervene, or whether the products are derived largely or entirely from nonplanar allylic intermediates related to **2** or **4**. Evidence as to the nature of the intermediates involved in addition reactions can in principle be obtained from studies employing optically active allenes. Thus, if planar allylic intermediates analogous to **3** (ions or radicals) are formed to any important extent from a dissymmetric allene prior to the product-forming steps, the products are expected to be racemic. If, however, the initially formed nonplanar intermediates **2** or **4** react to form products faster than they can interconvert to planar intermediates **3**, optically active products could result. Furthermore, the overall stereochemistry of addition can be determined from the relationship between the absolute configuration of the dissymmetric allene and the asymmetric adducts, if such are formed. Use of this stereochemical approach to probing the mechanism of addition in the case of electrophilic reactions is discussed later in the chapter.

The incidence of the various possible intermediates **1**, **2**, **3**, and **4** in a given reaction will clearly depend on the structure of the allene and the nature of the attacking reagent, and some attempt is made in this chapter to correlate the effect of structure on the orientation and stereochemistry of addition. Electrophilic, nucleophilic, and free-radical addition reactions are discussed separately in that order, and each section begins with a discussion of the reactions of the parent hydrocarbon, allene, and follows with a discussion of the reactions of substituted allenes.

I. ELECTROPHILIC ADDITION TO ALLENE

The products obtained on addition of hydrogen halides, water, halogens, sulfenyl halides, mercuric acetate, and other electrophiles to allene are summarized in Table 1. With the notable exception of the halogens and sulfenyl halides, the entering electrophile shows a pronounced selectivity for attack at the terminal carbon. This would indicate that the preferred intermediates are either vinylic cations or bridged ions that open by nucleophilic attack at the *central* carbon.

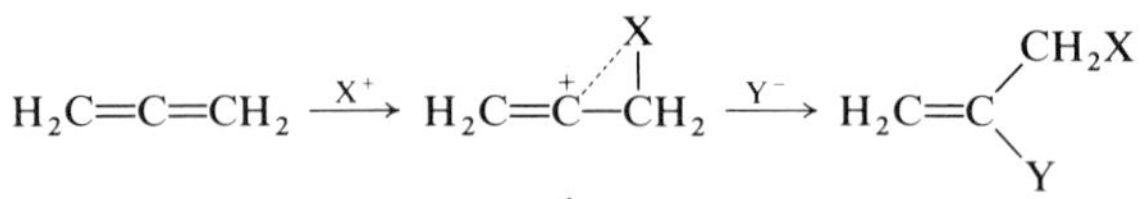

$$H_2C{=}C{=}CH_2 \xrightarrow{X^+} H_2C{=}\overset{+}{C}(\cdots X){-}CH_2X \xrightarrow{Y^-} H_2C{=}C(CH_2X)(Y)$$

1

The results obtained with the hydrogen halides are especially interesting because they illustrate the behavior of the vinylic cation **1**. This ion is presumably formed on protonation of both methylacetylene and allene, and accordingly both hydrocarbons give the same products upon reaction with hydrogen halides (2,3). These products are the expected mono- and diadducts, **5** and **6**, but a major proportion of the product is 1,3-dihalo-1,3-dimethyl-

TABLE 1
Products of Electrophilic Additions to Allene

Electrophile	Conditions	Products[a]	Ref.
H_2O	H_2SO_4	CH_3COCH_3	15
HF		$CH_3CF_2CH_3$	4
HCl	$-70°$[b]	$CH_2{=}C(Cl)CH_3$ (35)	2
		$CH_3CCl_2CH_3$ (54)	
H_3C Cl / Cl CH_3	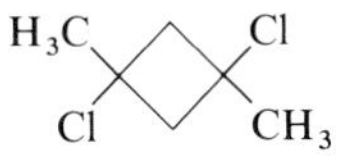 (8)	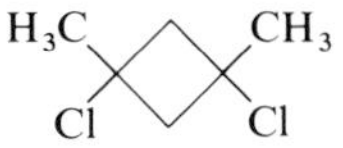 H_3C CH_3 / Cl Cl (3)	
HCl	$BiCl_3$, $-78°$	$CH_2{=}C(Cl)CH_3$ (60)	3
		$CH_3CCl_2CH_3$ (37)	
HBr	$-70°$	$CH_2{=}C(Br)CH_3$ (13)	2
		$CH_3CBr_2CH_3$ (35)	
H_3C Br / Br CH_3	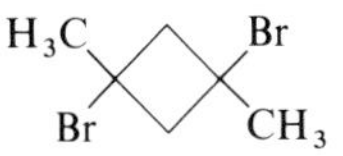 (44)	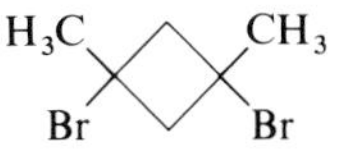 H_3C CH_3 / Br Br (8)	
HI	$-70°$	$CH_2{=}C(I)CH_3$ (6)	2
		$CH_3CI_2CH_3$ (94)	
HOCl		$CH_2{=}C(Cl)CH_2OH$ (10)	16
		$ClCH_2COCH_2Cl$ (15)	
		$ClCH_2COCH_2OH$ (50)	
$Hg(OAc)_2$	CH_3OH	$AcOHgCH_2COCH_2HgOAc$	5
B_2H_6		$[CH_2{=}CHCH_2BH_2]$[c,d]	6,7
R_2BH[e]	$130°$	$CH_2{=}CHCH_2BR_2$ (87)	8
		$CH_2{=}C(BR_2)CH_3$ (13)	
RSCl[f]	$-30°$, CH_2Cl_2	$CH_2{=}C(SR)CH_2Cl$[f]	3,10
Cl_2	$-30°$, $-70°$, CH_2Cl_2	$CH_2{=}C(Cl)CH_2Cl$ (40)	11,12,52
		$HC{\equiv}CCH_2Cl$ (60)	
Cl_2	AcOH	$CH_2{=}C(Cl)CH_2OAc$	11
Br_2	$-17°$, $CHCl_3$	$CH_2{=}C(Br)CH_2Br$	11
BrCl	$-30°$, CCl_4	$CH_2{=}C(Br)CH_2Cl$	11

[a] Percent composition in parenthesis.
[b] Up to 12% conversion.
[c] In solution, initial adduct oxidized to acetone (ref. 6).
[d] In the gas phase at 90–95°, allyldiborane is assumed to be the initial adduct: the isolated adducts were 1,2-trimethylenediborane and polymers (ref. 7).
[e] 4,4,6-Trimethyl-1,3,2-dioxaborinane, **13**.
[f] R, CH_3, C_6H_5, $2,4(NO_2)_2C_6H_3$, $CH_3CO(S)$; rearrangement of initial adduct observed for R = CH_3,C_6H_5.

cyclobutane **7** (2). Griesbaum has offered the suggestion that the cyclic product results from electrophilic attack of the initially formed vinylic cation **1** on the starting hydrocarbon to give other vinylic cations, **8** or **9**, which can cyclize and react further with the hydrogen halide (Scheme 1).* Hydrogen

$CH_2{=}C{=}CH_2 \overset{H^+}{\rightleftharpoons} [CH_3{-}\overset{+}{C}{=}CH_2] \overset{H^+}{\rightleftharpoons} CH_3C{\equiv}CH$

1

$CH_2{=}C{=}CH_2$ → $\left[CH_3{-}C(=CH_2){-}CH_2{-}\overset{+}{C}{=}CH_2\right]$ **8** → $\left[CH_3{-}\overset{+}{C}{-}CH_2{-}C(=CH_2){-}CH_2 \text{ (ring)}\right]$ → (X^-, HX) → **7**

X^- → $CH_3CX{=}CH_2$ **5** → (HX) → $CH_3CX_2CH_3$ **6**

$CH_3C{\equiv}CH$ → $\left[CH_3{-}C(=CH_2){-}CH{=}\overset{+}{C}{-}CH_3\right]$ **9** → $\left[CH_3{-}\overset{+}{C}{-}CH_2{-}C(CH_3){=}CH \text{ (ring)}\right]$ → (X^-, HX) → **7**

$CH_3{-}CX{-}CH_2{-}CX(CH_3){-}CH_2$ (cyclobutane)

7, X = Cl or Br
cis and *trans*

Scheme 1

iodide does not lead to cyclic products, however; but this is not surprising since a strong nucleophile such as iodide ion could perhaps compete effectively for the intermediate **1** and divert it totally to the adducts **5** and **6** (2). Hydrogen fluoride is reported to give 2,2-difluoropropane on reaction with allene (4), but no reference was made to the formation of cyclic products.

Methoxymercuration of allene, like hydrogen halide addition, gives products expected of electrophilic attack at the terminal carbon (5). The reaction cannot be arrested at the monoadduct stage **10** but proceeds to give

* There is reason to believe that the cycloaddition of **1** with allene or methylacetylene may actually be concerted. The rationale for this has been expressed by Woodward and Hoffmann in terms of the orbital symmetries of the participating components. [Address by R. B. Woodward on "The Conservation of Orbital Symmetry" at a symposium on *Orbital Symmetry Correlations in Organic Reactions*, January 7–9, 1969, Cambridge, England. See also R. B. Woodward and R. Hoffmann, Angew. Chemie., International Edn. *8*, 781 (1969)]. For a 2 + 2

the ketal **11**, which rapidly hydrolyzes under the reaction conditions to the ketone **12**.

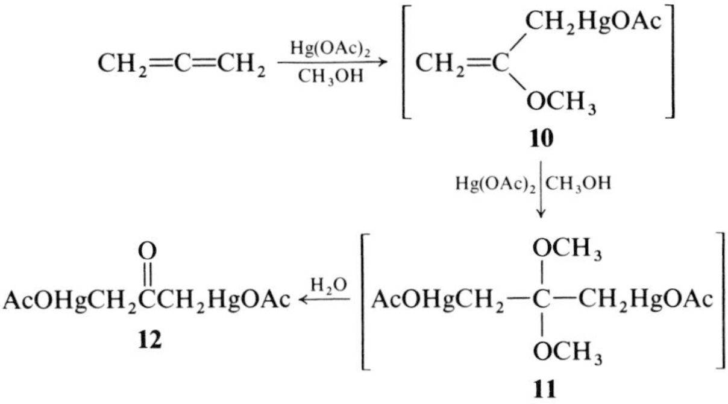

$$CH_2{=}C{=}CH_2 \xrightarrow[CH_3OH]{Hg(OAc)_2} \left[CH_2{=}C(CH_2HgOAc)(OCH_3) \right] \; \mathbf{10}$$

$$\downarrow Hg(OAc)_2,\ CH_3OH$$

$$\underset{\mathbf{12}}{AcOHgCH_2C({=}O)CH_2HgOAc} \xleftarrow{H_2O} \left[AcOHgCH_2{-}C(OCH_3)_2{-}CH_2HgOAc \right] \; \mathbf{11}$$

---

cycloaddition reaction to be symmetry allowed, the reacting bonds must necessarily approach each other in perpendicular planes (**I**) such that one π component reacts suprafacially (π^2s) and the other π component, which reacts antarafacially (π^2a). Since orbital overlap is generally

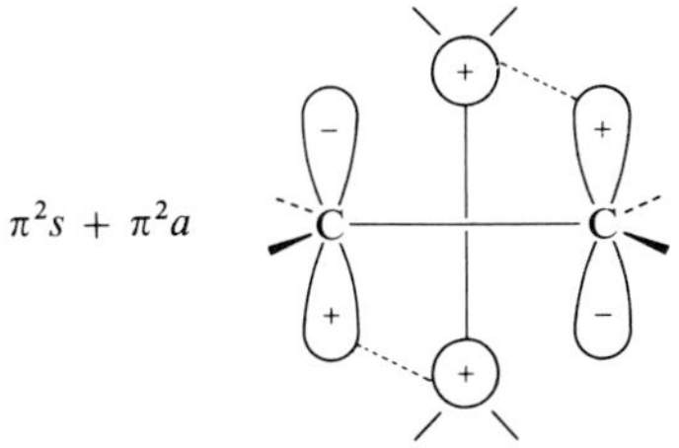

$\pi^2s + \pi^2a$

I

poor in this orientation, concerted cycloadditions of this type are likely to be energetically unfavorable and rarely observed. However, Woodward and Hoffmann have pointed out that the vinyl cation **II** has the advantage of possessing a vacant p orbital which can overlap favorably with the filled orbital of an approaching π bond and this interaction may lower the transition state energy sufficiently to allow the cycloaddition with allene (or methylacetylene) to proceed in a concerted manner.

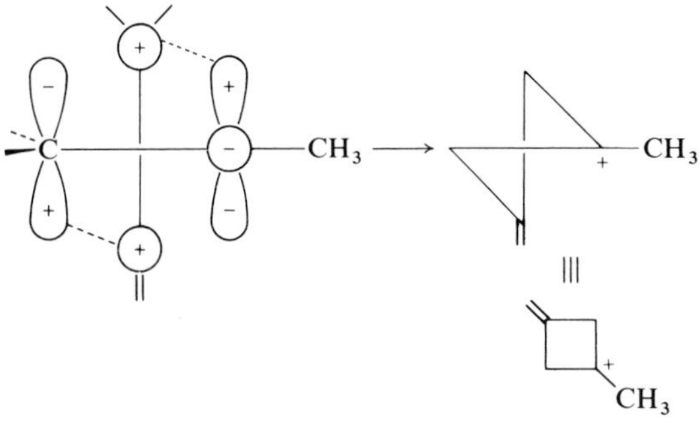

CH_3 ⟶ CH_3

≡

CH_3

Similarly, hydroboration of allene with diborane (6,7) leads to products derived from attachment of boron at the terminal position. The monofunctional borane **13** leads predominantly to **14**, which is the product of terminal attack, although some 13% of the product is derived from central attack, **15** (8). These reactions are included here as examples of electrophilic addition even though they are thought to proceed by way of a four-center transition state (9).

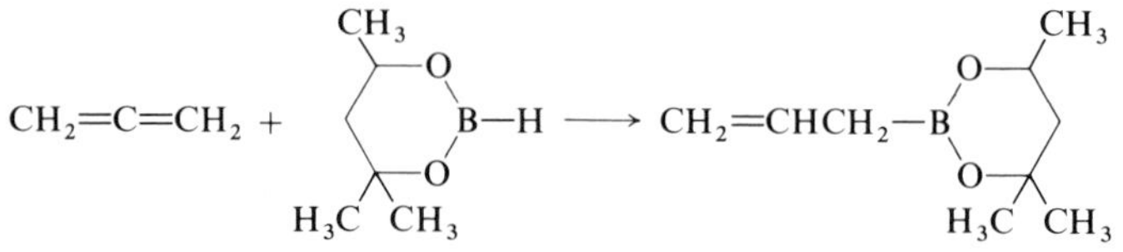

$$CH_2{=}C{=}CH_2 + \mathbf{13} \longrightarrow CH_2{=}CHCH_2{-}B(OCH(CH_3)CH_2C(CH_3)_2O)$$

13   **14** (87%)

$$CH_2{=}C(CH_3){-}B(OCH(CH_3)CH_2C(CH_3)_2O)$$

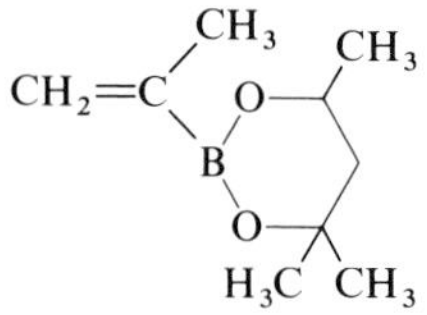

15 (13%)

In view of the observed orientation of addition of hydrogen halides, mercuric acetate and boranes to allene, it is perhaps surprising that addition of sulfenyl halides and halogens result in the *opposite* orientation. Jacobs and Johnson (3) first observed that 2,4-dinitrobenzenesulfenyl chloride reacts with allene to give a monoadduct corresponding to **16d**. The addition of several different sulfenyl chlorides to allene at low temperatures in methylene chloride has been studied by Mueller and Butler (10), and although the

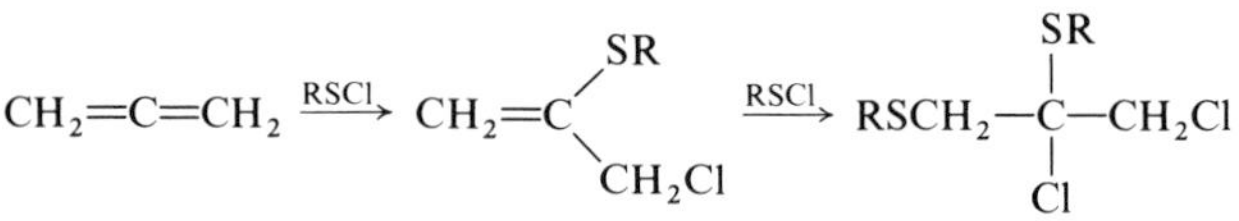

$$CH_2{=}C{=}CH_2 \xrightarrow{RSCl} CH_2{=}C(SR)CH_2Cl \xrightarrow{RSCl} RSCH_2{-}C(SR)(Cl){-}CH_2Cl$$

16a R, CH_3
b R, C_6H_5
c R, $CH_3(CO)S$
d R, 2,4-$(NO_2)_2C_6H_3$

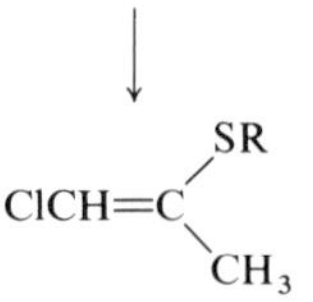

$$\downarrow$$

$$ClCH{=}C(SR)CH_3$$

17a, **b**, *cis* and *trans*

product mixtures obtained were complex the major kinetic product in each case was established as **16**. Although the adducts of methane- and benzenesulfenyl chloride addition, **16**a, **16**b, rearranged to more stable adducts **17**a and **17**b on warming to room temperature, the orientation in the initial adducts corresponds to net addition of the electrophile to the central carbon and the nucleophile to the terminal carbon.

Addition of halogens to allene has been studied by Peer (11) who observed that addition could be arrested at the monoadduct stage and that orientation of addition in nucleophilic solvents or with mixed halogens gave only products corresponding to *central* electrophilic attack.

$$CH_2{=}C{=}CH_2 \xrightarrow[HOAc]{Cl_2} CH_2{=}C(Cl)CH_2OAc$$

$$CH_2{=}C{=}CH_2 \xrightarrow[CCl_4]{BrCl} CH_2{=}C(Br)CH_2Cl$$

In the case of chlorine addition in inert solvents propargyl chloride was reported formed along with the expected product, 2,3-dichloropropene (11,12,13). It is not clear how the propargyl chloride arises except that its formation is partly suppressed by chlorination under strongly ionic conditions (12).

$$CH_2{=}C{=}CH_2 \xrightarrow[CH_2Cl_2]{Cl_2} \underset{(40\%)}{CH_2{=}C(Cl)CH_2Cl} + \underset{(60\%)}{HC{\equiv}CCH_2Cl}$$

$$CH_2{=}C{=}CH_2 \xrightarrow[(NaAlCl_4{-}KAlCl_4\ \text{melt})]{Cl_2} \underset{(88\%)}{CH_2{=}C(Cl)CH_2Cl} + \underset{(12\%)}{HC{\equiv}CCH_2Cl}$$

In a more recent investigation, Poutsma has shown that chlorine reacts with allene at $-30°$ under oxygen to give a complex mixture of products (52). Among these, 2,3-dichloropropene (23%) and propargyl chloride (20%) were positively identified. Low yields of dimeric products were also obtained, the structures of which were assigned as 2-chloro-1-hexen-5-yne (4.3%) and 2,5-dichloro-1-5,-hexadiene (1.5%). The most reasonable interpretation of the overall reaction involves initial electrophilic attack of chlorine to give an (allene-Cl)$^+$ species which may react both with chloride ion and with another molecule of allene to give the monomeric and dimeric products, respectively.

$$CH_2{=}C{=}CH_2 + Cl_2$$

$$\downarrow$$

$$CH_2{=}C\overset{\overset{Cl}{+}}{\text{———}}CH_2$$

$\xrightarrow{Cl^-}$ $CH_2{=}C(Cl)CH_2Cl$

$\xrightarrow[(-HCl)]{Cl^-}$ $CH{\equiv}CCH_2Cl$

$\xrightarrow{CH_2{=}C{=}CH_2}$ $CH_2{=}C(Cl)CH_2CH_2\overset{+}{C}{=}CH_2$

$\xrightarrow{Cl^-}$ $CH_2{=}C(Cl)CH_2CH_2C(Cl){=}CH_2$

$\xrightarrow[(-HCl)]{Cl^-}$ $CH_2{=}C(Cl)CH_2CH_2C{\equiv}CH$

Electrophilic additions of halogens, mercuric salts, and sulfenyl halides to alkenes are generally thought to proceed by way of bridged ion intermediates (14), in which case it may be more meaningful to discuss the orientation of addition in terms of the position of nucleophilic attack rather than the position of electrophilic attack. Thus ring-opening of intermediate episulfonium ions **18** and bromonium or chloronium ions **19** evidently occurs by nucleophilic attack at the methylene group, while mercurinium ions **20** open by nucleophilic attack at the vinylic carbon. A convincing reason for this difference in behavior has not been offered, and probably will not be forthcoming until we have a better understanding of the structure and charge distribution in the intermediate ions.

R–S bridged (+) H_2C—$C{=}CH_2$ **18**

X bridged (+) H_2C—$C{=}CH_2$ **19**

X = Cl, Br

AcO–Hg bridged (+) H_2C—$C{=}CH_2$ **20**

II. ELECTROPHILIC ADDITION TO SUBSTITUTED ALLENES

A. Hydrogen Halide Additions

There are several problems connected with the addition of hydrogen halides to allenes. For example, rearrangement of the starting allene and rearrangement of the initial adducts can severely complicate the reaction picture and, in HBr addition, it is difficult to completely suppress corresponding radical addition reactions. Because of these difficulties, it is hard to evaluate the literature and arrive at a self-consistent pattern of orientation. Yet the following conclusions emerge. Addition of reagents of the type HX to allene and monosubstituted allenes give products that are probably derived from vinylic cations (cf., Tables 1 and 2). More highly substituted

TABLE 2
Products of Addition of HX to Substituted Allenes

HX	Conditions	Allene	Products	Ref.
H_2O	H_2SO_4, $HgSO_4$	$RCH{=}C{=}CH_2$	$RCH_2C(=O)CH_3$	3,20–22
H_2O	H_2SO_4 (80%)	$RCH{=}C{=}CHCH_3$	$RCH_2C(=O)CH_2CH_3$	29
H_2O	H_2SO_4, $HgSO_4$ or HgO, $BF_3 \cdot Et_2O$	1,2-Cyclononadiene	$CH{=}CH{-}CH(OH)$ ring closed by $(CH_2)_6$	19,37b
CH_3OH	H_2SO_4	$(CH_3)_2C{=}C{=}CH_2$	$(CH_3)_2C{=}CHCH_2OCH_3$	3
CH_3OH	HgO, $BF_3 \cdot Et_2O$	$RCH{=}C{=}CH_2$	$RCH_2C(OCH_3)_2CH_3$	22
HCl	−78°	$RCH{=}C{=}CH_2$	$RCH{=}C(Cl)CH_3$ $RCH_2C(Cl_2)CH_3$	3,22
HCl	−78°	$(CH_3)_2C{=}C{=}CH_2$	$(CH_3)_2C(Cl)CH{=}CH_2$ $(CH_3)_2C{=}CHCH_2Cl$	3,25
HCl	−70°	$RCH{=}C{=}CHR$	$RCH{=}C(Cl)CH_2R$ $RCH{=}CHCH(Cl)R$ (mainly)	19,24
HCl	Et_2O	$CH_2{=}C{=}CHCN$	$CH_2{=}C(Cl)CH_2CN$	26
HBr	0°	$RCH{=}C{=}CH_2$	$RCH_2C(Br){=}CH_2$ $RCH{=}C(Br)CH_3$ (mainly)	20,27,29
HBr		$(CH_3)_2C{=}C{=}CH_2$	$(CH_3)_2C{=}CHCH_2Br$	29
HBr	−40°	$RCH{=}C{=}CHCH_3$	$RCH{=}C(Br)CH_2CH_3$ $RCH{=}CHCH(Br)CH_3$	24
HCl or HBr		$RCH{=}CH{-}CH{=}C{=}CH_2$	$RCH_2CH{=}CH{-}C(X){=}CH_2$	28
HOCl		$(CH_3)_2C{=}C{=}CH_2$	$(CH_3)_2C(OH)COCH_2Cl$ (38%)	16

allenes effect a clean reversal in orientation giving products indicative of allylic intermediates formed by *central* protonation.

As already mentioned, a major problem in addition of hydrogen halides is one of acid-catalyzed rearrangement of the allene to a more stable hydrocarbon. Extensive rearrangement of methylallene to 2-butyne, and 1,1-dimethylallene to isoprene, takes place under conditions of HCl addition

(3). This is not a serious complication in the case of methylallene since it was established that 2-butyne reacts more slowly than methylallene with HCl under the reaction conditions. The products of HCl addition may therefore be assumed to arise directly from methylallene by initial protonation at the terminal methylene group.

$$CH_3CH{=}C{=}CH_2 \underset{-78^\circ}{\overset{H^+}{\rightleftharpoons}} CH_3CH{=}\overset{+}{C}{-}CH_3 \xrightarrow[-78^\circ]{-H^+} CH_3C{\equiv}CCH_3$$

$$\downarrow Cl^-$$

$$CH_3CH{=}C(Cl)CH_3 \ \textit{cis and trans}$$

With 1,1-dimethylallene, the situation is more complex, but it is also more interesting. Rearrangement to isoprene occurs, and the isoprene formed reacts with hydrogen chloride to give the same adducts, **21** and **22**, as 1,1-dimethylallene. The product distributions are different, however, and more of the primary chloride **22** is formed from the allene than from isoprene. The implications of this result are difficult to assess reliably. It could be an artifact of allylic rearrangement of **21** to **22**, although this was checked and considered as insignificant (3). Or it could be that the addition products from 1,1-dimethylallene are not derived solely from planar allylic ions such as **23**, but are derived from nonplanar ions **24** and **25**, which precede the formation of **23** (Scheme 2). However, when correction is made for the adducts formed

$(CH_3)_2C{=}C{=}CH_2$ $\xrightarrow{H^+}$ **24** $[(H_3C)_2\overset{+}{C}{-}CH{=}CH_2]$ $\xrightarrow{Cl^-}$ (not formed) $(CH_3)_2C(Cl)CH{=}CH_2$ **21**

24 $\xrightarrow{\text{rotation}}$ **23** $[(H_3C)_2C{\cdots}CH{\cdots}CH_2]^+$ $\underset{-H^+}{\overset{H^+}{\rightleftharpoons}}$ $CH_2{=}C(CH_3){-}CH{=}CH_2$

23 $\xrightarrow{Cl^-}$ **21**; **23** $\xrightarrow{Cl^-}$ **22**

$(CH_3)_2C{=}C{=}CH_2$ $\xrightarrow{H^+}$ **25** $[(CH_3)_2C{=}CH{-}\overset{+}{C}H_2]$ $\xrightarrow{Cl^-}$ $(CH_3)_2C{=}CHCH_2Cl$ **22**

25 $\xrightarrow{\text{rotation}}$ **23**

Scheme 2

indirectly via isoprene, direct addition to 1,1-dimethylallene evidently leads predominantly or exclusively to the primary chloride **22** by way of the primary cation **25**. Addition of HBr to 1,1-dimethylallene is also reported to give the corresponding primary allylic bromide (17).

One may also consider the possibility that addition of hydrogen halides occurs by a termolecular mechanism of the type described in the work of Fahey (18), in which case the intermediate ions **23**, **24**, and **25** would serve only to explain the formation of isoprene.

$$(CH_3)_2C{=}C{=}CH_2 \xrightarrow{\text{H—Cl},\ \text{H—Cl}} (CH_3)_2C{=}CH{-}CH_2Cl$$

Addition of HCl to 1,3-disubstituted allenes also appears to proceed largely or exclusively by way of central electrophilic attack (19,24). Thus 1,2-cyclononadiene gives only 3-chlorocyclononene, whereas 1,3-dimethylallene gives 85% 4-chloro-2-pentene and 15% of 3-chloro-2-pentene (19). However, the configuration of the adducts does not appear to have been determined and the possibility of rearrangement of the allenes to conjugated dienes was not excluded.

$$\text{CH{=}C{=}CH}\,[(CH_2)_6] \xrightarrow[\text{BiCl}_3,\ -78^\circ]{\text{HCl}} \text{CH{=}CH{-}CHCl}\,[(CH_2)_6]$$

$$CH_3CH{=}C{=}CHCH_3 \xrightarrow[\text{BiCl}_3,\ -78^\circ]{\text{HCl}} \begin{cases} CH_3CH{=}CHCH(Cl)CH_3,\ 85\% \\ CH_3CH{=}C(Cl)CH_2CH_3,\ 15\% \end{cases}$$

Hydration reactions of allenes are difficult to evaluate. Several of these reactions utilize mercuric salts and strong acids as catalysts and under these conditions hydration probably involves initial oxymercuration followed by demercuration (cf. Table 2). This type of reaction is discussed further in the next section.

$$>C{=}C< \xrightarrow[H_2O]{HgSO_4} \left[-\underset{HO}{\overset{|}{C}}-\overset{Hg^+}{\underset{|}{C}}- \right] \xrightarrow[(-HgSO_4)]{H_2SO_4} -\underset{HO}{\overset{|}{C}}-\overset{H}{\underset{|}{C}}-$$

Even hydration reactions involving only acid catalysts are not straightforward; for example, hydration of 1,3-disubstituted allenes with 80%

sulfuric acid leads to ketones (23), and this result might be taken to indicate terminal electrophilic attack. If this is correct, then the orientation of hydration is opposite to that of hydrogen chloride addition, at least for 1,3-disubstituted allenes. If, however, hydration does occur by way of central electrophilic attack, then diadduct formation could lead to vicinal diols which would almost certainly undergo a pinacolic rearrangement under the reaction conditions. Ketone formation is therefore unhelpful in denoting the initial point of electrophilic attack.

$$RCH{=}C{=}CHR \xrightarrow{H_2O,\ H^+} \left[RCH{=}C(OH){-}CH_2R\right] \xrightarrow{H_2O,\ H^+} \left[RCH_2{-}C(OH)_2{-}CH_2R\right] \xrightarrow{-H_2O} RCH_2C({=}O)CH_2R$$

$$RCH{=}C(OH){-}CH_2R \longrightarrow RCH_2C({=}O)CH_2R$$

$$RCH{=}C{=}CHR \xrightarrow{H_2O,\ H^+} \left[RCH{=}CH{-}CH(OH)R\right] \xrightarrow{H_2O,\ H^+} RCH_2{-}CH(OH){-}CH(OH)R \xrightarrow{-H_2O} RCH_2C({=}O)CH_2R$$

B. Oxymercuration

In contrast to hydration and hydrogen halide addition, oxymercuration of allenes is remarkably well behaved. Rearrangement of the starting allene is not observed, and the adducts do not rearrange to more stable products under the reaction conditions. The adducts are therefore formed under kinetic control, and their structures are indicative of the effect of substitution on the orientation of addition.

The results of an investigation by Waters and Kiefer (5) on the orientation of addition of mercuric acetate in methanol to allene and methyl-substituted allenes are summarized in Table 3 together with other pertinent data. There is a clear change in orientation from exclusive terminal electrophilic attack on allene to exclusive central attack with increasing methyl substitution. Even with methylallene, 60% of the product corresponds to central electrophilic attack—the nucleophile entering at the more highly substituted terminal carbon.

There is considerable evidence pertaining to the mechanism of oxymercuration of alkenes (30) which shows that the transition state has appreciable carbonium-ion character (31). This and the fact that the stereochemistry of addition is largely *trans* has led to the proposition that the reaction intermediates are bridged mercurinium ions (cf. **4c**). In the case of

TABLE 3
Products of Oxymercuration of Allenes

Allene	Solvent	Electrophile	Adduct[a]		Ref.
$CH_2{=}C{=}CH_2$	CH_3OH	$Hg(OAc)_2$	$(AcOHgCH_2)_2C{=}O$		5
$CH_3CH{=}C{=}CH_2$	CH_3OH	$Hg(OAc)_2$	$AcOHgCH(CH_3){-}C(OCH_3)_2{-}CH_2HgOAc$[b]	(35)	5
			$CH_3CH(OCH_3){-}C(HgOAc){=}CH_2$	(60)	
$(CH_3)_2C{=}C{=}CH_2$	CH_3OH	$Hg(OAc)_2$	$(CH_3)_2C(OCH_3){-}C(HgOAc){=}CH_2$	(84)	5
$CH_3CH{=}C{=}CHCH_3$	CH_3OH	$Hg(OAc)_2$	$CH_3CH(OCH_3){-}C(HgOAc){=}CHCH_3$	*cis* (20) *trans* (80)	5
$CH_3CH{=}C{=}CHCH_3$	C_2H_5OH	$HgCl_2$	$CH_3CH(OC_2H_5){-}C(HgCl){=}CH_2$	(84)	19
$(CH_3)_2C{=}C{=}CHCH_3$	CH_3OH	$Hg(OAc)_2$	$(CH_3)_2C(OCH_3){-}C(HgOAc){=}CHCH_3$	(72)	5
			$(CH_3)_2C{=}C(HgOAc){-}CH(OCH_3)CH_3$	(18)	
$(CH_3)_2C{=}C{=}C(CH_3)_2$	CH_3OH	$Hg(OAc)_2$	$(CH_3)_2C(OCH_3){-}C(HgOAc){=}C(CH_3)_2$	(90)	5
1,2-Cyclononadiene	C_2H_5OH	$HgCl_2$	cyclononene ring: HgCl, OC_2H_5		19,37a
1,2-Cyclononadiene	C_2H_5OH or CH_3OH	$Hg(OAc)_2$	cyclononene ring: HgOAc, OR		35,37a

[a] Per cent yield in parenthesis.
[b] Isolated as the corresponding ketone.

oxymercuration of allenes the bridged intermediates would be vinylic mercurinium ions **26** which could conceivably gain allylic resonance stabilization by opening of the bridge and bond rotation to give **27**.

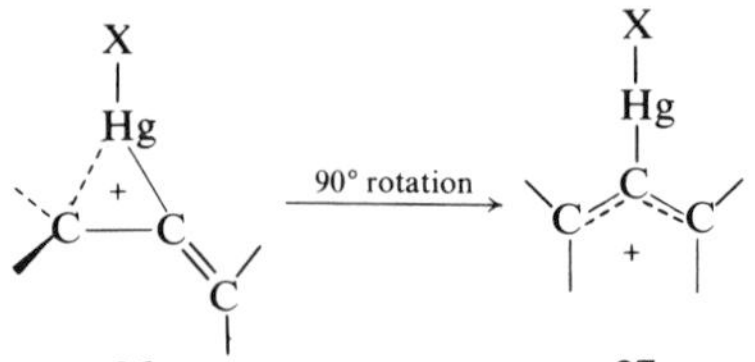

26 —90° rotation→ **27**

Compelling evidence that planar allylic ions analogous to **27** cannot be the sole intermediates in oxymercuration of allenes was obtained from an investigation of the stereochemistry of methoxymercuration of optically active 1,3-dimethylallene (32). Thus optically active adducts **28** and **29** were obtained on addition of mercuric acetate in methanol to (−)-1,3-dimethylallene **30**, and from the relationship between the configuration of the dissymmetric allene (33,34) and the configuration of the major asymmetric adduct **28**, the stereochemistry of addition was established as *trans*. This result rules out the exclusive intervention of allylic ions **27**, since these are expected to lead to racemic products. The observed stereochemistry supports the existence of product-forming intermediates such as **26** which retain their

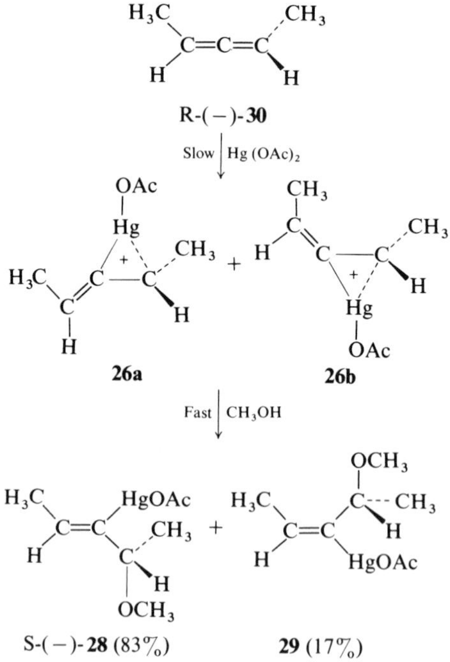

R-(−)-**30**

Slow ↓ $Hg(OAc)_2$

26a + **26b**

Fast ↓ CH_3OH

S-(−)-**28** (83%) + **29** (17%)

Scheme 3

dissymmetric configuration and open by nucleophilic attack at the more highly substituted carbon in a direction *trans* to the mercury bridge (Scheme 3).

The *cis–trans* product ratio in the oxymercuration of **30** is essentially independent of the structure of the alcohol solvent (Table 4) (35). Even in

TABLE 4
Products of Addition of Mercuric Acetate to 1,3-Dimethylallene in Alcohol Solvents (35)

$$CH_3CH{=}C{=}CHCH_3 \xrightarrow[2.\ NaCl]{1.\ Hg(OAc)_2,\ ROH} CH_3CH{=}C(HgCl)CH(OR)CH_3 + CH_3CH{=}C(HgCl)CH(OAc)CH_3$$

ROH	Acetate products, per cent	Ether products, per cent	*trans*:*cis*[a]
CH_3OH	—	100	83:17
CH_3OH[b]	35	65	83:17
CH_3OH[c]	65	35	83:17
$(CH_3)_2CHOH$	5	95	83:17
$(CH_3)_2CHOH$[c]	28	72	89:11
$(CH_3)_3COH$	25	75	86:14
CH_3COOH	100	—	83:17[d]

[a] Ratio refers to ether products.
[b] 0.5 *M* in sodium acetate.
[c] Saturated in sodium acetate.
[d] Ratio refers to acetate products.

t-butyl alcohol the proportion of *cis* and *trans* allylic ethers is not sensibly different from that in methanol. We regard this as evidence that attack of solvent on the bridged ions **26a** and **26b** is probably very rapid and does not directly determine the *cis–trans* product ratio (Scheme 3). This implies that the rate-limiting and product-determining step is the attack of mercuric acetate on the allene to give **26**. The preferential formation of the *trans* adduct **28** implies that the *trans* mercurinium ion **26a** is formed more rapidly and is therefore more stable than the *cis* ion **26b**. This is entirely reasonable since nonbonded interactions between *cis* alkyl substituents usually destabilize *cis* alkenes relative to their *trans* isomers. Also, steric interactions involving the acetoxymercuri group are expected to be negligible (36).

An interesting study by Bach (37) on the oxymercuration of optically active 1,2-cyclononadiene **31** strongly suggests that the stereospecificity of oxymercuration is dependent on the nature of the mercury electrophile. The highest level of activity retained in the allylic ether **33** formed by protodemercuration of the initial adduct **32** was obtained on the addition of ethylmercuric chloride, and the lowest on the addition of mercuric chloride. In general, the higher the electronegativity of the ligand X in the electrophile HgX^+, the lower the observed degree of stereospecificity.

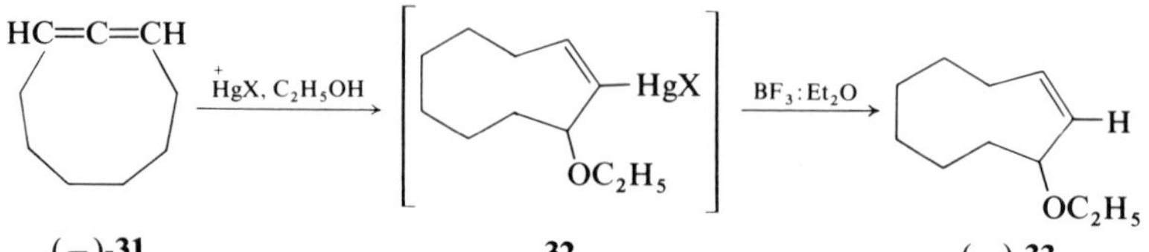

HC=C=CH — $\overset{+}{Hg}X$, C_2H_5OH → [—HgX, OC_2H_5] — $BF_3:Et_2O$ → —H, OC_2H_5

(−)-**31** **32** (−)-**33**

These results are considered to manifest the ease of interconversion of bridged and open allylic ions of type **26** and **27** as determined by the nature of the ligand X on mercury.

The possible involvement of bridged and open mercurinium ions **26** and **27** in the acid-catalyzed methanolysis of optically active 3-chloromercuri- and 3-acetoxy-mercuri-4-acetoxy-2-pentene, **34**, has been investigated (35). The solvolysis reaction (Scheme 4) was found to be *nonstereospecific*, which suggests that planar allylic ions **27** must participate to an important extent. Deoxymercuration to give 1,3-dimethylallene was not observed under the solvolysis conditions, and optically active ethers **28**, **29**, **35** and **36** readily racemized.

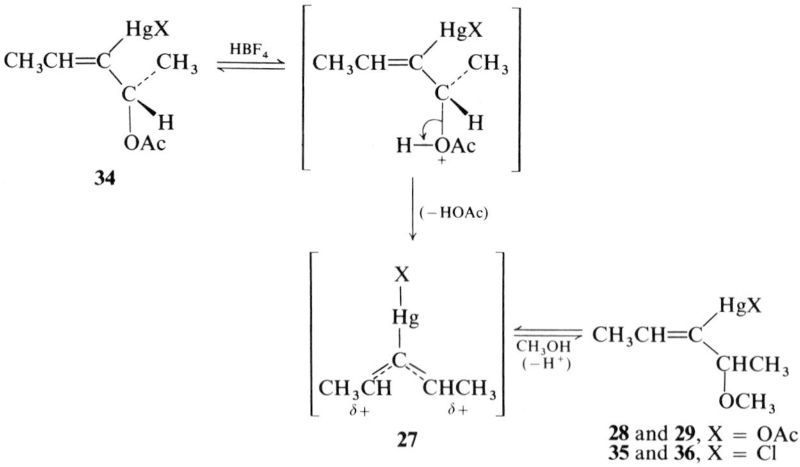

$CH_3CH{=}C(HgX)CH(H)(OAc)CH_3$ (**34**) $\underset{}{\overset{HBF_4}{\rightleftharpoons}}$ [$CH_3CH{=}C(HgX)CH(H)(\overset{+}{O}HAc)CH_3$] —(−HOAc)→ [X–Hg–C($CH_3\overset{\delta+}{CH}$)($\overset{\delta+}{CH}CH_3$)] (**27**) $\underset{CH_3OH\ (-H^+)}{\rightleftharpoons}$ $CH_3CH{=}C(HgX)CHCH_3(OCH_3)$

28 and **29**, X = OAc
35 and **36**, X = Cl

Scheme 4

Rate studies indicate that there is no significant participation by the neighboring vinylic mercury substituent in the solvolysis transition states. Dissymmetric intermediates such as **26** are, therefore, inaccessible by way of solvolysis of allylic derivatives **34** and related compounds, although **26** is demanded in the methoxymercuration reaction of 1,3-dimethylallene with mercuric acetate (Scheme 3).

C. Sulfenyl Halides

Sulfenyl chloride additions to allenes are reported to orient the sulfenyl group to the central allenic carbon and the chloride to a terminal position. The reactions in question are summarized in Table 5. The one apparent exception to the observed rule of central attack concerns the addition of

$$>C=C=C< \xrightarrow{RSCl} >C=C(SR)-C(Cl)<$$

2,4-dinitrobenzenesulfenyl chloride to 1,2-cyclononadiene. This reaction was originally reported to give a vinylic chloride (38) but, more recently, the adduct was reported to be an allylic chloride (39). One curious feature of the product is its lack of reactivity toward silver nitrate. It was this observation that first led Moore (38) to suggest that the product was a vinylic chloride, but it would seem that the inert character of allylic chlorine toward Ag^+ is also exhibited by the adduct of 2,4-dinitrobenzenesulfenyl chloride and 1,1-dimethylallene (39). Addition of dimethylthiomethylsulfonium fluoborate to cyclononadiene follows the same orientation as in sulfenyl halide addition, the electrophile (CH_3S^+) entering at the central carbon to give the *cis* isomer as the major adduct (40).

HC=C=CH (cyclononadiene) —RSCl→ (cyclononene with SR and Cl)

—$CH_3S\overset{+}{S}(CH_3)_2$ BF_4^-→ (cyclononene with SCH_3 and $\overset{+}{S}(CH_3)_2\bar{B}F_4$)

There is much evidence to suggest that sulfenyl halides add to alkenes by way of episulfonium ions (41,42) and in certain instances, episulfonium salts can be isolated (43). Jacobs et al. (44) have obtained evidence that a

TABLE 5
Products of Sulfenyl Halide Additions to Allenes

RSCl	Allene	Adducts[a]	Ref.
CH_3SCl	$CH_2{=}C{=}CH_2$	$CH_2{=}C(SR)CH_2Cl$	10
C_6H_5SCl	$CH_2{=}C{=}CH_2$	$CH_2{=}C(SR)CH_2Cl$	10
$CH_3COSSCl$	$CH_2{=}C{=}CH_2$	$CH_2{=}C(SR)CH_2Cl$	10
2,4-$(NO_2)_2C_6H_3SCl$	$CH_2{=}C{=}CH_2$	$CH_2{=}C(SR)CH_2Cl$	3
2,4-$(NO_2)_2C_6H_3SCl$	$(CH_3)_2C{=}C{=}CH_2$	$(CH_3)_2C{=}C(SR)CH_2Cl$	39
2,4-$(NO_2)_2C_6H_3SCl$	$(CH_3)_2C{=}C{=}CH_2$	$(CH_3)_2C{=}C(SR)CH_2Cl$ (40); $CH_2{=}C(CH_3){-}C(SR){=}CH_2$ (60)	45
$CH_3S\overset{+}{S}(CH_3)_2$ BF_4^-	$(CH_3)_2C{=}C{=}CH_2$	$(H_3C)_2C{=}C(SCH_3){-}CH_2{-}S^+(CH_3)_2$	40
2,4-$(NO_2)_2C_6H_3SCl$	1,2-Cyclononadiene	ring: $CH{=}C(SR){-}CHCl{-}(CH_2)_6$	39
$CH_3SS^+(CH_3)_2$ BF_4^-	1,2-Cyclononadiene	ring: $CH{=}C(SCH_3){-}CHS^+(CH_3)_2{-}(CH_2)_6$	40
$CH_3COSSCl$	$(CH_3)_2C{=}C{=}C(CH_3)_2$	$(CH_3)_2C{=}C(SR){-}C(CH_3){=}CH_2$	10

[a] Rearrangement products are not listed.

dissymmetric intermediate such as the vinylic episulfonium ion **37** must be involved in the addition of 2,4-dinitrobenzenesulfenyl chloride to optically active 2,2-dimethyl-3,4-hexadien-l-ol **38** since the cyclic adduct **39** was obtained optically active. This result could not arise if the adduct were derived from planar allylic ions.

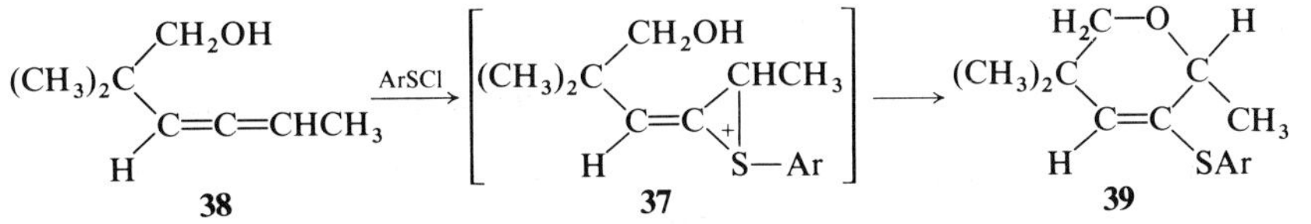

$$(CH_3)_2C(CH_2OH)-CH{=}C{=}CHCH_3 \ (\mathbf{38}) \xrightarrow{ArSCl} \left[\text{episulfonium ion } (\mathbf{37}):\ (CH_3)_2C(CH_2OH)-CH{=}C<\overset{+}{S}(Ar)>CHCH_3\right] \longrightarrow \mathbf{39}\ (H_2C{-}O{-}CH(CH_3){-}C(SAr){=}CH{-}C(CH_3)_2 \text{ ring})$$

Allenes in which the cumulated double bonds are differently substituted offer two different sites for attack by sulfenium ions. By analogy with the

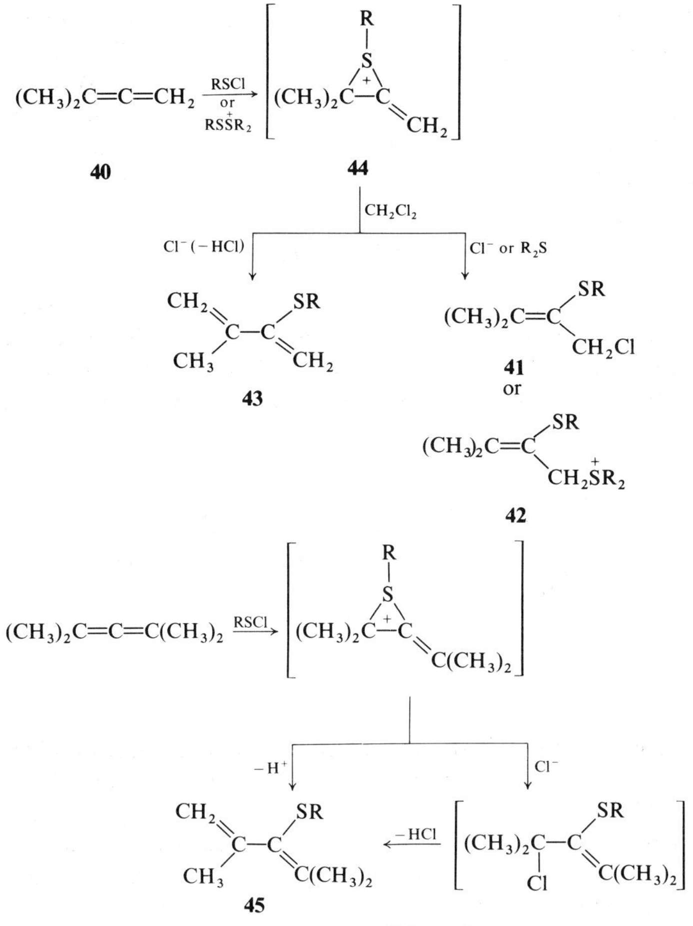

$$(CH_3)_2C{=}C{=}CH_2\ (\mathbf{40}) \xrightarrow[\text{RS}\overset{+}{\text{S}}\text{R}_2]{\text{RSCl or}} \left[(CH_3)_2C<\overset{+}{S}(R)>C{=}CH_2\right]\ (\mathbf{44})$$

44 in CH_2Cl_2:

- Cl^- ($-HCl$) → $CH_2{=}C(CH_3){-}C(SR){=}CH_2$ (**43**)
- Cl^- or R_2S → $(CH_3)_2C{=}C(SR)CH_2Cl$ (**41**) or $(CH_3)_2C{=}C(SR)CH_2\overset{+}{S}R_2$ (**42**)

$$(CH_3)_2C{=}C{=}C(CH_3)_2 \xrightarrow{RSCl} \left[(CH_3)_2C<\overset{+}{S}(R)>C{=}C(CH_3)_2\right]$$

- $-H^+$ → $CH_2{=}C(CH_3){-}C(SR){=}C(CH_3)_2$ (**45**)
- Cl^- → $\left[(CH_3)_2C(Cl){-}C(SR){=}C(CH_3)_2\right] \xrightarrow{-HCl}$ **45**

Scheme 5

observed orientation of oxymercuration of substituted allenes, one might anticipate that the more highly substituted double bond would be attacked preferentially. However, the addition of 2,4-dinitrobenzenesulfenyl chloride to 1,1-dimethylallene **40** gives the *primary* chloride **41** rather than the isomeric tertiary allylic chloride (39), although approximately 60% of the overall product mixture is the diene **43**. (45).

The same orientation is observed in the addition of $(CH_3)_2\overset{+}{S}SCH_3$ $\overset{-}{BF_4}$ to 1,3-dimethylallene giving **42**, but in this instance, no diene was formed (40). In contrast, the sole product of reaction of thioacetylsulfenyl chloride with tetramethylallene is the corresponding diene **45** (10). A reasonable course for these reactions is summarized in Scheme 5 wherein it is postulated that the more highly substituted double bond is attacked in conformity with related electrophilic reactions of **40**. The intermediate episulfonium ion **44** may either lose a proton to chloride ion (if present) to give the diene **43** or undergo S_N2' attack at the terminal methylene carbon to give the adducts **41** or **42** according to the nucleophile present. The mechanistic details of Scheme 5 have not, however, been experimentally verified.

D. Halogenation

Halogen addition can usually be arrested at the monoadduct stage and, without exception, the orientation corresponds to central electrophilic attack and terminal nucleophilic attack (Table 6). This is verified in the case of symmetrical reagents by reaction in the presence of competing nucleophiles (e.g., hydroxylic solvents).

$$>C{=}C{=}C< \xrightarrow[CH_3OH]{Br_2} >C{=}C(Br){-}C(Br)< \; + \; >C{=}C(Br){-}C(OCH_3)<$$

$$>C{=}C{=}C< \xrightarrow[CCl_4]{BrCl} >C{=}C(Br){-}C(Cl)<$$

The problem of allylic rearrangement in the first-formed adducts is a very real one and sometimes detracts from the true orientation. A study of bromine addition to methyl-substituted allenes by A. V. Fedorova (24) indicates that the kinetic products are generally those derived from attack at the more highly substituted double bond, and that subsequent rearrangement to more stable adducts frequently occurs. Therefore, with respect to orientation, halogen addition closely resembles oxymercuration in that the electro-

TABLE 6
Products of Halogen Additions to Substituted Allenes

Allene	Halogen	Solvent	Adducts	Ref.
$RCH{=}C{=}CH_2$	Br_2	CCl_4	$RCHBrCBr_2CH_2Br$	20
$RCH{=}C{=}CH_2$	Br_2	CCl_4	$RCHBrCBr{=}CH_2$ (mainly) $RCH{=}CBrCH_2Br$	28,51
$(CH_3)_2C{=}C{=}CH_2$	Br_2	CCl_4	$(CH_3)_2CBrCBr{=}CH_2$	28
$RCH{=}C{=}CHCH_3$	Br_2	$CHCl_3$, $-40°$	$RCHBrCBr{=}CHCH_3$ (mainly)	24
$CH_2{=}C{=}CHCN$	Br_2	CCl_4	$BrCH_2CBr{=}CHCN$	26
$CH_3CH{=}C{=}CHCH_3$ $CH_3CH{=}C{=}CHCH_3$	Br_2 $BrCl$	CH_3OH CH_3OH }	Br OCH$_3$ | | $CH_3CH{=}C{-}CHCH_3$[a]	32,48,49
$CH_3CH{=}C{=}CHCH_3$	Br_2	CCl_4	Br Br | | $CH_3CH{=}C{-}CHCH_3$[a]	32
$CH_3CH{=}C{=}CHCH_3$	$BrCl$	CCl_4	Br Cl | | $CH_3CH{=}C{-}CHCH_3$[a]	48
$CH_3CH{=}C{=}CHCH_3$ $CH_3CH{=}C{=}CHCH_3$ $CH_3CH{=}C{=}CHCH_3$ $CH_3CH{=}C{=}CHCH_3$ }	I_2 IBr ICl IN_3 }	CH_3OH }	I OCH$_3$ | | $CH_3CH{=}C{-}CHCH_3$	32,48
$CH_3CH{=}C{=}CHCH_3$	ICl	Pyridine	I Cl | | $CH_3CH{=}C{-}CHCH_3$[a]	48
1,2-Cyclononadiene	Br_2	CCl_4	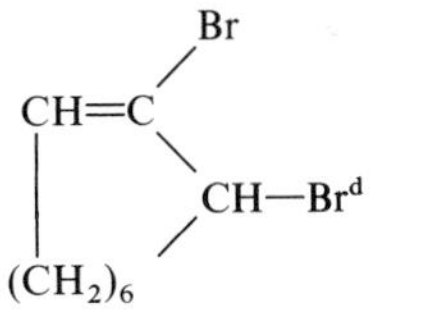 Br / CH=C | \\ | CH—Br[d] | / $(CH_2)_6$	48,50
1,2-Cyclononadiene	Br_2	CH_3OH	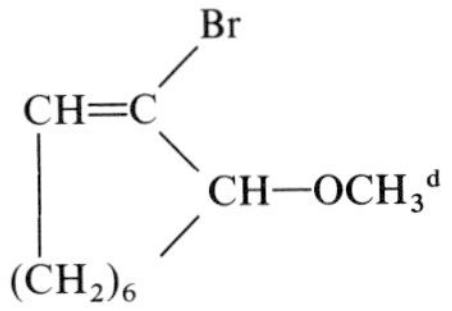 Br / CH=C | \\ | CH—OCH$_3$[d] | / $(CH_2)_6$	48
$(CH_3)_2C{=}C{=}CH_2$	Cl_2	$C_2F_3Cl_3$[b]	$CH_2{=}C(CH_3)C(Cl){=}CH_2$ (85) $CH_2{=}C(Cl)C(CH_3)_2(Cl)$ (2) $(CH_3)_2C{=}C(Cl)CH_2Cl$ (13)	52
$(CH_3)_2C{=}C{=}C(CH_3)_2$	Cl_2	$C_2F_3Cl_3$[b]	$(CH_3)_2C{=}C(Cl)C(CH_3){=}CH_2$[c]	52

[a] *Cis* and *trans* (mostly).

[b] Under oxygen to suppress radical reactions.

[c] Secondary products were also observed and are derived from reaction of HCl with tetramethylallene.

[d] *Cis* isomer; trans annular addition products are also obtained. (109).

phile enters at the central carbon and the nucleophile at the more highly substituted carbon. A similar orientation has been observed for the addition of nitrosyl chloride to 1,1-dimethylallene (46).

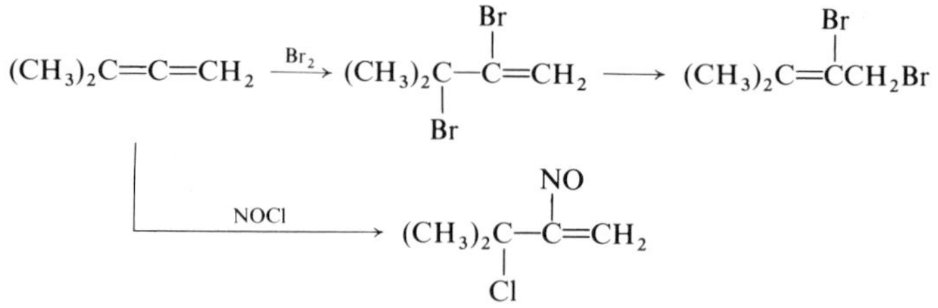

$$(CH_3)_2C{=}C{=}CH_2 \xrightarrow{Br_2} (CH_3)_2C(Br){-}C(Br){=}CH_2 \longrightarrow (CH_3)_2C{=}C(Br)CH_2Br$$

$$(CH_3)_2C{=}C{=}CH_2 \xrightarrow{NOCl} (CH_3)_2C(Cl){-}C(NO){=}CH_2$$

With respect to the stereochemistry of halogen aJdition, Jacobs et al. (44) found that optically active 2,2-dimethyl-3,4-hexadien-1-ol **38** gave inactive cyclic product **46** on reaction with bromine in carbon tetrachloride—from which it was inferred that a symmetrical intermediate or transition state was involved.

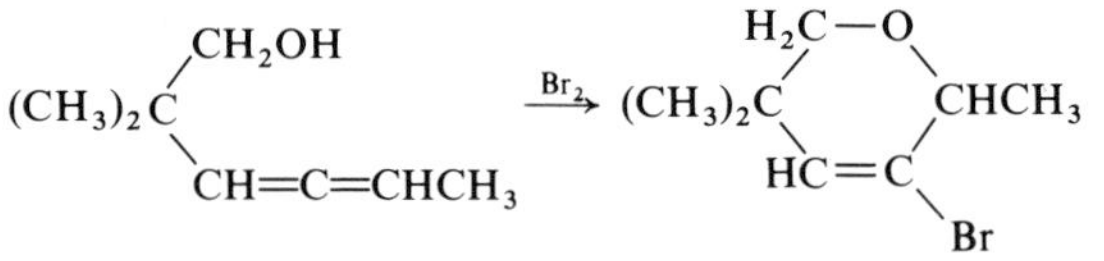

$$\underset{\mathbf{38}}{(CH_3)_2C(CH_2OH)CH{=}C{=}CHCH_3} \xrightarrow{Br_2} \underset{\mathbf{46}}{(CH_3)_2C\langle H_2C{-}O{-}CHCH_3{-}C(Br){=}CH\rangle}$$

In contrast bromine addition to optically active phenylallenecarboxylic acids **47** gave optically active bromolactones **48** with a high degree of stereospecificity (47).

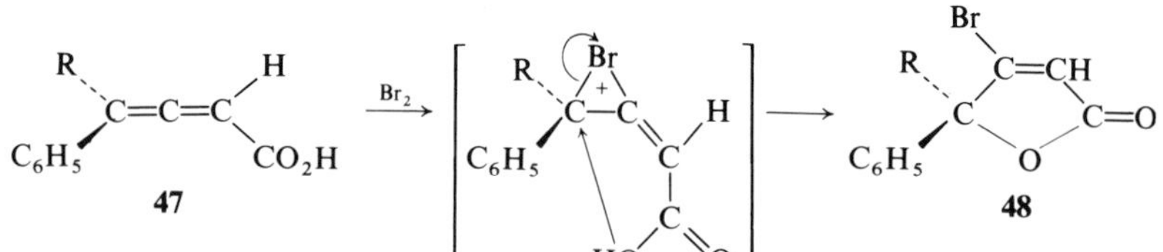

$$\underset{\mathbf{47}}{R(C_6H_5)C{=}C{=}C(H)CO_2H} \xrightarrow{Br_2} \left[\text{bromonium ion intermediate}\right] \longrightarrow \mathbf{48}$$

R = H, CH_3, C_2H_5, i-C_3H_7, t-C_4H_9

Addition of halogens to racemic and optically active 1,3-dimethylallene has been investigated from the standpoint of orientation and stereochemistry of reaction (32,48,49). Bromination of R-(−)-**30** with bromine or bromine chloride in methanol gave a mixture of optically active *cis* and *trans*-3-bromo-4-methoxy-2-pentenes **49** and **50**. The structure and configuration about the double bond and asymmetric carbon atom of the adducts was determined from their relationship to the known adducts of methoxymercuration, **28** and **29**, as shown in Scheme 6. In this way the stereochemistry of addition was established as *trans*.

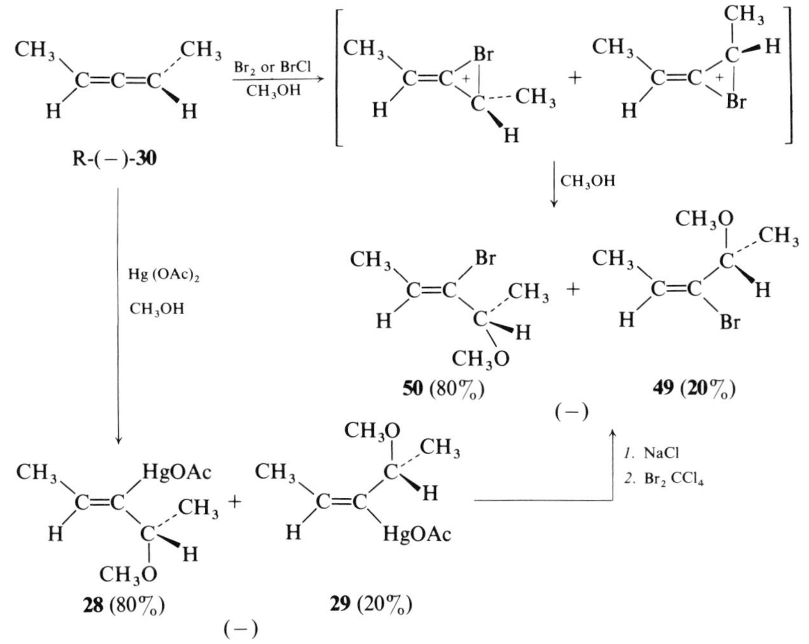

R-(−)-**30** $\xrightarrow[CH_3OH]{Br_2 \text{ or } BrCl}$ [bromonium ion intermediates] $\xrightarrow{CH_3OH}$ **50** (80%) + **49** (20%) (−)

R-(−)-**30** $\xrightarrow[CH_3OH]{Hg(OAc)_2}$ **28** (80%) + **29** (20%) (−) $\xrightarrow[\text{2. } Br_2\ CCl_4]{\text{1. NaCl}}$ **49**, **50**

Scheme 6

The degree of stereospecificity of bromination is not known with certainty, but it is at least as high as that in methoxymercuration (32). Bromination cannot then involve planar allylic ions to any important extent, and the products are more likely derived from nucleophilic attack on dissymmetric bromonium ion intermediates (Scheme 6).

Bromine addition to *R*-(−)-**30** in carbon tetrachloride gave a mixture of optically active *cis* and *trans* dibromopentenes, **51** and **52**, which were converted to the corresponding bromoethers, **49** and **50**, on reaction with silver nitrate in methanol. However, the latter reaction was accompanied with a change in the *cis–trans* ratio and extensive racemization with net inversion (Scheme 7). This observation is of some significance in that it implies that vinylic bromonium ions may not be formed reversibly or, what amounts to the same thing, the neighboring vinyl bromine does not participate in the silver-assisted methanolysis of the allylic bromine of **49** and **50**.

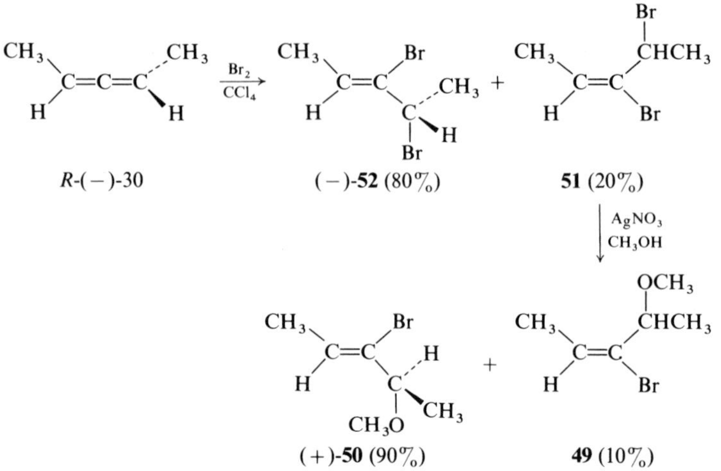

R-(−)-**30** $\xrightarrow[CCl_4]{Br_2}$ (−)-**52** (80%) + **51** (20%)

51 $\xrightarrow{AgNO_3,\ CH_3OH}$ (+)-**50** (90%) + **49** (10%)

Scheme 7

Addition of iodine, iodine bromide, and iodine chloride to R-(−)-**30** gave optically active iodoethers **53** and **54** in the ratio of 6:94. The stereochemistry of this reaction was also established as *trans* by the reaction sequence shown in Scheme 8 (32,48).

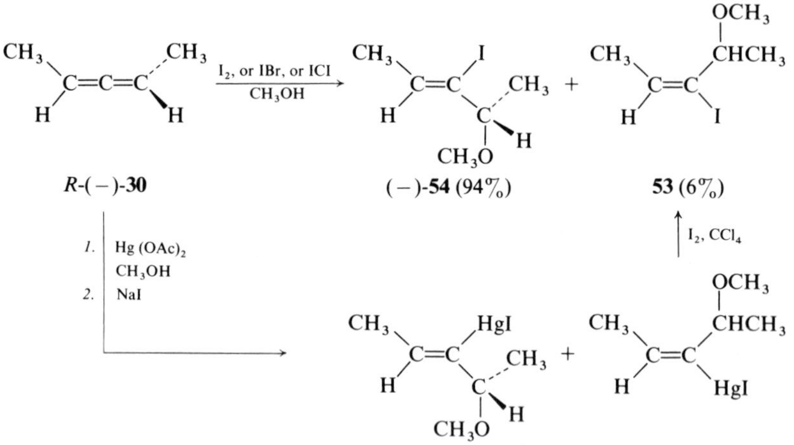

R-(−)-**30** $\xrightarrow[CH_3OH]{I_2,\ \text{or IBr, or ICl}}$ (−)-**54** (94%) + **53** (6%)

R-(−)-**30**: *1.* $Hg(OAc)_2$, CH_3OH; *2.* NaI → mercuric iodide ethers; I_2, CCl_4 → **53** (6%)

Scheme 8

The product composition of the mixture of iodoethers **54** and **53** was the same whether the iodinating agent was I_2, IBr, or ICl, but the magnitude of the rotation varied (Table 7). The optical rotations, hence the degree of stereospecificity, increased in the order $I_2 < IBr < ICl$. Evidently, the

TABLE 7
Product Distribution and Optical Rotations on Addition of Halogens to (−)-1,3-Dimethylallene in Methanol (48)

$CH_3CH{=}C{=}CHCH_3$, $[\alpha]_D^0$	Halogen, XY	$CH_3CH{=}CXCHCH_3$ (OCH_3) Per cent *cis*	Per cent *trans*	$[\alpha]_D$
−19.8	Br_2	31	69	−11.8
−19.8	BrCl	31	69	−11.1
−15.9	I_2	6	94	−4.1
−19.8	I_2	6	94	−5.5
−19.8	IBr	6	94	−7.8
−19.8	ICl	6	94	−13.4
−19.8	ICl	6	94	−12.5

stereospecificity of addition is dependent in part on the nature of the iodinating agent. Further investigation (45) has shown that the source of this difference in stereospecificity is the halide ion formed in the reactions in methanol. Halide ion effectively racemizes the starting allene in the order $I^- > Br^- > Cl^-$. The racemization sequence involves reversible halogen addition to the allene accompanied by S_N2-type racemization of the dihalo adduct.

$$CH_3CH{=}C{=}CHCH_3 + I_2 \underset{I^-}{\rightleftharpoons} CH_3CH{=}C(I){-}C(I)(CH_3)H$$

R-(−)-**30**

$$\Big\updownarrow I^-$$

$$R\text{-}(+)\text{-}\mathbf{30} \xleftarrow{I^-} CH_3CH{=}C(I){-}C(I)(CH_3)H$$

Bromination of 1,2-cyclononadiene in methanol gives a mixture of *cis* bromoethers **55** and **56** with **56** as the major product (Scheme 9) (48, 109). Formation of **55** is consistent with other electrophilic addition reactions of 1,2-cyclononadiene which give only *cis* adducts (e.g., oxymercuration). Models indicate that approach of an electrophilic reagent from within the ring is more hindered than from without (cf. **57**). Formation of **56** can be explained as the result of a transannular hydride shift occurring in the allylic

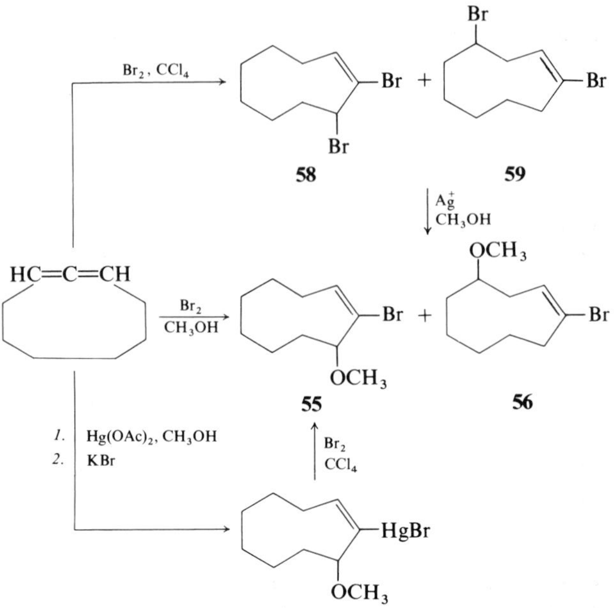

Scheme 9

cation initially formed by electrophilic attack of bromine at the central allenic carbon. The planarity or otherwise of the allylic intermediate is not yet known.

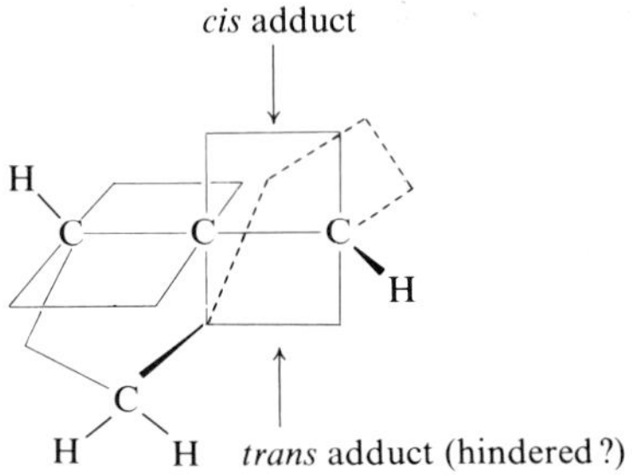

cis adduct

trans adduct (hindered?)

57

Addition of bromine to 1,2-cyclononadiene in carbon tetrachloride is reported to give a mixture of dibromides **58** and **59** (48,50,109). The homoallylic bromide **59** is evidently the product of transannular addition of bromine. It is interesting to note, however, that a 1:3 mixture of dibromides gave a 1:3 mixture of bromoethers **55** and **56** on reaction with silver fluoroborate in methanol (Scheme 9). Since the isomer distribution was unchanged, this might indicate that S_N1-type solvolysis of the dibromides does not lead to products of transannular hydride shifts.

Relative to bromination, chlorination of allenes is less straightforward.

Poutsma has described the reaction of chlorine with tetramethylallene in 1,1,2-trichlorotrifluoroethane in the presence of oxygen to suppress radical reactions (52). Under these conditions the chlorination product is 3-chloro-2,4-dimethyl-1,3-pentadiene, which is presumed formed by proton elimination from an intermediate chloronium ion.

$$(CH_3)_2C{=}C{=}C(CH_3)_2 \xrightarrow[C_2Cl_3F_3]{Cl_2 25°} \left[(CH_3)_3\overset{+}{C}\overset{Cl}{-}C{=}C(CH_3)_2\right] \xrightarrow{-H^+} CH_2{=}C(CH_3)-C(Cl){=}C(CH_3)_2$$

Formation of a diene on chlorination of tetramethylallene is reminiscent of the reaction of acetylthiosulfenyl chloride with tetramethylallene, which gives the diene **45** (42). The analogy between chlorination and sulfenyl halide addition is also evident in the reactions of 1,1-dimethylallene. Chlorination leads to the two possible dichloro adducts and to 2-chloro-3-methyl-1,3-butadiene as the major product (52). Likewise 2,4-dinitrobenzenesulfenyl chloride gives the monoadduct **41** and the diene **43** (Scheme 5) (45).

$$(CH_3)_2C{=}C{=}CH_2 \xrightarrow[C_2Cl_3F_3]{Cl_2,\ 25°} \underset{2\%}{(CH_3)_2C(Cl)-C(Cl){=}CH_2} + \underset{13\%}{(CH_3)_2{=}C(Cl)-CH_2Cl} + \underset{85\%}{CH_2{=}C(CH_3)-C(Cl){=}CH_2}$$

E. Addition of Chlorosulfonyl Isocyanate

Chlorosulfonyl isocyanate, $ClSO_2NCO$, is evidently an electrophilic reagent that adds to alkenes by a two-step dipolar cycloaddition reaction to

$$(CH_3)_2C{=}C{=}CH_2 \xrightarrow[Et_2O]{ClSO_2-\overset{\delta^-}{N}{=}\overset{\delta^+}{C}{=}O} \left[\begin{matrix}(CH_3)_2\overset{+}{C}-C({=}CH_2)\\ ClSO_2\overset{-}{N}-C{=}O\end{matrix}\right]_{\mathbf{60}} \rightarrow \underset{\mathbf{61}}{\begin{matrix}(CH_3)_2C-C{=}CH_2\\ R-N-C{=}O\end{matrix}}\ \ R = H, SO_2Cl \quad + \quad \underset{\mathbf{62}}{CH_2{=}C(CH_3)-C({=}CH_2)-C({=}O)-NH_2}$$

give β-lactams (53). Reaction of this reagent with variously substituted allenes has been reported (54). The major products are the expected β-lactams and conjugated dienes. These products may be considered formed by initial electrophilic attack of the reagent at the central allenic carbon to give a dipolar intermediate **60** which may cyclize to a β-lactam **61** or undergo proton transfer to give (after work-up) a conjugated diene **62**. Unsymmetrical allenes effectively add at the more highly substituted bond.

F. Hydroboration

Allenes react with diborane to give adducts that are not generally isolated but are oxidized to give either alcohols or ketones, depending on the position of the boron atom. A consistent picture of the orientation of addition using diborane is not evident, however. Allene reacts by terminal boron attack in both the gas phase (7) and in solution (6), while 1,2-cycloalkadienes and 2,3-nonadiene react by central attack to give ketones as the major products of oxidation (55); butylallene and phenylallene react to give 1,3 and 1-2 diols after oxidation, and these products unfortunately do not permit a clear choice as to the position of boron attack (6).

Studies involving diborane are difficult to interpret partly because of complications that arise in the oxidation step, and partly because addition to the trialkylborane stage involves three successive steps, each with different steric and electronic requirements.

$$\mathrm{R} + \mathrm{BH_3} \longrightarrow \mathrm{RBH_2} \xrightarrow{\mathrm{R}} \left(\begin{matrix}\mathrm{R}\\ | \\ \mathrm{H}\end{matrix}\right)_2\mathrm{BH} \xrightarrow{\mathrm{R}} \left(\begin{matrix}\mathrm{R}\\ | \\ \mathrm{H}\end{matrix}\right)_3\mathrm{B} \xrightarrow{[\mathrm{O}]} 3\left(\begin{matrix}\mathrm{R}\\ | \\ \mathrm{H}\end{matrix}\right)\!\!-\!\mathrm{OH}$$

(R = alkene)

One investigation that avoids these complications involves the hydroboration of allenes with the monofunctional borane **13** (8). The borane adducts proved to be relatively stable and were readily isolated and identified. The results are summarized in Table 8, which shows that terminal attack by boron is obviously preferred in allene and dimethylallene, but central attack is dominant in 1,3-dimethylallene, tetramethylallene, and 1,2-cyclononadiene.

Assuming that hydroboration involves a four-center transition state (9), then steric as well as electronic effects are likely to be important in determining the direction of addition. This is well illustrated in the case of 1,1-dimethylallene. The most nucleophilic of the two π bonds is undoubtedly the more highly substituted one, yet only 22% of the product involves addition

TABLE 8
Hydroboration of Allenes with 4,4,6-Trimethyl-1,3,2-dioxaborinane[a] (8)

Allene	Adducts[b,c]	
$CH_2{=}C{=}CH_2$	$CH_3C(BR_2){=}CH_2$ (13)	$CH_2{=}CHCH_2BR_2$ (87)
$(CH_3)_2C{=}C{=}CH_2$	$(CH_3)_2CH\,C(BR_2){=}CH_2$ (22)	$(CH_3)_2C{=}CCH_2BR_2$ (78)
$CH_3CH{=}C{=}CHCH_3$	$CH_3CH{=}C(BR_2)\,CH_2CH_3$ (81)	$CH_3CH{=}CHCH_2CH_2BR_2$ (19)
$(CH_3)_2C{=}C{=}C\,(CH_3)_2$	$(CH_3)_2C{=}C(BR_2){-}CH(CH_3)_2$ (100)	
$HC{=}C{=}CH$, ring $(CH_2)_6$	$CH{=}C(BR_2){-}CH$, ring $(CH_2)_6$ *cis* (85) *trans* (15)	

[a] Structure **13**.

[b] $-BR_2 = -B$ in a six-membered ring: $-B(O{-}CH(CH_3){-}CH_2{-}C(CH_3)_2{-}O)$

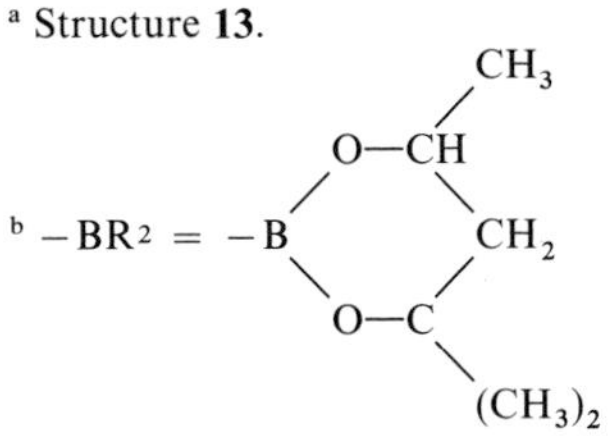

[c] Per cent yield in parenthesis.

to this bond. The major product involves boron attack at the terminal methylene, probably for the reason that steric congestion in the transition state **63** is at a minimum in this orientation.

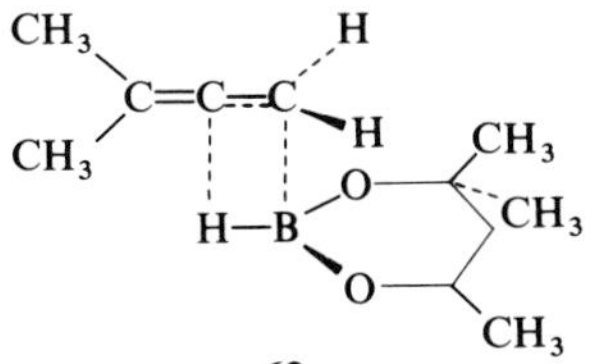

63

Other hydridic reagents react with allenes. For example, methyldichlorosilane is reported to form mono- and diadducts on addition to allenes in the presence of chloroplatinic acid (56). The structure of the monoadducts correspond to addition of hydrogen at the central *sp*-carbon and the silyl group at the least substituted terminal carbon.

$$C_3H_7CH{=}C{=}CH_2 \xrightarrow[H_2PtCl_6]{CH_3SiCl_2H} C_3H_7CH{=}CHCH_2SiCH_3Cl_2$$

$$C_3H_7CH{=}C{=}CHCH_3 \xrightarrow[H_2PtCl_5]{CH_3SiCl_2H} C_3H_7CH{=}CH\underset{\displaystyle CH_3}{\underset{|}{C}}HSiCH_3Cl_2$$

G. Summary of the Effect of Methyl Substitution

The influence of methyl substituents on the pattern of orientation is no doubt partly steric and partly electronic in origin, but it is not always easy to differentiate the two effects or to assess their relative importance. Nevertheless, the pronounced effect of methyl substituents on the orientation of addition is very probably related to the well-established fact that carbonium ion stability increases markedly with methyl substitution. Thus the change in orientation from terminal to central electrophilic attack with increasing methyl substitution surely reflects the increase in stability of secondary and tertiary carbonium ions relative to primary and vinyl carbonium ions. Likewise, a preference for attack at the more highly substituted of two cumulated bonds can be related to the stabilizing influence of the substituents in the incipient cation. The evidence obtained thus far does *not* support the formation of a delocalized allylic cation as the driving force for central attack. These arguments lose some credence in reactions which proceed through bridged ion intermediates since the position of electrophilic attack becomes obscure. Nevertheless the transition states probably have carbonium ion character and will thus be stabilized by electron accession from the methyl substituents.

Steric effects in these reactions are less well understood. A reagent approaching one double bond is coplanar with the substituents at the terminus of the other double bond, and this in principle could lead to unfavorable steric interactions. Steric effects of this type do not appear to be a dominating influence, however, as evidenced by the fact that the major products of halogenation, oxymercuration, and hydroboration of 1,3-dimethylallene are the *trans* adducts, even though the attacking reagent has thereby to approach in a direction *cis* to the terminal methyl group as opposed to a terminal hydrogen atom. This is not true in the bromination of cyclononadiene which does not lead to the formation of the *trans* isomer.

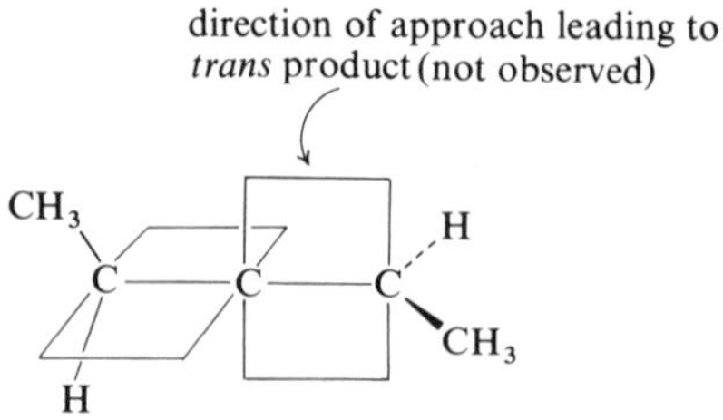

direction of approach leading to
trans product (not observed)

CH_3 H
C C C
CH_3
H

H. Perhaloallenes

Tetrafluoroallene is of particular interest in view of its seemingly anomolous behavior toward electrophilic reagents. While most perfluoroalkenes are notably resistant to attack by electrophilic reagents, tetrafluoroallene reacts smoothly and rapidly with HBr, HCl, HF, and excess chlorine to give high yields of monoadducts (57a). The orientation clearly spells electrophilic attack at the *central* carbon.

$$CF_2{=}C{=}CF_2 \xrightarrow{HX} XCF_2CH{=}CF_2$$

$$CF_2{=}C{=}CF_2 \xrightarrow{Cl_2} ClCF_2CCl{=}CF_2 \qquad X = F, Cl, Br$$

This result is contrary to the orientation observed for hydrogen halide additions to allene (Table 1), but the reactivity and the orientation of tetrafluoroallene can readily be explained as the result of stabilization of α-fluorocarbonium ions by electron release from fluorine to carbon. Accordingly, with increasing substitution of perfluoroalkyl groups, the reactivity

$$F_2C{=}C{=}CF_2 \xrightarrow{(HF)_x} \ddot{F}{-}\overset{F}{\underset{+}{C}}{-}CH{=}CF_2 \longleftrightarrow \overset{+}{F}{=}\overset{F}{C}{-}CH{=}CF_2 \text{ etc.} \xrightarrow{F^-} F_3CCH{=}CF_2$$

of perfluoroallenes towards electrophiles is sharply reduced. Thus, perfluoro(3-methyl-1,2-butadiene), $(CF_3)_2C{=}C{=}CF_2$, does not react with electrophiles (57b). Similarly, 1,2-perfluoropentadiene does not react with chlorine, hydrogen chloride and hydrogen bromide, but it reacts quantitatively at 20° with anhydrous hydrogen fluoride to give a *cis-trans* mixture of 2H-nona-fluoro-2-pentene (57c). Since the isomeric terminal alkene, 2H-nonafluoro-1-pentene was not observed, the suggestion has been made that the intermediate α-fluorocarbonium ion collapses to product faster than it can undergo bond rotation by 90° to form a delocalized ion that could lead to a mixture of 1-and 2-alkenes (57c).

$$C_2F_5CF{=}C{=}CF_2 \xrightarrow{(HF)_x} C_2F_5CF{=}C\begin{matrix} H \\ \overset{+}{C}F_2 \end{matrix} \overset{\text{rotn}}{\not\longrightarrow} C_2F_5CF\overset{\overset{H}{C}}{\underset{+}{\frown}}CF_2$$

$$\xrightarrow{F^-} \underset{\textit{cis} \text{ and } \textit{trans}}{C_2F_5CF{=}CHCF_3} \qquad \xrightarrow{F^-} \underset{\text{(not formed)}}{C_2F_5CF_2CH{=}CF_2}$$

Perchloroallene is also very reactive toward electrophilic reagents and the orientation of addition of hydrogen chloride is also contrary to that observed with allene itself (58). Halogen addition gives monoadducts, and the adduct formed from ICl could be considered to arise from initial electrophilic attack of iodine at the central carbon to give an α-chloro carbonium ion intermediate.

$$Cl_2C{=}C{=}CCl_2 \begin{cases} \xrightarrow[CH_2Cl_2,\ -75^\circ]{Br_2} Cl_2C{=}CBrCCl_2Br \\ \xrightarrow[CHCl_3,\ -75^\circ]{ICl} Cl_2C{=}CICCl_3 \\ \xrightarrow[AlCl_3,\ -75^\circ]{HCl} Cl_2C{=}CHCCl_3 \end{cases}$$

III. NUCLEOPHILIC ADDITION TO ALLENES

The following discussion of the addition of nucleophilic reagents to allenes includes the parallel reactions of isomeric alkynes. This is necessary because the basic character of nucleophilic reagents often promotes the rearrangement of allenes and alkynes, and it is not always obvious whether the products of nucleophilic addition are derived from the allene, the alkyne, or both. To make any reliable conclusions pertaining to addition, it is very important to know which of the equilibrating hydrocarbons is thermodynamically more stable and which is more reactive under the reaction conditions. The problems that arise when this information is lacking are illustrated in the reactions of nucleophilic reagents with the parent, allene.

A. Addition to Allene

The addition of ethanol to allene in the presence of potassium hydroxide was originally considered to be an example of anionic addition initiated by attack of ethoxide ion at the central carbon (59). However, the reaction is accompanied by rearrangement of allene to methylacetylene, and it is entirely possible that the adduct is formed from methylacetylene.

$$\left.\begin{array}{c} CH_2{=}C{=}CH_2 \\ \downarrow\uparrow \\ CH_3C{\equiv}CH \end{array}\right\} \xrightarrow[EtOH]{K^+OEt^-} CH_3{-}C(OEt){=}CH_2$$

The same ambiguity arises in the addition of sodium thiolates to allene, reported recently by Mueller and Griesbaum (60). Although mono- and diadducts were formed, equilibration of allene and methylacetylene also occurred. It was further shown that the products and their distribution were the same whether the starting hydrocarbon was allene or methylacetylene.

$$\left.\begin{array}{c} CH_2{=}C{=}CH_2 \\ \rightleftarrows \\ CH_3C{\equiv}CH \end{array}\right\} \xrightarrow[RSH]{Na^+\,{}^-SR,\ 85^\circ} \underset{90\%}{CH_3{-}C({=}CH_2){-}SR} + \underset{6\%}{CH_3{-}C(SR)_2{-}CH_3} + \underset{3\%}{CH_3CH{=}CHSR}$$

In an effort to resolve which hydrocarbon is more reactive, the related reaction of tetramethylallene with ethanethiol was studied. The finding that no detectable addition occurred suggests that allene may be similarly unreactive, in which case thiolate addition probably occurs predominantly or completely to methylacetylene (60). However, the pronounced retardation of nucleophilic addition by methyl substituents (61) makes it doubtful that the lack of reactivity of tetramethylallene can be reliably extrapolated to allene. The main conclusion is that nucleophilic additions to allene have not been demonstrated unambiguously, but if they occur, the nucleophile attacks the central carbon with high selectivity.

B. Addition to Substituted Allenes

High selectivity for central attack is demonstrated in nucleophilic additions to substituted allenes. Also, the reactivity of the allene is greatly enhanced by electron-withdrawing substituents. The products, however, do not always permit one to tell whether the reaction is under thermodynamic or kinetic control. The factors involved are well illustrated in the recent work of Stirling and co-workers (61–64), who have studied the reactions of arylsulfonylpropadienes **64** and arylsulfonylpropynes, **65** and **66**, with several different nucleophiles. Their work clearly shows that the overall course of these reactions depends not only on the stability and reactivity of the sulfones, but on the nucleophilic and base strength of the attacking nucleophile.

They first established that phenylsulfonylpropadiene **64a** is the thermodynamic product of base-catalyzed rearrangement of **65a** and **66a**. Conversion of **65a** to **64a** was found to be very rapid in the presence of triethylamine while that of **66a** to **64a** was comparatively slow (62).

$$\underset{\mathbf{65}}{ArSO_2CH_2C{\equiv}CH} \xrightarrow{(C_2H_5)_3N} ArSO_2CH{=}C{=}CH_2 \xleftarrow{(C_2H_5)_3N} \underset{\mathbf{66}}{ArSO_2C{\equiv}CCH_3}$$

64a Ar = C_6H_5
64b Ar = p-$CH_3C_6H_4$

Under weakly basic conditions (triethylamine) in the presence of a strong nucleophile (benzenethiol), both **64a** and **65a** reacted to give the same adduct, **67**, and it was presumed that, in the case of **65a**, rearrangement to the allene **64a** preceded addition (62). In contrast 1-phenylsulfonylpropyne **66a** gave the *cis* adduct **68**, which is the product expected of *trans* addition of benzenethiol to **66a**.

$$\underset{\textbf{64a}}{\phi SO_2CH{=}C{=}CH_2} \xrightarrow[(C_2H_5)_3N]{\phi SH,\ 20°} \underset{\textbf{67}}{\phi SO_2CH_2{-}C({=}CH_2)S\phi}$$

$$\underset{\textbf{65a}}{\phi SO_2CH_2C{\equiv}CH} \longrightarrow \textbf{64a}$$

$$\textbf{67} \xrightarrow{NaOCH_3} \underset{\textbf{69}}{(\phi SO_2)(H)C{=}C(CH_3)(S\phi)}$$

$$\underset{\textbf{66a}}{\phi SO_2C{\equiv}CCH_3} \xrightarrow[(C_2H_5)_3N]{\phi SH,} \underset{\textbf{68}}{(\phi SO_2)(H)C{=}C(S\phi)(CH_3)} \xrightarrow{NaOCH_3} \textbf{69}$$

The adducts **67** and **68** are apparently kinetic products since, in the presence of a strong base ($NaOCH_3$), they rearrange to the more stable *trans* adduct **69**.

The reactions of sodium benzenesulfinate in methanol with the phenylsulfonyl derivatives **64a**, **65a**, and **66a** took a similar course, and the reaction of **65a** is similarly presumed to occur by way of the allene.

$$\textbf{65a} \xrightarrow{\phi SO_2Na} \textbf{64a} \xrightarrow[CH_3OH]{\phi SO_2Na} \phi SO_2CH_2C({=}CH_2)SO_2\phi$$

$$\textbf{66a} \xrightarrow[CH_3OH]{\phi SO_2Na} (\phi SO_2)(H)C{=}C(SO_2\phi)(CH_3)$$

However, with more strongly basic nucleophiles (e.g., $NaOCH_3$) or with weaker nucleophiles (e.g., RCO_2Na) the *same* adducts **70** were obtained from all three isomers (62,64). Knowing that the allene isomer is both more stable and more reactive than **65** and **66**, these reactions most likely proceed by initial rearrangement of **65** and **66** followed by addition to **64** to give the adduct **70** (Scheme 10). The initial adduct is the kinetic product since, in methanol addition, increasing base concentration led to formation of the more stable adduct **71** (62). Clearly, a decrease in the ratio of nucleophilicity to basicity in the attacking nucleophile promotes base-catalyzed rearrangement of reactants and products.

$$\mathbf{65} \xrightarrow{\text{NaOR}} \underset{\mathbf{64}}{ArSO_2CH{=}C{=}CH_2} \xleftarrow{\text{NaOR}} \mathbf{66}$$

$$\downarrow \text{ROH, NaOR}$$

$$\underset{\mathbf{70}}{ArSO_2CH_2C(OR){=}CH_2} \xrightarrow{\text{NaOR}} \underset{\mathbf{71}}{(ArSO_2)(H)C{=}C(CH_3)(OR)}$$

$R = CH_3$, $COCH_3$, COC_6H_5 $\quad$ $Ar = C_6H_5$, $p\text{-}CH_3C_6H_4$, $p\text{-}CH_3C_6H_4$

Scheme 10

Reactions of sulfones **64**, **65**, and **66** with secondary amines represent an intermediate situation with regard to basicity versus nucleophilicity. Investigations by Stirling (61,63) showed that the same adduct **72**, which presumably is also the most stable of the possible adducts, was formed from each sulfone, even with a base as weak as dibenzylamine. This could mean that rearrangement precedes addition, as in the case of methanol and carboxylic acid addition. Kinetic studies on the addition of piperidine to the *p*-tolylsulfonyl derivatives **64b**, **65b**, and **66b** showed that the rates of adduct formation for **64b** and **65b** were essentially the same while the sulfone **66b** reacted appreciably slower (64). These results support the proposed rapid rearrangements of **65b** to **64b** followed by rate-controlling addition to **64b**, but do not distinguish between rate-controlling rearrangement or rate-controlling addition for **66b**. The latter alternative is favored by Stirling (64).

$$\underset{\mathbf{64}}{ArSO_2CH{=}C{=}CH_2} \xrightarrow[\text{Slow}]{R_2NH} \left[ArSO_2\overset{-}{C}H{\cdots}C(\overset{+}{N}HR_2){\cdots}CH_2 \right] \longrightarrow \underset{\mathbf{72}}{(ArSO_2)(H)C{=}C(CH_3)(NR_2)}$$

$$\underset{\mathbf{65}}{ArSO_2CH_2C{\equiv}CH} \xrightarrow[\text{rapid}]{R_2NH} \mathbf{64}$$

$$\underset{\mathbf{66}}{ArSO_2C{\equiv}CCH_3} \xrightarrow{?} \mathbf{65}$$

$$\underset{\mathbf{66}}{ArSO_2C{\equiv}CCH_3} \xrightarrow[?]{R_2NH} \left[ArSO_2\bar{C}{=}C(CH_3)(\overset{+}{N}R_2H) \right] \longrightarrow \mathbf{72}$$

A number of interesting nucleophilic addition reactions have been reported for 2,3-butadienoic acid and derivatives **73** ($X = CO_2C_2H_5$, CN). Most of these reactions are summarized in Table 9. Without exception,

TABLE 9
Products of Nucleophilic Addition to 2,3-Butadienoic Acid Derivatives and Related Compounds

Allene	Nucleophile	Adduct	Ref.
$H_2C{=}C{=}CHCO_2Et$	$EtOH(K_2CO_3)$	$CH_3C(OEt){=}CHCO_2Et$	65
$H_2C{=}C{=}CHCO_2Et$	ϕNH_2	$CH_3C(NH\phi){=}CHCO_2Et$	65
$H_2C{=}C{=}CHCO_2Et$	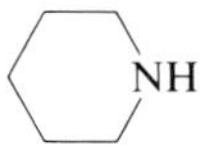 piperidine (cyclic NH)	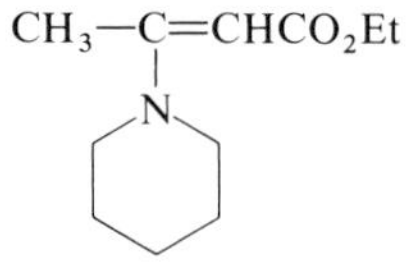 $CH_3{-}C(NC_5H_{10}){=}CHCO_2Et$ (N-piperidinyl)	65
$H_2C{=}C{=}CHCO_2Et$	$EtO{-}C_6H_4{-}NH_2$	$CH_3{-}C(NH{-}C_6H_4{-}OEt){=}CHCO_2Et$	65
$H_2C{=}C{=}CHCO_2Et$	NaN_3, H_2O, THF	$CH_3{-}C(N_3){=}CHCO_2Et$	67
$H_2C{=}C{=}CHCO_2Et$	$LiAlH_4$	$CH_2{=}CHCH_2CO_2Et$	65
$H_2C{=}C{=}CHCO_2H$	$LiAlH_4$	$CH_2{=}CHCH_2CO_2H$	65
$H_2C{=}C{=}CHCN$	ROH(RONa)	$CH_3C(OR){=}CHCN$	66
$H_2C{=}C{=}CHCN$	RCS_2H	$CH_3C(S{-}C({=}S){-}R){=}CHCN$	66
$CH_2{=}C{=}CHCN$	RR′NH	$CH_3{-}C(NRR'){=}CHCN$	66
$CH_2{=}C{=}CHCN$	H_2NNHR	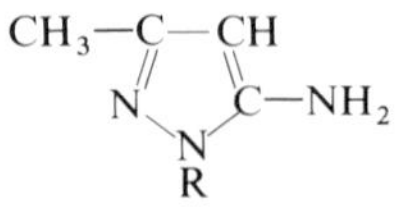 $CH_3{-}C{=}N{-}N(R){-}C(NH_2){=}CH$ (ring: 3-methyl-5-amino-1-R-pyrazole)	66

TABLE 9 (*continued*)

Allene	Nucleophile	Adduct	Ref.
$CH_2{=}C{=}CHCN$	$\phi NHNH_2$	$CH_3{-}C(=N{-}NH\phi)CH_2CN$	66
$CH_2{=}C{=}CHCN$	NH_2OH	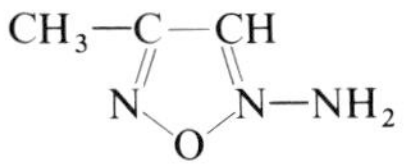 $CH_3{-}C{=}CH$ ring: N–O–N(–NH_2) (five-membered ring: CH_3–C=N–O–N(NH_2)=CH)	66
$CH_2{=}C{=}CHCN$	$CH_2(CO_2Et)_2$	$CH_3{-}C(=C(CO_2Et)_2){-}CH_2CN$	66
$R_2C{=}C{=}CHCN$	R_2NH	$R_2CH{-}C(NR_2){=}CHCN$ $R_2C{=}C(NR_2)CH_2CN$	70
$CH_2{=}C{=}C(R)CO_2H$	RMgBr	$CH_2{=}C(R)CH(R)CO_2H$	71,72
$CH_2{=}C{=}CHCOR$	EtMgBr	$CH_3CH_2C(CH_3){=}CHCOR$ (mostly) $CH_2{=}C(Et)CH_2COR$	74
$CH_2{=}C{=}CHCOR$	ROH (5% K_2CO_3)	$CH_3C(OR){=}CHCOR$ (mostly) $CH_2{=}C(OR)CH_2COR$	73
$RCH{=}C{=}CHCOR$	$RNHNH_2$	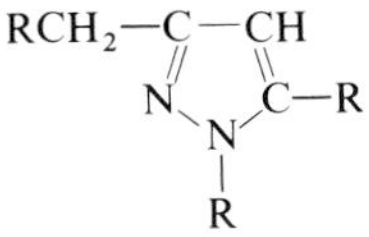 $RCH_2{-}C{=}CH$ ring: N–N(R)–C–R (pyrazole: RCH_2–C=N–N(R)–C(R)=CH)	69
$HO_2CCH{=}C{=}CHCO_2H$	$NaHSO_3$	$HO_2CCH_2{-}C(SO_3H){=}CHCO_2H$	67,68
$RCH{=}C{=}CHCH_2CH_2OH$	(NaOH)	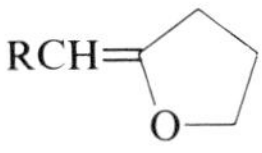 RCH= (2-tetrahydrofurylidene; five-membered ring with O)	75
$(CH_3)_2C{=}C{=}CHP(O)(OEt)_2$	ROH (base)	$(CH_3)_2CHC(R){=}CHP(O)(OEt)_2$	76
$(CF_3)_2C{=}C{=}C(CF_3)_2$	CH_3OH	$(CF_3)_2CHC(OCH_3){=}C(CF_3)_2$	77

nucleophilic attack occurs at the central allenic carbon and, apart from hydride addition, the major products are the thermodynamically stable adducts, which are usually β-substituted crotonic acid derivatives **74**.

$$\underset{\mathbf{73}}{H_2C{=}C{=}CHX} + \ddot{Y}H \longrightarrow \underset{\mathbf{74}}{CH_3{-}C(Y){=}CHX}$$

Intervention of acetylenic derivatives in these reactions is unlikely because equilibration studies have shown that base-catalyzed rearrangement of the 3-alkyne isomer to the corresponding allene is very facile (65,66) and that addition to the 2-alkyne isomer is relatively slow (65).

$$\mathrm{HC{\equiv}CCH_2CN \longrightarrow H_2C{=}C{=}CHCN}$$

$$\underset{\mathbf{76}}{\mathrm{HC{\equiv}CCH_2CO_2Et}} \xrightarrow{\mathrm{K_2CO_3}} \underset{\mathbf{75}}{\mathrm{CH_2{=}C{=}CHCO_2Et}} \xrightleftharpoons{\mathrm{K_2CO_3}} \underset{\mathbf{77}}{\mathrm{CH_3C{\equiv}CCO_2Et}}$$

$$\mathbf{75} \xrightarrow[\mathrm{EtOH}]{\mathrm{K_2CO_3}} \mathrm{CH_3{-}C(OEt){=}CHCO_2Et}$$

Thus in the case of ethyl 2,3-butadienoate **75** and its isomers **76** and **77** the tetrolic ester **77** is actually the most stable isomer but it is also the least reactive and, under mildly basic conditions (K_2CO_3), does not add ethanol. The allene isomer **75** reacts exothermically with ethanol under these conditions.

The exceptional reactivity of **75** toward nucleophilic reagents is most evident in its reaction with lithium aluminum hydride. The usual selectivity of the latter reagent for carbonyl bonds as opposed to carbon double bonds is reversed, and the product is ethyl vinylacetate **78**. In contrast acetylenic esters analogous to **77** are first reduced to acetylenic primary alcohols.

$$\underset{\mathbf{75}}{\mathrm{CH_2{=}C{=}CHCO_2Et}} \xrightarrow{\mathrm{LiAlH_4}} \underset{\mathbf{78}}{\mathrm{CH_2{=}CHCH_2CO_2Et}}$$

In the limited number of cases in which nucleophilic addition to allenes give products determined by kinetically controlled proton transfer the question is whether the intermediate carbanions attain a resonance-stabilized planar conformation. There is really no definitive information on this point, but there is one interesting reaction that may have some bearing on conformational changes in reaction intermediates and this concerns the addition of hydrazoic acid to butadienoic acid derivatives. Esters **75** and **79** are reported to react with excess sodium azide in aqueous tetrahydrofuran to give vinyl azides, **80** and **81**, respectively (78).

$$\mathrm{CH_2{=}C{=}C(R){-}CO_2C_2H_5} \xrightarrow[\mathrm{THF{-}H_2O}]{\mathrm{NaN_3}} \mathrm{CH_3{-}C(N_3){=}C(R)CO_2C_2H_5}$$

75 R = H **80** R = H

79 R = CH_3 **81** R = CH_3

This result is in distinct contrast to the behavior of the alkyne isomer **76** which undergoes 1,3-cycloaddition to give a vicinal triazole **82**.

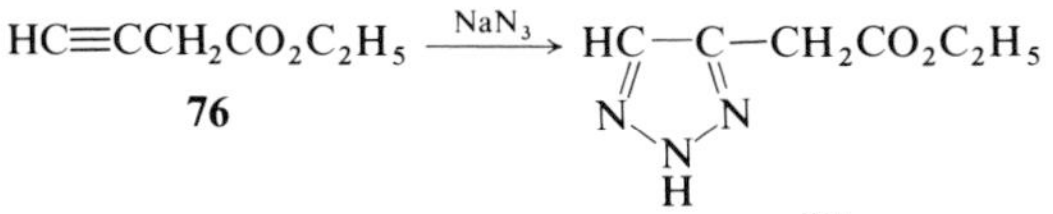

$$HC{\equiv}CCH_2CO_2C_2H_5 \xrightarrow{NaN_3} HC{-}C{-}CH_2CO_2C_2H_5$$

76 **82**

This difference in behavior prompted Harvey and Ratts (78) to make the interesting suggestion that the allylic anion **83**, initially formed from **75**, has time to rotate 90° about the 2–3 bond to form the planar anion **84**, and that this rotation prevents ring closure to the triazole because the orbitals at carbon and the terminal nitrogen of **84** are not properly oriented for sigma overlap. The orbitals in question are shown diagrammatically in structures **83** and **84** where the conformation and hybridization of the five-atom chain is drawn by analogy to the related conversion of diazoazides to cyclic pentazoles (79).

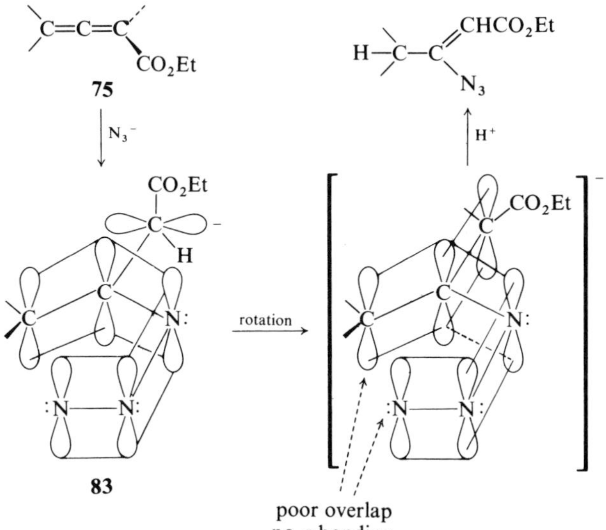

A corresponding orbital diagram for the intermediate ion derived from addition of azide to **76** shows that stereoelectronic requirements for ring closure are well satisfied.

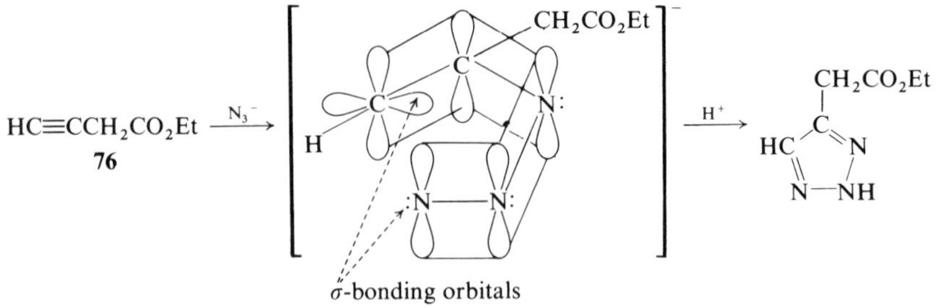

A second reaction that indicates different intermediates may participate in nucleophilic addition relates to the reaction of amines with cyanoallenes. Greaves and Landor (70) report that the isomeric enamines **85** and **86**, which are formed from amines and cyanoallenes, are *not* interconverted under the reaction conditions. Assuming the enamines are therefore kinetic products, the authors propose different mechanistic pathways for their formation. Paths A and B differ initially only in the direction of approach of the nucleophile to give either **87** or **88**. If proton transfer in **87** is fast relative to rotation to give **89**, the product will be **85** (path *A*). If rotation occurs, however, proton transfer can then give the more stable product **86** (path *B*).

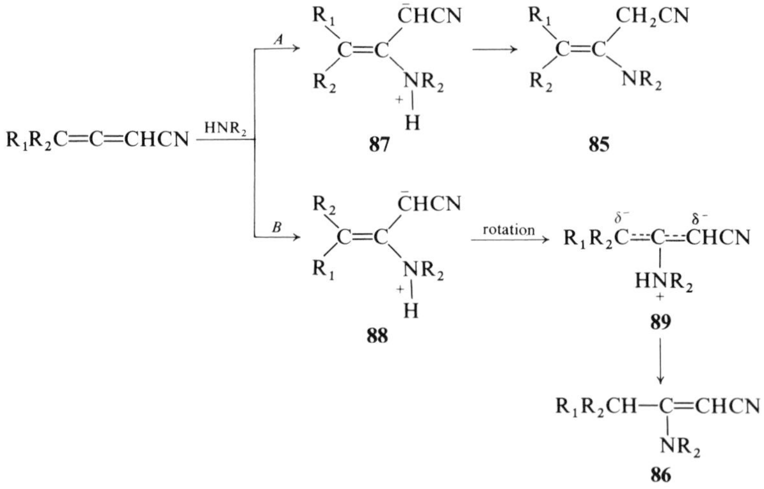

Without the activating influence of electron-withdrawing substituents, there is little or no addition of nucleophilic reagents to allenes. Mention has already been made that tetramethylallene is inert to thiols under basic conditions (60). Diphenylallene, however, does react with ethanol in the presence of sodium ethoxide to give the vinyl ether **90** and the allene dimer **91** (80).

Formation of the latter product is apparently unrelated to the nucleophilic addition reaction.

$$\phi_2C{=}C{=}CH_2 \xrightarrow[\text{NaOEt}]{\text{EtOH}} \phi_2C{=}C(CH_3)(OEt) + \text{(cyclobutane: } \phi_2\text{, =CH}_2\text{, =C}\phi_2)$$

90 **91**

C. Perhaloallenes

The remarkable reactivity of perfluoroallenes toward nucleophilic reagents and the interesting change in orientation observed with increasing substitution of perfluoroalkyl groups deserves special comment. Tetrafluoroallene reacts with reagents such as water, methanol and moist caesium fluoride under extremely mild conditions to give monadducts (57a,d). Spontaneous elimination or rearrangement of the initial adducts may occur, but the orientation of addition is such that the nucleophile enters at the *terminal* carbon.

$$CF_2{=}C{=}CF_2 \xrightarrow{H_2O} [CF_2{=}CHCF_2OH] \xrightarrow{-HF} CF_2{=}CHC(=O)F$$

$$CF_2{=}C{=}CF_2 \xrightarrow{CH_3OH} [CF_2{=}CHCF_2OCH_3] \longrightarrow CF_3CH{=}CFOCH_3$$

$$CF_2{=}C{=}CF_2 \xrightarrow[H_2O]{CsF} CF_2{=}CHCF_3$$

Similarly, perfluoro-1,2-pentadiene reacts readily with nucleophiles to produce monoadducts arising from attack at the terminal difluoromethylene group as well as products of retropropargylic rearrangement (57c).

$$C_2F_5CF{=}C{=}CF_2 \xrightarrow[\text{HCONH}_2]{\text{CsF, 20°}} C_2F_5CF{=}CHCF_3 \quad \textit{trans}$$

$$C_2F_5CF{=}C{=}CF_2 \xrightarrow[\text{100°, 0.1 mm}]{\text{anh. CsF}} \underset{96\%}{C_2F_5C{\equiv}CCF_3} + \underset{\textit{trans } 4\%}{C_2F_5CF{=}CHCF_3}$$

$$C_2F_5CF{=}C{=}CF_2 \xrightarrow[0°]{CH_3OH} \underset{\textit{trans } 77\%}{C_2F_5CF{=}CHCF_2OCH_3} + \underset{23\%}{C_2F_5C{\equiv}CCF_2OCH_3}$$

The orientation of nucleophilic addition to tetrafluoroallene and perfluoro-1,2-butadiene is clearly *opposite* to that observed in 2,3-butadienoic acid derivatives and related compounds previously discussed. This is not surprising since central nucleophilic attack in the case of the fluoroallenes would lead to α-fluorocarbanions, which are expected to be destabilized by electron repulsion and antibonding between the filled nonbonding orbital on carbon and the adjacent filled 2p orbitals on fluorine (105).

$$C_2F_5CF{=}C{=}CF_2 + F^- \longrightarrow [C_2F_5CF{=}\ddot{C}{-}CF_3]^- \xrightarrow{H_2O} C_2F_5CF{=}C(H)CF_3$$

$$C_2F_5CF{=}C{=}CF_2 + F^- \not\longrightarrow [C_2F_5CF{=}CF\ddot{C}F_2]^-$$

With increasing perfluoroalkyl substitution, the orientation changes, although the reactivity is still pronounced. The perfluoro analogs of 1,1-dimethylallene and tetramethylallene both react with nucleophiles to give monoadducts corresponding to *central* nucleophilic attack (57b,77).

$$(CF_3)_2C{=}C{=}CF_2 \xrightarrow[0\text{–}20^\circ]{CH_3OH} (CF_3)_2CHC({=}CF_2)OCH_3$$

$$(CF_3)_2C{=}C{=}CF_2 \xrightarrow[HCONH_2]{CsF} (CF_3)_2CHCF{=}CF_2$$

The observed change in orientation adds to the increasing body of evidence that perfluoralkyl groups are better able to stabilize a carbanion center than α-fluoro substituents (105). In this respect, it is noteworthy that the reaction of perfluoro(3-methyl-1,2-butadiene) does not give rise to $(CF_3)_2C{=}C(OCH_3)CHF_2$ in methanol corresponding to the formation of

$$(CF_3)_2C{=}C{=}CF_2 \overset{CH_3OH}{\not\longrightarrow} (CF_3)_2C{=}C(\bar{:C}F_2)OCH_3 \longrightarrow (CF_3)_2C{=}C(OCH_3)CHF_2 \text{ (not observed)}$$

$$(CF_3)_2C{=}C{=}CF_2 \xrightarrow{CH_3OH} (CF_3)_2\ddot{\bar{C}}{-}C({=}CF_2)OCH_3 \longrightarrow (CF_3)_2CHC({=}CF_2)OCH_3$$

an intermediate (or transition state) with carbanion character at the terminal difluoromethylene position. This suggests that the intermediate carbanion is probably nonplanar with charge localized at the trifluoromethyl-substituted carbon.

IV. FREE RADICAL ADDITION TO ALLENES

There are relatively few documented examples of free-radical addition reactions of allenes, particularly substituted allenes. Most of the work pertains to allene itself and, until recently, the factors influencing the product distributions were not clearly understood and the position of initial attack by free radicals was a point of controversy. The problem seems to be that the addition products are not always formed under conditions where kinetic control in the formation of intermediate radicals prevails. The validity of relating product distributions to the selectivity of free radical attack at the terminal and central positions of allene is, therefore, open to question. The situation is well illustrated by the reactions of HBr and thiols with allene and is described below in some detail.

A. Addition to Allene

1. Hydrogen Bromide

The addition of HBr to allene under free-radical conditions has been reported by several groups of workers (81–84). Photoinduced reactions in the gas phase over the temperature range 25–150° are reported to give 2-bromopropene **5** as the major 1:1 adduct (81–83). Little or no 3-bromopropene **94** is formed under these conditions. This result has been interpreted in various ways. Kovachic and Leitch (81) considered that the product reflects the greater stability of the allylic radical **92** relative to the vinylic radical **93** without regard for whether **92** was planar or nonplanar. Abell and

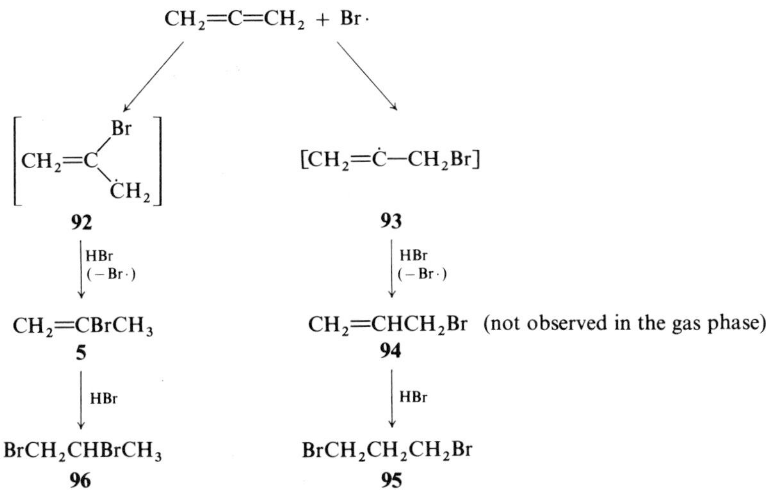

Anderson (83) have suggested that initial attack occurs at the terminal position but that the vinyl radical **92** so formed rapidly rearranges to the resonance-stabilized allylic radical, which then leads to product. Reticence in accepting initial central attack of Br· stems largely from the fact that other electrophilic radicals, $CF_3\cdot$ and $CCl_3\cdot$, selectively attack the terminal allenic carbon, and it is not obvious why bromine atoms should behave exceptionally. However, as discussed shortly, there appears to be no correlation between the polar character of the attacking radical and the point of initial attack.

When, however, all the experimental evidence from a variety of experimental conditions is examined, a more involved picture of the HBr reaction emerges. Foremost, the ratio of products (mono and diadducts) derived from terminal and central attack, $[T]/[C]$, depends critically on the phase of the reaction, the reaction temperature, and the relative concentrations of reactants (82,84).

$$\frac{[T]}{[C]} = \frac{\mathbf{94} + \mathbf{95}}{\mathbf{5} + \mathbf{96}}$$

Thus, larger amounts of terminal attack products, **94** and **95**, are formed in liquid-phase reactions than in gas-phase reactions. Terminal attack is also promoted by lower temperatures and increasing reactant ratios, [HBr]/[allene]. A careful investigation by Heiba and Haag (84) on the effect of HBr concentration on the product compositions of reactions in pentane solution at -78° initiated by γ-radiation has helped to clarify the reaction sequence. The effect is shown schematically in Figure 1. At high HBr concentrations, the $[T]/[C]$ ratio is constant and corresponds to the statistical

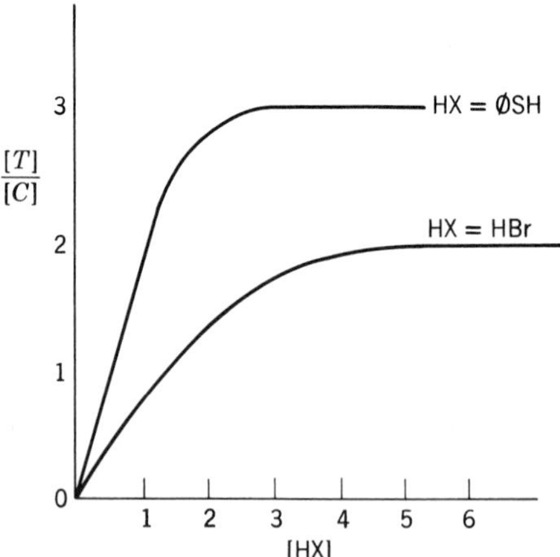

Figure 1. The effect of [HX] concentration on the ratio of products of terminal and central attack on the free radical addition of HX to allene.

ratio of 2 expected for indiscriminant attack of Br· at the central and terminal positions. At low HBr concentrations, the $[T]/[C]$ ratio is approximately linear with [HBr]. These results are accommodated by the sequence of

$$H_2C{=}C{=}CH_2 + Br\cdot \underset{k_{-1}}{\overset{k_1}{\rightleftharpoons}} [BrCH_2{-}\dot{C}{=}CH_2]\ (\mathbf{93}) \xrightarrow[HBr]{k_3} BrCH_2{-}CH{=}CH_2\ (\mathbf{94})$$

$$H_2C{=}C{=}CH_2 + Br\cdot \xrightarrow{k_2} \left[H_2\dot{C}{-}C(Br){=}CH_2\right]\ (\mathbf{92a}) \xrightarrow{\text{rapid rotation}} \left[H_2C{\overset{\cdot\cdot}{=}}C(Br){\overset{\cdot\cdot}{-}}CH_2\right]^{\cdot}\ (\mathbf{92b}) \xrightarrow[HBr]{k_4} H_3C{-}C(Br){=}CH_2\ (\mathbf{5})$$

Scheme 11

Scheme 11 from which eq. 1 may be derived relating the relative rates of terminal and central attack to the HBr concentration,

$$\frac{d[T]}{d[C]} = \frac{k_1 k_3[\mathrm{HX}]}{k_2(k_{-1} + k_3[\mathrm{HX}])} \tag{1}$$

(HX = HBr or HSR)

The important point about Scheme 11 is that the vinyl radical **93** is formed *reversibly* while the allylic radical **92a** is formed *irreversibly*. At high HBr concentrations, the hydrogen abstraction step (k_3) dominates over bromine elimination (i.e., $k_3[\mathrm{HBr}] \gg k_{-1}$) and eq. 1 simplifies to

$$\frac{d[T]}{d[C]} = \frac{k_1}{k_2} \tag{2}$$

At low HBr concentrations, the reverse situation applies, and eq. 1 reduces to

$$\frac{d[T]}{d[C]} = \frac{k_1 k_3[\mathrm{HBr}]}{k_2 k_{-1}} \tag{3}$$

These limiting situations are observed experimentally (Fig. 1).

To appreciate that Scheme 11 is eminently reasonable, the thermodynamics of the situation should be examined. Based on reasonable estimates for the dissociation energies of the bonds involved in the attack of Br· at the terminal and central positions of allene, it can be calculated that *both* modes of addition are exothermic, probably by about 10 kcal/mole or more. Once the allylic intermediate **92a** is formed, it may be expected to rapidly undergo bond rotation by 90° to form the planar radical **92b**, thereby gaining approximately 12 kcal of resonance stabilization. This makes elimination of Br· from **92b** more strongly endothermic and of higher activation energy than the comparable elimination from the vinylic radical **93**. Consequently *reversible* attack of bromine atoms would appear to be energetically feasible at the terminal position only.

The temperature effect on product composition now becomes understandable since this should have the greatest influence on processes of high activation energy. Increasing temperature should mostly enhance k_{-1} which results in decreasing amounts of terminal attack products.

In summary, then, bromine atoms apparently attack allene indiscriminantly, although the products do not necessarily reflect this because terminal attack is reversible.

2. *Thiols*

Free-radical additions of thiols to allene have also been studied extensively. These reactions are generally uncomplicated by competing ionic processes; they are usually initiated by irradiation, azobisisobutyronitrile (AIBN), or peroxides, and the major products are mono- and diadducts, **97**, **98**, **99**, and **100**.

$$CH_2{=}C{=}CH_2 \xrightarrow{RSH} \underset{\mathbf{97}}{RSCH_2CH{=}CH_2} \xrightarrow{RSH} \underset{\mathbf{98}}{RSCH_2CH_2CH_2SR}$$

$$CH_2{=}C{=}CH_2 \xrightarrow{RSH} \underset{\mathbf{99}}{CH_3{-}C(SR){=}CH_2} \xrightarrow{RSH} \underset{\mathbf{100}}{CH_3{-}CH(SR)CH_2SR}$$

Terminal attack of thiyl radicals leads to products **97** and **98** while central attack gives **99** and **100** (85–87). The monoadduct **99** is often undetected for the reason that it reacts with thiols to give the diadduct **100** more rapidly than it is formed.

The proportions of products derived from terminal and from central attack appear to depend to some extent on the nature of the attacking thiyl radical RS·, but in general the selectivity favors terminal attack (Table 10). However, the product compositions in thiol additions are influenced by temperature and reactant concentrations in much the same way as in HBr addition. Increasing temperature favors central attack (85,88),* and increasing thiol concentration favors terminal attack (88). The product ratios, $[T]/[C]$, in the addition of benzenethiol to allene were found to vary with the benzenethiol concentration as shown in Figure 1, which closely resembles the dependence of $[T]/[C]$ on [HBr] in HBr addition to allene. The reaction

* The higher percentage of central attack reported in ref. 87 for benzenethiol addition may reflect the higher temperatures used compared to refs. 85 and 88.

TABLE 10
Orientation of Addition of Thiols to Allene

Thiol, RSH	Terminal attack, $RSCH_2CH=CH_2 + RSCH_2CH_2CH_2SR$ (mole %)	Central attack, $RSCH_2CH(CH_3)SR$ (mole %)	Ref.
H_2S[a]	93	2	90
CH_3SH[b]	90	10	85
CH_3CH_2SH[c]	81	19	89
CH_3CH_2SH[d]	84	16	86
$CH_3CH_2CH_2SH$[e]	84	16	87
C_6H_5SH[f]	80	20	88
C_6H_5SH[e]	65	35	87
C_6H_5SH[b]	81	19	85

[a] Mole ratio H_2S/allene = 10; initiated by UV and AIBN.
[b] 17°.
[c] Mole ratio allene/thiol = 1:2 at 17°.
[d] 67°; AIBN.
[e] Mole ratio allene/thiol = 12:18 at 82°.
[f] 20°.

sequence of Scheme 12 is consistent with this data, and the relative rates of product formation by terminal and central attack is accordingly expressed by eq. 1. At high benzenethiol concentrations, reversibility of terminal attack is negligible and the product ratios reflect the relative rates of terminal and central attack.

$$\frac{[T]}{[C]} = \frac{k_1}{k_2} = 3$$

The benzenethiyl radical is, therefore, somewhat more selective than bromine atoms and shows some preference for terminal attack.

Increasing temperature and low benzenethiol concentrations lead to an increase in central attack by favoring the elimination of PhS· from the vinyl radical **101** over hydrogen abstraction to give **97**. Irreversible formation of

$$CH_2=C=CH_2 + \phi S\cdot \underset{k_{-1}}{\overset{k_1}{\rightleftharpoons}} [\phi SCH_2\dot{C}=CH_2]\ (\mathbf{101}) \xrightarrow[\phi SH]{k_3} \phi SCH_2CH=CH_2\ (\mathbf{97}) \text{ etc.}$$

$$CH_2=C=CH_2 + \phi S\cdot \xrightarrow{k_2} [\dot{C}H_2-C(S\phi)=CH_2]\ (\mathbf{102a}) \xrightarrow{\text{rapid rotation}} [CH_2\dot{=}C(S\phi)\dot{=}CH_2]\ (\mathbf{102b}) \xrightarrow[\phi SH]{k_4} CH_3-C(S\phi)=CH_2\ (\mathbf{99}) \text{ etc.}$$

Scheme 12

the allylic radical **102a** is considered to result from its rapid conversion to the planar form **102b** which is resonance-stabilized and unlikely to reverse to allene for energetic reasons. In all important respects, the situation is analogous to HBr addition.

3. Trimethyltin Hydride

Addition of trimethyltin hydride to alkenes is a free-radical chain reaction initiated by AIBN and other free-radical sources to give trimethyltin $(CH_3)_3Sn\cdot$, as the chain transfer radical. The addition of trimethyltin hydride to allenes has been investigated by Kuivila, Rahman, and Fish (91), and the product compositions in this reaction may closely reflect kinetic control of the formation of intermediate radicals **103** and **105**. This may be deduced from the thermodynamics of the propagation steps. The first step is probably thermoneutral or slightly endothermic based on a carbon–tin bond energy of 56 kcal/mole (92) and a value of 54–59 kcal/mole required to break one allene double bond.* The second step, however, is almost certainly highly exothermic based on a tin–hydrogen bond energy of 50 kcal and a carbon–hydrogen bond energy of 85 kcal (allylic) or 104 kcal (vinylic) (91,94). The activation energy for this step is probably close to zero; from the rate expression of eq. 1 and Scheme 13 it follows that $k_3[\mathrm{HX}] \gg k_{-1}$ where

$$CH_2{=}C{=}CH_2 + Sn\cdot \xrightarrow{k_1} [Sn{-}CH_2{-}\dot{C}{=}CH_2]\ (\mathbf{103}) \xrightarrow[k_3]{SnH} SnCH_2CH{=}CH_2\ \ \mathbf{104}\ (55\%)$$

$$CH_2{=}C{=}CH_2 + Sn\cdot \xrightarrow{k_2} [\dot{C}H_2{-}C(Sn){=}CH_2]\ (\mathbf{105}) \xrightarrow[k_4]{SnH} CH_3{-}C(Sn){=}CH_2\ \ \mathbf{106}\ (45\%)$$

$Sn \equiv (CH_3)_3Sn$

Scheme 13

HX is $HSn(CH_3)_3$. The product ratios are therefore expressed by the relative rates of terminal and central attack.

$$\frac{[T]}{[C]} = \frac{k_1}{k_2}$$

The actual product composition observed in trimethyltin hydride addition to allene was 55% of **104** and 45% of **106**, corresponding to terminal and central attack, respectively. For indiscriminant attack the per cent com-

* The bond-dissociation energy $C{=}C{=}C \rightarrow C{=}\dot{C}{-}\dot{C}$ is probably less than the average value of 63 kcal for an isolated double bond since the heat of hydrogenation of allene to propane is 70 kcal, whereas that of propene to propane is 30 kcal (cf. ref. 93).

position should approach 66% of **104** and 33% of **105**, corresponding to a $[T]/[C]$ ratio of 2. The fact that this ratio is actually less than 2 reflects some preference of the trimethyltin radical for central attack.

4. *Miscellaneous*

A limited amount of data exists on the addition of tetrahalomethanes to allene. The photoinduced addition of trifluoromethyl iodide to allene was reported by Haszeldine and co-workers to give only the monoadduct corresponding to terminal attack of the trifluoromethyl radical (95). The kinetics of the reaction have been studied by Meunier and Abell (96) who have confirmed that the sole product is that of terminal attack.

$$CF_3I + CH_2{=}C{=}CH_2 \longrightarrow CF_3CH_2CI{=}CH_2$$

The related reaction of bromotrichloromethane with allene is also reported to give only products of terminal attack (87,88). Since the orientation did not alter with temperature over the range -10 to 135°, the orientation probably reflects that the trichloromethyl radical is highly selective and favors terminal attack.

$$CCl_3Br + CH_2{=}C{=}CH_2 \xrightarrow[\text{or ROOR}]{h\nu} Cl_3CCH_2CBr{=}CH_2 \xrightarrow{CCl_3Br} Cl_3CCH_2CBr_2CH_2CCl_3$$

Tetrafluorohydrazine adds homolytically to allene (97) but the products give no information concerning the position of initial attack of the NF_2 radical.

$$H_2C{=}C{=}CH_2 + N_2F_4 \longrightarrow \left[CH_2{=}\overset{\overset{\displaystyle NF_2}{|}}{C}{-}CH_2NF_2 \right] \longrightarrow FCH_2{-}\overset{\overset{\displaystyle NF}{||}}{C}{-}CH_2NF_2$$

However, the photolysis of tetrafluorohydrazine in the presence of allene gives products of substitution and addition (98). The attacking radicals under these conditions are presumably fluorine atoms which evidently lead to adducts of central attack exclusively.

$$CH_2{=}C{=}CH_2 + N_2F_4 \xrightarrow{h\nu} CH_2{=}C(F)CH_2NF_2 + HC{\equiv}CCH_2NF_2$$

Thermal decomposition of S_2F_{10} leads to $SF_5\cdot$, which adds to alkenes under pressure. The initially formed adduct of S_2F_{10} and allene reacts further and obscures the initial orientation of addition (99).

From the addition of phosphine to allene under free-radical conditions, only one monoadduct, isopropenylphosphine, was isolated (100). Unfortunately the yield was low and much of the initial adduct polymerized under the reaction conditions which precluded any valid conclusions to be drawn concerning the selectivity of phosphine addition.

$$CH_2{=}C{=}CH_2 + PH_3 \xrightarrow[53\text{ atm}, 77^\circ]{h\nu} CH_2{=}C\begin{matrix} PH_2 \\ CH_3 \end{matrix}$$

Finally free-radical addition of diethylchloramine to allene reportedly gives products of terminal attack of the diethylammonium radical (101).

B. Conclusions

If one is to make any valid comparisons of the selectivities of various radical reagents for terminal versus central attack, the products of the reactions in question should be formed under conditions of kinetic control. The available data that meet this requirement are limited but are summarized in Table 11.

TABLE 11
Selectivity in Free Radical Attack of Allene

Radical	Terminal attack, per cent	Central attack, per cent	Ref.
$(CH_3)_3Sn\cdot$	55	45	91
$Br\cdot$	66	33	84
$\phi SH\cdot$	75	25	88
$CF_3\cdot$	100	0	92
$CCl_3\cdot$	100	0	87,88
$CH_3\cdot$	100	0	96

The data of Table 11 indicate no striking difference in reactivity of the terminal and central positions toward bromine, trimethyltin, benzenethiyl radicals, even though the polar characteristics of these radicals are widely different.* The nucleophilic trimethyltin radical does show a small preference for central attack ($[T]/[C] < 2$), but the electrophilic radicals react indiscriminately ($Br\cdot$) or by preferential terminal attack. Even more striking is the fact that radicals of the type $CX_3\cdot$ show a high selectivity for terminal attack regardless of the nature of X (F, Cl, or H).† Since this would indicate

* Fluorine atoms, however, appear to differ from bromine atoms in showing a strong preference for central attack (cf. ref. 98).

† It has been argued on the basis of relative reactivities of allenes and substituted allenes toward $CH_3\cdot$ and $CF_3\cdot$ that the relatively nucleophilic methyl radical attacks the central carbon of allenes (102). This has been shown to be incorrect by the isolation of $CH_3CH_2C(I){=}CH_2$ as the sole product of photoaddition of CH_3I to allene (96).

there is no clear correlation between the polar nature of the attacking radical and the orientation of addition, we may infer that polar effects are unimportant and that vinylic and nonplanar allylic radical intermediates **107** and **108** are closely similar in energy.

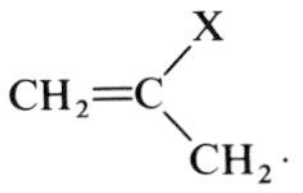

$$CH_2{=}C(X)CH_2\cdot \qquad CH_2{=}\dot{C}{-}CH_2X$$

107 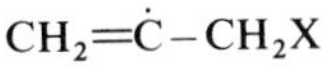 **108**

C. Additions to Substituted Allenes

The effect of methyl substitution on the orientation of free-radical additions to allenes has been studied for only two reactions—ethanethiol addition (86) and trimethyltin hydride addition (91). Qualitatively, the results of these two investigations clearly indicate that central attack becomes progressively more important with increasing substitution at the terminal carbons. The selectivities observed are summarized below; the numbers beneath the carbons represent the per cent of product derived from attack at that position.

ethanethiol addition:

allene	1,2-butadiene	3-methyl-1,2-butadiene
$CH_2{=}C{=}CH_2$	$CH_3CH{=}C{=}CH_2$	$(CH_3)_2C{=}C{=}CH_2$
43.6 12.8 43.6	6.3 48.2 45.5	100

trimethyltin hydride addition:

allene	1,2-butadiene	3-methyl-1,2-butadiene
$H_2C{=}C{=}CH_2$	$CH_3CH{=}C{=}CH_2$	$(CH_3)_2C{=}C{=}CH_2$
27.5 45 27.5	0 86 14	100

2,3-pentadiene	2-methyl-2,3-pentadiene
$CH_3CH{=}C{=}CHCH_3$	$(CH_3)_2C{=}C{=}CHCH_3$
100	100

The simplest explanation for these results is that selectivity is primarily determined by the steric and electronic effects of methyl substituents and not by the nature of the attacking radical. Steric effects serve to decrease the rate of terminal attack while polar effects serve to stabilize secondary and tertiary radical centers promoting central attack. The stability sequence of radicals derived from the attack of radical X· on allenes may therefore be written as

$$CH_2{=}\dot{C}{-}CH_2X \sim \dot{C}H_2{-}CX{=}CH_2 \text{ (nonplanar)}$$
$$CH_3CH{=}\dot{C}{-}CH_2X < CH_3\dot{C}H{-}CX{=}CH_2 \text{ (nonplanar)}$$
$$(CH_3)_2C{=}\dot{C}CH_2X \ll (CH_3)_2\dot{C}{-}CX{=}CH_2 \text{ (nonplanar)}$$

The nature of the products obtained in trimethyltin hydride additions to allenes resulting from central attack reflects the intervention of planar resonance-stabilized allylic radicals. Thus products of hydrogen abstraction at *both ends* of the allyl system were observed in additions to 1,2-butadiene 3-methyl-1,2-butadiene, and 2-methyl-2,3-pentadiene.

$$CH_3CH{=}C{=}CH_2 \xrightarrow{Sn\cdot} \left[CH_3CH \overset{\overset{\displaystyle Sn}{|}}{\underset{}{\cdots C \cdots}} CH_2\right]^{\cdot} \xrightarrow{SnH} \underset{63\%}{CH_3CH{=}C(Sn)CH_3} + \underset{37\%}{CH_3CH_2C(Sn){=}CH_2}$$

$$(CH_3)_2C{=}C{=}CH_2 \xrightarrow{Sn\cdot} \left[(CH_3)_2C \overset{\overset{\displaystyle Sn}{|}}{\cdots C \cdots} CH_2\right]^{\cdot} \xrightarrow{SnH}$$

$$\underset{73\%}{(CH_3)_2C{=}C(Sn)CH_3} + \underset{27\%}{(CH_3)_2CH{-}C(Sn){=}CH_2}$$

$$(H_3C)_2C{=}C{=}C(CH_3)H \xrightarrow{Sn\cdot} \left[(CH_3)_2C \overset{\overset{\displaystyle Sn}{|}}{\cdots C \cdots} CHCH_3\right]^{\cdot} \xrightarrow{SnH}$$

$$\underset{45\%}{(CH_3)_2C{=}C(Sn)CH_2CH_3} + \underset{55\%}{(CH_3)_2CH{-}C(Sn){=}CHCH_3}$$

Although bridged or nonplanar radical intermediates cannot be rigorously excluded, it seems more reasonable that the products are formed by hydrogen abstraction by planar allylic radicals. There is a definite preference for abstraction at the primary position. This may be compared with ethanethiol addition to 1,2-butadiene and 3-methyl-1,2-butadiene where exclusive abstraction at the primary position was observed (86). The greater selectivity exhibited by allylic radicals in abstracting hydrogen from thiols may be due to the fact that the S—H bond is stronger than the Sn—H bond. The transition state for hydrogen abstraction may then be expected to be relatively far along the reaction coordinate in thiol addition reflecting considerable C—H bond making and S—H bond breaking. The preferred transition state will then be that leading to the more stable product, which in this case is presumably the more highly substituted alkene.

Intervention of symmetrical allylic free-radical intermediates is also manifest in the reaction of *t*-butyl hypochlorite with 1,3-dimethylallene (106). The reaction is fairly complex in that it gives products of both free-radical substitution and addition by "spontaneous" initiation since the product distribution is the same whether or not the reaction mixture is irradiated (107). The major addition products are the *cis* and *trans* dichloro adducts

$$\underset{\mathbf{30}}{CH_3CH{=}C{=}CHCH_3} \xrightarrow[\text{ether}]{t\text{-BuOCl}} \underset{5\%}{CH_3C{\equiv}CCH(Cl)CH_3} + \underset{33\%}{CH_3CH{=}C(Cl)CH{=}CH_2} + \underset{\substack{\mathbf{109}\ (30\%) \\ cis \text{ and } trans}}{CH_3CH{=}C(Cl)CH(Cl)CH_3} + \underset{\substack{\mathbf{110}\ (28\%) \\ cis \text{ and } trans}}{CH_3CH{=}C(OBu\text{-}t)CH(Cl)CH_3}$$

109 and the *cis* and *trans* *t*-butoxychloro adducts **110**. The structure of **110** is that expected of a free-radical chain addition reaction in which *t*-butoxy radicals attack the *central* allenic position of **30** (Scheme 14). Furthermore, optically active **30** gave *racemic* **110**. which strongly suggests that, if dissymmetric allylic intermediates such as **111a** are initially formed, bond rotation to give symmetrical free radicals **111b** occurs more rapidly than chlorine-atom abstraction to give **110**.

$$\underset{R\text{-}(-)\text{-}\mathbf{30}}{CH_3(H)C{=}C{=}C(H)CH_3} \xrightarrow{t\text{-BuO}} \underset{\mathbf{111a}}{\left[CH_3CH{=}C(OBu\text{-}t)\dot{C}(H)CH_3\right]} \xrightarrow[\text{rotation}]{\text{rapid}} \underset{\mathbf{111b}}{\left[CH_3HC{\cdots}C(OBu\text{-}t){\cdots}CHCH_3\right]^{\cdot}} \xrightarrow[(-t\text{-BuO}\cdot)]{t\text{-BuOCl}} \underset{(\pm)\text{-}110}{CH_3CH{=}C(OBu\text{-}t)CH(Cl)CH_3}$$

Scheme 14

Racemic dichloro adduct **109** was also formed from *R*-(−)-**30**, and a mechanism by which it may be formed is represented in Scheme 15. Homolysis of *t*-butyl hypochlorite, which normally does not occur unless irradiated, is assisted by concomitant covalent bond formation with **30** and ether solvation of the incipient radicals (107–108). The racemic adduct **109** is considered to result from chlorine-atom abstraction by symmetrical allylic free-radical intermediates **112**.

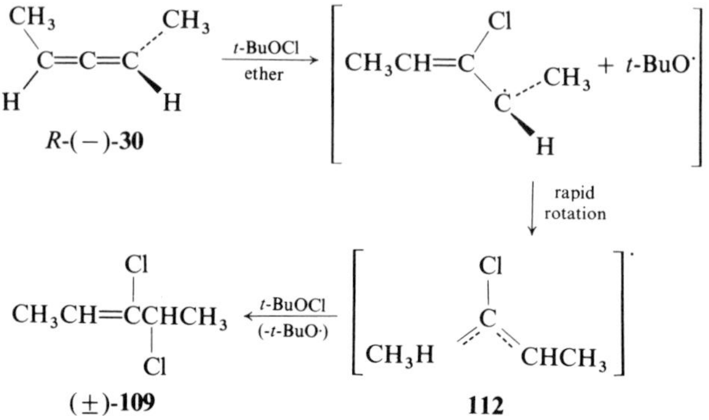

$CH_3(H)C{=}C{=}C(CH_3)H$ *R*-(−)-**30** $\xrightarrow[\text{ether}]{t\text{-BuOCl}}$ $[CH_3CH{=}C(Cl)\dot{C}(CH_3)H + t\text{-BuO}^{\cdot}]$ $\xrightarrow{\text{rapid rotation}}$ $[CH_3H\ C(Cl)\ CHCH_3]^{\cdot}$ **112** $\xrightarrow[(-t\text{-BuO}\cdot)]{t\text{-BuOCl}}$ $CH_3CH{=}C(Cl)CH(Cl)CH_3$ (±)-**109**

Scheme 15

Another point of interest concerning this reaction is the recovery of unreacted allene (**30**) of retained optical purity which implies that allylic free-radicals **111** and **112** are formed *irreversibly*. It also means that attack at the terminal allenic position is negligible since hypochlorite addition products of this orientation were not observed and reversible addition would be expected to result in partial or complete racemization of **30**.

The reaction of *N*-bromosuccinimide with 1,1-disubstituted allenes in refluxing carbon tetrachloride is also reported to give substitution and additions products, presumably by a radical process (103). The structure of the products indicates that bromine attacks the central carbon to give allylic intermediates which lead to the two possible monoadducts of the starting allene as well as to a conjugated diene.

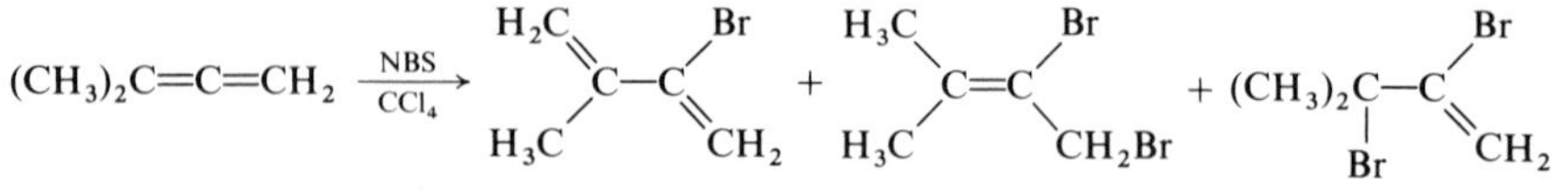

$$(CH_3)_2C{=}C{=}CH_2 \xrightarrow[CCl_4]{NBS} (H_2C{=})(H_3C)C{-}C(Br){=}CH_2 + (H_3C)_2C{=}C(Br)CH_2Br + (CH_3)_2C(Br){-}C(Br){=}CH_2$$

Finally, addition of *p*-toluenesulfonyl iodide to 1,3-dimethylallene, phenylallene, and allene is reported to occur rapidly under free radical conditions to give 1:1 adducts in which the iodine enters at the central position (104).

$$RCH{=}C{=}CHR + ArSO_2I \xrightarrow{h\nu} RCH{=}C(I)CHR(SO_2Ar)$$

D. Perhaloallenes

There is as yet little information on the reactivity and orientation of free radical addition reactions of perfluoroallenes. Tetrafluoroallene polymerizes under conditions that could support radical addition of other reagents, thereby precluding their study (57a). Perfluoro-1,2-pentadiene does not polymerize readily, however, and a study of its photochemical chlorination and hydrobromination has been made (57c). The major product of photochlorination of perfluoro-1,2-pentadiene with excess chlorine is the tetrachloride adduct **113**. There can be little doubt that this is a radical addition reaction since no reaction takes place in the dark at room temperature.

$$C_2F_5CF{=}C{=}CF_2 + Cl_2 \xrightarrow{h\nu} \underset{\textbf{113}\ (95\%)}{C_2F_5CFClCCl_2CClF_2}$$

Photolysis of hydrogen bromide in the presence of perfluoro-1,2-pentadiene gives a complex mixture of monoadducts that can be rationalized as arising from attack of a bromine atom at the central allenic carbon with subsequent bond rotation to give stabilized allylic radicals **114** which abstract hydrogen to give the major observed products **115** and **116** (Scheme 16).

$$HBr \xrightarrow{h\nu} H\cdot + Br\cdot$$

$$C_2F_5CF{=}C{=}CF_2 + Br\cdot \longrightarrow \left[C_2F_5FC\overset{\overset{Br}{|}}{\dot{C}}CF_2\right] \ \textbf{114}$$

$$\textbf{114} \xrightarrow{HBr} \underset{\textbf{115}}{C_2F_5CHFCBr{=}CF_2} + \underset{\substack{\textbf{116}\\ cis\ \text{and}\ trans}}{C_2F_5CF{=}CHBrF_2}$$

Scheme 16

In summary, the available evidence on free-radical addition reactions of substituted allenes, including perhaloallenes, indicates that attack at the central position is the preferred orientation to give adducts derived from resonance-stabilized allylic intermediate free radicals.

References

1. Several comprehensive reviews on the chemistry of allenes have been published. See, for example, (a) K. Griesbaum, *Angew. Chem., Intern. Ed.*, **5**, 933 (1966); (b) D. R. Taylor, *Chem. Rev.*, **67**, 317 (1967); (c) M. V. Maurov and V. F. Kucherov, *Russ. Chem. Rev.*, **36**, 233 (1967); (d) A. A. Petrov and A. V. Fedorova, *Russ. Chem. Rev.*, **33**, 1 (1964); (e) H. Fischer, in *Alkenes*, S. Patai, Ed., Interscience, New York, 1964, pp. 1060–1083.
2. (a) K. Griesbaum, *J. Am. Chem. Soc.*, **86**, 2301 (1964); (b) K. Griesbaum, *Angew. Chem., Intern. Ed.*, **3**, 697 (1964); (c) K. Griesbaum, W. Naegele, and G. G. Wanless, *J. Am. Chem. Soc.*, **87**, 3151 (1965).
3. T. L. Jacobs and R. N. Johnson, *J. Am. Chem. Soc.*, **82**, 6397 (1960).
4. P. R. Austin, U.S. Patent 2,585,529 (Feb. 12, 1952); *Chem. Abstr.*, **46**, 3799 (1952).
5. W. L. Waters and E. F. Kiefer, *J. Am. Chem. Soc.*, **89**, 6261 (1967).
6. S. Gorsano, *Atti Acad. Naz. Lincei, Rend., Classe Sci. Fis., Mat. Nat.*, **34**, 430 (1963).
7. H. H. Lindner and T. Onak, *J. Am. Chem. Soc.*, **88**, 1886 (1966).
8. R. H. Fish, *J. Am. Chem. Soc.*, **90**, 4435 (1968).
9. H. C. Brown, *Hydroboration*, W. A. Benjamin, New York, 1962.
10. W. H. Mueller and P. E. Butler, *J. Org. Chem.*, **33**, 1533 (1968).
11. H. G. Peer, *Rec. Trav. Chim.*, **81**, 113 (1962).
12. W. H. Mueller, P. E. Butler, and K. Griesbaum, *J. Org. Chem.*, **32**, 2651 (1967).
13. J. Van Leeuwen, Brit. Pat. 908,219; *Chem. Abstr.*, **58**, 6693 (1963).
14. P. D. B. de la Mare and R. Bolton, *Electrophilic Additions to Unsaturated Systems*, Elsevier, New York, 1966.
15. G. Gustavon and N. Demjanoff, *J. Prakt. Chem.*, **38**, 201 (1888).
16. F. W. Smirnoff, *J. Russ. Phys. Chem. Soc.*, **35**, 854 (1903); **36**, 1184 (1904); *Chem. Zentr.*, **75**, 576 (1904); **76**, 344 (1905).
17. A. V. Fedorova and A. A. Petrov, *J. Gen. Chem. USSR*, **31**, 3273 (1961).
18. R. C. Fahey and D. J. Lee, *J. Am. Chem. Soc.*, **89**, 2780 (1967); R. C. Fahey, in *Topics in Stereochemistry*, Vol. 3, E. Eliel and N. Allinger, Eds., Wiley-Interscience, New York, 1968.
19. R. K. Sharma, B. A. Shoulder, and P. D. Gardner, *J. Org. Chem.*, **32**, 241 (1967).
20. M. Bouis, *Ann. Chim. (Paris)*, [10] **9**, 402 (1928).
21. W. H. Carothers and G. J. Berchet, *J. Am. Chem. Soc.*, **55**, 1628 (1933).
22. G. F. Hennion and J. J. Sheehan, *J. Am. Chem. Soc.*, **71**, 1964 (1949).
23. A. V. Fedorova and A. A. Petrov, *J. Gen. Chem.*, **31**, 3273 (1961).
24. A. V. Fedorova, *J. Gen. Chem.*, **33**, 3508 (1963).
25. I. W. Kondakov, *J. Russ. Phys. Chem. Soc.*, **21**, 36 (1889); *Chem. Zentr.*, **61**, 311 (1890).
26. P. Kurtz, H. Gold, and H. Disselnkötter, *Annalen*, **624**, 1 (1959).
27. M. Bouis, *Compt. Rend.*, **184**, 1563 (1927).
28. J. Grimaldi, A. Cozzone, and M. Bertrand, *Bull. Soc. Chim. France*, **1967**, 2723.
29. A. V. Fedorova and A. A. Petrov, *J. Gen. Chem.*, U.S.S.R, **32**, 1740 (1962).
30. This subject has been thoroughly reviewed by W. Kitching, *Organometal. Chem. Rev.*, **3**, 61 (1968).

31. See for example, J. Halpern and H. B. Tinker, *J. Am. Chem. Soc.*, **89**, 6427 (1967), on the kinetics of oxymercuration of alkenes. Also, the extensive studies by Kreevoy and coworkers on the kinetics and mechanism of deoxymercuration of alkoxymercuri iodides are very pertinent to the discussion of the mechanism of the forward reaction. See M. M. Kreevoy and M. A. Turner, *J. Org. Chem.*, **30**, 373 (1965) and references to earlier work cited therein.
32. W. L. Waters, W. S. Linn, and M. C. Caserio, *J. Am. Chem. Soc.*, **90**, 6741 (1968).
33. W. L. Waters and M. C. Caserio, *Tetrahedron Letters*, **1968**, 5233.
34. W. M. Jones and J. M. Walbrick, *Tetrahedron Letters*, **1968**, 5229.
35. W. S. Linn, W. L. Waters, and M. C. Caserio, presented at the 157th National Meeting of the American Chemical Society, Minneapolis, April 1969.
36. F. R. Jensen and L. H. Gale, *J. Am. Chem. Soc.*, **81**, 6337 (1959).
37. (a) R. D. Bach, *J. Am. Chem. Soc.*, **91**, 1771 (1969); (b) *Tetrahedron Letters*, **1968**, 5841.
38. W. R. Moore and R. C. Bertelson, *J. Org. Chem.*, **27**, 4182 (1962).
39. T. L. Jacobs and R. Macomber, *J. Org. Chem.*, **33**, 2988 (1968).
40. D. Ralston, E. Laserna, and M. C. Caserio, unpublished results, University of California, Irvine, 1969.
41. W. A. Thaler, W. H. Mueller, and P. E. Butler, *J. Am. Chem. Soc.*, **90**, 2069 (1968).
42. W. H. Mueller and P. E. Butler, *J. Am. Chem. Soc.*, **90**, 2075 (1968).
43. D. J. Pettitt and G. K. Helmkamp, *J. Org. Chem.*, **29**, 2702 (1964).
44. T. L. Jacobs, R. Macomber, and D. Zunker, *J. Am. Chem. Soc.*, **89**, 7001 (1968).
45. W. L. Waters, M. Findlay, and M. C. Caserio, unpublished results, University of California, Irvine, 1968.
46. G. E. Heinsohn and H. G. Kuivila, *J. Org. Chem.*, **33**, 2165 (1968).
47. K. Shingu, S. Hagishita, and M. Nakagawa, *Tetrahedron Letters*, **1967**, 4371.
48. L. R. Byrd, W. L. Waters, and M. C. Caserio, paper presented at the 157th National Meeting of the American Chemical Society, Minneapolis, April 1969.
49. F. Acree, Jr. and F. B. Laforge, *J. Org. Chem.*, **5**, 430 (1940).
50. D. K. Wedegaertner and M. J. Millam, *J. Org. Chem.*, **33**, 3943 (1968).
51. M. Lespieau and Wiemann, *Bull. Soc. Chim.*, **45**, 632 (1929); *Chem. Abstr.*, **23**, 4925.
52. M. L. Poutsma, *J. Org. Chem.*, **33**, 4080 (1968).
53. The chemistry of chlorosulfonyl isocyanate has been reviewed by H. Ulrich, *Chem. Rev.*, **65**, 369 (1965).
54. E. J. Moriconi and J. F. Kelly, *J. Am. Chem. Soc.*, **88**, 3657 (1966); *J. Org. Chem.*, **33**, 3036 (1968).
55. D. Devaprabhakara and P. D. Gardner, *J. Am. Chem. Soc.*, **85**, 1458 (1963).
56. A. V. Fedorova, M. D. Stadnichutz, and A. A. Petrov, *Dokl. Akad. Nauk SSSR*, **145**, 837 (1962).
57a. R. E. Banks, R. N. Haszeldine, and D. R. Taylor, *J. Chem. Soc.*, **1965**, 978; *Proc. Chem. Soc.*, **1964**, 121.
57b. R. E. Banks, A. Braithwaite, R. N. Haszeldine and D. R. Taylor, *J. Chem. Soc.* (*C*), **1969**, 996.
57c. idem., ibid., **1969**, 454.
57d. R. E. Banks, W. R. Deem, R. N. Haszeldine and D. R. Taylor, ibid., **1966**, 2051.
58. A. Roedig and B. Heinrich, *Chem. Ber.*, **100**, 3716 (1967).
59. A. Faworsky, *J. Prakt. Chem.* [2], **44**, 208 (1891).
60. W. H. Mueller and K. Griesbaum, *J. Org. Chem.*, **32**, 856 (1967).
61. S. T. McDowell and C. J. M. Stirling, *J. Chem. Soc.*, *(B)*, **1967**, 351.
62. C. J. M. Stirling, *J. Chem. Soc.*, **1964**, 5856.
63. C. J. M. Stirling, *J. Chem. Soc.*, **1964**, 5863.

64. G. D. Appleyard and C. J. M. Stirling, *J. Chem. Soc.*, *(C)*, **1967**, 2686.
65. G. Eglinton, E. R. H. Jones, G. H. Mansfield, and M. C. Whiting, *J. Chem. Soc.*, **1954**, 3197.
66. P. Kurtz, H. Gold, and H. Disselnkötter, *Annalen*, **624**, 1 (1959).
67. J. M. van der Zanden, *Rec. Trav. Chim.*, **54**, 289 (1935). But see ref. 68.
68. E. R. H. Jones, G. H. Mansfield, and M. C. Whiting, *J. Chem. Soc.*, **1954**, 3208.
69. M. Bertrand, J. Elguero, R. Jacquier, and J. Le Gras, *Compt. Rend.*, *(C)*, **262**, 782 (1966).
70. P. M. Greaves and S. R. Landor, *Chem. Commun.*, **1966**, 322.
71. J. H. Wotiz and J. S. Matthews, *J. Am. Chem. Soc.*, **74**, 2559 (1952); J. H. Wotiz and H. E. Merrill, *J. Am. Chem. Soc.*, **80**, 866 (1958); see also ref. 72.
72. J. Klein, *Tetrahedron*, **20**, 465 (1964).
73. M. Bertrand and J. Le Gras, *Compt. Rend.*, **260**, 2926 (1965).
74. M. Bertrand and J. Le Gras, *Bull. Soc. Chim. France*, **1962**, 2136.
75. F. Bohlmann, P. Herbst, and H. Gleinig, *Chem. Ber.*, **94**, 948 (1961).
76. A. N. Pudovik, N. G. Khusainova, and I. M. Aladzheva, *J. Gen. Chem., USSR*, **34**, 2484 (1964).
77. Yu. A. Cheburkov, Yu. E. Aronov, and I. L. Krunyants, *Izv. Akad. Nauk SSSR, Ser. Khim.*, **1966**, 582–3; *Chem. Abstr.*, **65**, 8740 (1966).
78. G. R. Harvey and K. W. Ratts, *J. Org. Chem.*, **31**, 3907 (1966).
79. See J. D. Roberts, *Notes on Molecular Orbital Calculations*, W. A. Benjamin, New York, 1961, p. 131.
80. P. Beltrame, D. Pitea, A. Marzo, and M. Simonetta, *J. Chem. Soc.*, *(B)*, **1967**, 71.
81. D. Kovachic and L. C. Leitch, *Can. J. Chem.*, **39**, 363 (1961).
82. K. Griesbaum, A. A. Oswald, and D. N. Hall, *J. Org. Chem.*, **29**, 2404 (1964).
83. P. I. Abell and R. S. Anderson, *Tetrahedron Letters* (49), **1964**, 3727.
84. E. I. Heiba and W. O. Haag, *J. Org. Chem.*, **31**, 3814 (1966).
85. K. Griesbaum, A. A. Oswald, E. R. Quiram, and W. Naegele, *J. Org. Chem.*, **28**, 1952 (1963).
86. T. L. Jacobs and G. E. Illingworth, Jr., *J. Org. Chem.*, **28**, 2692 (1963).
87. H. J. van der Ploeg, J. Knotnerus, and A. F. Bickel, *Rec. Trav. Chim.*, **81**, 775 (1962).
88. E. I. Heiba, *J. Org. Chem.*, **31**, 776 (1966).
89. A. A. Oswald, K. Griesbaum, D. N. Hall, and W. Naegele, *Can. J. Chem.*, **45**, 1173 (1967).
90. K. Griesbaum, A. A. Oswald, E. R. Quiran, and P. E. Butler, *J. Org. Chem.*, **30**, 261 (1965).
91. H. G. Kuivila, W. Rahman, and R. H. Fish, *J. Am. Chem. Soc.*, **87**, 2835 (1965).
92. S. J. W. Price and A. E. Trotman-Dickenson, *Trans. Faraday Soc.*, **54**, 1630 (1958).
93. G. B. Kistiakowsky, J. R. Ruhoff, H. A. Smith, and W. E. Vaughan, *J. Am. Chem. Soc.*, **58**, 146 (1936); **57**, 876 (1935).
94. J. A. Kerr, *Chem. Rev.*, **66**, 465 (1966).
95. R. N. Haszeldine, K. Leedham, and B. R. Steele, *J. Chem. Soc.*, **1954**, 2040.
96. H. G. Meunier and P. I. Abell, *J. Chem. Phys.*, **71**, 1430 (1967).
97. G. N. Sausen and A. L. Logothetis, *J. Org. Chem.*, **32**, 2261 (1967).
98. C. L. Bumgardner and K. G. McDaniel, *J. Am. Chem. Soc.*, **91**, 1032 (1969).
99. M. Tremblay, *Can. J. Chem.*, **43**, 219 (1965).
100. H. Goldwhite, *J. Chem. Soc.*, **1965**, 3901.
101. R. S. Neale, *J. Am. Chem. Soc.*, **86**, 5340 (1964).
102. A. Rajbenbach and M. Szwarc, *Proc. Roy. Soc. (London)*, **A251**, 1266 (1959); A. Rajbenbach and M. Szwarc, *J. Am. Chem. Soc.*, **79**, 6343 (1957); A. P. Stefani, L. Heak, and M. Szwarc, *J. Am. Chem. Soc.*, **83**, 4732 (1961).
103. M. Bertrand and R. Marvin, *Acad. Sci.*, Marseille, 1107 (1963).
104. W. E. Truce and G. C. Wolf, *Chem. Commun.*, **1969**, 150.

105. D. Holtz and A. Streitwieser Jr. in *Progress in Physical Organic Chemistry*, Vol. 8, S. G. Cohen, A. Streitwieser Jr., and R. W. Taft, Eds. Wiley–Interscience, New York, N.Y. In Press. Chapter on *A Critical Evaluation of the Concept of Fluorine Hyperconjugation.*
106. L. R. Byrd and M. C. Caserio, unpublished results, University of California, Irvine, 1969.
107. For examples and discussion of spontaneous initiation in alkyl hypochlorite reactions, see C. Walling, L. Heaton and D. D. Tanner, *J. Am. Chem. Soc.*, **87**, 1715 (1965).
108. The effects of solvent in alkyl hypochlorite reactions are discussed by C. Walling and P. J. Wagner, *J. Amer. Chem. Soc.*, **86**, 3368 (1964).
109. The author is indebted to Professor C. B. Reese, Cambridge University, for personally communicating in advance of publication the results of his work with M. S. Baird and A. Shaw on the bromination of 1,2-cyclononadiene. This work has established that the product initially thought to be *trans*-2,3-dibromocyclononene (cf. refs. 48, 50) is actually *cis*-1,4-dibromocyclononene **59**.

Note Added in Proof. Free-radical addition of hydrogen bromide to allene, methylallene, 1,1-dimethylallene, 1,3-dimethylallene and tetramethylallene in the gas phase is reported to involve irreversible addition of atomic bromine to the *center* carbon of the allenic system; only in the case of 1,3-dimethylallene was any product detected of *terminal* attack by bromine atoms (6%) [R. Y. Tien and P. I Abell, *J. Org. Chem.*, **35**, 956 (1970)]. The major product of photo-induced addition of trifluoromethyl iodide to 1,3-dimethylallene has been identified as 3-iodo-4-trifluoromethyl-2-pentene arising from *terminal* attack by trifluoromethyl radicals (106). Nmr and uv studies of 1,3-dimethylallene and tetramethylallene in $FSO_3H\text{-}SbF_5$ at $-70°$ have established that stable allylic cations are formed by protonation at the *central* *sp*-hybridized carbon (C. U. Pittman, Jr., *Chem. Commun.*, **1969**, 122). Electrophilic addition of mercuric acetate in acetic acid to 1,1,4,4-tetraphenyl-3-bromo-1,2-butadien-4-ol has been reported [F. Toda and K. Akagi, *Tetrahedron*, **25**, 3795 (1969)] and further work has been published on nucleophilic additions to substituted allenes [G. D. Appleyard and C. J. M. Stirling, *J. Chem. Soc.*, (C), **1969**, 1904; M. V. Mavrov, E. S. Voskanyan and V. F. Kucherov, *Tetrahedron*, **25**, 3277 (1969)].

Selective Oxidations with Selenium Dioxide

ROBERT A. JERUSSI

Chemical Laboratory, General Electric Research and Development Center, Schenectady, New York

I. INTRODUCTION

Selenium dioxide has been used as a reagent in organic chemistry since Riley's pioneering work which first appeared in 1932 (1). Several reviews have been published, the most extensive of which are those by Rabjohn (2) and Waitkins and Clark (3). This author's intention was to focus on those transformations particularly characteristic of the reagent and, in so doing, to outline the current mechanistic information available for each process. Thus a general review of selenium dioxide oxidations was not intended.

Latimer (4) lists the E° for the Se—H_2SeO_3 couple (eq. 1) as -0.740 V; a value about half that for MnO_2, Cr_2O_7, or PbO_2. This potential is great enough to explain all the oxidations of selenium dioxide. In fact, the difficulty

$$Se + 3H_2O{=}H_2SeO_3 + 4H^+ + 4e^- \qquad (1)$$

is explaining why the reagent does not normally carry out certain oxidations, for example, an aldehyde to an acid. The latter reaction does sometimes occur when a methyl attached to an aromatic ring is oxidized, presumably passing through the aldehyde. However, when an aliphatic aldehyde is oxidized, the α-dicarbonyl compound is the product rather than a carboxylic acid. This is very likely due to dione formation being a lower activation energy process compared to acid formation.

This author considers the three major selective oxidative processes for selenium dioxide as the transformation of a ketone or an aldehyde to an α-dione, allylic oxidation, and the conversion of a monoketone or a 1,4-diketone to an α,β-unsaturated ketone or to an ene-dione. Three other types of oxidation appear to be unique and of good enough yield to warrant discussion; the conversion of alcohols to ketones or aldehydes, olefins to *vic*-diols, and acetylenes to α-diones. Oxidations of other classes of compounds such as aromatics to quinones (2), sulfur-containing compounds (2), and phenols (2) have not received general acceptance by organic chemists nor have they been the subject of mechanistic study. Therefore they are not discussed in this chapter.

II. *alpha*-DIONE FORMATION

A. Oxidation of Ketones and Aldehydes—Scope of the Reaction

In the first paper published by Riley (1) on selenium dioxide oxidations, the conversion of a monocarbonyl compound having an adjacent methylene unit to an α-dicarbonyl compound was discussed. This first report indicated that both ketones and aldehydes would react to give reasonable yields of product (eqs. 2 and 3). With the unsymmetrical compound, ethylmethyl ketone, both possible oxidation products were isolated and the oxidation of acetophenone and of phenylacetaldehyde gave the same product, phenylglyoxal (eq. 4).

$$CH_3COCH_3 \longrightarrow CH_3COCHO \quad 60\% \tag{2}$$

$$CH_3CH_2CH_2CHO \longrightarrow CH_3CH_2COCHO \quad \text{ca. } 45\% \tag{3}$$

$$\left.\begin{matrix} C_6H_5COCH_3 \\ C_6H_5CH_2CHO \end{matrix}\right\} \longrightarrow C_6H_5COCHO \tag{4}$$

The reagent has been documented three times in *Organic Syntheses*: the first to produce phenylglyoxal from acetophenone in 70% yield (5), glyoxal, isolated as the bisulfite adduct in 70% yield, from paraldehyde (6), and cyclohexa-1,2-dione from cyclohexanone in 60% yield (7). Excellent yields can be obtained if the ketone being oxidized contains structural features which prevent side or competing reactions from occurring. Thus *A*-nor-allobetulone-3 and *A*-nor-4,4-dimethylcholest-5-en-3-one, both of which have the monoketone unit, partial structure **1**, are oxidized to the diketone **2** in 87 and 92%, respectively (8). Several substituted desoxybenzoins can be oxidized to the corresponding benzils in almost quantitative yields (9). A compilation of the yields obtained from a number of ketones and aldehydes is given in the review by Rabjohn (2).

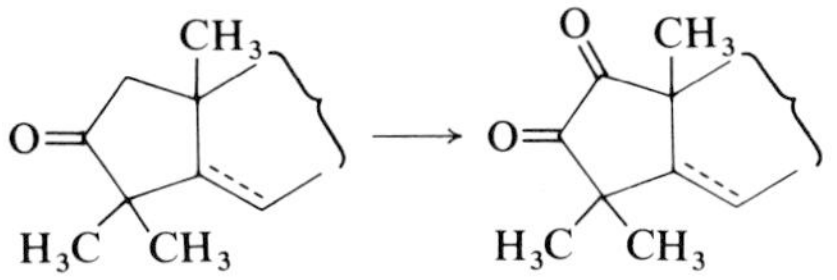

B. Kinetics and Mechanism

Mel'nikov and Rokitskaya (10–16) studied the mechanism of the oxidation in a series of papers. They found that dialkyl esters made from selenium dioxide and primary alcohols decompose at 300° to give aldehydes (10). They also studied the kinetics of the oxidation of aldehydes and ketones to dicarbonyl compounds by plotting the amount of selenium deposited against time and found the reaction to be bimolecular (11–13). They considered the rate as a measure of the degree of enolization since the rates of oxidation seemed to parallel the ability of the ketones to enolize. In analogy with the reactions of alcohols, they proposed that the carbonyl compounds oxidize through their enol form which interacts with selenium dioxide to give an ester which further reacts to product (11,12) (eq. 5).

$$2\,RCH{=}\overset{OH}{C}CH_3 + SeO_2 \longrightarrow Se(RCHCOCH_3)_2 + H_2O \qquad (5)$$

Duke (17) studied the kinetics of the oxidation of acetone and found that the rate law contained the concentrations of acetone, selenious acid, and hydrogen ion. As an intermediate, Duke proposed **3**, formation of which is rate determining.

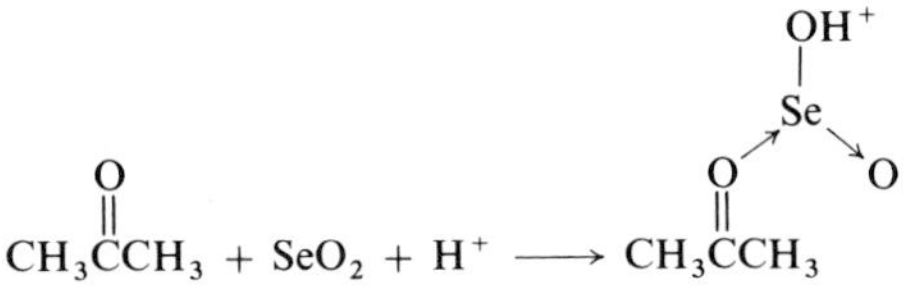

$$CH_3\overset{\overset{O}{\|}}{C}CH_3 + SeO_2 + H^+ \longrightarrow CH_3\overset{\overset{O\rightarrow Se(OH^+)\rightarrow O}{\|}}{C}CH_3$$

3

However, the most critical study of the reaction was made by Corey and Schaefer who studied the oxidation of desoxybenzoin in 70% acetic acid at 89.2° (18). They found the reaction to be second order: first order in ketone and first order in selenium (IV) reagent and to be catalyzed by added strong acid. Various *p*-substituted desoxybenzoins with substituents in the benzyl or benzoyl moiety were studied. Electron-supplying substituents in the benzoyl group increased the rate of reaction ($\rho = -0.56$) while the effect in the benzyl group was to decrease the rate ($\rho = +0.25$). This is what one would expect in an acid-catalyzed enolization process. The reaction was found to be catalyzed also by acetate ion, a maximum being reached when the concentration of acetate was equivalent to that of selenium dioxide and the substrate. The electronic effects with sodium acetate were the reverse of those found for acid catalysis, electron-withdrawing groups attached to the benzoyl part increased the rate of reaction ($\rho = +0.77$) while those on the benzyl group decreased the reaction rate ($\rho = -0.47$).

The rate was found to be insensitive to free-radical initiators and inhibitors. Benzoin as an intermediate was ruled out since the rate constant for its oxidation was over 1/20 that of desoxybenzoin.

The reaction exhibited an isotope effect; the $k\text{H}/k\text{D}$ for α,α-dideuterio-desoxybenzoin was 6.0 and for the oxalic acid catalyzed reaction, it was 5.8. In the latter stages of the reaction some exchange of deuterium with the solvent was noted. The $k\text{H}/k\text{D}$ for acetate catalysis was initially about 7.6 but fell off as the reaction proceeded.

1. The Ketone Versus the Enol as Intermediates

Corey and Schaefer (18) conclude from their results that in the acid-catalyzed oxidation the enol can be excluded as an intermediate. Since the concentration of selenium dioxide appears in the rate expression, enolization cannot be rate determining. Fast reversible enolization, followed by slow reaction of the enol with oxidant, should have exchanged out the deuterium prior to reaction and $k\text{H}/k\text{D} = 1$.

The mechanism proposed involves attack of an electrophile, $HSeO_2^+$, or $H_3SeO_3^+$, and the nucleophile, H_2O, on desoxybenzoin, **4**, in a slow step to give an enol selenite ester, **5**. The latter rearranges in a series of fast steps to an α-selenium (II) ketoester, **6**, and finally to benzil, **7**. The mechanism for acetate ion catalysis is similarly formulated where the interaction of ketone, selenious acid, and acetate give the anion of the enol ester, **8**, in the rate-determining step (19). This is converted to the enol selenite ester, **5**, by protonation which then decomposes as diagrammed previously.

Finally, the authors point out that there are probably cases of selenium dioxide oxidation where the enol is involved, e.g., ketones that are highly enolic may possibly react directly to give the enol selenite ester.

$$\underset{\mathbf{4}}{C_6H_5C(=O)CH_2C_6H_5} + HSeO_2^{+} + H_2O \underset{}{\overset{\text{slow}}{\rightleftharpoons}} \underset{\mathbf{5}}{C_6H_5C(OSeO_2H)\!=\!CHC_6H_5} \xrightarrow{\text{fast}}$$

$$\underset{\mathbf{6}}{C_6H_5C(=O)CH(OSeOH)C_6H_5} \xrightarrow{\text{fast}} \underset{\mathbf{7}}{C_6H_5C(=O)C(=O)C_6H_5}$$

$$\mathbf{4} + H_2SeO_3 + CH_3CO_2^{-} \overset{\text{slow}}{\rightleftharpoons} \underset{\mathbf{8}}{C_6H_5C(O\text{—}SeO_2^{-})\!=\!CHC_6H_5} \xrightarrow{CH_3CO_2H} \mathbf{5}$$

The proposal that the enol is not involved in the oxidation has been questioned by Best, Littler, and Waters (20) who found the rate of oxidation of cyclohexanone by chromic acid to be slower than its enolization rate at the same acidity. They point out that the rate of formation of an enol should exhibit the same primary isotope effect as any other reaction that involves the rupture of an α-carbon–hydrogen bond in a ketone. Enolization is pictured as involving two steps, the first fast (eq. 6) since a proton is exchanging between oxygen atoms, and the forward reaction of the second step (eq. 7) as being slow since a carbon–hydrogen bond is broken. Hence a reaction proceeding through an enol which is slower than the enolization step should occur less readily with a deuterated ketone until the unreacted ketone has attained isotopic equilibrium with the solvent through exchange.

$$RCOCH_2R' + H_3O^{+} \rightleftharpoons R\overset{+}{C}(OH)CH_2R' \tag{6}$$

$$R\overset{+}{C}(OH)CH_2R' + H_2O \rightleftharpoons RC(OH)\!=\!CHR' \tag{7}$$

Wiberg (21) has pointed out that the steady-state concentration of an enol being generated from a deuterated ketone will be reduced by a factor equal to the isotope effect for the enolization. The reduced enol concentration will decrease the rate of the next step and thus the overall rate until the ketone exchanges. This will be the case where the enolization rate is faster than the overall oxidation by about the same order of magnitude as the isotope effect.

Recently Rocek and Riehl have found that the chromic acid oxidation of several ketones proceeds at a rate slower than the enolization rate and have also proposed a mechanism proceeding through the enol (22). For isobutyrophenone and chlorocyclohexanone, they were able to experimentally verify

Wiberg's (21) conclusion that higher concentrations of chromic acid should make oxidation and enolization rates about equal and thus have the reaction approach zero-order in chromic acid. As the chromic acid concentration was increased, the reaction rate approached the enolization rate.

Hence it is possible to have the oxidation of a ketone be slower than the rate of enolization and still pass through the enol. Waters (23) has suggested that the selenium dioxide oxidation involves direct interaction of the enol and oxidant to give the α-keto selenium (II) intermediate, **9**, which in turn decomposes in a slow step. The important point is that the enolization rate

9

for desoxybenzoin can be faster than its oxidation rate by selenium dioxide and still be oxidized through its enol form, but the rate cannot be much faster and still agree with Corey and Schaefer's data (18). Unfortunately, the enolization rate for desoxybenzoin under the same conditions used for its oxidation is not available. Fischer, Packer, and Vaughn, however (24), measured the bromination of desoxybenzoin in 87% acetic acid containing 0.631*M* sodium acetate at 25, 40, and 55°, and reported the results as apparent bimolecular rate constants (*k* bromination of ketone/[acetate]). They determined an activation energy of 20.0 kcal/mole and log *A* in the Arrhenius equation of 9.4. Using these figures, this author has calculated the rate at 89.2° to be 21.5×10^{-4} liter mole^{-1} sec^{-1}. A value of 6.06×10^{-4} was found in 70% acetic acid containing 0.685 *M* sodium acetate at 89.2° (18) for the selenium dioxide oxidation and only a small increase (ca. 12%) was observed by switching from 70 to 90% acetic acid in an uncatalyzed reaction. Therefore, ignoring the small difference in acetate concentration and the difference in solvent composition, a *k* enolization/*k* oxidation for desoxybenzoin of ca. 3.5 is obtained. Although inherent in this calculation is the problem of comparing first and second order reactions, it may indicate that the enolization of desoxybenzoin is not very fast compared to its oxidation by selenium dioxide.

Using the rate constants for bromination at 55° obtained for various substituted desoxybenzoins with the substituent located in the *para* position of the benzyl group (24), this author has made a plot using the *sigma* values for the particular substituents given by Jaffé (25). A ρ value equal to $+1.3$ is obtained indicating that electron-withdrawing groups increase the rate of reaction in the presence of sodium acetate. This is the reverse of that found for the selenium dioxide reaction catalyzed by sodium acetate, and may mean that different intermediates are present in each reaction.

Oxidation of desoxybenzoin at room temperature for three days in 80% deuterioacetic acid and 20% deuterium oxide gave a 38% yield of benzil (26). Recovery of the unreacted starting ketone followed by deuterium analysis indicated the incorporation of no deuterium. The recovery procedure used was mild and the substrate did not come in contact with water. Therefore the oxidation does occur under conditions which do not catalyze enolization and this result does argue for the molecular mechanism.

From the preceding discussion it is apparent that the general features of the mechanism of diketone formation are known. However, the published information to date does not allow a clear-cut decision between the ketone or the enol as the actual intermediate.

III. ALLYLIC OXIDATIONS

A. Alkenes

1. Discovery and Scope

Schwenk and Borgwardt (27) appear to have been the first to demonstrate an allylic oxidation with selenium dioxide. In general, the reagent will oxidize an allylic methylene to an alcohol (eq. 8), which will be isolated as an acetate or ether if the reaction is run in acetic anhydride or anhydrous

$$\mathrm{RCH{=}CHCH_2R'} \longrightarrow \mathrm{RCH{=}CHCHOHR'} \tag{8}$$

alcohols (28). With water present, and depending on the amount of oxidant, the alcohol first formed may be further oxidized to a carbonyl group. Both cyclic and acyclic olefins undergo the oxidation.

After several publications on the oxidation of olefins (29), Guillemonat collected his observations in one report (30). He found that oxidation always occurred *alpha* to the double bond and that with both cyclic and acyclic trisubstituted olefins, oxidation took place at the most substituted end of the double bond. In addition, he found that methylene [—CH_2—] oxidized in preference to methyl [—CH_3] which in turn oxidized preferentially to methine [—CH—] (with a vertical bond drawn above the CH). 1,2-Disubstituted olefins very often gave products which resulted from reaction at both allylic positions. Terminal olefins gave primary alcohols with a shift of the double bond.

2. *Mechanism*

The mechanism of this oxidation has received considerable attention and an important contribution has been made by Wiberg and Nielsen (31) who oxidized (+)-1-*p*-menthene (carvomenthene), **10**, in 57% yield to (+)-*p*-menth-6-en-2-one (carvotanacetone), **11**, with 44–55% retention of asymmetry. The only other carbonyl containing compound present was phellandral, **12**, detected in 8% yield. No *p*-menth-1-en-3-one (piperitione), **13**, was evident from gas chromatographic analysis. That (+)-*p*-menth-6-en-2-one did indeed have the structure shown in **11** and not its mirror image was shown by relating the relative configuration of **10** to **11** through (+)-limonene. Thus, the asymmetric carbon atoms in **10** and **11** have the same configuration and oxygen was introduced at the allylic methylene in **10** rather than at the double bond.

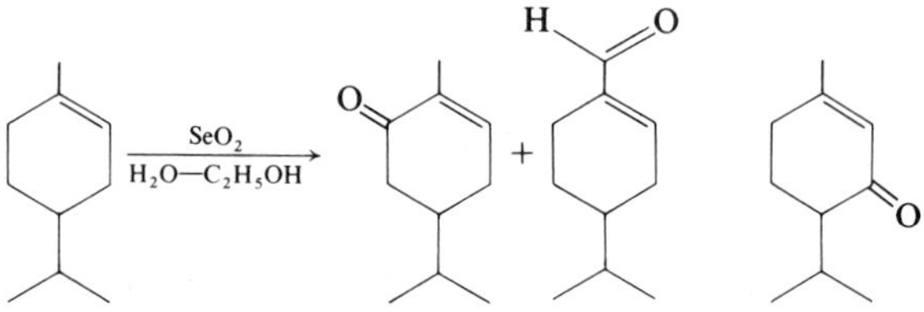

$$\mathbf{10} \xrightarrow[H_2O-C_2H_5OH]{SeO_2} \mathbf{11} + \mathbf{12} \qquad \mathbf{13}$$

The authors suggest only two mechanisms can be operative; a direct insertion reaction and an addition–elimination mechanism. They reject the former based on Guillemonat's observations (30) that oxidation occurs preferentially at the most substituted end of the double bond and that a terminal olefin gives a primary allylic alcohol. The mechanism proposed is formulated as follows; attack by protonated selenious acid (water–ethanol solvent) at the least substituted end of the double bond to give the intermediate carbonium ion, **14**, which loses a proton to give the allylic seleninic acid, **15**. Intermediate **15** can decompose via several pathways: SN2′ to give **11** with retention, SN2 to give **16** with inversion and SN1 to give a racemic product. A predominance of SN2′ displacement would account for the observed stereochemical result.

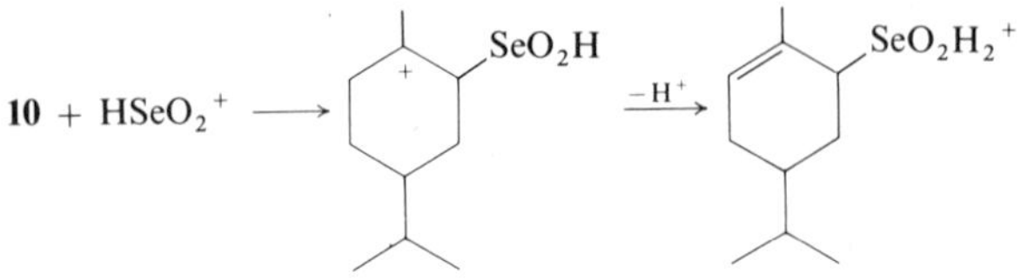

$$\mathbf{10} + HSeO_2^+ \longrightarrow \mathbf{14}\ (SeO_2H) \xrightarrow{-H^+} \mathbf{15}\ (SeO_2H_2^+)$$

$H_2O \longrightarrow$ **15** (H_2O, $SeO_2H_2^+$) — SN2′ → HO-substituted menthene → **11** Retention; SN2 → OH-substituted menthene → **16** Inversion; SN1 → carbonium ion $\xrightarrow{1.\ H_2O;\ 2.\ \text{oxid.}}$ **11** + **16** Racemic mixture

Mousseron and Jacquier (32) oxidized (+)-4-methylcyclohexene, **17**, in acetic anhydride–acetic acid to give 5-methyl-2-cyclohexen-2-yl acetate. Subsequent reactions which did not affect the configuration of the methyl

17 $\xrightarrow[CH_3CO_2H,\ (CH_3CO)_2O]{SeO_2}$ CH_3OCO–(methylcyclohexene, CH_3) $\xrightarrow{1.\ OH^-;\ 2.\ H_2\text{—}Pt;\ 3.\ CrO_3}$ **18**

group gave (−)-3-methylcyclohexanone, **18**, with over 80% retention of asymmetry. Again an SN2′ displacement by acetate ion could account for this result.

Nelson (33) has studied the same system as Wiberg and Nielsen and found racemic product in acetic acid–acetic anhydride, but retention in several aqueous dioxane solvent systems. He proposed a similar reaction mechanism to account for the results but prefers addition of the oxidant to the double bond through oxygen rather than selenium.

It was also shown (33) that free-radical intermediates were not present in an oxidation reaction of (+)-1-*p*-menthene since added acrylonitrile did not polymerize. The oxidation of α-pinene, **19**, was reinvestigated and in addition to the major products of the reaction, myrtenol, **20**, and myrtenal, **21**, which had been previously reported (34), Nelson was able to identify *p*-cymene, **22**. He considers the formation of **22** to be due to the rearrangement of the initially formed carbonium ion **23**. It has already been pointed out why the carbonium ion rearrangement may not be a favorable process (31) and thus not lead to a major product.

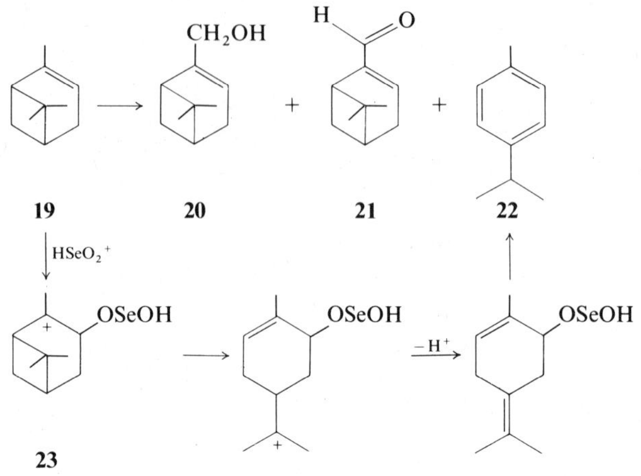

19 20 21 22

$HSeO_2^+$

OSeOH OSeOH $-H^+$ OSeOH

23

Schaefer and Horvath (35) studied the oxidation of 1,3-diphenylpropene in 99% acetic acid at 115°, a reaction which gives 1,3-diphenyl-2-propen-1-ol acetate in high yield. When competitive experiments were run using a limited amount of selenium dioxide with equimolar mixtures of the unsubstituted 1,3-diarylpropene and a substituted 1,3-diarylpropene, the results were indicative of an attack on the olefinic bond by an electrophilic species, possibly $HSeO_2^+$. Thus, the ratio

$$[C_6H_5CH_2CH{=}CHC_6H_5]/[C_6H_5CH_2CH{=}CHAr]$$

is 0.56, 1.76, and >13.5 when Ar is *p*-$CH_3OC_6H_4$—, *p*-ClC_6H_4—, and mesityl, respectively. The latter result indicates that the oxidation can be slowed up by sterically protecting the double bond. These ratios are in general agreement with the results of Nelson (33) who found a negative ρ value of -2.34 when he studied the kinetics of oxidation of various ring-substituted α-methylstyrenes.

When 3-deuterio-1,3-diphenylpropene **24** containing 0.87 atoms of deuterium per molecule was oxidized (35), 1,3-diphenyl-2-propen-1-ol **25** was produced containing 0.67 atoms of deuterium. This gives a kH/kD of 3.1 at 115°. When **25** was reduced to 1,3-diphenyl-2-propan-1-ol **26**, and the latter oxidized to 1,3-diphenylpropan-1-one **27**, half the deuterium was lost. Since the possibility of olefin isomerization before oxidation or rearrangement during isolation was ruled out by control experiments, the results with the deuterated substrate indicate that the two benzylic positions of **24** become equivalent during the oxidation.

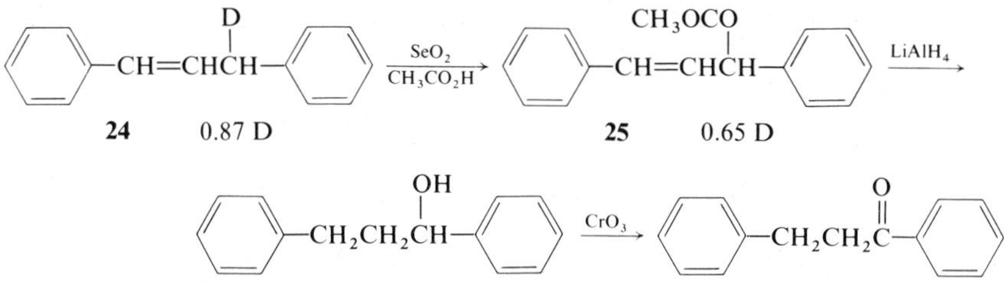

$$C_6H_5CH{=}CHCD(H)C_6H_5 \xrightarrow[CH_3CO_2H]{SeO_2} C_6H_5CH{=}CHCH(OCOCH_3)C_6H_5 \xrightarrow{LiAlH_4}$$

24 0.87 D **25** 0.65 D

$$C_6H_5CH_2CH_2CH(OH)C_6H_5 \xrightarrow{CrO_3} C_6H_5CH_2CH_2C(=O)C_6H_5$$

26 0.67 D **27** 0.32 D

The mechanism proposed involves interaction of oxidant and olefin to form an allylic selenium(II) ester (eq. 9) which ionizes to a symmetrical 1,3-diarylcation (eq. 10). Attack of acetic acid statistically at both ends of the cation gives the product (eq. 11).

$$C_6H_5CH{=}CHCH_2C_6H_5 + HSeO_2{}^+ \longrightarrow C_6H_5CH(\ddot{O}SeOH)CH{=}CHC_6H_5 \qquad (9)$$

$$C_6H_5CH(\ddot{O}SeOH)CH{=}CHC_6H_5 \longrightarrow C_6H_5\overset{+}{\overline{CHCHCH}}C_6H_5 + H\ddot{S}eO_2{}^- \qquad (10)$$

$$C_6H_5\overset{+}{\overline{CHCHCH}}C_6H_5 + CH_3CO_2H \longrightarrow C_6H_5CH(OCOCH_3)CH{=}CHC_6H_5 \qquad (11)$$

Since the selenium(II) ester formed in eq. 9 is benzylic, the preference for ionization (SN1) rather than SN2′ attack as found by Wiberg and Nielsen is understandable.

In a more recent paper Schaefer, Horvath, and Klein (36) extended these results. They pointed out that the intermediate suggested by Wiberg and Nielsen, for example, the seleninic acid, **15**, which contains a C—Se bond is very likely a stable compound as are other alkyl seleninic acids. Thus it probably would not undergo solvolysis as well as a selenium(II) ester type intermediate which contains an O—Se bond.

Suga, Sugimoto, and Matsuura (37) oxidized (+)-*p*-menth-3-ene **28** in acetic acid–acetic anhydride, and after working up the product by steam

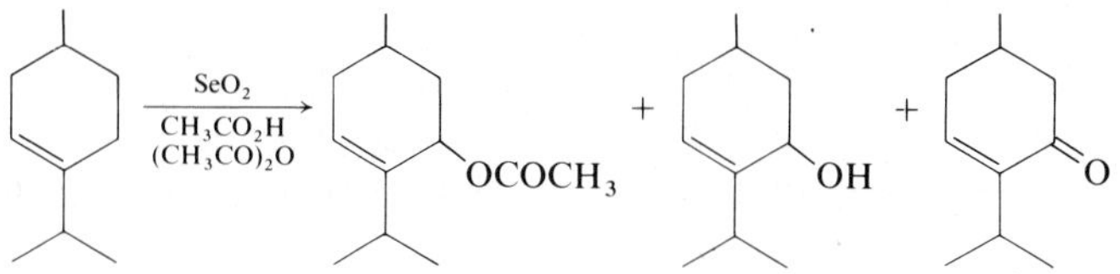

$$\mathbf{28} \xrightarrow[\substack{CH_3CO_2H\\(CH_3CO)_2O}]{SeO_2} \mathbf{29}\ (OCOCH_3) + \mathbf{30}\ (OH) + \mathbf{31}\ (=O)$$

28; **29** 35%; **30** 18%; **31** 8%

distillation, identified *p*-menth-3-en-5-yl acetate **29**, *p*-menth-3-en-5-ol **30** and *p*-menth-3-en-5-one **31** all racemic. No product resulting from oxidation at the 2-position was detected.

In view of the racemization observed in acetic acid with the menthene system by two groups (33,37) Schaefer, Horvath, and Klein reinvestigated the oxidation of (+)-4-methylcyclohexene (**17**) using that solvent. The initially formed allylic acetate was hydrogenated and the acetate group removed by lithium aluminum hydride. Oxidation of the resultant alcohol gave 3-methylcyclohexanone (**18**) that was almost completely racemized.

Therefore the solvent has an important role in the oxidation coupled with the nature of the selenium(II) intermediate. In acetic acid–acetic anhydride, a good ionizing solvent, a secondary intermediate such as **32** dissociates to an allylic cation and gives racemization. In less ionizing solvent systems **32** would undergo SN2′ in preference to SN2 due to steric reasons and thus give retention. However, when the intermediate is primary **33**, the SN2 reaction predominates and the basis for Guillemonat's observation that

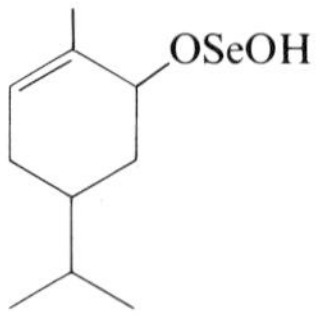

OSeOH

32

$$\begin{array}{c} \text{OSeOH} \\ | \\ \text{RCH{=}CHCH}_2 \end{array}$$

33

oxidation of a terminal olefin produces a primary product with a shift of the double bond is explained.

When one mole of selenium dioxide and two moles of 1,3-diphenylpropene (35) were reacted in acetic acid at room temperature, a white crystalline material slowly precipitated. An infrared spectrum revealed the presence of acetate ester and combustion analysis indicated the presence of selenium in the ratio of about one selenium atom to two 1,3-diarylpropyl acetate groups. Structure **34** was suggested for the compound. When **34** was heated in acetic acid under conditions used for the oxidation of 1,3-diphenylpropene, about 7% of the theoretical amount of selenium was obtained, whereas in the same time, the normal oxidation reaction gave 53% selenium. Pyrolysis of **34** at 300° gave a mixture of **34**, 1,3-diphenylpropene, and 1,3-diphenyl-2-propen-1-yl acetate. These experiments do indicate a pyrolytic pathway can be operative as well as a solvolysis step in selenium dioxide oxidations, but in view of the yields of selenium obtained, pyrolysis is not a major pathway in the oxidation of 1,3-diphenylpropene.

However, pyrolysis can be a major pathway in the oxidation of cyclohexene (31) where the reaction of two moles of olefin with one mole of selenium dioxide in acetic acid–acetic anhydride at steam bath temperatures gave no oxidized product. Pyrolysis of the nonsteam volatile material gave 3-acetoxycyclohexene in 39% yield based on SeO_2. The intermediate in this reaction is considered to be **35**.

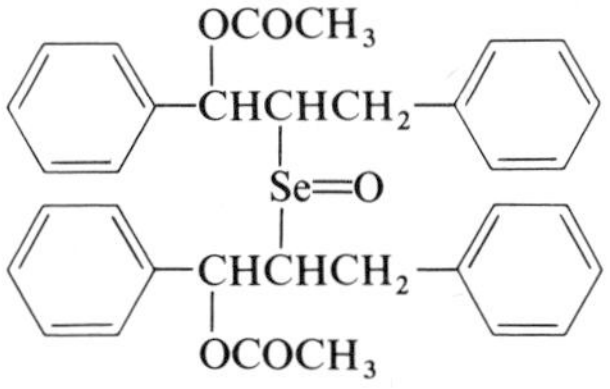

34

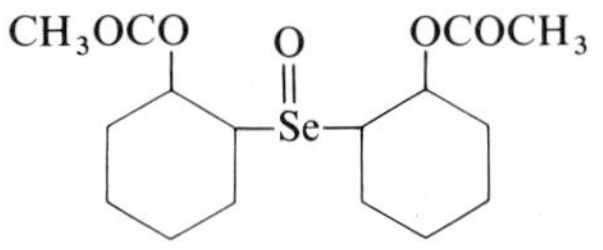

35

B. Benzylic Type Positions

1. Utility of the Reaction

Methyl or methylene attached to an aromatic nucleus oxidize to aryl carbonyl compounds and in certain cases high yields can be obtained (2). For instance, diphenylmethane can be oxidized to benzophenone in 87% yield (38) and fluorene to fluorenone in 68% yield (39). The reaction has been very useful in the oxidation of methyl groups attached to nitrogen heterocycles. Some examples are the oxidation of 2,3,8-trimethylquinoline **36** to 3,8-dimethyl-2-quinoline aldehyde (**37**) in 82% yield (40) and the oxidation of 2-methylpyridine to a mixture of 2-pyridylaldehyde and 2-pyridylcarboxylic acid (41).

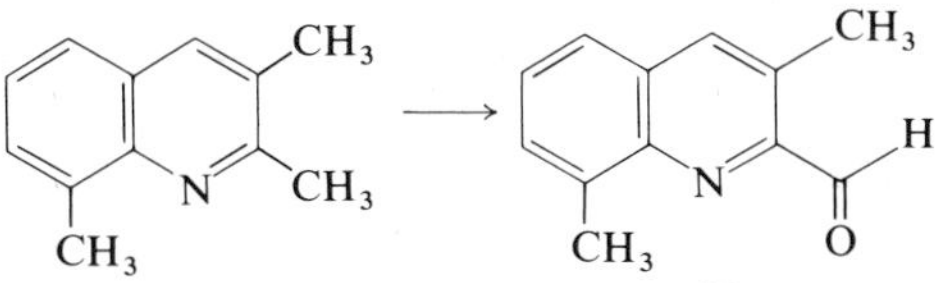

36 → **37**

2. Mechanism

Several mechanisms have been proposed for the oxidation. Waters (23) has suggested attack on an isomeric double bond as in **38** to give the selenium(II) ester, **39**, which rearranges as shown to product **40**. Corey and Schaefer (18) have proposed **42** as a possible intermediate in the oxidation of 7-methylquinoline **41**. Intermediate **42** isomerizes and rearranges to selenium(II) ester, **43**, which goes to the aldehyde, **44**. Both mechanisms appear plausible especially due to the known ability of 2- and 4-methyl pyridines and quinolines to undergo condensation reactions (42,43). Also, the specific oxidation of the 2-methyl group in **36** rather than the 3- or

8-group where an isomeric structure cannot be written is another indication that such a pathway may be operative.

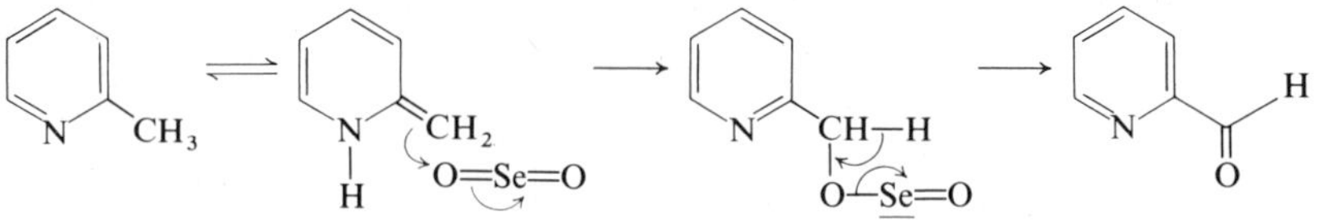

38 39 40

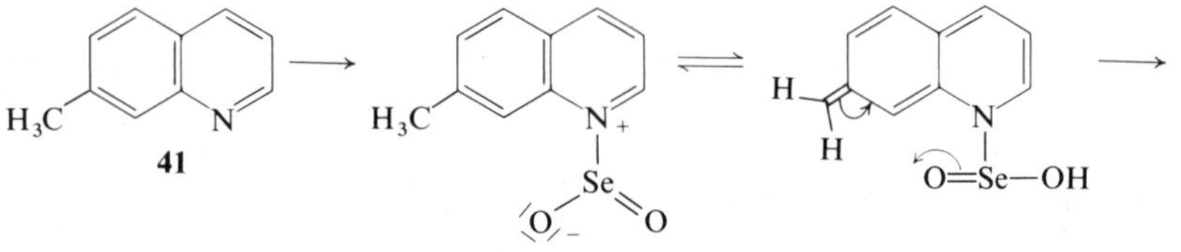

41 42

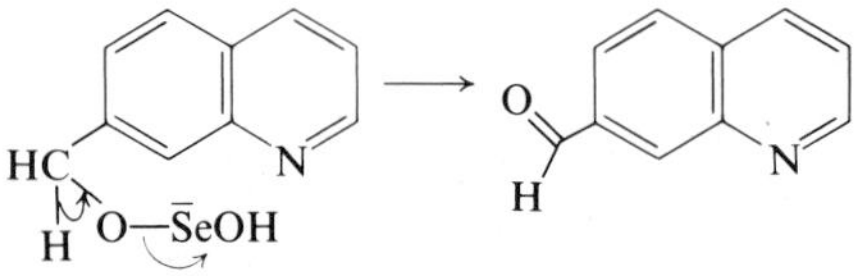

43 44

Jerchel, Heider, and Wagner (44) have made an extensive study of the oxidation of various methyl-substituted pyridines and quinolines. Using 1.5 moles of selenium dioxide per mole of methyl group, they were able to oxidize the methyl to carboxyl groups in yields greater than 50% in both pyridine or dioxane at 100–115°. 3-Methylpyridine or 3-methylquinoline did not react and in fact the former could be used as a solvent for the oxidation. 2,4-Dimethylpyridine **45** was oxidized to the 2-methylpyridine-4-carboxylic acid **46** under these conditions. They also oxidized the methyl groups in pyridine *N*-oxides and in *N*-*n*-butylpyridinium chlorides. With 2-methylpyridine *N*-oxides, they isolated the aldehyde even with excess oxidant and considered the hydrate to be stabilized by hydrogen bonding to the *N*-oxide, structure **47**. Oxidation was thought to occur through the carbanion structure, **48**.

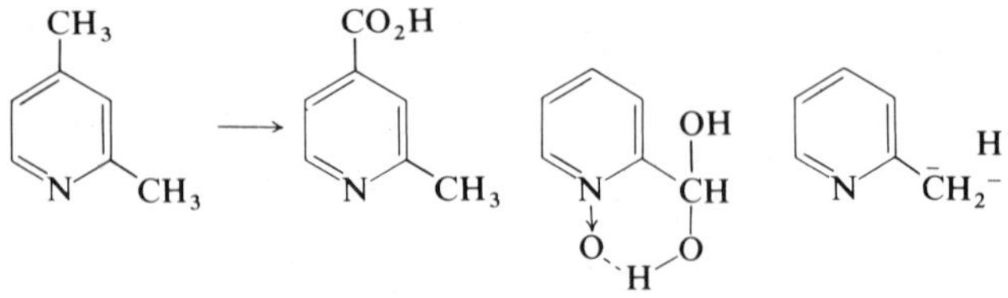

45 46 47 48

The 3-methylpyridine must be considered the same type as a methyl attached to a phenyl ring, and it is difficult to rationalize a pathway for either of these involving an isomeric structure. Oxidations of this type require higher temperatures and may involve free radicals.

IV. α,β-DEHYDROGENATION OF CARBONYL COMPOUNDS

A. Historical

1. 1,4-Diketones

In Riley's third paper on selenium dioxide oxidation (45) the dehydrogenation of diethylsuccinate to a mixture of the di- and half ester of maleic acid was reported (eq. 12). Armstrong and Robinson (46) were able to utilize the reaction to convert 2,5-hexanedione to 1,2-diacetylethylene in

$$2CH_3CH_2OCOCH_2CH_2COOCH_2CH_3 \xrightarrow[170°]{SeO_2} CH_3CH_2OCOCH{=}CHCOOCH_2CH_3 + \underset{\text{ca. } 40\%}{HOCOCH{=}CHCOOCH_2CH_3} \quad (12)$$

15% yield. Riley (47) showed that the presence of two carbonyl groups was unnecessary and that an aryl group would act as a second carbonyl unit (eq. 13). In fact, in the same paper 1,2-diphenylethane was reported to give

$$C_6H_5CH_2CH_2CO_2CH_2CH_3 \xrightarrow{SeO_2} \underset{8\%}{C_6H_5CH{=}CHCO_2CH_2CH_3} \quad (13)$$

stilbene in 33% yield, so that the requirement of even one carbonyl group is unnecessary. However, since this type of reaction has been used chiefly with ketones, that aspect of the reaction is discussed.

2. Monoketones

In 1947 Schwenk and Stahl (48) reported the discovery that selenium dioxide oxidation of a 12-keto steroid **49** produced the $\Delta^{9,11}$-12 ketone,

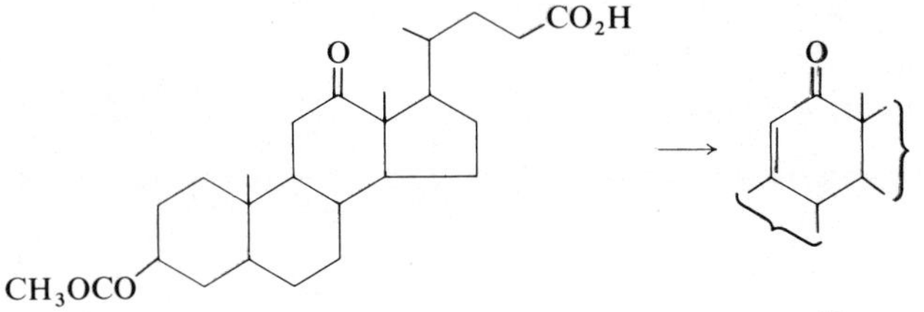

CO_2H O O CH_3OCO →

49 **50**

partial structure **50**, not the 11,12-diketone. Thus, dehydrogenation can occur without the presence of two straddling activating groups. In 1956 two groups reported that selenium dioxide introduced a double bond at the 1,2-position in either a 5α-3-keto steroid or Δ^4-3-keto steroid, partial structures **51** and **52**, when the reaction was carried out in refluxing tertiary alcohols (49). These discoveries spurred the use of selenium dioxide in steroid chemistry and a number of examples have already been cited (50).

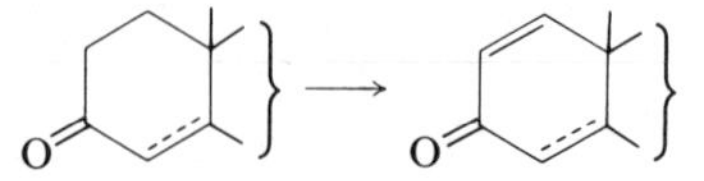

51 → **52**

B. Kinetics and Mechanism

1. 1,4-Diketones

Three kinetic studies have been made of this dehydrogenation reaction. Schaefer (51) oxidized 1,2-dibenzoylethane **53** to *trans*-1,2-dibenzoylethylene **54** and found that under catalysis by *p*-toluene sulfonic acid, the reaction was first order in substrate and first order in oxidant. This was also true of acetate catalysis at low concentration of base. In the absence of a catalyst, however, the dependence of the rate on the oxidant was one-half order, perhaps due to a series of equilibria which preceded the rate-determining step.

$$H_2SeO_3 + H^+ \rightleftharpoons H_3SeO_3^+ \tag{14}$$

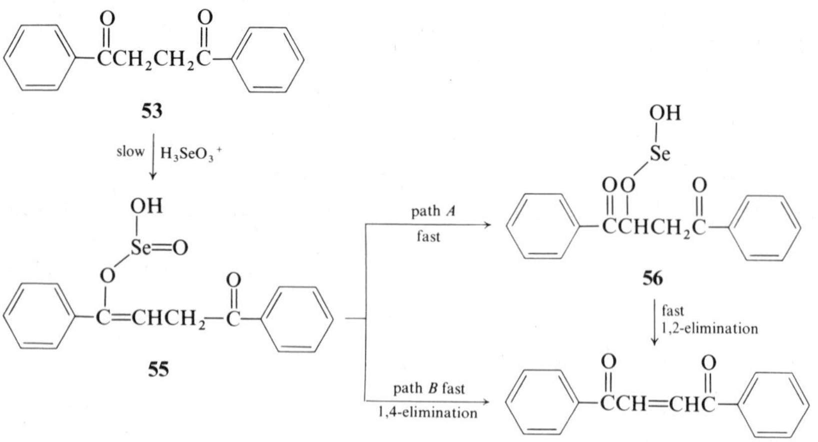

$C_6H_5COCH_2CH_2COC_6H_5$ **53**

slow ↓ $H_3SeO_3^+$

$C_6H_5C(OSe(=O)OH)=CHCH_2-COC_6H_5$ **55**

path *A* fast → $C_6H_5COCH(OSe-OH)CH_2COC_6H_5$ **56**

fast 1,2-elimination ↓

path *B* fast 1,4-elimination →

$C_6H_5COCH=CHCOC_6H_5$ **54**

The reaction exhibited a deuterium isotope effect of 6.5 (initial) when 1,1,2,2-tetradeuterio-1,2-dibenzoylethane was oxidized at 90°. Schaefer also showed that the biselenite ion, $HSeO_3^-$, is very likely not involved in the reaction since it does not oxidize acetone. 1,2-Dibenzoylethylene **54** is also oxidized by selenious acid, but at 1/30 the rate of starting dione.

With these facts the following mechanism was proposed for acid catalysis: (*1*) production of the oxidant, $H_3SeO_3^+$, by protonation of selenious acid (eq. 14); (*2*) attack of the oxidant on the substrate **53** to give an enol selenite ester **55**; and (*3*) decomposition of **55** to the product via one of two pathways. Path *A* involves rearrangement of **55** to the α-selenium(II) keto ester **56** and then to product by a 1,2-elimination. In Path *B* **55** proceeds directly to product by a 1,4-elimination. Path *A* is essentially the same mechanism and intermediates already proposed for α-dione formation (18).

The enol as an important intermediate is rejected for the same reasons as in α-dione formation, the deuterium isotope effect and the presence of the oxidant in the rate expression.

Banerji, Barton, and Cookson (52) studied the oxidation of a series of stereoisomeric methyl 3,6-dioxoeudesmanoates, two of which are **57** and **58**. The relative rates of oxidation of these diketones were measured and it was found that for any pair of stereoisomeric ketones, those with the 4,5-hydrogen atoms *cis* **57** are oxidized faster than those with a *trans* relationship **58** to the corresponding methyl 3,6-dioxoeudesm-4-enoate, **59**.

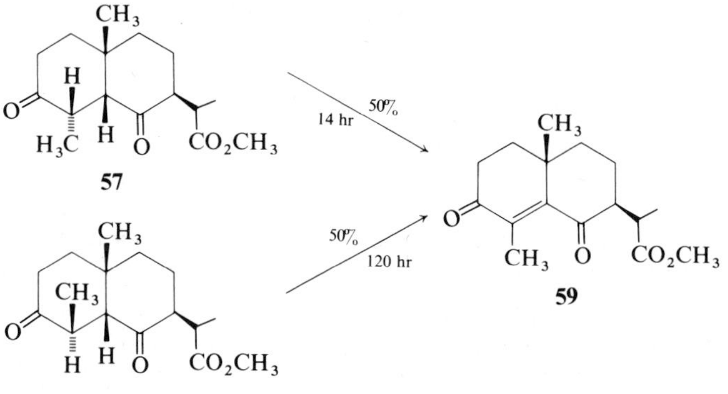

CH_3 H H_3C H O CO_2CH_3

57

14 hr 50%

CH_3 CH_3 H H O CO_2CH_3

58

50% 120 hr

CH_3 CH_3 O CO_2CH_3

59

The rates of bromination of the various diketones gave the same type of result; the *cis* were faster than the *trans* compounds. Thus, possibly enolization was involved in the selenium dioxide oxidation and the *cis* compounds simply enolized faster than the *trans*. In attempting to obtain information on the effect of selenious acid ($K_1 = 2.4 \times 10^{-3}$) in catalyzing enolization, the polarographic rate of isomerization of **57** to **58** was studied using chloroacetic acid ($K = 1.36 \times 10^{-3}$). The rate increased with increasing acid concentration, but the rate of oxidation was about 15 times faster than isomerization at equivalent acid concentration. It was suggested that these results could be explained either by the oxidation proceeding through direct attack of selenium dioxide on the ketone or that in the isomerization study, reprotonation of the enol **60** to give back **57** is faster than protonation to give **58**. The

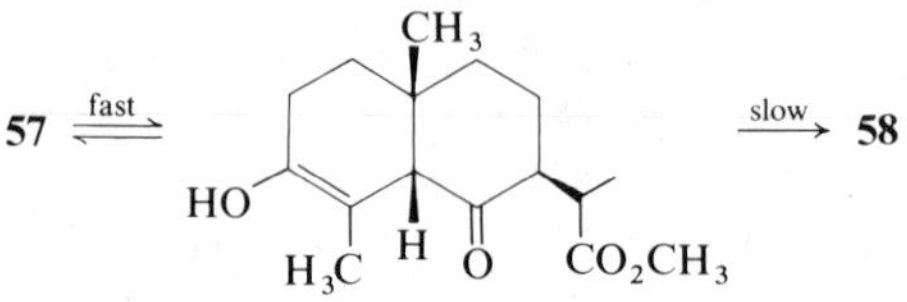

57 $\overset{\text{fast}}{\rightleftharpoons}$ [structure **60**: CH_3, HO, H_3C, H, O, CO_2CH_3] $\xrightarrow{\text{slow}}$ **58**

60

latter would explain the slow isomerization rate as compared to oxidation.

Banerji, Barton, and Cookson (52) also found that in buffered medium, the oxidation rate was first order in both selenium dioxide and diketone. They suggest that both possible mechanisms may be involved; attack of selenium dioxide on the diketone form of the *cis* compounds and attack on the enol of the *trans* isomers. Barnes and Barton (53) had previously indicated that a cyclic transition state such as partial structure **61** or **62** may be involved to account for the faster rate for the *cis* compounds.

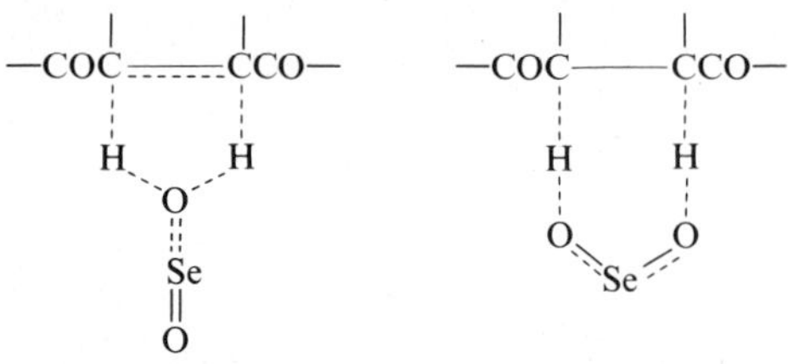

[structure **61**: —COC—CCO—, H, H, O, Se, O] [structure **62**: —COC—CCO—, H, H, O, Se, O]

61 **62**

Schaefer (51) gives some evidence to indicate that if molecular attack is involved, selenium dioxide does not remove the two hydrogens simultaneously as the *cis–trans* results of Barton might indicate. He reasons that if simultaneous removal is operative, the deuterium isotope effect calculated by a comparison of the deuterium analysis of partially deuterated starting material and the deuterium analysis of the product should be equal to that measured kinetically from the tetradeuterated starting dione. However, if the process is stepwise, the loss of the second hydrogen occurs from a high energy intermediate and should occur in a less discriminatory fashion. This should give a lower calculated comparison isotope effect than the kinetic one. The comparison isotope effect was found to be 5.4 which is well below the 6.5 initial value found kinetically. However, the initial kinetic isotope effect decreased slowly due to exchange with solvent and the 5.4 comparison value was obtained after a three-hour reaction, enough time so that some of the starting material may have equilibrated with the solvent. Thus the amount of deuterium in the product would be less than it should be and the amount of hydrogen would be greater, a combination which would reduce the isotope effect. Therefore it appears that the conclusion drawn from this result would be valid only if the deuterated substrate either did not undergo exchange or did so very slowly.

Of the two possibilities, the present author favors the 1,4-elimination from the enol selenite ester, **55**, since this intermediate contains a doubly activated methylene unit. The latter simply requires an enolization to give the half ester of the di-enol, **63**, which can decompose to product via bond migration, the driving force being the reduction of the selenium. The possibility of 1,2-elimination from the α-selenium(II) keto ester, **56**, appears less likely in view of the fact that the alternate product of its decomposition, 1,4-diphenyl-1,2,3-trioxobutane could not be detected (51).

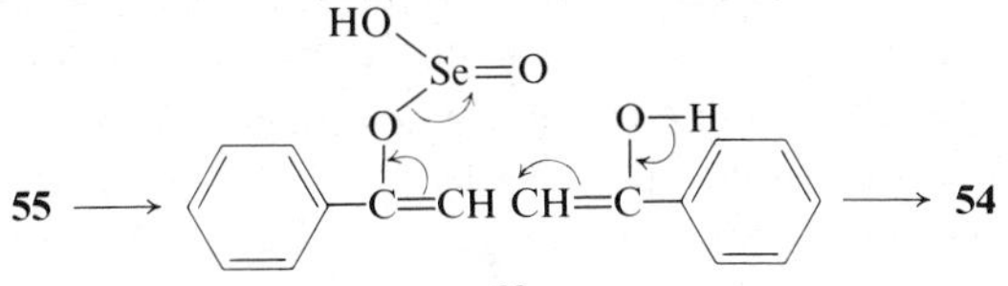

55 ⟶ [structure] ⟶ **54**

63

2. *Monoketones*

The kinetics of dehydrogenation of an α,β-unsaturated ketone were studied by Langbein (54). He obtained a second-order rate constant for the Δ^1-dehydrogenation of cortisoneacetate, **64**, to **65** from a plot which contained the concentrations of ketone and selenium dioxide. Langbein pictures

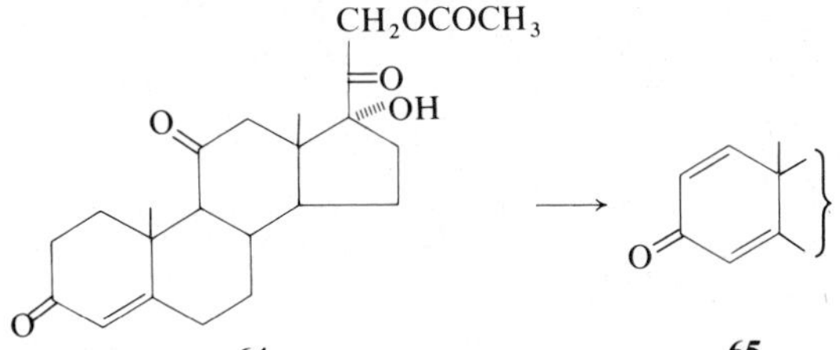

64 ⟶ **65**

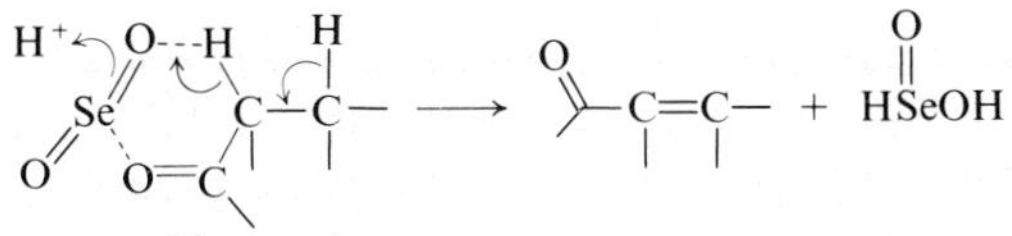

[structure] ⟶ $-C=C-$ + HSeOH

66

a common intermediate, similar to **56**, formed by direct attack of the oxidant on the ketone, which decomposes to form all the possible oxidation products. However, for α,β-dehydrogenation, he considers the more plausible path as one that does not involve carbon–oxygen bond formation, as in **66**.

The reaction of a 5β-3-keto steroid gives the Δ^4-3-ketone while a 5α-3-keto compound gives the Δ^1-3-ketone (49). Thus, the reaction appears to follow the direction of enolization. With the 5α-3-keto steroid, where there are two hydrogens at C-1, the α-hydrogen (axial) is stereoselectively removed (55). Since the enolization of a 5α-3-keto steroid, the 2β-hydrogen is preferentially lost due to more efficient overlap in the transition state (56), the net process appears to be a *trans* diaxial loss of hydrogen, thereby paralleling the dichlorodicyanoquinone (57) and *Bacillus sphaericus* dehydrogenation (58). The same study (55) indicated that some Δ^4-ketone was obtained from the 5α starting ketone and that this was not produced by rearrangement of the Δ^1 compound.

A mechanism for the dehydrogenation of a monoketone may be written which involves either 1,4-elimination from **67** or 1,2-elimination from **68**. These intermediates are similar to **55** and **56**, but without a second carbonyl group to activate the *beta* position and therefore should be less prone to undergo elimination. In addition, **68** is an α-ketoselenium(II) ester similar to **6** which has been proposed as an intermediate in α-dione formation (Section II). Another path, which circumvents the difficulty inherent in **68**, would be direct attack on the allylic position in the enol **69** by selenium dioxide to remove hydride ion. This would be similar to the mechanism proposed for the same type of oxidation by dichlorodicyanoquinone (57) and *Bacillus sphaericus* (59) and would explain the preference for loss of the

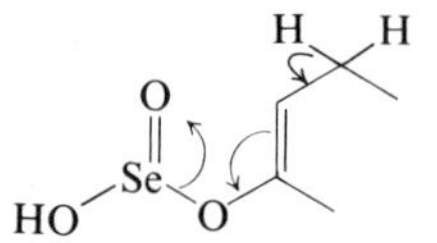

67 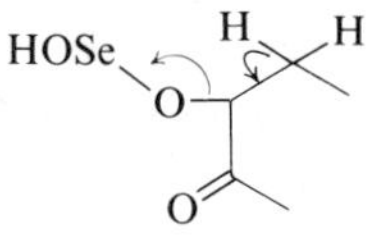 **68**

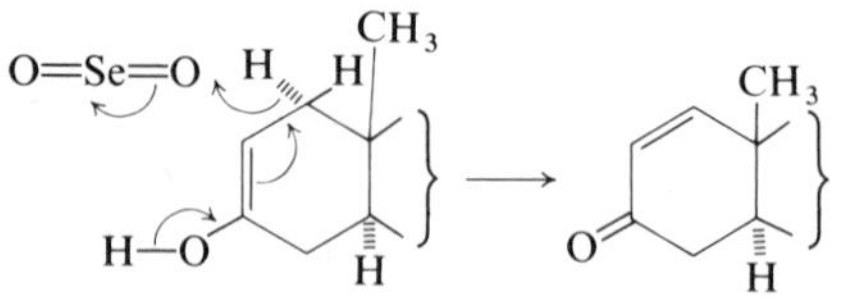

69

1α hydrogen since removal of the 1β hydrogen would require the oxidant to attack from the most hindered side due to the C-19 methyl group.

Unfortunately an important question cannot be answered; for example, why do some monoketones give α-diones and others α,β-unsaturated ketones? This appears to be partially a solvent effect. Tertiary alcohols are normally used to carry out the dehydrogenation reaction (49), but the reaction can be effected in acetic acid (48) or in aromatic solvents (60). α-Diones are generally produced using ethyl alcohol (1) or dioxane (7). The nature of this solvent effect has not been elucidated. Corey and Schaefer (18) have pointed out that a steric factor may be involved; shielding of an α-methylene unit by a nearby large substituent often favors the formation of the α,β-unsaturated ketone.

V. ALCOHOL OXIDATION

In Riley's third paper (45) he reported the oxidation of primary alcohols to aldehydes. Ethyl alcohol was oxidized to oxalyl aldehyde (eq. 15) and benzyl alcohol to benzaldehyde (eq. 16).

$$CH_3CH_2OH \xrightarrow[200^\circ]{SeO_2} HCOCHO \quad 41\% \tag{15}$$

$$C_6H_5CH_2OH \xrightarrow[SeO_2]{reflux} C_6H_5CHO \quad ca.\ 40\% \tag{16}$$

In the former the dialdehyde is very likely produced via acetaldehyde. Attempts to oxidize *n*-propyl and *n*-butyl alcohols gave only traces of oxidation.

Mel'nikov and Rokitskaya (10) prepared dialkylselenite esters which decompose when heated at 300° to aldehydes (eq. 17),

$$2ROH + SeO_2 \longrightarrow ROSeO_2R + H_2O \xrightarrow{300^\circ} 2RCHO + Se + H_2O \tag{17}$$

where R = methyl, ethyl, propyl, butyl, *iso*-butyl, and *iso*-amyl radicals.

Weygand, Kinkel, and Tietjen (61) were able to prepare the cyclic selenite ester, **70**, from *ortho*-phthalalcohol and decompose it thermally to *ortho*-phthalaldehyde, **71**.

Kaufmann and Spannuth (62) prepared a series of dialkyl selenites from C_{10}–C_{18} alcohols and were able to decompose them to the corresponding aldehydes in yields of about 90% by simply heating the ester somewhat above its melting point under vacuum and irradiating with ultraviolet light.

70 → **71**

Allylic alcohols are quite easily oxidized. Several examples have been recorded by Rabjohn (2). In fact, Nelson (33) found that when an allylic position was oxidized in aqueous dioxane, the ratio of selenium dioxide/olefin had to be as low as 0.125 to get the allylic alcohol. Higher ratios gave mixtures of alcohol and ketone. Stagno d'Alcontres and Lo Vecchio (63) found that allyl alcohol and propargyl alcohol reacted exothermally to give the corresponding aldehydes in 18 and 27% yields, respectively. They also oxidized cinnamic alcohol to cinnamaldehyde in 59% yield, but were unable to oxidize 1-phenyl-5-hydroxymethyl-1,2,3-triazole **72**. With furfuryl alcohol **73** a small quantity of the aldehyde was obtained and with 3-phenyl-5-hydroxymethylisoxazole **74** they were able to prepare and decompose the selenite ester, but obtained no aldehyde.

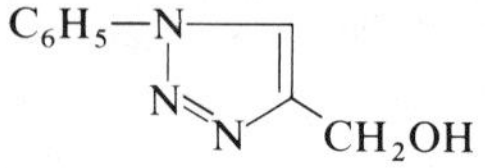

C_6H_5—N, N, N, CH_2OH **72** O, CH_2OH **73** C_6H_5, N, O, CH_2OH **74**

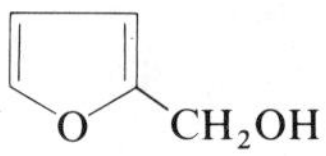

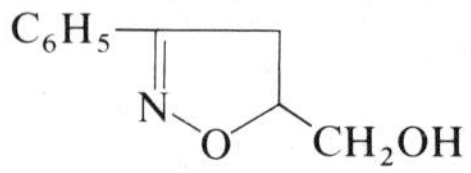

Jerchel and Heck (64) have studied the oxidation of 2-, 3-, and 4-hydroxymethylpyridines and found that the 2- and 4-substituted compounds easily underwent oxidation to the aldehyde in yields of ca. 90% at 80–90° in both dioxane or pyridine using one-half mole of selenium dioxide per mole of substrate. However, the 3-hydroxymethylpyridine did not react under these conditions, but did give a 60% yield of the aldehyde when reacted for 10 min at 160° without solvent. With a selenium dioxide–alcohol ratio of 1, only aldehyde was obtained in solution with the 2- and 4-hydroxymethyl compounds, but without solvent, the acids were obtained.

They explain these results as follows. Without solvent, the half ester, **75**, or diester, **76**, forms depending on the amount of selenium dioxide present.

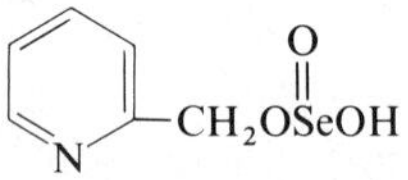

N, —CH_2OSeOH (Se=O) **75** N, —$CH_2OSeOCH_2$— (Se=O), N **76**

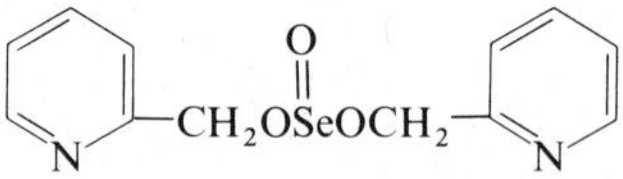

Both decompose at elevated temperatures, **75** to a carboxylic acid and **76** to the aldehyde. In the presence of solvent the esters do not form. Instead the 2- and 4-hydroxymethylene compounds react through their carbanion form, **77**. Since this carbanion cannot be stabilized for the 3-hydroxymethyl compound, it does not react.

Jerchel and Heck's explanation appears reasonable. Selenium dioxide, however, cannot directly attack the double bond of **77** since that would lead to the acid. Rather, esterification of the hydroxy group, structure **78**, would give an intermediate which would decompose to aldehyde.

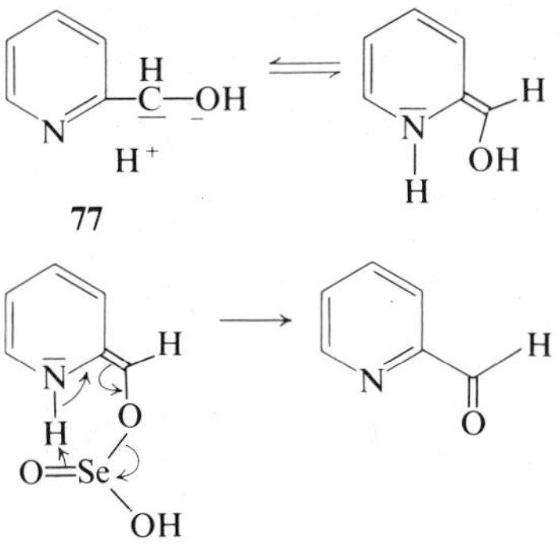

77

78

VI. OXIDATION OF OLEFINS AND ACETYLENES

In his second paper on selenium dioxide oxidations Riley (65) reported the oxidation of ethylene to oxalyl aldehyde (eq. 18).

$$CH_2{=}CH_2 + SeO_2 \xrightarrow{110\text{–}120^\circ} HCOCHO \quad 82\% \tag{18}$$

In a later paper he reported the oxidation of stilbene to benzil in 86% yield (45) (eq. 19).

$$C_6H_5CH{=}CHC_6H_5 \xrightarrow{SeO_2} C_6H_5COCOC_6H_5 \quad 86\% \tag{19}$$

Although the reaction appears to be restricted to nonalkyl substituted olefins, since one might expect allylic oxidation in such a case (see Section III), Riley (45) was able to oxidize propene to pyruvaldehyde (eq. 20).

$$CH_3CH{=}CH_2 + SeO_2 \longrightarrow CH_3COCHO \quad 19\% \tag{20}$$

Recently Olson (66) has reported the oxidation of ethylene in acetic acid with or without sodium acetate to give primarily bis-(β-acetoxy ethyl)selenide **79** in 35–40% yield, whereas in acetic acid containing aqueous hydrochloric acid a 45% yield of ethylene glycol diacetate **80** and a 50% yield of ethylene glycol monoacetate **81** were obtained. He has outlined a rather complicated

$$CH_3OCOCH_2CH_2SeCH_2CH_2OCOCH_3$$
79

$$CH_3OCOCH_2CH_2OCOCH_3$$
80

$$HOCH_2CH_2OCOCH_3$$
81

mechanism for the formation of **80** which proceeds through the nonclassical ion, **82**.

$$HO\overset{+}{Se}(OCOCH_3)OH + CH_2{=}CH_2 \longrightarrow (HO)_2Se(OCOCH_3)CH_2{-}CH_2^{+} \longrightarrow (HO)_2\overset{+}{Se}{-}CH_2CH_2OCOCH_3 \longrightarrow \mathbf{82} \longrightarrow \mathbf{80}$$

A reasonable mechanism, at least in systems containing water where the diketone is the product, is one which involves formation of a selenium(II) ester carbonium ion **83**. This would be rapidly converted to α-hydroxyacetaldehyde **84**, which would suffer further oxidation to the product.

The oxidation of acetylene (45) also gives oxalyl aldehyde (eq. 21). The most likely mechanism would appear to be hydration to acetaldehyde, followed by normal oxidation to the dione.

$$HC{\equiv}CH \xrightarrow[40\text{–}50°]{SeO_2} HCOCHO \tag{21}$$

$$H_2C{=}CH_2 + O{=}Se{=}O \longrightarrow \overset{+}{C}H_2CH_2O\overline{Se}{=}O \ (\mathbf{83}) \xrightarrow{H_2O} HOCH_2CH(H){-}O{-}Se{=}O \longrightarrow HOCH_2\overset{O}{\overset{\|}{C}}H \ (\mathbf{84}) \xrightarrow{SeO_2} H\overset{OO}{\overset{\|\|}{CC}}H$$

Acknowledgment

Helpful discussion and assistance in calculations of kinetic processes in Section II were kindly provided by Dr. Arnold Factor of this Laboratory.

* NOTE ADDED IN PROOF: An excellent Chapter on Selenium dioxide oxidations has recently been published in which many experimental suggestions are given (67).

References

1. H. L. Riley, J. F. Morley, and N. A. C. Friend, *J. Chem. Soc.*, **1932**, 1875.
2. N. Rabjohn, *Organic Reactions*, Vol. 5, Wiley, New York, 1949, ch. 8.
3. G. R. Waitkins and C. W. Clark, *Chem. Rev.*, **36**, 235 (1945).
4. W. M. Latimer, *Oxidation Potentials*, 2nd ed., Prentice-Hall, Englewood, N.J., 1952, ch. 6.
5. H. A. Riley and A. R. Gray, *Organic Syntheses*, Coll. Vol. 2, Wiley, New York, 1943, p. 509.
6. A. R. Ronzio and T. D. Waugh, *Organic Syntheses*, Coll. Vol. 3, Wiley, New York, 1955, p. 438.
7. C. C. Hach, C. V. Banks, and H. Diehl, *Organic Syntheses*, Coll. Vol. 4, Wiley, New York, 1963, p. 229.
8. R. Hanna and G. Ourisson, *Bull. Soc. Chim. France*, **1961**, 1945.
9. H. H. Hatt, A. Pilgrim, and W. J. Hurran, *J. Chem. Soc.*, **1936**, 93.
10. N. M. Mel'nikov and M. S. Rokitskaya, *J. Gen. Chem. (USSR)*, **7**, 1532 (1937); *Chem. Abstr.*, **31**, 8502 (1937).
11. N. M. Mel'nikov and M. S. Rokitskaya, *J. Gen. Chem. (USSR)*, **7**, 2728 (1937); *Chem. Abstr.*, **32**, 2903 (1938).
12. N. M. Mel'nikov and M. S. Rokitskaya, *J. Gen. Chem. (USSR)*, **8**, 1369 (1938); *Chem. Abstr.*, **33**, 4194 (1939).
13. N. M. Mel'nikov and M. S. Rokitskaya, *J. Gen. Chem. (USSR)*, **9**, 1158 (1939); *Chem. Abstr.*, **34**, 1233 (1940).
14. N. M. Mel'nikov and M. S. Rokitskaya, *J. Gen. Chem. (USSR)*, **9**, 1808 (1939); *Chem. Abstr.*, **34**, 3676 (1940).
15. N. M. Mel'nikov and M. S. Rokitskaya, *J. Gen. Chem. (USSR)*, **10**, 1439 (1940); *Chem. Abstr.*, **35**, 2400 (1941).
16. N. M. Mel'nikov and M. S. Rokitskaya, *J. Gen. Chem. (USSR)*, **10**, 1713 (1940); *Chem. Abstr.*, **35**, 3226 (1941).
17. F. R. Duke, *J. Am. Chem. Soc.*, **70**, 419 (1948).
18. E. J. Corey and J. P. Schaefer, *J. Am. Chem. Soc.*, **82**, 918 (1960).
19. J. P. Schaefer, *J. Am. Chem. Soc.*, **84**, 717 (1962).
20. P. A. Best, J. S. Littler, and W. A. Waters, *J. Chem. Soc.*, **1962**, 822.
21. K. B. Wiberg, *Oxidation in Organic Chemistry*, Part A, Academic Press, New York, 1965, ch. II, p. 179.
22. J. Roček and Sr. A. Riehl, *J. Am. Chem. Soc.*, **89**, 6691 (1967).
23. W. A. Waters, *Mechanism of Oxidation of Organic Compounds*, Wiley, New York, 1964, pp. 94, 95.
24. A. Fischer, J. Packer, and J. Vaughn, *J. Chem. Soc.*, **1963**, 226.
25. H. H. Jaffé, *Chem. Rev.*, **53**, 191 (1953).
26. J. E. White, Ph.D. dissertation, University of Arizona, Tucson, Arizona, 1967.
27. E. Schwenk and E. Borgwardt, *Ber.*, **65**, 1601 (1932).
28. K. Alder and G. Stern, *Ann.*, **504**, 205 (1933).
29. A. Guillemonat, *Compt. Rend.*, **200**, 1416 (1935); **201**, 904 (1935); **205**, 67 (1937); **206**, 1126 (1938).
30. A. Guillemonat, *Ann. Chim. (Paris)*, **11**, 143 (1939).
31. K. B. Wiberg and S. D. Nielsen, *J. Org. Chem.*, **29**, 3353 (1964).
32. M. Mousseron and R. Jaquier, *Bull. Soc. Chim. France*, **1952**, 467.
33. C. H. Nelson, Ph.D. thesis, Clark University, Worcester, Mass., 1966.
34. G. Dupont, W. Zacharewicz, and R. Dulou, *Compt. Rend.*, **198**, 1699 (1934); T. Matsuura and K. Fujita, *J. Sci. Hiroshima Univ., Ser. A*, **15**, 277 (1951); *Chem. Abstr.*, **48**, 3932 (1954).
35. J. P. Schaefer and B. Horvath, *Tetrahedron Letters* (30), **1964**, 2023.

36. J. P. Schaefer, B. Horvath, and H. P. Klein, *J. Org. Chem.*, **33**, 2647 (1968).
37. T. Suga, M. Sugimoto, and T. Matsuura, *Bull. Chem. Soc. Japan*, **36**, 1363 (1963).
38. J. J. Postowsky and B. P. Lugowkin, *Ber.*, **68**, 852 (1935).
39. G. M. Badger, *J. Chem. Soc.*, **1941**, 535.
40. A. Burger and L. R. Modlin, *J. Am. Chem. Soc.*, **62**, 1079 (1940).
41. H. Henze, *Ber.*, **67**, 750 (1934); W. Borsche and H. Hartmann, *Ber.*, **73**, 839 (1940).
42. H. S. Mosher, in *Heterocyclic Compounds*, Vol. 1, R. C. Elderfield, Ed., Wiley, New York, 1950, ch. 8.
43. R. C. Elderfield, in *Heterocyclic Compounds*, Vol. 4, R. C. Elderfield, Ed., Wiley, New York, 1953, ch. 1.
44. D. Jerchel, J. Heider, and H. Wagner, *Ann.*, **613**, 153 (1958).
45. S. Astin, A. C. C. Newman, and H. L. Riley, *J. Chem. Soc.*, **1933**, 391.
46. K. F. Armstrong and R. Robinson, *J. Chem. Soc.*, **1934**, 1650.
47. S. Astin, L. de V. Moulds, and H. L. Riley, *J. Chem. Soc.*, **1935**, 901.
48. E. Schwenk and E. Stahl, *Arch. Biochem.*, **14**, 125 (1947).
49. C. Meystre, H. Frey, W. Voser, and A. Wettstein, *Helv. Chim. Acta*, **39**, 734 (1956); S. A. Szpilfogel, T. A. P. Posthumus, M. S. deWinter, and D. A. vanDorp, *Rec. Trav. Chim.*, **75**, 475 (1956).
50. R. Owyang, in *Steroid Reactions*, C. Djerassi, Ed., Holden-Day, San Francisco, 1963, ch. 5.
51. J. P. Schaefer, *J. Am. Chem. Soc.*, **84**, 713 (1962).
52. J. C. Banerji, D. H. R. Barton, and R. C. Cookson, *J. Chem. Soc.*, **1957**, 5041.
53. C. S. Barnes and D. H. R. Barton, *J. Chem. Soc.*, **1953**, 1419.
54. G. Langbein, *J. Prakt. Chem.*, **18**, 244 (1962).
55. R. A. Jerussi and D. Speyer, *J. Org. Chem.*, **31**, 3199 (1966).
56. E. J. Corey and R. A. Sneen, *J. Am. Chem. Soc.*, **78**, 6269 (1956).
57. H. J. Ringold and A. Turner, *Chem. Ind. (London)*, **1962**, 211.
58. M. Hayano, H. J. Ringold, V. Stefanovic, M. Gut, and R. I. Dorfman, *Biochem. Biophys. Res. Commun.*, **4**, 454 (1961).
59. H. J. Ringold, M. Hayano, and V. Stefanovic, *J. Biol. Chem.*, **238**, 1960 (1963); R. Jerussi and H. J. Ringold, *Biochemistry*, **4**, 2113 (1965).
60. H. J. Ringold, G. Rosenkranz, and F. Sondheimer, *J. Org. Chem.*, **21**, 239 (1956); B. F. McKenzie, V. R. Mattox, L. L. Engel, and E. C. Kendall, *J. Biol. Chem.*, **173**, 271 (1948).
61. F. Weygand, K. G. Kinkel, and D. Tietjen, *Ber.*, **83**, 394 (1950).
62. H. P. Kaufmann and D. B. Spannuth, *Ber.*, **91**, 2127 (1958).
63. G. Stagno d'Alcontres and G. Lo Vecchio, *Gazz. Chim. Ital.*, **90**, 337 (1960).
64. D. Jerchel and H. E. Heck, *Ann.*, **613**, 180 (1958).
65. H. L. Riley and N. A. C. Friend, *J. Chem. Soc.*, **1932**, 2342.
66. P. H. Olson, *Tetrahedron Letters* (19), **1966**, 2053.
67. E. N. Trachtenberg, in *Oxidation*, Vol. 1, R. L. Augustine, Ed., Dekker, New York, 1969. ch. 3.

Triazoline Decomposition

PETER SCHEINER *

Mobil Research and Development Corporation, Princeton, New Jersey

* Present address: York College of the City University, 158–11 Jewell Ave, Flushing, N.Y.

I. INTRODUCTION

A. General

Since their discovery in 1912 (1), the decomposition of Δ^2-1,2,3-triazolines (hereafter triazolines) has attracted the attention of organic chemists. In part, interest has stemmed from the facile observation of decomposition; triazolines expel nitrogen under rather mild conditions. The synthetic potential of these reactions was recognized in the earliest studies (1), but useful procedures have only recently emerged. By means of triazoline intermediates olefins may be selectively oxidized to azomethine, carbonyl, or aziridine compounds. Paralleling synthetic developments, mechanistic information has been obtained. To an appreciable extent, this information allows a rational interpretation of the diverse decomposition paths. However, several important mechanistic questions remain unresolved.

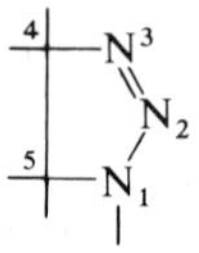

Unlike *vicinal*-triazoles, the triazoline system is nonaromatic. Triazolines might therefore be expected to exhibit chemical behavior characteristic of aliphatic azo compounds. The additional, basic ring-nitrogen (N_1) gives rise to reaction paths not encountered in simpler systems. Both homolytic and heterolytic fragmentations occur. Thus, by virtue of their constitution, triazolines provide an interesting model for examining the chemistry of the azo linkage. Three types of triazoline decomposition have been described: thermal, acid and base catalyzed, and photodecomposition. This chapter surveys current knowledge of these reactions and attempts to relate them to broader aspects of reactivity.

B. Preparation of Triazolines

Numerous stable triazolines have been characterized, and many others have been generated *in situ*. Four preparative reactions are known, but only the azide–olefin and the diazoalkane–imine cycloadditions are generally applicable.

1. Azide–Olefin Addition (2)

Organic azides are 1,3-dipoles (2a), and as such they undergo cycloaddition with olefins. Most triazolines have been obtained by this route. In

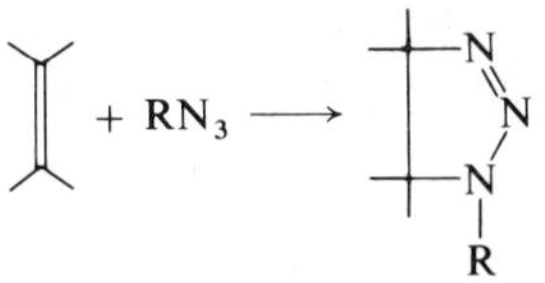

$$\text{>C=C<} + RN_3 \longrightarrow \text{triazoline (N-R)}$$

contrast to simple olefins, Alder and Stein (3) noted a striking rate enhancement for addition to strained double bonds, mainly norbornene derivatives. Other strained olefins (e.g., *trans*-cycloalkenes) display similar high reactivity (4); the reaction has therefore been utilized for derivatization (5) and quantitative analysis (6) of such compounds. Although strain unquestionably promotes addition, it is not a necessary condition for triazoline formation. Recent work has shown that adducts may be obtained with unstrained linear olefins, conjugated dienes (7), and simple fluorinated olefins (8). Steric (9) and electronic factors dramatically affect the rate of addition. Exceptional reactivity has been reported for enamines (10) and vinyl ethers (11). Although aryl and benzyl azides have most frequently been employed, the reaction is not limited with regard to the azide component. Satisfactory additions have been described using phosphoryl azides (12), trimethylsilyl azide, α-styryl azide (13), and low molecular weight alkyl azides (14). The scope of azide–olefin addition is clearly quite broad.

Mechanistic studies with norbornene and substituted phenyl azides (15,16) show that the reaction occurs in a concerted manner (*cis* addition). Consequently the olefin geometry is preserved in the triazoline adduct (17,18). Furthermore, the addition to unsymmetrically substituted olefins exhibits a marked orientational specificity. The direction of addition is controlled by electronic rather than steric factors. Thus aryl azide-styrene adducts (**1**) are exclusively 1,5-diaryltriazolines (19); the same orientation is found in additions to enamines (10), vinyl ethers (11) (**2**) and simple olefins (7) (**3**). The related cycloaddition of azides and acetylenes, to give 1,2,3-triazoles, does not show a comparable degree of orientational specificity (20).

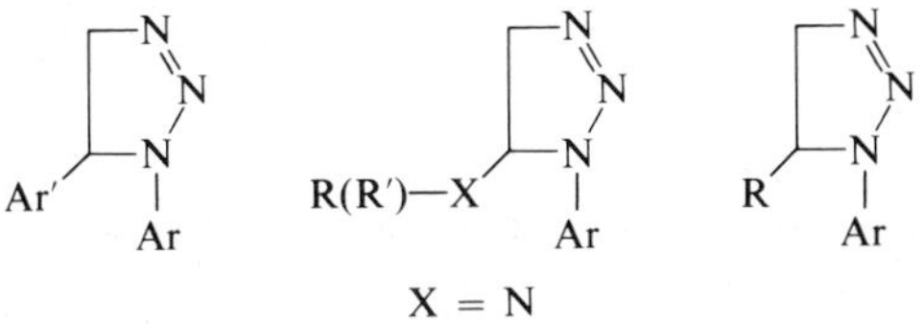

Structures: **1** (Ar′ at C-5, N-1–Ar); **2** (R(R′)—X at C-5, N-1–Ar; X = N, X = O); **3** (R at C-5, N-1–Ar)

1 **2** **3**

For predictive purposes azide addition to electron-rich double bonds may be considered to occur in a Markownikoff sense. Electrophilic attack by the terminal azido nitrogen to generate the more stable carbonium ion correctly predicts the course of reaction. (This useful rule must not be viewed as a mechanistic description.) Electron-deficient double bonds, such as acrylonitrile (21), add azides in the opposite sense, giving 1,4-disubstituted triazolines.

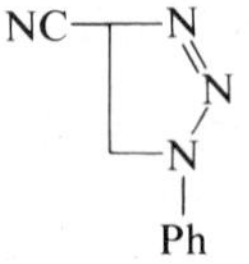

NC, N, N, N, Ph

2. *Diazoalkane–Imine Addition*

A second 1,3-dipolar cycloaddition, diazomethane and anils, has been successfully used to prepare 1,5-diaryltriazolines. Discovered by Mustafa (22), the correct orientation of the addition was established in later work (19). Rather than a Δ'-1,2,4-triazoline, the product from diazomethane and

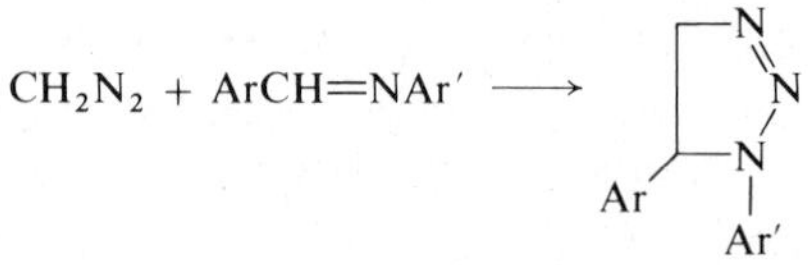

$$CH_2N_2 + ArCH{=}NAr' \longrightarrow$$

N, N, N, Ar, Ar′

benzalaniline was identical to the Δ^2-1,2,3-triazoline obtained by phenyl azide–styrene cycloaddition. Kinetic studies have revealed marked substituent effects on the rate of addition (23), and the reaction is catalyzed by small amounts of water (24). Unfortunately few extensions of the reaction have as yet been reported. Preparation of the unusual triazoline **4** (25), however, indicates that extensions are indeed feasible.

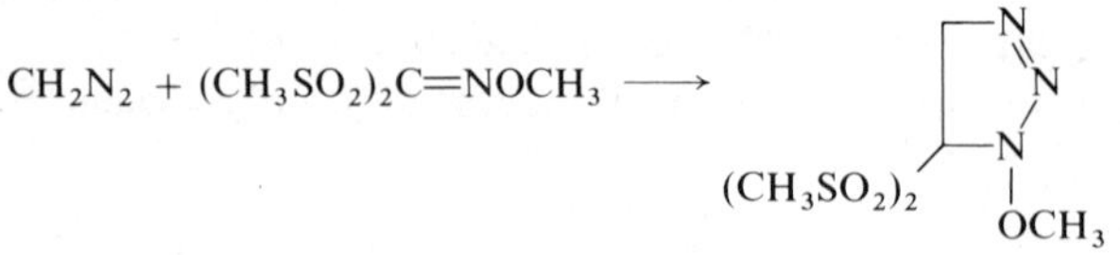

$$CH_2N_2 + (CH_3SO_2)_2C{=}NOCH_3 \longrightarrow$$

N, N, N, $(CH_3SO_2)_2$, OCH_3

4

3. *Arylazoaziridine Rearrangement and Hydrocyclization*

An iodide and thiocyanate catalyzed rearrangement of 1-arylazoaziridines to triazolines has been described by Heine and Tomalia (26). Formally analogous to vinyl cyclopropane–cyclopentene isomerization, the reaction more closely resembles other ionic ring expansions of aziridines bearing unsaturated substituents (27). Although the starting materials are hazardously

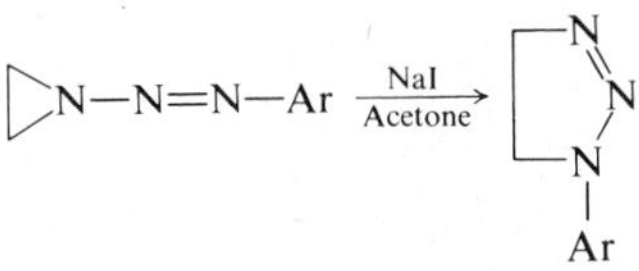

explosive, the rearrangement constitutes the only presently known route to 1-aryltriazolines unsubstituted at positions 4 and 5.

Catalytic hydrogenation of α-azidodiphenylacetonitrile gave a 5-triazolone derivative (**5**, 72%) rather than the anticipated hydrogenolysis product, α-aminodiphenylacetonitrile (28). The reaction was found to be quite general for disubstituted α-azidoacetic acid derivatives, and a number

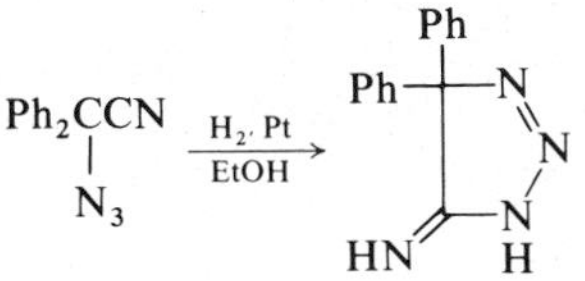

of authentic 5-triazolones were obtained in this way. It should be noted that the chemical literature abounds with incorrect "5-triazolone" structures (29). Quite early Curtius (30) recognized these compounds as aromatic 5-hydroxytriazoles, but to the present day various authors have persisted in misnaming them (31). The unfortunate effect of preserving incorrect, "traditional" tautomeric assignments has been pointed out by J. D. Watson (32).

II. THERMAL DECOMPOSITION

A. Thermal Stability

Triazolines as a class vary widely in thermal stability. Some, such as those resulting from addition of picryl azide (33), arylsulfonyl azides (34), and cyanogen azide (35) to olefins, have not been isolated. Reaction of the azide and olefin in these instances, even at low temperatures, is accompanied by nitrogen evolution. Evidence, both structural and kinetic (36–38), clearly shows that the nitrogen arises from decomposition of transient triazoline intermediates. Direct displacement of nitrogen on interaction with the olefin and unimolecular azide decomposition (nitrene formation) (39) have been considered and rejected. The rate of decomposition of these triazolines exceeds their rate of formation, thus precluding isolation.

1-Acyltriazolines are somewhat more stable. Although the benzoyl azide–norbornene adduct (**6a**) was not isolated from solution, its existence was demonstrated (16) by NMR spectroscopy. At 40°, **6a** decomposed slowly

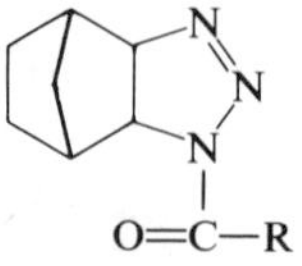

6 a R = Ph
b R = OC_2H_5, OCH_3

when compared to the corresponding 1-benzenesulfonyl compound. 1-Carbalkoxytriazolines (**6b**) have been isolated and characterized, but they too decompose on attempted purification (40,41).

In general, 1-aryl- and 1-alkyltriazolines are moderately stable. Protected from light, they may be stored indefinitely under refrigeration. All, however, decompose (nitrogen evolution) at elevated temperatures. Some years ago it was observed that 1-*p*-nitrophenyl triazolines are thermally less stable than the corresponding 1-phenyl compounds (42). Unpublished work (43) on the decomposition of triazolines (**7**) in phenetole (171°) agrees with this observation. Relative to phenyl (X = H), the following first-order rates

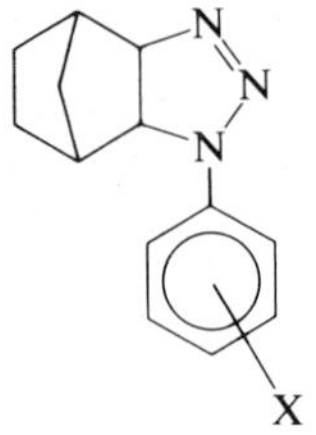

7

for nitrogen evolution were measured: *p*-NO_2, 6.8; *m*-NO_2, 3.8; *m*-Br, 2.2; *m*-CH_3, 0.9. The brief foregoing survey supports the contention that electron-withdrawing substituents on N_1 promote thermal instability.

B. Products

1. Early Investigations

The analogy between the thermal decompositions of triazolines and 1-pyrazolines (**8**) was first drawn by Wolff (1). Whereas pyrazolines were known to produce cyclopropanes by expulsion of nitrogen, the thermolysis of triazoline **9** reportedly gave 1,2-diphenylaziridine. (Unfortunately, experimental detail was not included in the report.) Later Alder and Stein

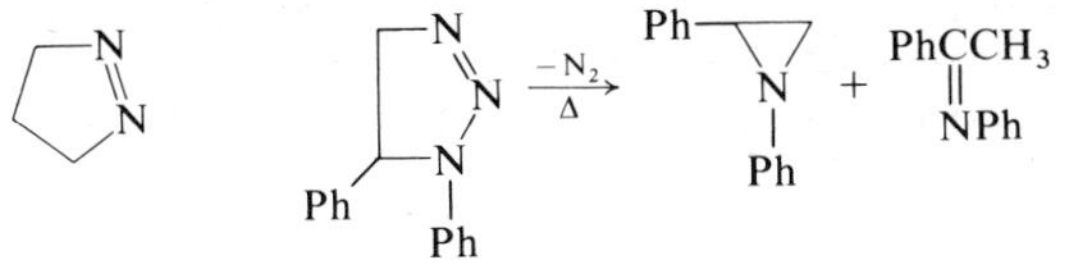

8 **9** **10** **11**

(44) suggested the formation of acetophenone anil (**11**) as well as **10**, but a more recent reinvestigation (23) found no evidence for the aziridine (or hydrolysis products derived from it), and another report has described only an 81% yield of the anil (45). Thus conflicting reports still surround this earliest example of triazoline pyrolysis. On the other hand, thermolysis of the related triazoline **12** in heptane solution afforded the corresponding aziridine (26). To some extent divergent results such as these probably reflect

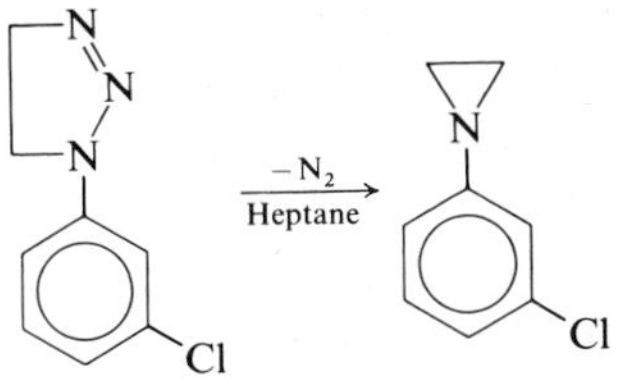

12

differing reaction conditions. Secondary reactions may also be anticipated, particularly with unstable aziridine products.

According to Chattaway and Parkes (46), aryl azide–toluquinone adducts decomposed to aziridines in high yield (eq. 1). But an entirely different thermolysis path was attributed (47) to the triazoline formed from naphthoquinone and phenyl azide (eq. 2). Here the product was formulated as an anil,

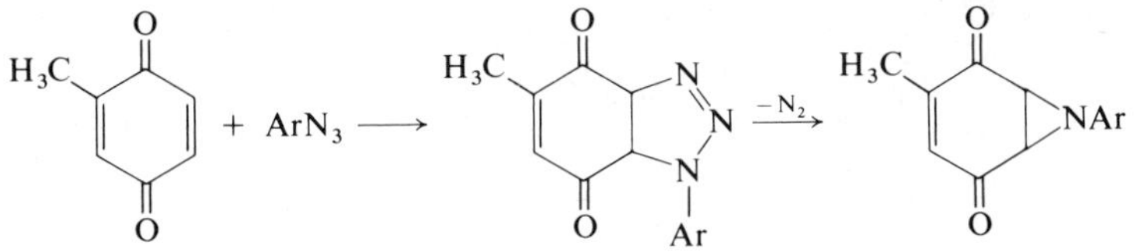

$$+ ArN_3 \longrightarrow \xrightarrow{-N_2} \quad (1)$$

$$+ PhN_3 \longrightarrow \xrightarrow{-N_2} -CH{=}NPh \quad (2)$$

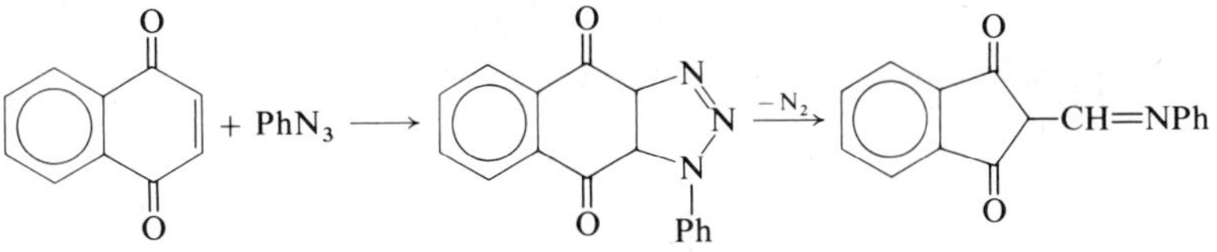

resulting from an unusual ring-contracting acyl migration. More recent evidence favors the former interpretation (42), although the naphthoquinone–phenyl azide reaction has not been reinvestigated.

A possible example of triazoline decomposition appears in the work of Curtius (48). The alleged aziridine product from carbamoyl azide and ethyl fumarate might have formed via a triazoline intermediate. It is perhaps more

$$C_2H_5O_2C\text{–CH=CH–}CO_2C_2H_5 + N_3CONH_2 \xrightarrow{-N_2} H_2NCON\langle(CO_2C_2H_5)_2 \qquad :\ddot{N}CONH_2\ (\mathbf{13})$$

likely that reaction occurred by addition of carboxamido nitrene (**13**) to the double bond, a reaction of the type investigated by Lwowski and associates (49).

2. *Structure and Products*

Until 1933 no clear picture of triazoline decomposition had emerged. In that year Alder and Stein (44) reported the results of an extensive study relating triazoline structure to the decomposition products. Two types of product were found, imines and isomeric aziridines. However, depending on triazoline structure, markedly different product distributions were obtained. Bicyclic triazolines (adducts of azide and bicycloalkenes) gave mixtures of aziridines and imines, with the former predominating. But only imine products (anils) were detected from triazolines derived from linear or cyclic

Ph (triazoline) $\xrightarrow[150^\circ]{-N_2}$ NPh (aziridine) + NPH (imine) 4 : 1

CH_3CO, H_3C, N–Ph triazoline $\xrightarrow[140^\circ]{-N_2}$ $CH_3C(=NPh)CH_2COCH_3$

${}_n(CH_2)$ triazoline, N–Ar $\xrightarrow[\Delta]{-N_2}$ ${}_n(CH_2)$C=NAr

$n = 3, 5, 6$

olefins. Although these experiments were conducted without solvent, similar results have been reported for thermolysis in dioxane. Aryl azide–indene adducts, for example, afforded anils of 1-indanone (50).

Without benefit of gas chromatography and spectroscopic techniques, minor isomeric products may easily escape detection. Thus recent studies suggest that Alder's generalization concerning cyclic and linear triazolines, that is, exclusive imine formation, is an oversimplification. Under rather different conditions (315°), pyrolysis of various triazolines produced appreciable amounts of aziridine (13). Furthermore, with azide–cycloalkene adducts the product distributions were sensitive to ring-size (Table 1).

TABLE 1
Pyrolysis of Cyclic Triazolines, 315°[a]

$_n(CH_2)$ triazoline (N-Ar) → n	$_n(CH_2)$ C=NAr (imine)	+ $_n(CH_2)$ aziridine NAr
3	79%	21%
5	46	54
6	45	55

[a] Ref. 13.

The important role played by geometrical factors is apparent from the data of Henery-Logan and Clark (51). In this work decomposition was effected by heating phenyl azide and excess olefin at moderate temperatures. The triazoline adducts were not isolated. Their presence was demonstrated kinetically, and they may also be inferred from related studies under similar conditions in which the triazolines were isolated. Contrasting with the high-temperature pyrolyses (Table 1), these reactions displayed marked selectivity for either aziridine or anil products (Table 2).

The results with *cis*- and *trans*-cyclooctene are particularly notable. Following earlier proposals (see below) the authors considered triazoline decomposition a heterolytic process; rupture of the N_1—N_2 bond produces a diazonium zwitterionic intermediate (**14**).* Formation of imine, then, most

* Intermediate **14** has also been termed a "diazonium betaine." See ref. 38.

TABLE 2
Decomposition of Phenyl Azide in Cyclic Olefins[a]

Olefin	Temp. (°C)	Product[b]	Yield (%)
	42	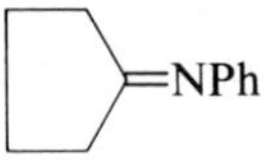 =NPh	75
	80	 NPh	79
	80	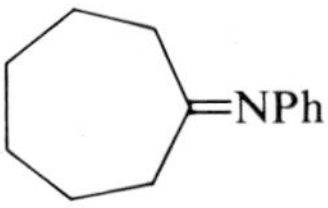 =NPh	80
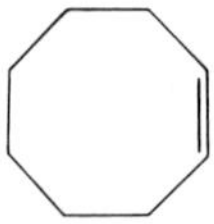	80	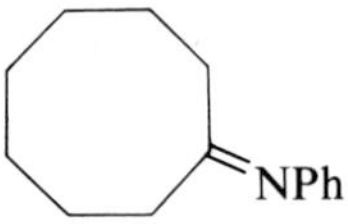 NPh	87
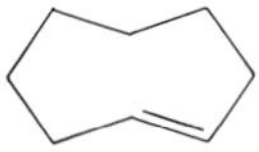	80	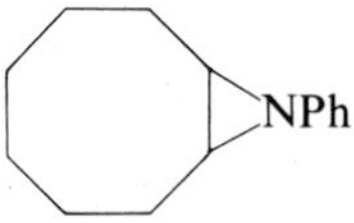 NPh	85

[a] Ref. 51.
[b] Other product(s) not characterized.

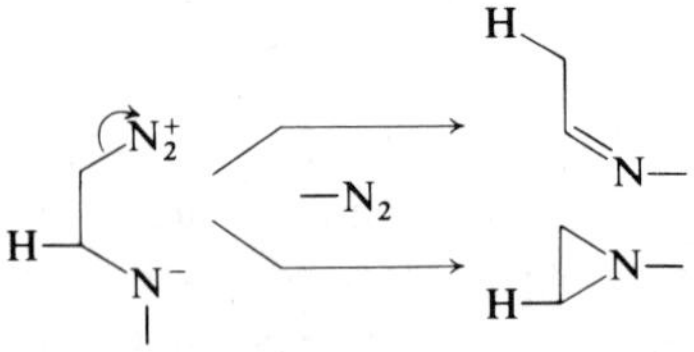

N_2^+ H— N^- —N_2 H N— H N—

14

probably occurs by a 1,2-hydride shift with involvement of the negative charge on nitrogen (N_1). In common with other nucleophilic rearrangements, a *trans*-orientation of the diazonium and hydride moieties would be preferred for migration, i.e., rear-side displacement. Using reasonable assumptions regarding ring conformations, the authors found that only the zwitterion from the *cis*-triazoline (*cis*-cyclooctene) may readily attain this orientation. On the other hand, *transoid* displacement of nitrogen by the amide anion to give aziridine is conformationally feasible for the *trans* isomer. Hence the observed products.

Hydride migration and aziridine ring-closure are therefore concerted with nitrogen expulsion. Were this not the case, either isomer would afford the same product. But the *necessity* for rear-side displacement is not entirely clear, as shown by work with bicyclic triazolines.

Bicyclic triazolines have received considerable attention. Three products were obtained from **15** (160°, no solvent) (16), and similar results are reported for the corresponding 1-*p*-bromophenyl compound decomposed in phenetole solution (52). The structure of the unidentified isomer from **15** is suggested

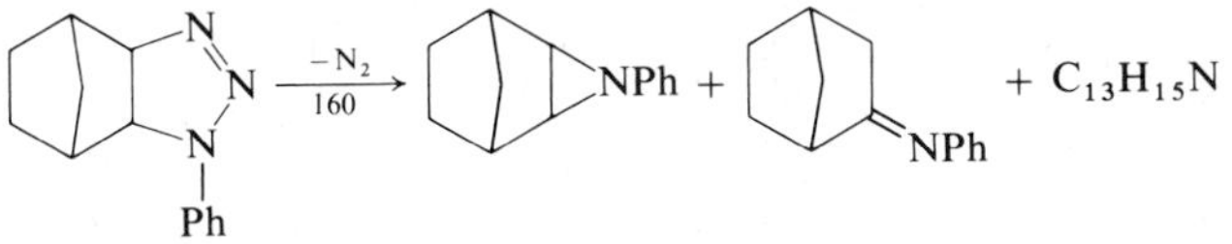

N N N Ph $\xrightarrow[160]{-N_2}$ NPh + NPh + $C_{13}H_{15}N$

15 5:2:1

by the identification of a Wagner-Meerwein rearrangement product from triazoline **16** (53). Although alkyl shifts frequently occur in the formation of imine product, the limited extent (5–12%) of Wagner-Meerwein rearrangement in the norbornyl system is somewhat surprising. Considerable skeletal rearrangement might be anticipated from the proposed diazonium intermediate. However, formation of rearranged product necessitates a formal 1,4-prototropic shift. Since this process may be difficult, or require an

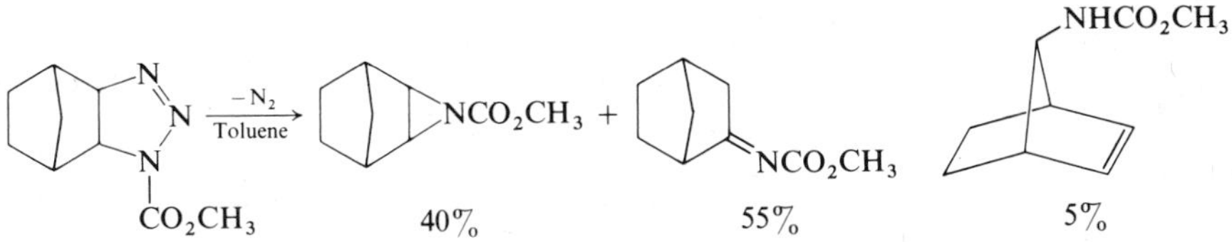

N N N CO_2CH_3 $\xrightarrow[Toluene]{-N_2}$ NCO_2CH_3 + NCO_2CH_3 $NHCO_2CH_3$

16 40% 55% 5%

$\bar{N}R$ N_2^+ $\xrightarrow{-N_2}$ $\bar{N}R$ + → NR + NR

NHR = NHR

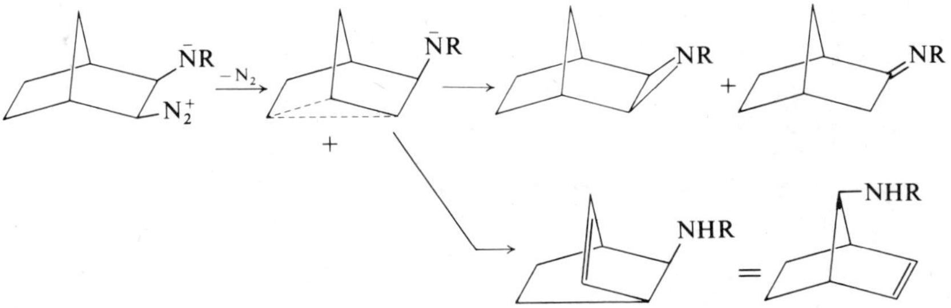

additional basic species, it is not unreasonable that the rates of aziridine and imine formation are considerably faster. In the norbornyl system rear-side displacement of nitrogen would most easily be effected by the *transoid* $C_6—C_1$ bond; products may therefore derive from a short-lived, nitrogen-deficient intermediate.

Arylsulfonyl azides (16,34,38) and picryl azide (54) react directly with bicyclic olefins to produce aziridines and, to a lesser extent, imines. An unusual stereochemical result, the formation of *endo*-aziridine, has been described by Zalkow and co-workers (55) for certain of these reactions. With benzoyl azide and norbornene, the initial product is the expected aziridine (16,56). The latter undergoes facile ring-expansion, thereby accounting for earlier reports of oxazoline formation during decomposition (57).

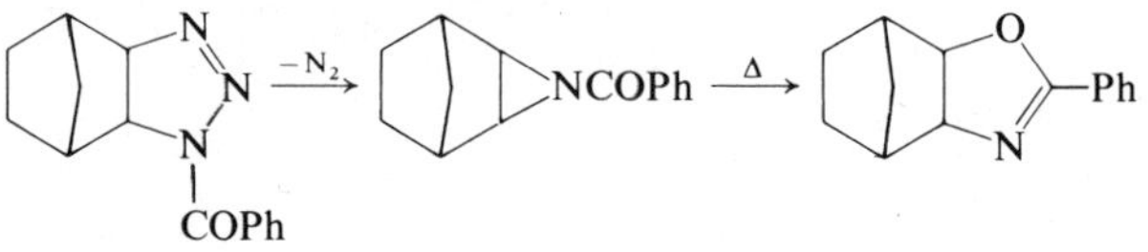

N N N COPh $\xrightarrow{-N_2}$ NCOPh $\xrightarrow{\Delta}$ O N Ph

Monotriazolines formed from norbornadiene and azides thermally decompose by two distinctive paths. With reactive azides (58,59) the initially formed aziridines (**17**) are unstable. In a manner similar to norbornadiene

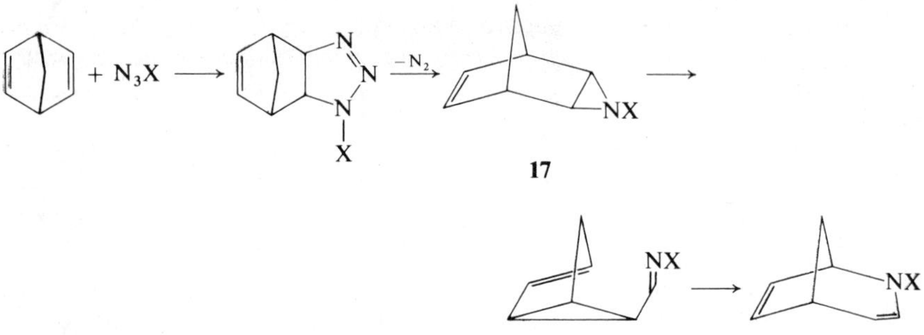

+ N_3X ⟶ N N N X $\xrightarrow{-N_2}$ NX **17** ⟶ NX ⟶ NX **18**

X = SO_2Ph, CN

oxide (60) they undergo extensive electrocyclic rearrangement leading to the observed products (**18**). Monoadducts of phenyl azide, on the other hand, fragment without loss of nitrogen. Retro Diels-Alder reactions have been reported (16,61).

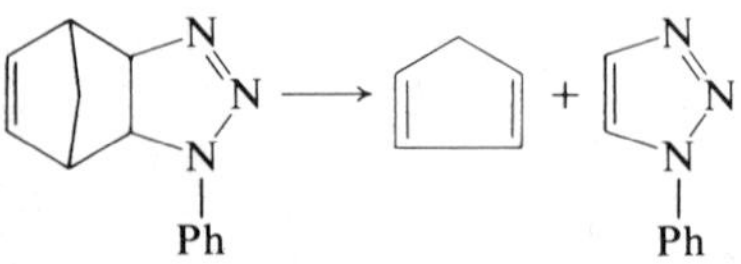

N N N Ph ⟶ + N N N Ph

Regrettably, a systematic investigation of the effect of reaction variables (temperature, solvent polarity) has not as yet appeared. From Logothetis' results (62) it is indeed clear that the imine/aziridine ratio is quite sensitive to reaction conditions. Fused-ring triazolines (**19**) were prepared or generated *in situ* by intramolecular cyclization of olefinic azides. Decomposition at various temperatures, in several solvents and the vapor phase, gave rise to

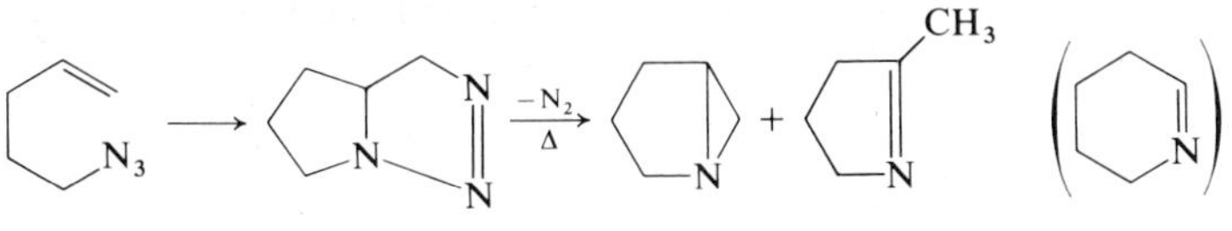

19

widely differing product distributions, although the azomethine compounds predominated in all cases. Regarding solvent polarity, a notable kinetic effect was observed. First-order rates for nitrogen evolution increased 10- and 20-fold when nitromethane and aqueous diglyme were substituted for toluene as solvents. In light of this solvent effect, and the relative migratory aptitudes of groups in imine formation, the polar mechanism involving a diazonium zwitterionic intermediate was proposed.

In general, triazolines thermally decompose to aziridines and imines. Their behavior in this respect is analogous to 1-pyrazolines; pyrolysis of these compounds gives cyclopropanes and isomeric olefins (63). As Alder correctly pointed out, bicyclic triazolines favor aziridine formation (44,64), but mixed products are the rule. Certain triazolines, however, decompose exclusively to aziridines. Rather than the previously reported anils (21), triazolines **20** were found to produce only the corresponding aziridines (65). Similar thermolytic behavior was attributed to triazolines of type **21** (66),

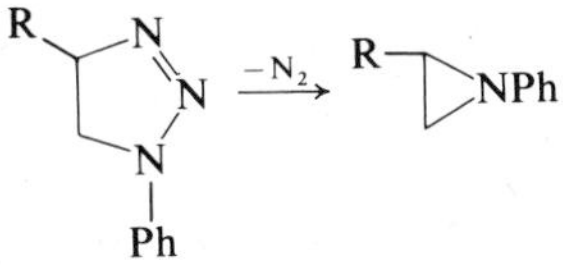

20a R = CN
b R = CO_2CH_3

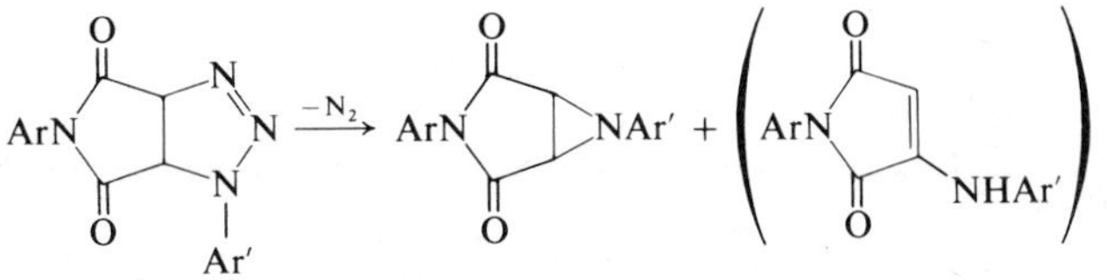

21 **22**

although the major aziridine product was subsequently shown to be contaminated with trace amounts of enamine **22** (67), an anil tautomer. It is noteworthy that in these highly selective decompositions, and perhaps those described by Chattaway and Parkes (46) (see above), the triazoline is substituted at the 4-position with an acyl or nitrile group. Kinetic results (Section II-C-2) indicate the operation of a "special" decomposition path in these instances.

3. *Cyanogen Azide*

A useful new technique for oxidizing olefins has been developed by Marsh and Hermes (35). Cyanogen azide (**23**), conveniently generated by eq. 3, reacts rapidly with olefins at room temperature and below. Mechanistic studies show that the reaction proceeds through an unstable triazoline intermediate. With most linear and cyclic olefins the predominant and sometimes

$$NaN_3 + ClCN \longrightarrow NaCl + \underset{\mathbf{23}}{N_3CN} \tag{3}$$

exclusive products are alkylidene cyanimides (imines); lesser amounts of 1-cyanoaziridines may also form. The cyanimides are readily hydrolyzed to carbonyls (35,39), thus providing a convenient, rapid procedure for the oxidation of olefins to ketones. The sequence is illustrated for cyclopentene.

$$\text{cyclopentene} + N_3CN \longrightarrow \left[\text{triazoline (N–CN)}\right] \xrightarrow{-N_2} \text{cyclopentylidene}{=}NCN \xrightarrow[H_2O]{Ag^+} \text{cyclopentanone}{=}O + Ag_2NCN$$

As a consequence of the orientational specificity of azide addition the oxidation is selective; the more highly substituted olefinic position is oxidized. Terminal olefins give methyl ketones. With olefins such as isobutene the intermediate imine is **24** (methyl migration) rather than **25**, affording

$$(\overset{1}{C}H_3)(CH_3)\overset{2}{C}{=}\overset{3}{C}H_2 + N_3CN \xrightarrow{-N_2} \underset{\mathbf{24}}{\overset{1}{C}H_3\overset{2}{C}(=NCN)\overset{3}{C}H_2CH_3} \qquad \underset{\mathbf{25}}{\overset{1}{C}H_3\overset{2}{C}H(CH_3)\overset{3}{C}H{=}NCN}$$

methyl ethyl ketone after hydrolysis. Thus cyanogen azide follows the orientation principle observed generally for the addition of azides to unsymmetrical olefins (Section I-B-1). Picryl azide reacts in an analogous manner (33).

4. 5-Amino- and 5-Alkoxytriazolines

Despite the presumably incorrect structures proposed, 5-aminotriazolines were first prepared by Alder (44,68). Cycloalkenes heated with excess phenyl azide gave the compounds, termed by Alder "N_4-compounds." Later, Fusco and co-workers, and others (10,11) demonstrated the exceptional reactivity of enamines to azide addition. Hence the "N_4-compounds" probably were formed as shown for cyclooctene. Other Shiff bases

+ PhN_3 → ; $-N_2$; NPh ; PhNH ; Ph ; PhN_3 ; NHPh

capable of enamine tautomerism have been employed (69,70), but most 5-aminotriazolines have now been obtained by direct addition to enamine compounds. Similarly, 5-alkoxytriazolines result from azide addition to vinyl ethers (11,17).

Depending on structure and reaction conditions, 5-aminotriazolines thermally decompose in various ways. Elimination of amine to give 1,2,3-triazoles is characteristic of such triazolines bearing 4-acyl and 4-sulfonyl groups (71). The same reaction may be induced by acid catalysis. (In the absence of 5-amino or 5-alkoxy substituents the triazoline ring is resistant to aromatization. Triazoles have been obtained from 4-chlorosulfonyl triazolines (72) and by the action of strong oxidants (47,65a).) Normal decomposition, that is, nitrogen expulsion, is observed with other 5-aminotriazolines. In tetralin solution, for example, amidines are formed in high yield by hydrogen migration, while alkyl migration with ring-fused triazolines gives ring-contracted amidine products (73). *N*-Phosphorylated amidines have also been prepared (74). An additional fragmentation mode is discussed below (Section II-B-5).

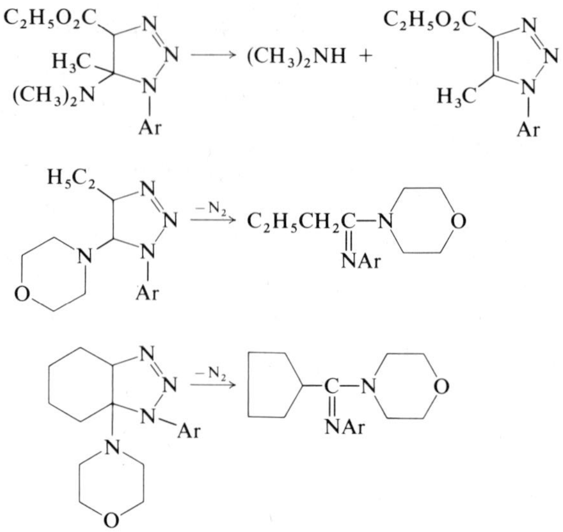

Thermolysis of 5-alkoxytriazolines is similar to that of the 5-amino compounds. Both aromatization (alcohol elimination) and amidate formation have been observed (11,17,37,75,76). Decomposition with rearrangement leading to amidine and amidate products is typically encountered in triazoline thermolysis, that is, imine formation, but unlike other triazolines aziridines have not been detected from these compounds. In one instance it has been shown that the corresponding aziridine is stable under the decomposition conditions (Section IV-B).

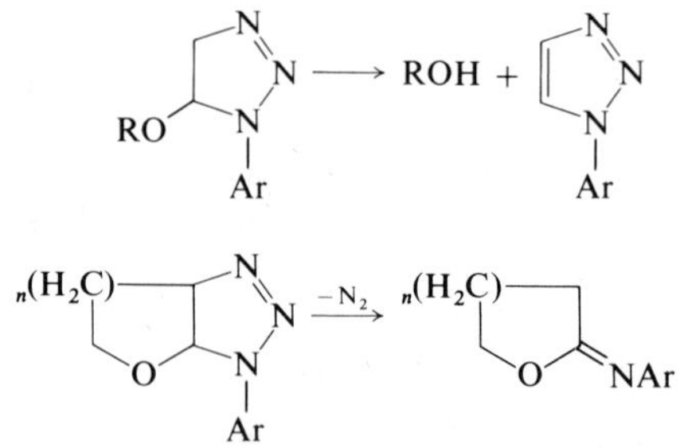

5. *Diazoalkane Formation*

The usual course of triazoline thermolysis involves expulsion of nitrogen; the remaining fragment is converted to imine and aziridine products (path *a*). Fusco and associates have called attention to a second type of fragmentation (77). Cleavage by path *b* produces diazoalkane and imine. Although this decomposition was first noted in the reaction of arylsulfonyl azides and linear enamines, it appears to be of wider scope (70).

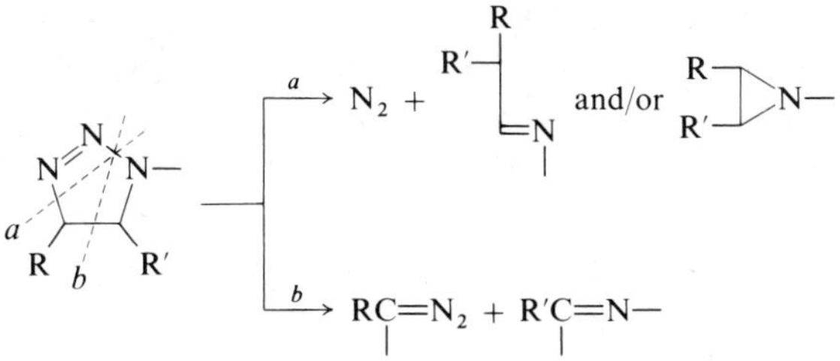

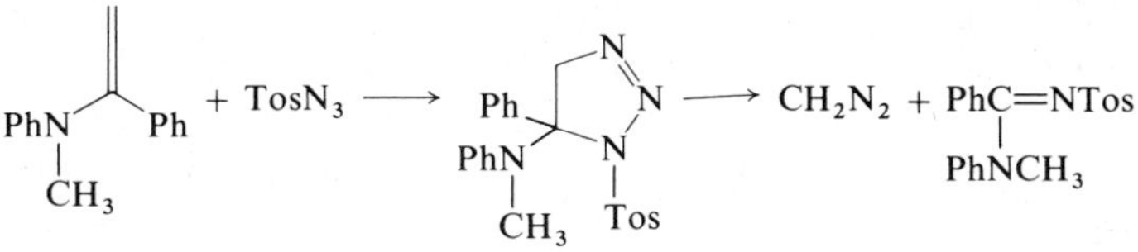

An interesting example occurs in the pyrolysis of fluorinated triazoline **26** (8). Decomposition (250–325°) over glass beads gave aziridine, but difluorodiazomethane was obtained on pyrolysis over a nickel surface. Depending

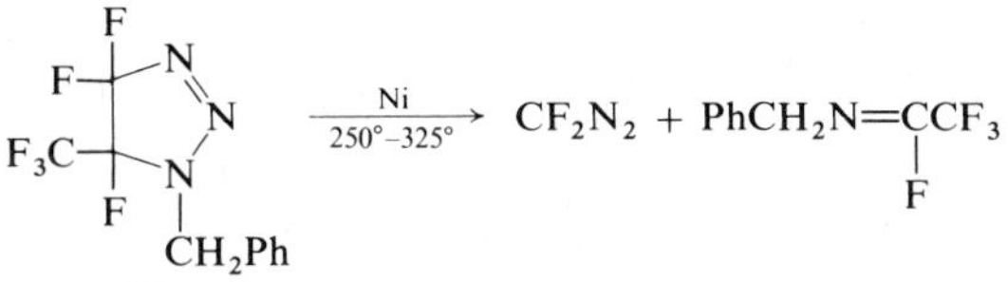

$$\mathbf{26} \xrightarrow[250^\circ\text{–}325^\circ]{\text{Ni}} CF_2N_2 + PhCH_2N{=}C(F)CF_3$$

on conditions, the reaction may prove to be quite general. Baldwin et al. have invoked diazoalkane formation to account for an unusual result (see Section II-C-1), and the unexplained formation of *endo*-aziridines (55) (e.g., **27**) might well involve this path.

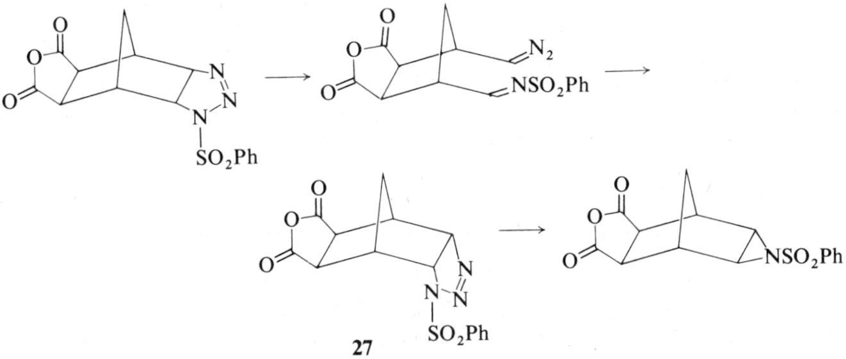

27

The reaction has been used as a preparative method for α-diazocarbonyl compounds (78). Quite probably it is also operative in the numerous reactions of active methylene compounds with arylsulfonyl azides. Diazoalkane formation via triazoline decomposition may thus be considered a "diazo transfer" process; Regitz (79) has cited many examples. Formally the reaction is a cycloelimination, the reverse of diazoalkane-imine addition (Section I-B-2). As yet, detailed mechanistic information is not available.

C. Mechanisms

A priori, four paths appear reasonable for the usual course of triazoline thermal decomposition. Concerted cycloelimination with no transient intermediate (eq. 4) accounts for aziridine, but is unsatisfactory with regard to the imine product. (In several instances imine was shown to be a primary product; it did not arise by rearrangement of aziridine (13,37).) Furthermore, solvent

(4)

N_2 + (5)

N_2 + (6)

N_2^+ N^- (7)

and substituent effects argue against this path. On similar grounds homolytic cleavage (eq. 5) does not appear *typical* of triazoline thermolysis. Available data support a polar mechanism. Zwitterionic intermediates, either a 1,3-dipole (eq. 6) or a diazonium zwitterion (eq. 7), or both, are implicated.

1. Trapping Experiments

First proposals tentatively supported decomposition through a 1,3-dipolar intermediate (57, cf. ref. 2a), but this supposition was based on an incorrect structure for the product obtained from triazoline **28** decomposed in the presence of phenyl isothiocyanate. Originally formulated as a 1,3-cycloadduct, the product was later found to be **29** (80), a structure not

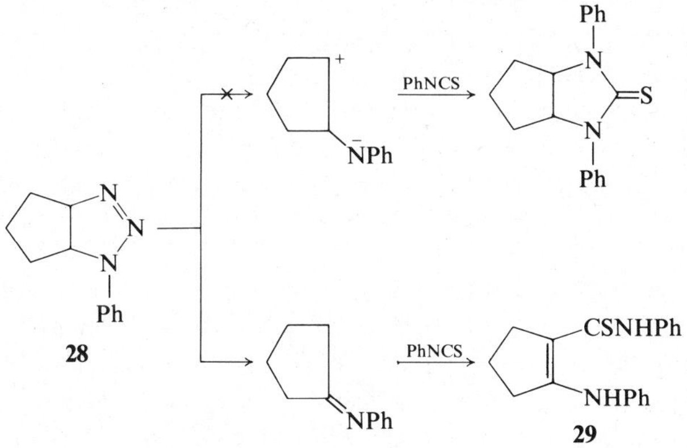

28

29

derived from the proposed 1,3-dipolar intermediate. Indeed, the normal decomposition product, cyclopentanone anil, also gave **29** when heated with phenyl isothiocyanate (81).

In related work Baldwin and co-workers (80) reinvestigated the decomposition of the phenyl azide–norbornene adduct in phenyl isocyanate. The major product was the tricyclic lactam **30** and not a previously proposed structure formed by trapping the 1,3-dipolar intermediate. Baldwin's interpretation is illustrated. According to this view, decomposition via a diazonium zwitterion produces a δ-diazoanil; the latter reacts in a plausible

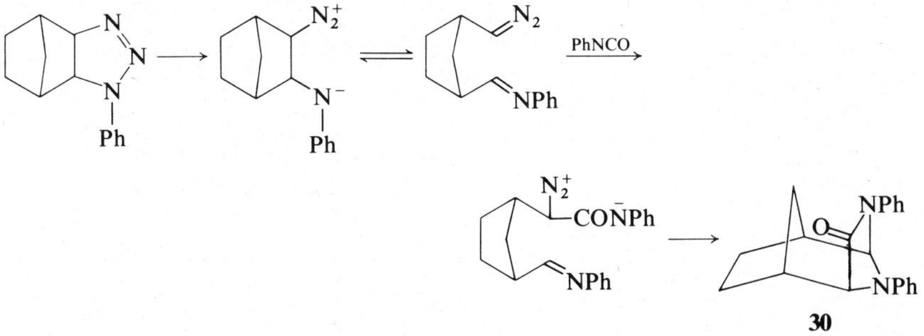

30

manner (82) with phenyl isocyanate. The involvement of a zwitterionic precursor to the diazoanil (diazoalkane formation) is open to question. It has also been pointed out (81) that the reaction may be more complex since decomposition is induced by phenyl isocyanate. Nevertheless, these results provide no evidence for a 1,3-dipolar intermediate. If formed at all, the lifetime of such a species has been too short for bimolecular, trapping reactions.

2. *Kinetic Experiments*

The weight of evidence supports N_1–N_2 heterolysis leading to an unstable diazonium zwitterionic intermediate. Once formed, the intermediate achieves stabilization by paths anticipated for an alkyldiazonium ion. Nucleophilic displacement of nitrogen by the adjacent amide anion gives aziridine; displacement by a migrating group with involvement of the negative charge on nitrogen produces imine. Unassisted loss of nitrogen to a 1,3-dipole and ring-closure back to the triazoline may also occur, but evidence for these paths is lacking.

The recognized instability of triazolines bearing electron-withdrawing substituents at N_1 may thus be attributed to stabilization of developing negative charge in the transition state (**31**). (Arguments based on ground-state stability (2a,16) are less persuasive.) First-order rates for nitrogen evolution have been reported in several instances (41,62,65b). Heterolysis is

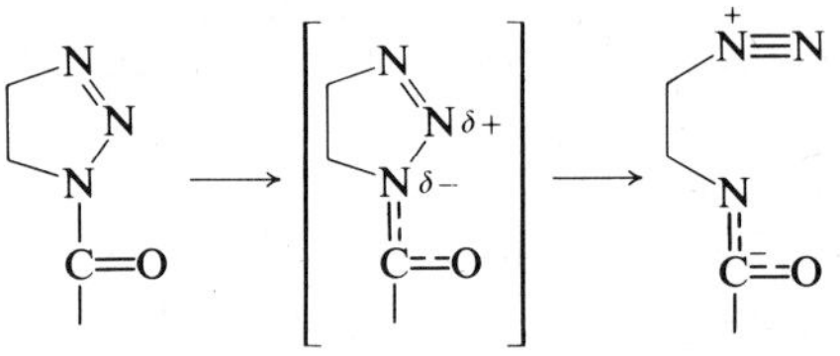

31

also indicated by the solvent effect; increased polarity enhances the rate of decomposition (41,62). Regarding imine formation, the migratory aptitudes of various groups parallel those found in alkyldiazonium ion rearrangements. The order is somewhat similar to other reactions that involve migration to electron-deficient centers; for example, pinacol–pinacolone rearrangement, Bayer-Villiger oxidation (62).

Direct kinetic evidence for the zwitterionic intermediate was obtained by Berlin (83). Apart from nitrogen, the sole decomposition product from **32** was the phosphoramidate **34**. While previous studies had found first-order

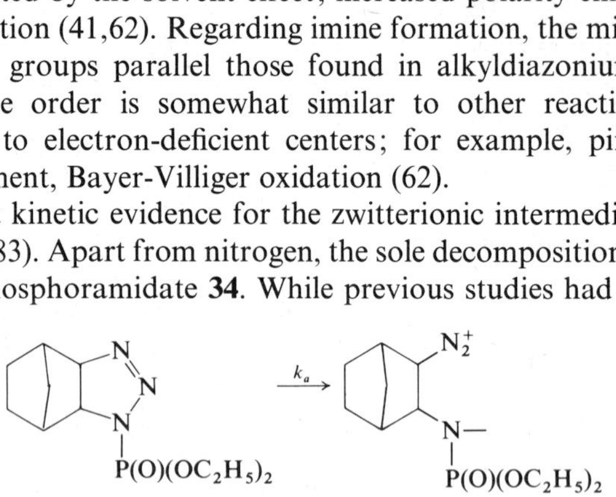

$P(O)(OC_2H_5)_2$ **32** $\xrightarrow{k_a}$ $P(O)(OC_2H_5)_2$ **33**

33 $\xrightarrow{k_b}$ N_2 + $NP(O)(OC_2H_5)_2$ **34**

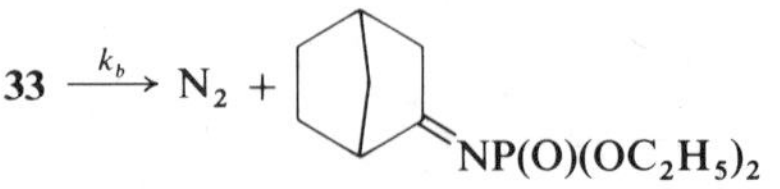

rate dependence on triazoline, in this instance careful examination of the data (N_2 evolution) revealed significant deviation from simple first-order kinetics. Rather, the rate of decomposition was precisely described by two consecutive first-order reactions. Thus the results require a two-step process, with nitrogen expulsion in the second step. *Direct* formation of a 1,3-dipole was therefore excluded, and the reaction was interpreted in terms of intermediate **33**.

An elegant application of computer techniques allowed evaluation of the individual rate constants (k_a, k_b) at several temperatures. Activation parameters were then calculated (Table 3). The large negative entropy of

TABLE 3
Rate Constants and Activation Parameters.[a] Decomposition of **32** in Diglyme (118°)

	k_a (min^{-1})	k_b (min^{-1})
	4.60×10^2	16.3×10^2
E_a (kcal/mole)	30.9	14.0
$\Delta S\ddagger$	5.53	−34.9

[a] Ref. 83.

activation for the second step is unexpected. To account for its magnitude and sign the authors suggest product formation via a six-membered transition state involving the phosphoryl group. Exclusive production of the imine (**34**) differs from the thermolytic behavior of related compounds and appears to require product formation in an exceptional manner. Solvent effects on the rate of decomposition accorded with the postulated mechanism.

Although most triazoline decompositions probably occur through a zwitterionic intermediate, Huisgen and co-workers have convincingly demonstrated a second thermolytic path. A series of *para*-substituted phenyl azide-methyl acrylate adducts was investigated (65b). Aziridine products were formed in a simple first-order process. Contrary to previous experience,

CO_2CH_3 N N N X $\xrightarrow[85^\circ]{-N_2}$ X—N CO_2CH_3

the rate of nitrogen evolution was essentially independent of polarity in several solvents (dielectric constants 2.5–35). Electron-withdrawing groups on the phenyl ring did not facilitate decomposition in the usual manner. Relative rates for the *p*-nitro- and *p*-methoxyphenyl triazolines, for example, were 1.16 and 4.48, respectively, and the data were not correlated by σ-constants. Comparison with other (unspecified) 4-substituted triazolines revealed a strong rate-accelerating effect due to the 4-carbomethoxy group. These results are clearly inconsistent with rate-determining formation of a zwitterionic species.

An attractive alternative proposed by the authors is homolytic decomposition to a spin-coupled (singlet) diradical (**35**). This proposal accounts

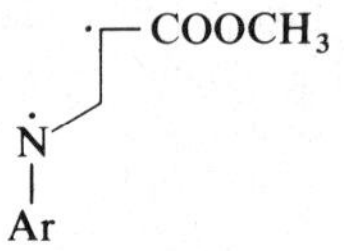

COOCH$_3$

N

Ar

35

for the noted 4-carbomethoxy effect as well as the failure of electron-withdrawing substituents to enhance the rate. Triazoline photodecomposition, a free-radical process, exhibits a similar specificity for aziridine formation.

3. Conclusions

Two decomposition paths have been discussed: heterolytic and homolytic. The distinction may in fact arise from the descriptive formalism. Consideration of the initially formed transients in either case (**36,37**) suggests that they might be viewed as resonance contributors to a single intermediate. Substituents would then alter the relative contributions of **36** and **37**.

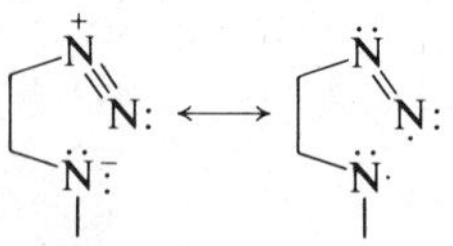

36 **37**

4-Alkyl groups, for example, would increase the polar component (heterolytic decomposition), whereas 4-carbalkoxy or nitrile substituents would diminish its importance, imparting radical character to the reaction. These arguments are particularly applicable to the prior transition state.

III. ACID- AND BASE-INDUCED DECOMPOSITION

A. Acid Decomposition

Triazolines are rapidly and quantitatively decomposed by acids, including carboxylic acids. Vigorous nitrogen evolution occurs on acid treatment providing a simple qualitative test for the triazoline system as well as a convenient volumetric method for molecular weight. Detailed investigations of the reaction have not been described, but the diversity of examples from the literature is sufficient for generalization.

1. Mechanism

Protonation of the basic triazoline nitrogen (N_1) produces a 2-aminoalkyldiazonium ion. The role of the basic site is evident from the comparative behavior of triazoline **38** and 1-pyrazoline **39**. On treatment with

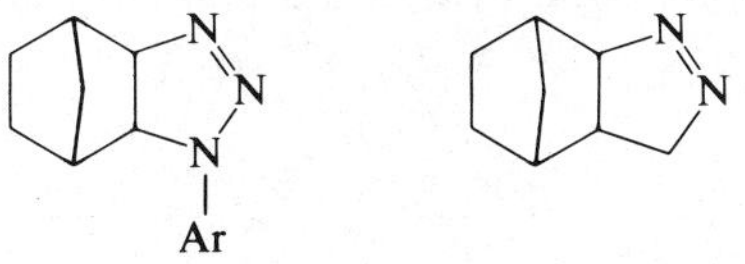

38 **39**

dilute acid, **38** immediately expels one equivalent of nitrogen; under identical conditions no nitrogen loss from **39** is observed (84). (The single reported isolation of a triazoline hydrobromide salt (72) is apparently erroneous.)

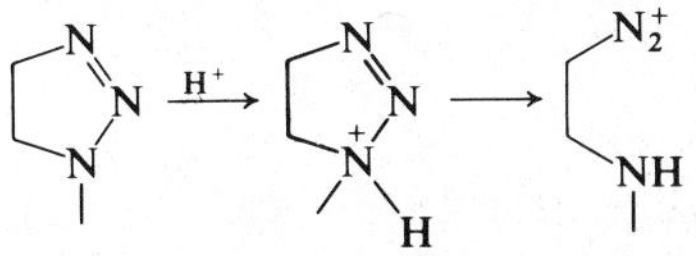

Stabilization of the diazonium ion is achieved in the anticipated manner; loss of nitrogen and nucleophilic substitution at the incipient or actual electron-deficient center. Triazoline acid decomposition thus closely resembles nitrous acid deamination reactions, notably the semipinacolic rearrangement of 2-amino-alcohols. Multiple reaction paths are characteristic of such "carbonium ion" reactions. The actual course followed depends on both structural features and reaction conditions.

2. *Displacement; Internal and External Nucleophiles*

Displacement of the diazonium nitrogen by external nucleophiles or by the adjacent amino group has been reported. Triazoline **40** afforded *N*-methyl benzylic acid amide (85), and **41** gave β-chloro-*N*-ethylaniline hydrochloride (26). Although these products appear to arise by external nucleophilic

$$\mathbf{40} \xrightarrow[-N_2]{\text{dil. HCl}} Ph_2C(OH)CONHCH_3 \qquad \mathbf{41} \xrightarrow[-N_2]{\text{conc. HCl}} PhNHCH_2CH_2Cl\cdot HCl$$

substitution (H_2O, Cl^-), another path is possible. Acid-induced cleavage of an initially formed aziridine could account for the products. Indeed, unambiguous examples of aziridine formation have been reported (86) in dilute acetic acid, but because of the facile acid-cleavage of aziridines it is usually difficult to distinguish these paths. Failure to recognize this duality has had unfortunate results. Based on the observed acid decomposition product (**44**), structure **42** was assigned to the adduct of phenyl azide and dimethylketene

42

$$\mathbf{43} \xrightarrow[-N_2]{\text{2N HOAc}} [\text{aziridine}] \longrightarrow (CH_3)_2C(NHPh)CO_2CH_3\ (\mathbf{44})$$

acetal (87). Direct displacement of nitrogen by water in **42** was assumed. Overwhelming orientational evidence (Section I-B-1) however, indicates **43** for the correct structure of the adduct. The decomposition therefore probably occurred through the illustrated aziridine intermediate. (For another example see ref. 2a.)

3. *Displacement with Group Migration*

Rearrangements characteristic of alkyldiazonium ions are encountered with triazolines. Treatment of **45** with dilute sulfuric acid gave aniline and cyclopentanone via a 1,2-hydride shift, while alkyl migration produced hexahydrobenzaldehyde from **46** (44). As expected, the norbornene–phenyl

$$\mathbf{45} \xrightarrow[-N_2]{\text{dil. } H_2SO_4} [\text{NHPh, OH}] \longrightarrow PhNH_2 + \text{cyclopentanone}$$

$$\mathbf{46} \xrightarrow[-N_2]{\text{dil. } H_2SO_4} [\text{CH(OH)NHPh}] \longrightarrow PhNH_2 + \text{CHO}$$

azide adduct gave Wagner-Meerwein rearranged product, *syn*-7-anilino-*exo*-2-hydroxynorbornane (86%) (88), a compound also obtained by acid hydrolysis of the corresponding aziridine (16). Acid decomposition of the *endo*-dicyclopentadiene–phenyl azide adduct was re-examined by Funakubo (89). Contrary to earlier reports, rearranged products were found.

In common with related 1,2-nucleophilic rearrangements, competition between migrating groups is determined by both electronic factors (migratory aptitudes) and steric effects (90). The importance of the latter is apparent from results obtained with *trans*- and *cis*-1,5-diphenyl-4-methyltriazolines (88). Acid decomposition afforded, in differing ratios, the same products in either case. 2-Phenylpropionaldehyde resulted from phenyl migration, and

$$\xrightarrow[-N_2]{\text{dil. } H_2SO_4} PhNH_2 + PhCH(CH_3)CHO + PhCOCH_2CH_3$$

	$PhCH(CH_3)CHO$	$PhCOCH_2CH_3$
trans	87%	4%
cis	38%	58%

hydride shift produced propiophenone. Although phenyl usually migrates in preference of hydrogen, as observed with the *trans*-compound, hydrogen migration predominated with the *cis*-isomer. Models reveal the ground and transition state conformations for H-migration in the *cis*-triazoline to be sterically more favorable than those for phenyl migration. The effect is quite similar to that described in the deamination of certain aminoalcohols (91). Aziridines were shown not to be involved in the above reactions.

In view of the mechanism it is hardly surprising that other electrophilic reagents [e.g., Cu(II)] capable of coordination with N_1 also bring about decomposition.

B. Base-catalyzed Isomerization

The effect of base on triazolines has been examined by Huisgen and coworkers (65a). Triazolines bearing acyl, carbalkoxy, and nitrile groups at the 4-position readily isomerized to ring-opened diazo compounds in the presence of triethylamine. Compound **47**, for example, gave methyl 2-diazo-3-anilinopropionate (**48**). In addition to spectroscopic evidence, structure **48**

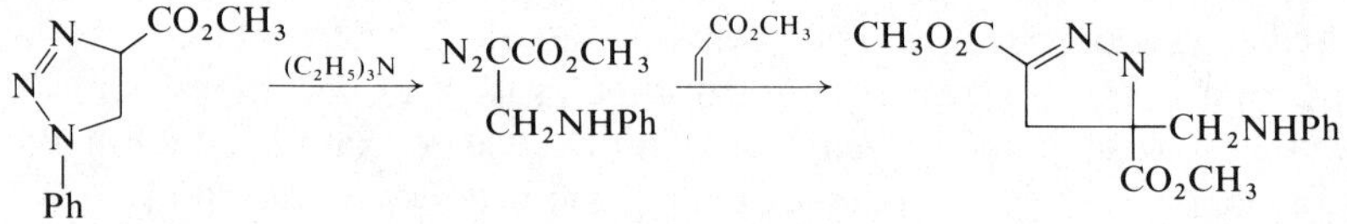

CO_2CH_3 (triazoline, N-Ph) $\xrightarrow{(C_2H_5)_3N}$ $N_2CCO_2CH_3$ / CH_2NHPh $\xrightarrow{CH_2=CHCO_2CH_3}$ CH_3O_2C (pyrazoline) $-CH_2NHPh$, CO_2CH_3

47 **48**

was confirmed by further cycloaddition reactions and by pyrolysis to methyl β-anilinoacrylate. Pyrolysis of the triazoline (**47**) gave aziridine.

The isomerization is reversible, as indicated by the establishment of equilibria in certain instances. The closest chemical analogy appears to be

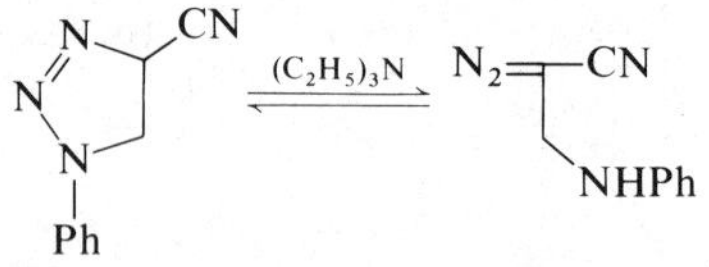

CN (triazoline, N-Ph) $\underset{}{\overset{(C_2H_5)_3N}{\rightleftharpoons}}$ $N_2=$ $-CN$ / NHPh

71% 29%

the base-catalyzed Dimroth rearrangement of 5-amino and 5-hydroxy-1,2,3-triazoles (92). A similar mechanism was formulated.

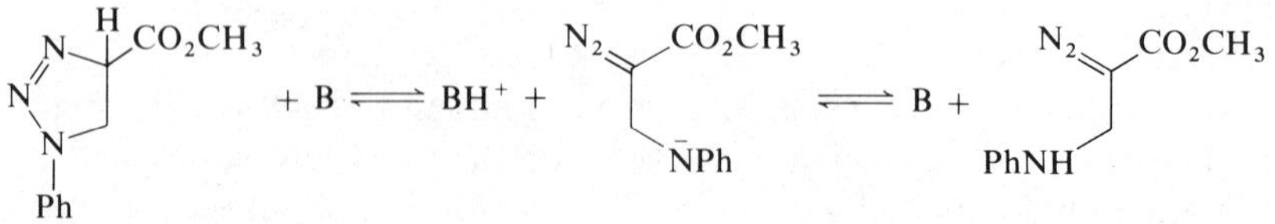

H, CO_2CH_3 (triazoline, N-Ph) + B $\rightleftharpoons$ BH^+ + N_2, CO_2CH_3, $\bar{N}Ph$ $\rightleftharpoons$ B + N_2, CO_2CH_3, PhNH

IV. PHOTODECOMPOSITION

The photolability of aliphatic azo compounds (93) suggests that triazolines too should photolyze readily. As a class this behavior is observed. The distinctive feature of photodecomposition is its high selectivity. In contrast to thermolysis high yields of aziridine are usually obtained.

A. Spectra, Photoefficiency, and Solvents

The effect of substituents at N_1 on the long wavelength (>240 mμ) absorption maximum is shown in Table 4. The tabulated maxima are actually aggregates with discernible shoulders. The data show the triazoline chromophore to be sensitive to N_1 substitution, presumably by conjugation of the

TABLE 4
Ultraviolet Absorption of Triazolines[a,b]

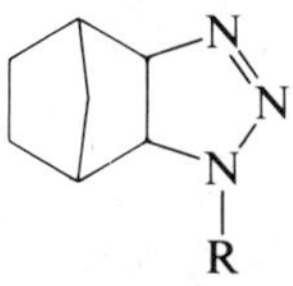 R	λ_{max}, mμ	ε_{max}
—Ph	309	10,700
—C(=CH_2)Ph	274	6,670
—CH_2Ph	270	4,860
—Si$(CH_3)_3$	257	3,580
—$CO_2C_2H_5$	244	3,320

[a] Ref. 13.
[b] Ethanol and acetonitrile solvents.

azo π-system through the nonbonding electron pair. Diazoamino compounds display a similar effect (94). With the exception of chlorinated solvents (see below), solvent polarity does not significantly affect the position of λ_{max}. Unlike typical azo compounds, the weak absorption corresponding to the n,π^* transition has not been detected. It is likely that this transition lies beneath the more intense π,π^* bands centered at the maximum. The spectra of aryl azides have been interpreted in this manner (95).

Irradiation in the region of λ_{max} causes nitrogen evolution. Quantum yield determinations for several 1-phenyltriazolines lie close or equal to unity, a value characteristic of aliphatic azo compounds. Nitrogen evolution ceases abruptly when the irradiation is interrupted. The reaction may therefore be characterized as a highly efficient, nonchain photoprocess.

A useful aspect of photodecomposition is its insensitivity to solvent. Any inert and transparent material may be employed. Diverse solvents such as cyclohexane, carbon disulfide, methyl isocyanate, and 20% aqueous dimethylformamide have been used without affecting the product distribution (13,16). Chlorinated solvents, however, are exceptional and appear to induce undesirable side reactions. Triazoline spectra in such solvents show marked changes; charge-transfer complexation may be involved (96).

B. Products

Although thermolysis usually gives mixtures of imine and aziridine isomers, triazoline photolysis displays a higher selectivity for aziridines. The phenomenon appears general; an analogous photolytic preference for small-ring formation has been described with 1-pyrazolines (97a), 3-*H*-pyrazoles (97b), and 4-*H*-1,2,3-triazenes (98). Aromatic 1,2,3-triazoles, however, do not produce small-ring products (99). The comparative results in Table 5 are

TABLE 5
Photo and Thermal Decompositions[a]

	Photodecomposition		Thermal Decomposition	
Triazoline	% Aziridine	% Imine	% Aziridine	% Imine
Cyclopentane-fused triazoline, N-Ar	94	6	21	79
Cycloheptane-fused triazoline, N-Ar	89	11	—	80
n-C_4H_9-substituted triazoline, N-Ar	88	12	64	36

[a] Refs. 13 and 51.

illustrative. Imine formation is significantly less important in photolysis than in thermolysis. Additional examples of this selectivity are cited in refs. 16, 52, and 56.

An indication of the synthetic potential of the method is evident from results obtained with triazoline **49** (100). Thermally, **49** was quantitatively converted to the iminolactone, a reaction common to triazolines of this type (11,36,37,75). Photolysis, on the other hand, gave 67% (isolated) of the

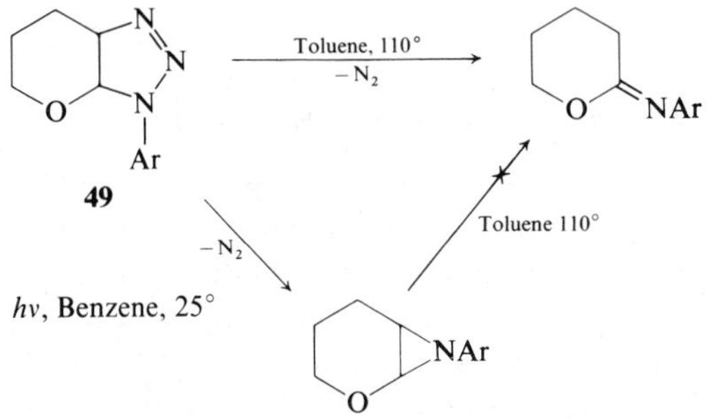

Toluene, 110° $-N_2$

49

$h\nu$, Benzene, 25° $-N_2$

Toluene 110°

aziridine. The unusual nature of the aziridine suggested that it might have been a precursor to the imine during thermolysis. However, the aziridine was found to be completely stable under the thermolytic conditions. While the iminolactone resulted from a usual thermal path (N_1—N_2 heterolysis followed by hydride shift), photolysis clearly occurred by a different route.

Photodecomposition has been utilized to prepare aziridines. The method is particularly useful where the desired aziridine is thermally unstable or sensitive to polar reagents. Vinylaziridines have been conveniently prepared, as were several other unusual aziridine structures (13,52,101). Synthesis in this manner is exceptionally simple; azide–olefin cycloaddition followed by photolysis. Conventional techniques (102) frequently require lengthy preparations of the starting materials, and the excellent newer procedures, such as the methods of Hassner (103a) and Swern (103b), are not applicable to *N*-arylaziridines. When the triazoline is available, photodecomposition is a useful synthetic route to the aziridine.

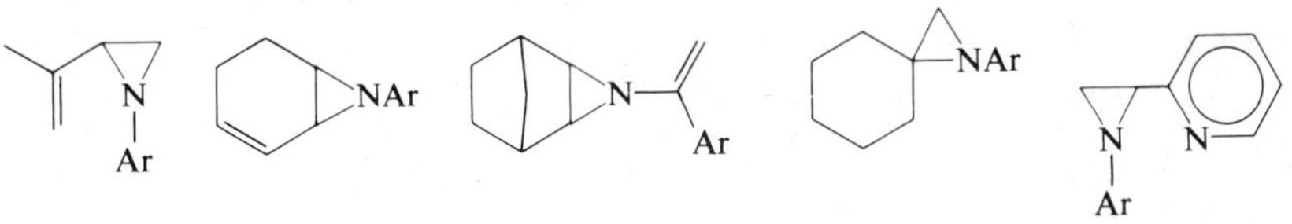

Additional photoreactive groups may inhibit aziridine formation.

Regardless of the nature of substituent X, triazolines of type **50** photolyze quantitatively to the corresponding aziridines (13,16). The *m*- and *p*-nitro-

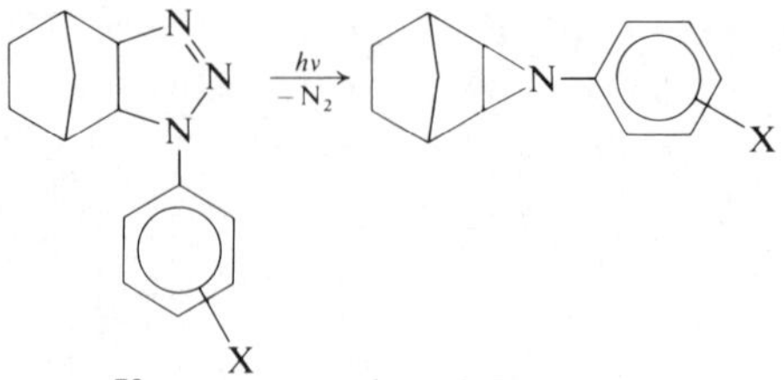

50

X = H; *m,p*-Br; *m,p*-Cl; *m,p*-OCH_3; *m,p*-CH_3; *p*-$CO_2C_2H_5$

phenyl compounds are exceptions; complex mixtures containing only traces of aziridine result from these compounds. The generally unsatisfactory photolytic behavior of 1-nitrophenyltriazolines is not due to the ground state electronegativity of the nitro group, cf. *p*-$CO_2C_2H_5$ and *p*-NO_2. Rather, competing processes associated with the electronically excited nitro group (104) are probably responsible.

Photoinduced side reactions complicate the photolysis of **51** (Ar = *p*-BrC_6H_4) (13). Under identical conditions, the oxa analog (**52**) gave 96% of aziridine while **51** produced a highly colored mixture from which

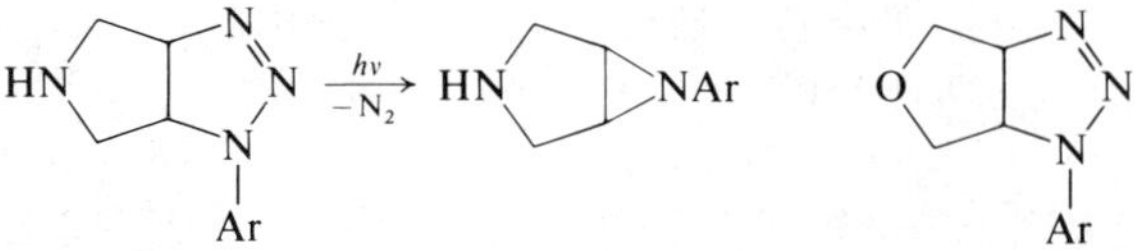

51 61% **52**

61% of the aziridine was isolated. Prolonged irradiation increased tar formation. These observations are suggestive of photodisplacement reactions, as discussed by Havinga (105).

C. Mechanism

The electronic transition(s) responsible for photodecomposition has not been defined. A π^* azo configuration formed by promotion of either *n* or π electrons is most likely. The often dramatic differences observed in the photo and thermal reactions indicates the operation of chemically distinct paths. Heterolytic cleavage accounts best for the (usual) thermal process. Photolysis of azo compounds, on the other hand, is homolytic, giving rise to nitrogen and free radicals. In agreement with the noted insensitivity to solvent, triazoline photodecomposition also appears to be a homolytic process.

The photodecomposition of isomeric triazolines **53** and **54** has been investigated (106). With excellent material balance, three products were obtained. The product distributions, however, varied significantly under direct and sensitized photolysis conditions (Table 6).

TABLE 6
Stereochemistry of Triazoline Photodecomposition[a]

		Products, %		
		trans-2-methyl-3-phenyl-1-phenylaziridine (Ph, CH$_3$, N–Ph)	cis-2-methyl-3-phenyl-1-phenylaziridine (Ph, CH$_3$, N–Ph)	PhC(=NPh)CH$_2$CH$_3$
53, *trans*	Direct[b]	66	22	12
	Photosens.[c]	42	54	4
54, *cis*	Direct[b]	17	65	18
	Photosens.[c]	60	36	4

[a] Ref. 106.
[b] 313 mμ, benzene.
[c] 366 mμ, benzophenone in benzene.

The results shown in Table 6 for direct photolysis were unaffected by variations in initial concentration, temperature, and solvent. Similarly, efficient triplet energy acceptors had negligible effect on either the product distributions or the quantum yields ($\Phi_{N_2} = 1.0$). Direct photolysis occurred with predominant, though not exclusive, retention of triazoline geometry in the aziridine product. Solution-phase photolysis of 1-pyrazolines to cyclopropanes exhibits the same geometrical specificity (97,107).

A simple free-radical path accommodates the results (Fig. 1). The ineffectiveness of triplet quenchers (piperylene and oxygen) indicates reaction

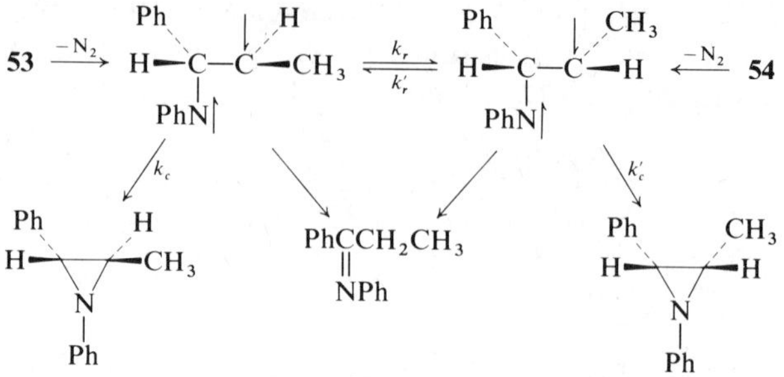

Figure 1.

from an excited singlet state. At the wavelengths of light employed the departing nitrogen must also be singlet. Therefore, in accordance with spin-conservation, the intermediate diradicals are depicted as spin-paired. The observed predominant retention of geometry may then be attributed to the rapidity of ring-closure as compared to rotation, $k_c > k_r$, $k'_c > k'_r$. Calculations based on this model reveal that closure occurred about 1·8 times more frequently than rotation ($k_c/k_r = 1.8$); a k_c/k_r value of 0.1 has been found for 1,4-diradicals (108). The radical coupling (ring-closure) is extremely rapid, precluding chemical interaction with the solvent.

In terms of Figure 1 photosensitization would be expected to produce 1,3-diradicals with initially unpaired electron spins (triplets). Prior to aziridine formation, in this case, a single spin inversion is required. Experimentally, similar product mixtures were obtained from either triazoline (Table 6). It therefore appears that spin inversion is appreciably slower than σ-bond rotation, allowing time for rotational equilibrium to be approached (18).

The only imine product, propiophenone anil, may also result from the 1,3-diradical by a hydrogen shift. Hydrogen and alkyl migrations have been amply documented in 1,3-diradical systems (109). It is noteworthy that in the acid-induced (ionic) decomposition of these triazolines (Section III-A-3) both phenyl and hydrogen migrations were observed. Regarding photolytic imine formation, a stereoelectronic requirement has been suggested (13,106). The extent of imine produced correlates with the degree of alkyl substitution at C_4 and with rotational freedom about the C_4—C_5 triazoline bond.

Although the experimental results are consistent with the indicated path (Fig. 1), a mechanism involving synchronous elimination of nitrogen and aziridine formation (**55**) deserves consideration. Such a process would completely preserve triazoline geometry in the aziridine product. The appreciable rotation (and imine formation) found with **53** and **54** on direct photolysis argue against concerted cycloelimination. Nevertheless, to the extent that the new σ bond is forming as nitrogen departs, **55** has validity. The data do

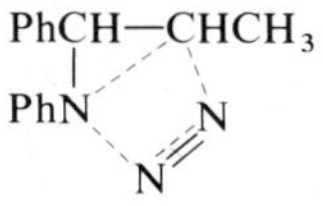

PhCH—CHCH$_3$
PhN N
N

55

not allow assessment of the extent of transitional bonding. It therefore seems best to consider the reaction in terms of 1,3-diradical intermediates.

The selectivity of triazoline photodecomposition is a consequence of its homolytic nature. Short-lived 1,3-diradicals are formed with their ends in close proximity. Aziridine formation follows. Huisgen has demonstrated (Section II-C-2) that, in certain instances, thermal radical decomposition may occur. Although photolysis represents the homolytic extreme, varying degrees of radical character may be anticipated for thermolysis.

V. CONCLUSION

The increasing availability of triazolines makes them attractive materials for study and synthetic purposes. The thermolytic and photolytic behavior of these compounds resembles that of related cyclic azo structures. But the additional nitrogen introduces reactions unique to the triazoline system. Among the several decompositions discussed, the cyanogen azide reaction (Section II-B-3), diazo transfer (Section II-B-5) and photodecomposition are notable. These new reactions provide excellent, selective synthetic procedures.

References

1. L. Wolff, *Ann.*, **394**, 30, 60 (1912).
2. For brief reviews see: (a) R. Huisgen, R. Grashey, and J. Sauer, in *The Chemistry of Alkenes*, S. Patai, Ed., Interscience, New York, 1964, pp. 835ff; (b) J. H. Boyer and F. C. Canter, *Chem. Rev.*, **54**, 42 (1954).
3. K. Alder and G. Stein, *Ann.*, **485**, 211 (1931).
4. K. Ziegler, H. Wilms, H. Sauer, L. Bruns, H. Froitzheim-Kuhlhorn, and J. Schneider, *Ann.*, **567**, 1 (1950); **589**, 122 (1954); A. T. Blomquist, L. H. Liu, and J. C. Bohrer, *J. Am. Chem. Soc.*, **74**, 3643 (1952).
5. W. E. Parham, W. T. Hunter, R. Hanson, and T. Lahr, *J. Am. Chem. Soc.*, **74**, 5646 (1952).
6. A. A. Danish and R. E. Lidov, *Anal. Chem.*, **22**, 702 (1950).
7. P. Scheiner, *Tetrahedron*, **24**, 349 (1968).

8. W. R. Carpenter, A. Haymaker, and D. W. Moore, *J. Org. Chem.*, **31**, 789 (1966).
9. K. Alder and G. Stein, *Ann.*, **515**, 165 (1935).
10. R. Fusco, G. Bianchetti, and D. Pocar, *Gazz. Chim. Ital.*, **91**, 849 (1961); M. E. Munk and Y. K. Kim, *J. Am. Chem. Soc.*, **86**, 2213 (1964); G. Nathansohn, E. Testa, and N. Dimola, *Experientia*, **18**, 57 (1962).
11. R. Huisgen, L. Mobius, and G. Szeimies, *Chem. Ber.*, **98**, 1138 (1965).
12. K. D. Berlin and L. A. Wilson, *Chem. Commun.*, **1965**, 280.
13. P. Scheiner, *Tetrahedron*, **24**, 2757 (1968).
14. P. A. S. Smith, J. M. Clegg, and J. LaKritz, *J. Org. Chem.*, **23**, 1595 (1958).
15. P. Scheiner, J. H. Schomaker, S. Deming, W. J. Libbey, and G. P. Nowack, *J. Am. Chem. Soc.*, **87**, 306 (1965).
16. R. Huisgen, L. Mobius, G. Mueller, H. Stangl, G. Szeimies, and J. M. Vernon, *Chem. Ber.*, **98**, 3992 (1965).
17. R. Huisgen and G. Szeimies, *Chem. Ber.*, **98**, 1153 (1965).
18. P. Scheiner, *J. Am. Chem. Soc.*, **88**, 4759 (1966).
19. G. D. Buckley, *J. Chem. Soc.*, **1954**, 1850.
20. R. Huisgen, R. Knorr, L. Mobius, and G. Szeimies, *Chem. Ber.*, **98**, 4014 (1965).
21. S. M. Gurvich and A. P. Terent'ev, *Sb. Statei Obsch. Khim., Akad. Nauk SSSR*, **1**, 401 (1953); *Chem. Abstr.*, **49**, 1047 (1955).
22. A. Mustafa, *J. Chem. Soc.*, **1949**, 234.
23. P. K. Kadaba and J. O. Edwards, *J. Org. Chem.*, **26**, 2331 (1961).
24. P. K. Kadaba, *Tetrahedron*, **22**, 2453 (1966); P. K. Kadaba and N. F. Fanin, *J. Heterocycl. Chem.*, **4**, 301 (1967).
25. H. J. Backer, *Rec. Trav. Chim.*, **69**, 1223 (1950).
26. H. W. Heine and D. A. Tomalia, *J. Am. Chem. Soc.*, **84**, 993 (1962).
27. H. W. Heine, *Angew. Chem. Intern. Ed. Engl.*, **1**, 528 (1962).
28. K. Hohenlohe-Oehringen, *Monatsh.*, **89**, 557, 562, 597 (1958).
29. O. Dimroth and H. Aickelein, *Ber.*, **39**, 4390 (1906); T. Curtius and J. Thompson, *Ber.*, **39**, 4140 (1906).
30. T. Curtius and E. Welde, *Ber.*, **40**, 1197 (1907).
31. R. Neidlein, *Ber.*, **97**, 3476 (1964); R. Gompper, *Ber.*, **90**, 374 (1957).
32. J. D. Watson, *The Double Helix*, Atheneum, New York, 1968, pp. 190ff.
33. A. S. Bailey, J. J. Mercer, and J. E. White, *Chem. Commun.*, **1965**, 4.
34. J. E. Franz, C. Osuch, and M. W. Dietrich, *Tetrahedron Letters*, **1963**, 837; *J. Org. Chem.*, **29**, 2922 (1964); A. C. Oehlschlager and L. H. Zalkow, *J. Org. Chem.*, **28**, 3303 (1963).
35. F. D. Marsh and M. E. Hermes, *J. Am. Chem. Soc.*, **86**, 4506 (1964).
36. A. S. Bailey and J. E. White, *J. Chem. Soc., (B)*, **1966**, 819.
37. I. Brown and O. E. Edwards, *Can. J. Chem.*, **43**, 1266 (1965).
38. A. C. Oehlschlager and L. H. Zalkow, *J. Org. Chem.*, **30**, 4205 (1965).
39. A. G. Anastassiou, *J. Am. Chem. Soc.*, **90**, 1527 (1968).
40. P. Scheiner, *J. Org. Chem.*, **30**, 7 (1965).
41. A. C. Oehlschlager, P. Tillman, and L. H. Zalkow, *Chem. Commun.*, **1965**, 596.
42. G. Caronna and S. Palazzo, *Gazz. Chim. Ital.*, **82**, 292 (1952).
43. P. Scheiner and S. Deming, unpublished results.
44. K. Alder and G. Stein, *Ann.*, **501**, 1 (1933).
45. R. Huisgen, *Angew. Chem. Intern. Ed. Engl.*, **2**, 565 (1963).
46. F. D. Chattaway and G. D. Parkes, *J. Chem. Soc.*, **127**, 1307 (1925).
47. L. Wolff and R. Hercher, *Ann.*, **399**, 274 (1913).
48. T. Curtius and W. Dorr, *J. Prakt. Chem.*, **125**, 425 (1930).
49. For leading references see W. Lwowski, *Angew. Chem. Intern. Ed. Engl.*, **6**, 897 (1967).

50. J. Jaz, E. D. deHault, and R. Navette, *Tetrahedron Letters*, **1965**, 2751; cf. P. Walker and W. A. Waters, *J. Chem. Soc.*, **1962**, 1632.
51. K. R. Henery-Logan and R. A. Clark, *Tetrahedron Letters*, **1968**, 801.
52. P. Scheiner, *J. Org. Chem.*, **30**, 7 (1965).
53. A. C. Oehlschlager, P. Tillman, and L. H. Zalkow, *Chem. Commun.*, **1965**, 596.
54. A. S. Bailey and J. J. Wedgwood, *J. Chem. Soc., (C)*, **1968**, 682.
55. L. H. Zalkow and C. D. Kennedy, *J. Org. Chem.*, **28**, 3309 (1963); A. C. Oehlschlager and L. H. Zalkow, *Chem. Commun.*, **1966**, 5.
56. L. H. Zalkow, A. C. Oehlschlager, G. A. Cabat, and R. L. Hale, *Chem. Ind. (London)*, **1964**, 1556.
57. R. Huisgen, *Angew. Chem. Intern. Ed. Engl.*, **2**, 565 (1963).
58. J. E. Franz and C. Osuch, *Chem. Ind. (London)*, **1964**, 2058; A. C. Oehlschlager and L. H. Zalkow, *J. Org. Chem.*, **30**, 4205 (1965).
59. A. G. Anastassiou, *J. Org. Chem.*, **31**, 1131 (1966).
60. J. Meinwald, S. S. Labana, and G. H. Wahl, *Tetrahedron Letters*, **1965**, 1789.
61. K. Alder and W. Trimborn, *Ann.*, **566**, 58 (1950).
62. A. L. Logothetis, *J. Am. Chem. Soc.*, **87**, 749 (1965).
63. T. L. Jacobs, in *Heterocyclic Compounds*, Vol. 5, R. C. Elderfield, Ed., Wiley, New York, 1957, p. 76.
64. K. Alder and G. Stein, *Ann.*, **485**, 211 (1931); P. Scheiner and W. R. Vaughan, *J. Org. Chem.*, **26**, 1923 (1961).
65. (a) R. Huisgen, G. Szeimies, and L. Mobius, *Chem. Ber.*, **99**, 475 (1966); (b) G. Szeimies and R. Huisgen, *Chem. Ber.*, **99**, 491 (1966).
66. A. Mustafa, S. M. A. D. Zayed, and S. Khattab, *J. Am. Chem. Soc.*, **78**, 145 (1956); S. J. Davis and C. S. Rondestvedt, *Chem. Ind. (London)*, **1956**, 845.
67. W. I. Awad, S. M. A. R. Omran, and F. Nagieb, *Tetrahedron*, **19**, 1591 (1963).
68. K. Alder and G. Stein, *Ann.*, **515**, 165, 185 (1935).
69. G. Bianchetti, P. D. Croce, D. Pocar, and G. G. Gallo, *Rend. Ist. Lombardo Sci Lettere*, **A99**, 296 (1965); *Chem. Abstr.*, **65**, 15367 (1966); R. D. Burpitt and V. W. Goodlett, *J. Org. Chem.*, **30**, 4308 (1965).
70. A. C. Ritchie and M. Rosenburger, *J. Chem. Soc., (C)*, **1968**, 227.
71. R. Fusco, G. Bianchetti, D. Pocar, and R. Ugo, *Gazz. Chim. Ital.*, **92**, 1040 (1962).
72. C. S. Rondostvedt and P. K. Chang, *J. Am. Chem. Soc.*, **77**, 6532 (1955).
73. R. Fusco, G. Bianchetti, and D. Pocar, *Gazz. Chim. Ital.*, **91**, 933 (1961).
74. K. D. Berlin and L. A. Wilson, *Chem. Ind. (London)*, **1965**, 1522.
75. J. E. Franz, M. W. Dietrich, A. Henshall, and C. Osuch, *J. Org. Chem.*, **31**, 2847 (1966); D. L. Rector and R. H. Harmon, *ibid.*, **31**, 2837. (1966).
76. R. Scarpata, D. Sica, and A. Lionetti, *Gazz. Chim. Ital.*, **93**, 30 (1963).
77. R. Fusco, G. Bianchetti, D. Pocar, and R. Ugo, *Ber.*, **96**, 802 (1963).
78. J. Kucera and Z. Arnold, *Tetrahedron Letters*, **1966**, 1109; M. Rosenberger, P. Yates, J. B. Hendrickson, and W. Wolf, *Tetrahedron Letters*, **1964**, 2285.
79. M. Regitz, *Angew. Chem. Intern. Ed. Engl.*, **6**, 733 (1967).
80. J. E. Baldwin, G. V. Kaiser, and J. A. Romersberger, *J. Am. Chem. Soc.*, **87**, 4114 (1965); **86**, 4509 (1964).
81. R. Huisgen, R. Grashey, J. M. Vernon, and R. Kunz, *Tetrahedron*, **21**, 3311 (1965).
82. J. C. Sheehan and P. T. Izzo, *J. Am. Chem. Soc.*, **70**, 1985 (1948); **71**, 4059 (1949).
83. K. D. Berlin, L. A. Wilson, and L. M. Raff, *Tetrahedron*, **23**, 965 (1967).
84. N. S. Zefirov, P. P. Kadzauskas, and Yu. K. Yurev, *Zh. Obsh. Khim.*, **36**, 23 (1966).
85. K. Hohenlohe-Oehringen, *Monatsh.*, **89**, 588 (1958).

86. K. Alder and R. Ruhrman, *Ann.*, **566**, 1 (1950); K. Alder, W. Gunzl, and K. Wolff, *Ber.*, **93**, 809 (1960).
87. R. Scarpati and D. Sica, *Gazz. Chim. Ital.*, **93**, 942 (1963).
88. P. Scheiner, unpublished results.
89. E. Funakubo, I. Moritani, H. Taniguchi, T. Yamamoto, and S. Tsuchiya, *Ber.*, **96**, 2035 (1963).
90. D. J. Cram, in *Steric Effects in Organic Chemistry*, M. S. Newman, Ed. John Wiley and Sons, New York, 1956, pp. 250ff.
91. E. L. Eliel, *Stereochemistry of Carbon Compounds*, McGraw-Hill Book Co., New York, 1962, pp. 149, 154–156.
92. O. Dimroth, *Ann.*, **373**, 336 (1910); B. R. Brown, D. L. Hammick, and S. G. Heritage, *J. Chem. Soc.*, **1953**, 3820.
93. J. G. Calvert and J. N. Pitts, *Photochemistry*, Wiley, New York, 1966, pp. 452, 462ff.
94. V. O. Lukashevich and E. S. Lisitsyna, *Dokl. Akad. Nauk SSSR*, **172**, 345 (1967).
95. A. Reiser, G. Bowes, and R. J. Horne, *Trans. Faraday Soc.*, **62**, 3167 (1966).
96. W. J. Lautenberger, E. N. Jones, and J. G. Miller, *J. Am. Chem. Soc.*, **90**, 1110 (1968).
97. (a) T. V. VanAuken and K. L. Rinehart, *J. Am. Chem. Soc.*, **84**, 3736 (1962); **82**, 5251 (1960); (b) G. L. Closs, W. A. Boll, H. Heyn, and V. Dev, *J. Am. Chem. Soc.*, **90**, 173 (1968).
98. E. M. Burgess and L. McCullagh, *J. Am. Chem. Soc.*, **88**, 1581 (1966), and references therein to thermolysis.
99. E. M. Burgess, R. Carithers, and L. McCullagh, *J. Am. Chem. Soc.*, **90**, 1923 (1968).
100. P. Scheiner, *J. Org. Chem.*, **32**, 2022 (1967).
101. P. Scheiner, *J. Org. Chem.*, **32**, 2628 (1967).
102. For reviews see: P. E. Fanta, in *Heterocyclic Compounds with Three and Four-Membered Rings*, Part I, A. Weissberger, Ed., Interscience, New York, 1964, pp. 524–575; P. A. Gembitskii, N. M. Loim, and D. S. Zhuk, *Uspekhi Khemii*, **1966**, 229; English translation, *Russian Chemical Reviews*, No. **2**, 105 (1966).
103. (a) A. Hassner and C. Heathcock, *Tetrahedron*, **20**, 1037 (1964); *J. Org. Chem.*, **29**, 3640 (1964); (b) T. A. Foglia and D. Swern, *J. Org. Chem.*, **32**, 75 (1967).
104. R. Hurley and A. C. Testa, *J. Am. Chem. Soc.*, **89**, 6917 (1967); **88**, 4330 (1966).
105. For leading references see: E. Havinga, R. O. deJongh, and M. E. Kronenberg, *Helv. Chim. Acta.*, **50**, 2550 (1967).
106. P. Scheiner, *J. Am. Chem. Soc.*, **90**, 988 (1968).
107. C. G. Overberger, R. E. Zangara, and J-P. Anselme, *J. Org. Chem.*, **31**, 2046 (1966); D. E. McGreer and W. S. Wu, *Can. J. Chem.*, **45**, 461 (1967).
108. L. K. Montgomery, K. Schueller, and P. D. Bartlett, *J. Am. Chem. Soc.*, **86**, 622 (1964).

Asymmetric Selection via Elimination*

STANLEY I. GOLDBERG

Department of Chemistry, University of South Carolina, Columbia, South Carolina

I. INTRODUCTION AND DEFINITION OF ASYMMETRIC SELECTION

This discussion is concerned with elimination reactions that embody rather special stereochemical features. It turns out that these special stereochemical features are by no means new; in fact, they have been recognized for a long time although they have been discussed, as have many developing concepts in chemistry, in various ways at different times. It also turns out that these special stereochemical features are not unique to elimination reactions, but that they are in fact much better known in connection with a wide variety of addition reactions. Reactions and processes which manifest these special stereochemical features have been referred to and variously defined as asymmetric inductions, asymmetric transfers, asymmetric syntheses (both absolute and ordinary), asymmetric transformations, asymmetric destructions,

* Supported by the Petroleum Research Fund of the American Chemical Society.

asymmetric hydrogenations, reductions, oxidations, etc. Clearly, these names have been used to emphasize and call attention to one or more particular stereochemical aspect associated with a given constitutional change. After repeated exposure to the terminology one may develop an intuitive cognizance for these special stereochemical features, but this is usually attended with an annoying feeling of vagueness because of the lack of precise, comprehensive definition. The difficulty in formulating a secure definition is not exactly a problem of precision but rather one more of boundaries; that is to say, the problem is, and always has been, simply that of deciding which reactions to include or rather how to formulate one's definition to include those reactions one believes ought to be included. The problem of boundaries is brought about by the fact that, while special recognition of certain stereochemical features is desirable and justifiable, it is actually somewhat artificial because those special stereochemical features of interest here all fall under the general and precisely formulated concept of stereoselectivity. The vagueness, therefore, has been introduced by the practice of assigning specialized names and definitions to reactions that actually embody various manifestations of the same stereochemical concept. Nevertheless, this practice may be justified on the grounds that it serves to emphasize important or novel aspects of various reactions.

For the present purpose we want to define asymmetric selectivity in terms of a set of criteria which will allow us to single out certain reactions for special attention. Since all of the reactions of potential interest fall under the general concept of stereoselectivity, we need only set conceptual boundaries for separating those of special interest. This may be done in the same way as stereospecific reactions are singled out for special recognition from the general classification of stereoselective reactions. The latter category is all inclusive. It denotes reactions that give or may give two or more stereoisomeric products which are formed in unequal amounts. For example, the dehydrochlorination of **1**, which gives a mixture of diastereomeric products, the (*E*)- and the (*Z*)-olefins* (**2** and **3**, respectively) is a simple case of a stereoselective process if, and only if, the two olefins are formed in unequal amounts. In thermodynamic terms a stereoselective reaction is a consequence of kinetic control. That is to say it arises out of reaction rate differences due to inequalities in the free energies of the ground states and/or the transition states.† In the illustrative example shown in Figure 1 the transition states

* The recently adopted (1) *Chemical Abstracts* convention for specification of double bond configurational isomers. For the reader who does not yet feel "at home" with this convention there will be no ambiguity introduced in this discussion if he merely equates (*E*) and (*Z*) to the more familiar but obsolete designations, *trans* and *cis*, respectively.

† See Mislow (2) for an elegant systematic treatment of the various types of energy relationships constituting the fundamental origins of stereoselective reactions.

leading to the diastereomeric products must be of different free energy. Indeed, the two transition states themselves must bear a diastereomeric relationship to one another and are, therefore, by definition of different free energy. The greater the difference in free energies of the transition states of a given reaction, the greater the stereoselectivity of that reaction.

(1) —HCl→ (2) + (3)

Figure 1. When stereoisomeric products are formed in unequal amounts, then the reaction by which they are formed is stereoselective.

For reasons of emphasis special recognition is given to some stereoselective reactions involving stereoisomeric substances as reactants. When, with everything else equal, stereoisomeric materials exhibit a reaction difference—usually in the ratio of their products—then such reactions are said to be stereospecific. It is important to note that these reactions are first and always stereoselective, and that they are singled out of the general category by means of the set of boundaries stated above. The diastereomeric olefins, **2** and **3**, may be used to serve as a simple illustration. Suppose each is caused to undergo addition of the elements of the molecule A_2. The various possibilities are idealized in Figure 2.

Addition of A_2 to **2** in a *syn* manner gives an equal mixture (racemic modification) of the dissymmetric *threo* enantiomers. This mixture is also obtained from the diastereomeric olefin **3** but only via *anti* addition of A_2, while the nondissymmetric *erythro* product **6**—a *meso* form—is also provided by **2** and **3** but only through *anti* addition of A_2 to the former and *syn* addition to the latter. Thus under an equal set of reaction circumstances—say, *syn* addition of A_2—the stereoisomeric substances **2** and **3** react differently, constituting stereospecific reactions. This difference may be accounted for terms of the free energies ($\Delta G\ddagger$) of the relevant transition states. It can be readily seen, even with the simplified representations of Figure 2, that the transition states involved in *syn* addition of A_2 to **2** and *syn* addition to **3** are neither identical nor related as object and mirror image (enantiomeric). They possess, in fact, a diastereomeric relationship; and they must, therefore, differ in free energy. The difference in free energy ($\Delta\Delta G\ddagger$) may lead to a significant difference in reaction rates which would be reflected in the product ratios. This is, of course, the same principle upon which all stereoselective reactions may be interpreted. It is important to emphasize this point, for we want to develop it further in formulating the definition of asymmetric selectivity. For that purpose we may again refer to Figure 2 considering only the *syn* addition of A_2 to **2** which gives an *equal* mixture of

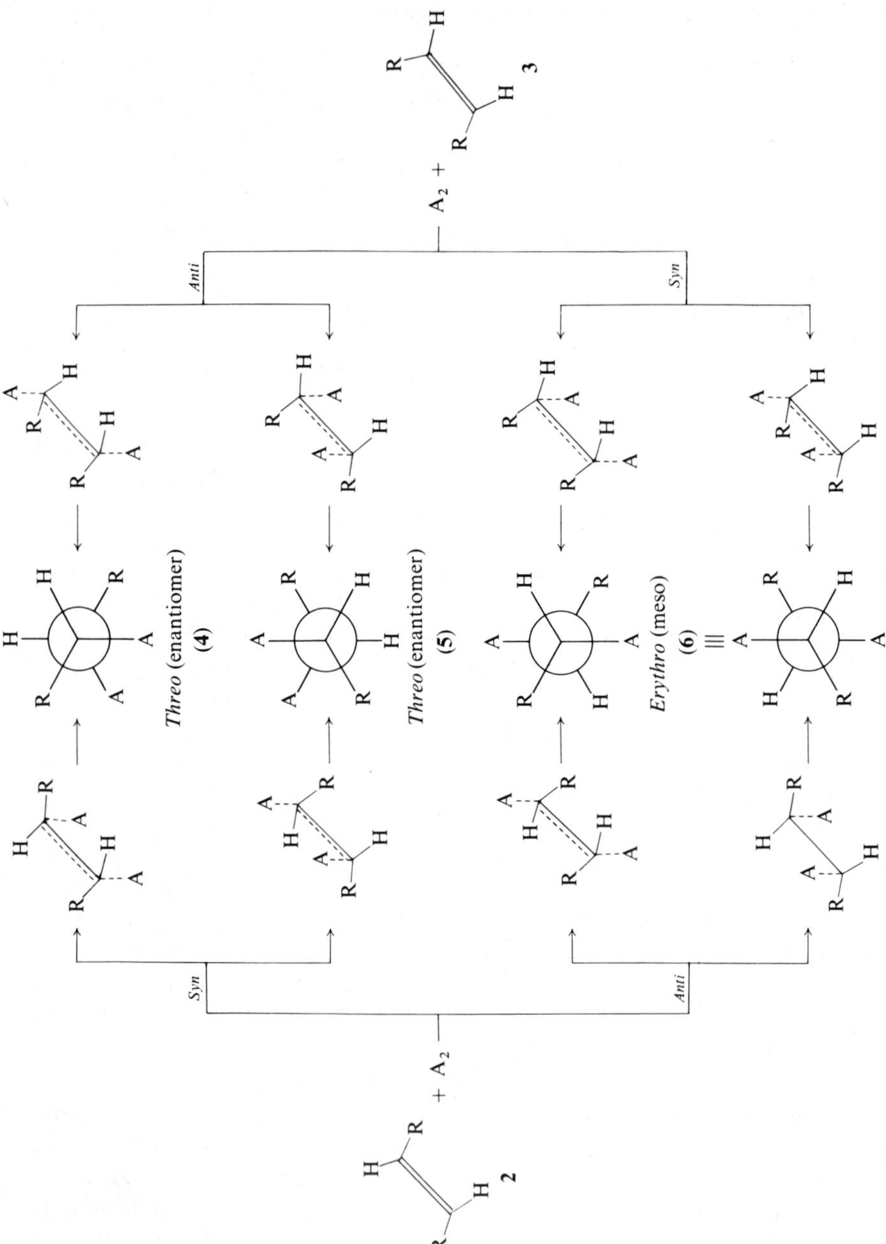

Figure 2. Stereospecific reactions illustrated by *syn* and *anti* additions of A_2 to the diastereomeric olefins, **2** and **3**.

the *threo* enantiomers, **4** and **5**. Clearly, we should expect an equal mixture because the corresponding transition states have an object-mirror image or enantiomeric relationship and possess, therefore, the same free energy. With a $\Delta\Delta G\ddagger$ equal to zero there can be no rate difference, and the two products must accumulate in equal amounts. Reactions that give equal amounts of enantiomers or racemic modifications are well known in many different forms and need not be further illustrated. The intriguing question is, however, what may be done so that such a reaction will give *unequal* amounts of enantiomers? The answer is that a dissymmetric influence must be brought to bear on the reaction so that the transition states leading to the enantiomeric products can no longer be in an enantiomeric relationship. This is the only way to have $\Delta\Delta G\ddagger$ not equal to zero. In other words, something must interact so that the transition states possess a diastereomeric relationship. This may be brought about in several different ways. For example, in the *syn* addition of A_2 to **2**, one enantiomeric form of an interacting dissymmetric solvent may be used. The two solvated transition states would then be diastereomerically related, and could, in principle, lead to enantiomer enrichment.† The addition may also be one that takes place in the presence of an enzyme where the reactants must "fit" onto the dissymmetric enzyme surface. Here, too, the transition states would possess a diastereomeric relationship. Enzymatic reactions generally give rise to such large $\Delta\Delta G\ddagger$ values that one enantiomer is formed to the exclusion of the other. Reactions such as these are of sufficient interest that they should be relegated to a special category of stereoselective processes: processes of *Asymmetric Selection*; those that give *enantiomers* in *unequal* amounts.*

It will be noted that although this definition of asymmetric selectivity is fairly precise it perhaps suffers from the fact that it is also fairly restrictive. It does not include a number of reactions that ordinarily one might intuitively wish to relegate to this category. For example, if R in **2** is dissymmetric then *syn* addition of A_2 would proceed through diastereomeric transition states and give products in unequal amounts. The same would be true if A_2 contains a dissymmetric grouping. But these situations do not give enantiomers. The products they provide (Fig. 3) are diastereomers. Such reactions are, therefore, merely stereospecific processes in which new dissymmetric groupings are formed. Although they may be closely related to processes of

† Actual experimental demonstrations of precisely this type of asymmetric selectivity have been provided by recent work at the University of South Carolina (2a): Two similar free radical coupling reactions carried out in the presence of an optically active solvent gave, in each case, a measurable enantiomeric enrichment among the dissymmetric products.

* An alternative designation, suggested by Professor George Wahl (Dept. of Chemistry, North Carolina State University) is *enantiomeric selection*. Perhaps an even more concise and apt designation would be *enantioselection*.

asymmetric selection, they should be distinguished from reactions in that category.

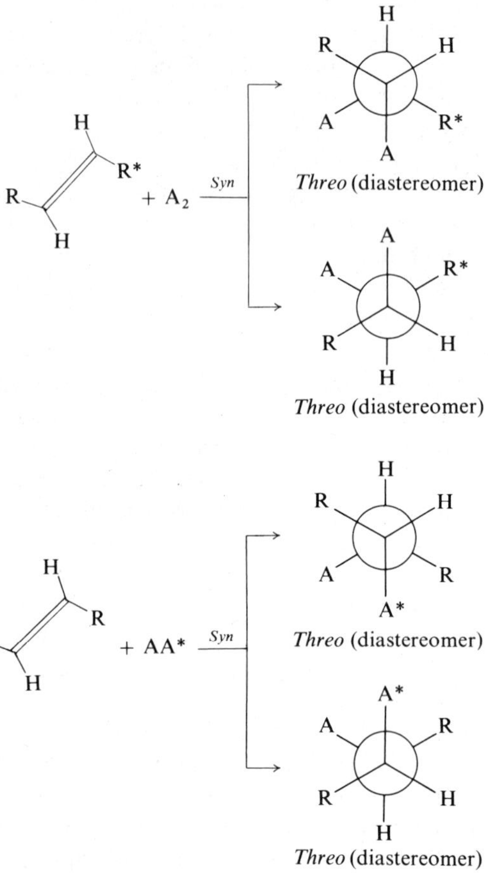

Figure 3. When either R* or A* is a dissymmetric grouping, the *syn* addition products are not enantiomers; they are diastereomerically related.

Nevertheless, there are borderline cases which include processes that *indirectly* lead to unequal mixtures of enantiomers. In such cases unequal mixtures of diastereomers are initially formed; they are then transformed in a subsequent step to *unequal* mixtures of enantiomers. An example (3) may be seen in the synthesis of optically active lupinine (**7**), illustrated in Figure 4. Sodium borohydride treatment of (−)-1-menthoxycarbonyl-1 (10)-dehydroquinolizidine (**8**) gave exclusive *syn* addition to yield the mixture of diastereomers, **9a** and **9b**. Removal of the dissymmetric menthyl group by the action of lithium aluminium hydride on the diastereomeric mixture provided an unequal mixture of the enantiomeric lupinines (**7**) in which the dextrorotatory form (**7a**) predominated.

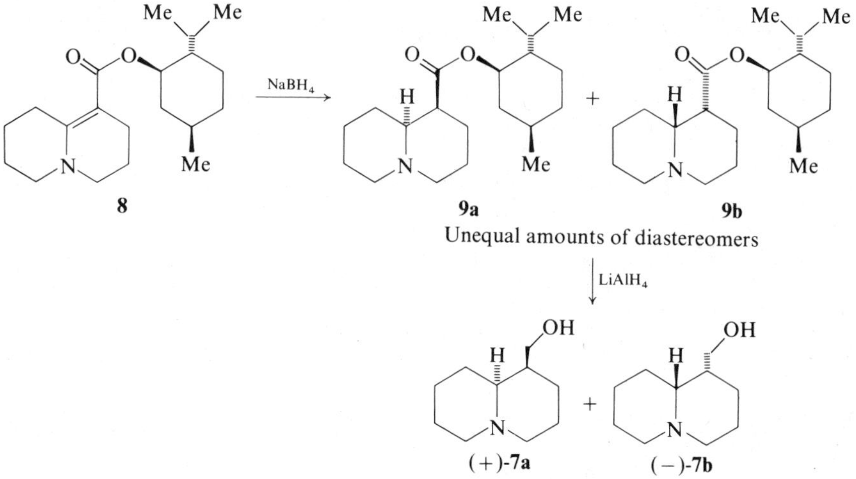

8 **9a** **9b**
Unequal amounts of diastereomers

(+)-**7a** (−)-**7b**
Unequal amounts of enantiomers

Figure 4. An example of asymmetric selection in which the enantiomeric enrichment is indirectly obtained.

Other examples from this category that may be cited (Table 1) are additions to dissymmetric glyoxalates (**10**), which gave diastereomeric adducts, **11** and **12**, initially, but these were then hydrolyzed to enantiomeric mixtures, **13** and **14**, containing a predominant form (4–8). Similarly, catalytic hydrogenation of dissymmetric *β*-methylcinnamates (**15**) were shown to give initially diastereomerically related *β*-phenylbutyrates (**16** and **17**) which provided an enantiomerically enriched mixture of *β*-phenylbutyric acids (**18** and **19**) upon hydrolysis (9). Finally, in a different application of the same principle, use of the dissymmetric hydroboration reagent, (−)-*sym*-tetraisopinocamphenyldiborane (**20**), in the treatment of a series of nondissymmetric (*Z*)-olefins gave diastereomerically related hydroboration products (**21** and **22**). In each case these mixtures were converted to unequal enantiomeric alcohols (**23** and **24**) (10).

Thus, although these processes do not involve a reaction that is enantiomerically selective, it is seen that they do combine a sequence consisting of a diastereomerically selective step with a stereospecific step that leads overall to enantiomers in unequal mixture.

It should also be noted that there are reactions that are not just enantiomerically selective in products, but that some may be enantiomerically selective in reactants as well. These reactions have been included in the

TABLE 1
Additional Examples of Processes that Give Enantiomeric Enrichment Indirectly

Dissymmetric reactant	⟶	Unequal amounts of diastereomers	⟶	Unequal amounts of enantiomers
10	R′MgX ⟶	11 + 12	Hydrolysis ⟶	13 + 14
15	H_2/Cat. ⟶	16 + 17	Hydrolysis ⟶	18 + 19
20	⟶	21 + 22	OH^- / H_2O_2 ⟶	23 + 24

categories of kinetic resolution or asymmetric destruction; and since they do provide an unequal mixture of enantiomers, such reactions also would be considered as processes of asymmetric selection. An example may be found in the treatment of the racemic mixture of (−)-(*S*)- and (+)-(*R*)-4-methylcyclohexenes, **25a** and **25b**, with a limited amount of the optically active hydroboration agent, (−)-*sym*-tetraisopinocamphenyldiborane (**20**). Evidently the transition states for the reactions between (−)-(*S*)- **25a** and **20** possess lower free energy than those for (+)-(*R*)-**25b** and **20** since (−)-(*S*)-**25a** was hydroborated faster, leaving an enrichment (partial resolution) of (+)-(*R*)- **25b** in the recovered starting material (11).

Stereoselective processes, therefore, that give either directly or indirectly unequal mixtures of enantiomers constitute the special category we wish to recognize and designate as processes of asymmetric selection.

II. ASYMMETRIC SELECTION VIA ELIMINATION

As noted above, there are many examples known of addition reactions or processes which meet the requirements of asymmetric selectivity. The vast majority of these reactions have been cited and discussed in a number of recent review articles (12–14). In comparison the number of examples embodying bona fide asymmetric selectivity in elimination reactions is very small. This condition is most probably a reflection of the fact that, compared to elimination reactions, addition reactions offer a much greater variety of experimental systems. Nevertheless, when given sufficient thought it is not difficult to devise a number of projected experimental systems each of which is capable, in principle, of allowing observation of asymmetric selection during an elimination process. Indeed, this realization along with the increasing growth of methods and techniques for preparation and determination of configurational and conformational isomers may be taken as strong indications of increasing activity in this area. While it is the major purpose of this article to present a review of work already reported in this field, an account of some additional possibilities shall be included as well. The organization of this discussion is based, solely for reasons of convenience, on the carbon skeleton of the main group involved in the elimination.

A. Even-Membered Rings

Conceptually, asymmetric selection may be observed during elimination of a chiral molecule, AH, from any even-membered ring that meets the following requirements. The systems, as represented in Figure 5, are substituted with a chiral group A and an achiral group R. These substituents are located on the ring so that the ring atoms to which they are attached remain achiral. Thus for cyclobutane [(**26**), $n = 0$] they are disposed 1,3-, for cyclohexane [(**26**), $n = 1$], 1,4-, etc. The relative configurations (*cis* or *trans*) of the two substituents are fixed and known in each instance. The total number of individual configurational isomers (with only two configurational possibilities within the chiral group A) of a given system is, therefore, four: *cis*-enantiomers and *trans*-enantiomers.

R
$(H_2C)_n$ $(CH_2)_n$
A
26
R
$(H_2C)_n$ $(CH_2)_n$
H
A
27a
R
$(H_2C)_n$ $(CH_2)_n$
28a
R
$(H_2C)_n$ $(CH_2)_n$
H
A
27b
R
$(H_2C)_n$ $(CH_2)_n$
28b

Figure 5. Asymmetric selection in even-membered rings via elimination of the chiral molecule AH (adapted from ref. 30).

Elimination of AH, therefore, from a given configurational isomer of **26** must proceed through diastereomerically related transition states, **27a** and **27b**—by whatever way one chooses to represent the transition states—to enantiomerically related olefins, **28a** and **28b**, directly. If the $\Delta\Delta G\ddagger$ of the transition states is large enough to provide a detectable enrichment of one enantiomer, this is most readily determined by the observation of optical activity in the product. Indeed, a measure of the degree of asymmetric selectivity is obtained by calculation of the optical yield or enantiomeric purity.*

* These equivalent quantities may be defined as the product of the ratio of observed specific rotation to the specific rotation of the pure enantiomer and 100.

While no experiments corresponding to this concept have been reported for any cyclobutane systems (**26**, $n = 0$), and although cyclooctane and larger ring systems (**26**, $n = 2$) are complicated by the fact that the double bond may form in (**E**) and the (**Z**) configurations (see discussion of this point under Section II-D), a number of experiments demonstrating the correctness of the concept have been carried out with a variety of cyclohexane systems.

1. A = p-Tolylsulfinyl

The bonding sequence corresponding to **26** with $n = 1$, R = methyl, and A = *p*-tolylsulfinyl represents 4-methylcyclohexyl *p*-tolyl sulfoxide, which may exist in four configurationally isomeric forms: *trans*-enantiomers and *cis*-enantiomers. Goldberg and Sahli (15,16) prepared the individual *trans*-enantiomers, **29a** and **29b**. Each sulfoxide was directly obtained by treatment of enantiomerically pure menthyl *p*-toluenesulfinate, **30a** and **30b** (17,18) with 4-methylcyclohexylmagnesium chloride. As shown in Figure 6 the

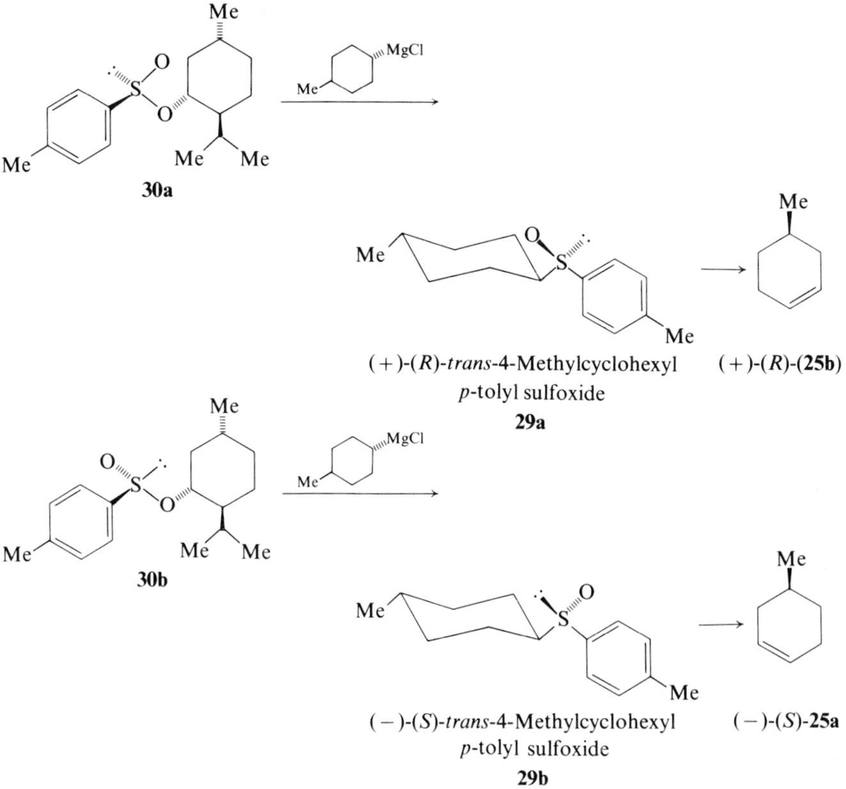

30a

(+)-(*R*)-*trans*-4-Methylcyclohexyl *p*-tolyl sulfoxide **29a**

(+)-(*R*)-(**25b**)

30b

(−)-(*S*)-*trans*-4-Methylcyclohexyl *p*-tolyl sulfoxide **29b**

(−)-(*S*)-**25a**

Figure 6. Preparation and pyrolysis of (+)-(*R*)-*trans*- and (−)-(*S*)-*trans*-4-methylcyclohexyl *p*-tolyl sulfoxides, **29a** and **29b** (adapted from ref. 16).

absolute configurational details of each sulfoxide were assigned. The configurations at sulfur followed directly from the known absolute sulfur configuration in each sulfinate ester (18), and from the fact that displacement of alkoxy from sulfinate by the action of Grignard reagent takes place with complete inversion (19). The *trans* configuration of the 4-methylcyclohexyl group in each sulfoxide was shown to result as a consequence of thermodynamic control during formation of the Grignard reagent from magnesium and either *trans*- or *cis*-4-methyl-1-chlorocyclohexane.*

Pyrolysis of each enantiomeric sulfoxide manifested asymmetric selection in that the olefinic product, 4-methylcyclohexene, was obtained optically active. Thus (+)-(*R*)-*trans*-4-methylcyclohexyl *p*-tolyl sulfoxide (**29a**) gave 4-methylcyclohexene enriched in the dextrorotatory (*R*)-enantiomer (**25b**), and the levorotatory (*S*)-enantiomer (**25a**) was the predominant form obtained from pyrolysis of the (−)-(*S*)-sulfoxide (**29b**). Since the absolute configurational details of each sulfoxide and of each predominant olefin were known it was possible to correlate the pyrolyses results in terms of absolute transition state topologies using the generally accepted (20), five-centered, transition state for pyrolytic *syn*-elimination of sulfoxides. This analysis is illustrated in Figure 7 with the (+)-(*R*)-sulfoxide

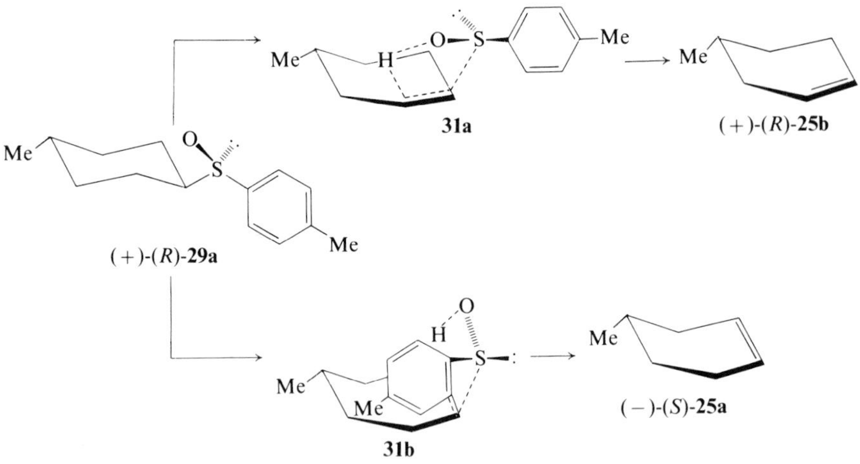

Figure 7. Transition-state analysis of results of pyrolytic *syn* elimination of (+)-(*R*)-*trans*-4-methylcyclohexyl *p*-tolyl sulfoxide (**29a**) (adapted from ref. 16).

(**29a**). The two diastereomeric transition states are represented by **31a** and **31b**. It is seen that in **31b** serious nonbonded interactions between the *p*-tolyl group and the cyclohexyl hydrogens are present, while the *p*-tolyl group in

* See ref. 16 for details as well as an account of previous work in support of this conclusion.

31a is relatively free of such interactions. This simple consideration indicates **31a** to be the lower energy pathway, and the experimentally observed preponderance of (+)-(*R*)-**25b** is therefore readily rationalized. Furthermore, from consideration of these transition-state models with their inherently large difference in degree of nonbonded interactions, one would expect intuitively a relative large $\Delta\Delta G\ddagger$. That this view is essentially correct is indicated by the exceptionally high optically yields observed for this pyrolytic process: pyrolysis at 250° gave 40% conversion to olefin with *42% optical yield*; at 226° olefin was obtained in 33% conversion, *optical yield was 51%*; and finally, pyrolysis at 200° provided the olefin in 15% conversion and *70% optical yield.*

2. *A = N-Methyl-N-phenyl-N-oxoamino*

Berti and Bellucci (21) carried out preparation and pyrolysis of optically active stereomers of an amine oxide system corresponding to **26**, $n = 1$, and A = *N*-methyl-*N*-phenyl-*N*-oxoamino. Catalytic hydrogenation of the enamine **32**, obtained from treatment of the diethyl ketal of 4-methylcyclohexanone with *N*-methylaniline, gave a mixture of *cis*- and *trans*-*N*-methyl-*N*-phenyl-4-methylcyclohexylamines, **33** and **34**. Fractional crystallizations of the hydrochlorides provided, however, only one of the diastereomers, and this was tentatively assigned *cis* (**33**) because its density and refractive index were each found to be higher than the corresponding values determined from the mixture of *cis* and *trans* amines. Oxidation of the isolated amine gave its amine oxide which was resolved with (−)-dibenzoyltartaric acid to provide the *cis*-enantiomeric amine oxides (**35a** and **35b**) with the dextrorotatory form in higher enantiomeric purity.

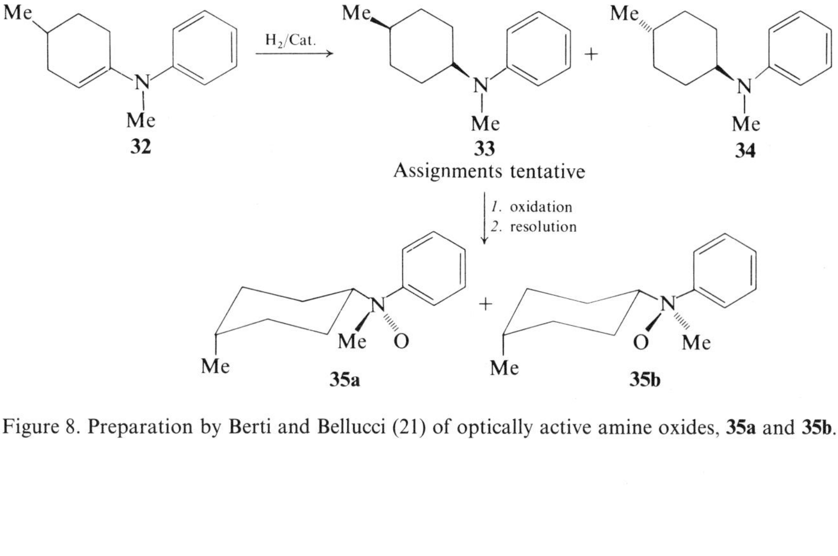

Figure 8. Preparation by Berti and Bellucci (21) of optically active amine oxides, **35a** and **35b**.

Pyrolysis of the (+)-amine oxide at 110° gave (+)-(*R*)-4-methylcyclohexene (**25b**) in 60% conversion and in 32% optical yield. (−)-(*S*)-4-Methylcyclohexene (**25a**) was obtained as the predominant enantiomer from pyrolysis of the levorotatory amine oxide of lower enantiomeric purity. If these results are analyzed in terms of a transition-state model that parallels that for pyrolytic *syn*-elimination in sulfoxides (20), and such a model is in fact generally accepted for amine oxide pyrolysis (22), then it is possible to deduce the absolute configurations at nitrogen. This analysis is presented in the subsequent section in connection with another amine oxide system where configurational details of the 4-methylcyclohexyl moiety were more reliably determined.

3. A = N-Methyl-N-neopentyl-N-oxoamino

In a study which had as one of its objectives the attainment of high degrees of asymmetric selectivity Goldberg and Lam (23) reasoned that the two groups attached to nitrogen of the amine oxide that are not directly involved in the elimination should possess a maximum difference in "bulkiness." This notion was based on the validity of the five-centered, planar transition state for pyrolytic *syn*-elimination of amine oxides (22). It appeared to be essentially correct, for the sulfoxide system (**29**) provided very high degrees of asymmetric selectivity. Since amine oxide and sulfoxide eliminations may be considered to be described by very similar transition-state models, the exceptionally high degrees of asymmetric selectivity observed with **29** may be ascribed to the large difference in "bulkiness" between the *p*-tolyl group and the electron pair. Accordingly, all four stereoisomeric *N*-methyl-*N*-neopentyl-4-methylcyclohexylamine oxides, in which the relevant groups were methyl and neopentyl, were individually prepared as indicated in Figure 9.

Evidently the same equilibrium mixture of azomethine compounds [(**37a**) $\rightleftharpoons$ (**37b**) $\rightleftharpoons$ (**37c**)] was obtained from reaction of pure *trans*-4-methylcyclohexylamine (**36**) (24) and pivaldehyde as was produced from treatment 4-methylcyclohexanone with neopentylamine, for in each case reduction (either by lithium aluminum hydride or hydrogenation over palladium on charcoal) gave the same mixture of secondary amines (**38** and **39**). The two were separated by preparative gas–liquid chromatography, and the stereochemical identifications were made according to the shapes of the signal envelopes of the cyclohexyl protons in the nuclear magnetic resonance spectra determined from each isomer (25). This correlation was also made with each of the tertiary amines, prepared by *N*-methylation of **38** and **39**. The amine oxides, **40** and **41**, obtained from each of the tertiary amines, also gave rise to nuclear magnetic resonance spectra that were consistent with the

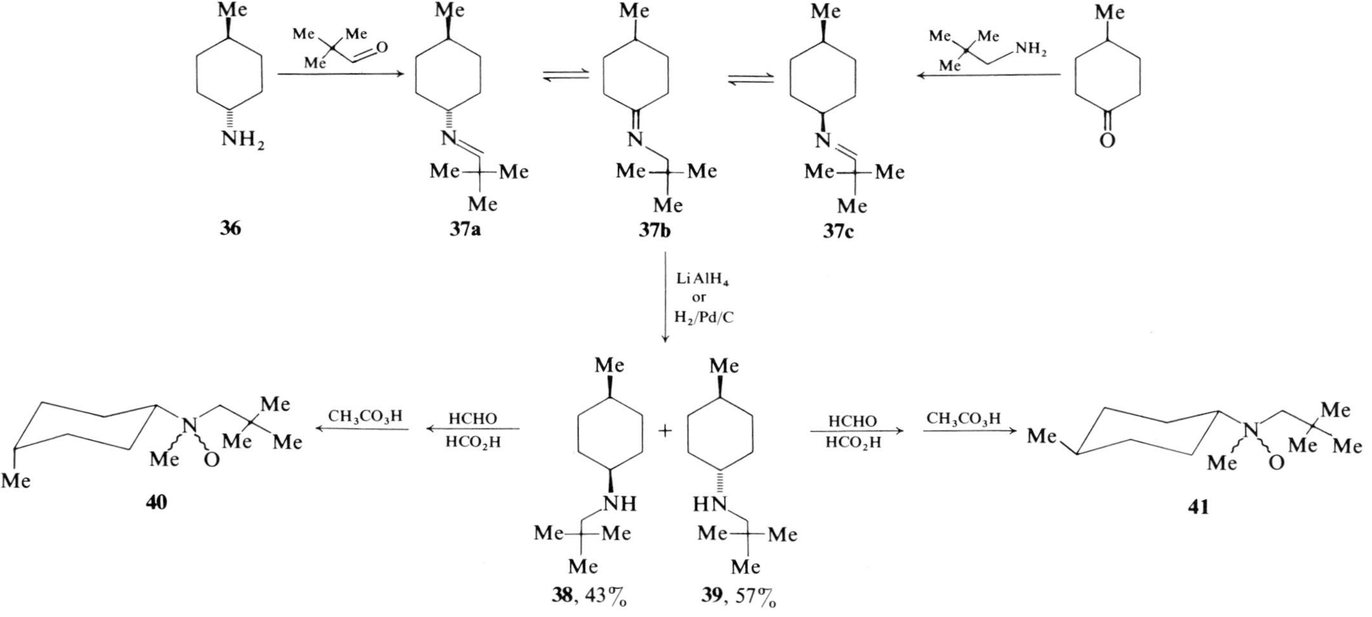

Figure 9. Preparation (23) of *cis*- and *trans*-amine oxides, **40** and **41**.

stereochemical identification. Resolution, while difficult, was at least partially accomplished with (−)-dibenzoyltartaric acid, giving each of the four possible configurational isomers with unknown enantiomeric purity. Each of these optically active amine oxides exhibited asymmetric selectivity in its pyrolysis, for the presence of a predominant enantiomeric 4-methylcyclohexene was evident by the optical activity displayed by the olefin isolated from each pyrolyzate. These data are presented in Table 2.

TABLE 2
Pyrolyses of Isomeric *N*-methyl-*N*-neopentyl-4-methylcyclohexyl Amine Oxides (23)

Amine oxide			Olefin	
Isomer	$[\alpha]_D$	Pyrolysis temperature	Isomer	$[\alpha]_D$
(−)-*cis*	−2.9°	120°	(−)-(*S*)-**25a**	−10°
(−)-*cis*	−2.9°	100°	(−)-(*S*)-**25a**	−10°
(−)-*cis*	−2.9°	85°	(−)-(*S*)-**25a**	−12°
(+)-*cis*	+2.9°	120°	(+)-(*R*)-**25b**	+7.8°
(+)-*cis*	+2.9°	100°	(+)-(*R*)-**25b**	+9.1°
(+)-*cis*	+2.9°	85°	(+)-(*R*)-**25b**	+13°
(−)-*trans*	−0.7	120°	(+)-(*R*)-**25b**	+2.0°
(−)-*trans*	−0.7°	100°	(+)-(*R*)-**25b**	+3.0°
(+)-*trans*	+2.4°	120°	(−)-(*S*)-**25a**	−5.8°

Although the unexpected high degree of asymmetric selectively did not materialize, probably because of very incomplete resolutions, enough data were obtained to allow assignment of absolute configuration to the nitrogen of each of the isomeric amine oxides. The deductions were based on analyses using the five-centered, coplanar transition state (22), and the argument may be illustrated in Figure 10, using the (−)-*trans*-amine oxide as a representative example.

The two contending sets of diastereomeric transition states are represented by **42a** and **42b**, with the (*S*) configuration at nitrogen, and by **43a** and **43b**, with the (*R*) configuration at nitrogen. In each set, **42a** and **43a** are judged to be the more favorable since the bulky neopentyl group is involved in much less nonbonded interactions with the cyclohexyl hydrogens than it is in **42b** and **43b**. Since the (−)-*trans*-amine oxide gave (+)-(*R*)- **25b** as the predominant enantiomeric 4-methylcyclohexene, its nitrogen configuration must be (*S*) as in **42a** and not (*R*) as in **43b** because the former leads to the

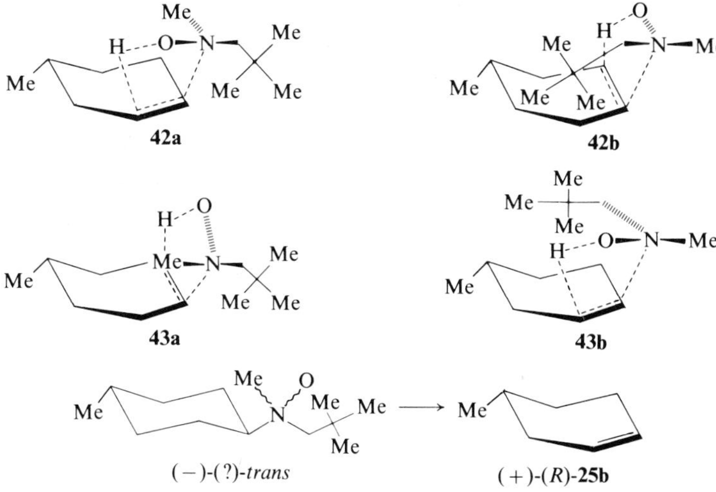

Figure 10. Transition state analysis (23) leading to assignment of (*S*) configuration at nitrogen in the (−)-*trans*-amine oxide.

experimentally observed (+)-(*R*)-**25b** while the latter predicts the unobserved (−)-(*S*)-**25a**.

In this way the nitrogen configuration for each of the stereoisomeric amine oxides was deduced (Fig. 11).

It may be noted as well that should the tentatively assigned *cis* configuration for the 4-methylcyclohexyl group in the Berti and Bellucci (21) amine oxides **35a** and **35b** be confirmed, than by a similar transition state

(−)-(*S*)-*cis*-**46a** → (−)-(S)-**25a** ← (+)-(*R*)-*trans*-**44b**

(−)-(*S*)-*trans*-**44a** → (+)-(*R*)-**25b** ← (+)-(*R*)-*cis*-**46b**

Figure 11. Configurational details of the stereoisomeric amine oxides, and the predominant enantiomeric 4-methylcyclohexene obtained on pyrolysis in each case (23).

analysis (Fig. 12), the nitrogen configuration in each may be established: the dextrorotatory form, which gave enrichment of (+)-(*R*)-**25b**, would be (*S*), **35b** (Fig. 8), and **35a** (Fig. 8) would be assigned to the levorotatory enantiomer.*

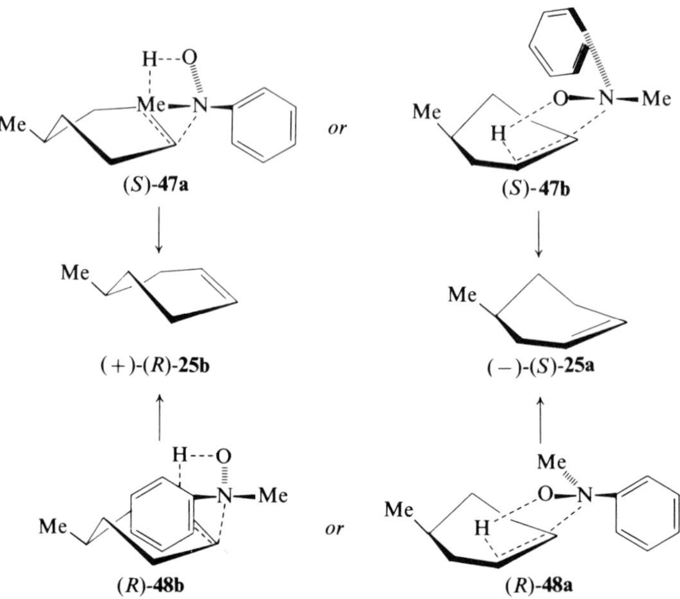

Figure 12. Transition state analysis leading to assignments of the (*S*) nitrogen configuration (**47a**) in the (+)-*cis*-amine oxide (**35b**) and the (*R*) nitrogen configuration (**48a**) in the (−)-*cis*-amine oxide (**35a**) of Berti and Bellucci (21).

4. A = Methylphenylphosphinyl

An interesting situation emerged from the preliminary work (27) carried out on the pyrolysis of the optically active phosphine oxide system corresponding to **26**, $n = 1$, with A = methylphenylphosphinyl. The dextrorotatory phosphine oxide, probably (*R*)-*trans*-**49**,† was prepared by displacement [inversion of phosphorus configuration (28)] of the (−)-menthoxy group in (−)-menthyl (*S*)-methylphenylphosphinate (**50**) through the action of 4-methylcyclohexylmagnesium chloride (16), as illustrated in Figure 13.

* It is amusing to note that as a consequence of the priority rules (26) **35b** and **46b**, which actually possess the same configurational sense at nitrogen, must be given opposite configuration specifications.

† Although combustion analysis showed this material to be constitutionally pure, the failure to induce its crystallization may have admitted the question of its diastereomeric purity.

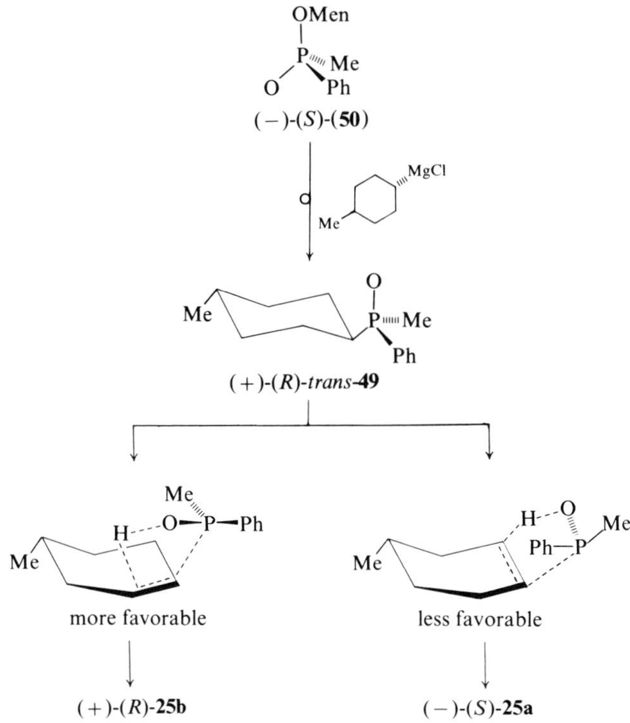

Figure 13. Preparation and pyrolysis (27) of (+)-(*R*)-*trans*-methylphenyl-4-methylcyclohexylphosphine oxide (**49**).

Unlike corresponding sulfoxide and amine oxide systems, remarkably high temperatures were required to effect pyrolysis of **49** to 4-methylcyclohexene. Thus at 460° only a trace (less than 1% conversion) of the olefin was formed; and, with pyrolysis temperatures of 505 and 527°, 4-methylcyclohexene was formed in 13 and 17% conversion, respectively. Even more remarkable was the fact that the olefin isolated from the 505° pyrolyzate was optically active with $[\alpha]_D$ − 2.0°, showing that the processes took place with asymmetric selectivity giving (−)-(*S*)-**25a** as the more abundant enantiomer in 1.9% optical yield.

The interesting feature of these results lies in the fact that, with the absolute configurational details of the (+)-phosphine oxide as shown in **49**, transition state analysis (Fig. 13) with the five-centered, coplanar model, as used for the sulfoxides and amine oxides, leads to prediction of enantiomeric enrichment of (+)-(*R*)-**25b**, but (−)-(*S*)-**25a** was shown experimentally to be the enriched olefin.

These results (27) appear to suggest that phosphine oxides pyrolyze in a manner different from that of sulfoxides and amine oxides. Indeed, the dramatically higher pyrolysis temperature required for the phosphine oxide as compared to the sulfoxide and amine oxide systems would appear to support this view, but until the configurational details given in **49** are corroborated any interpretation must be considered as tentative.

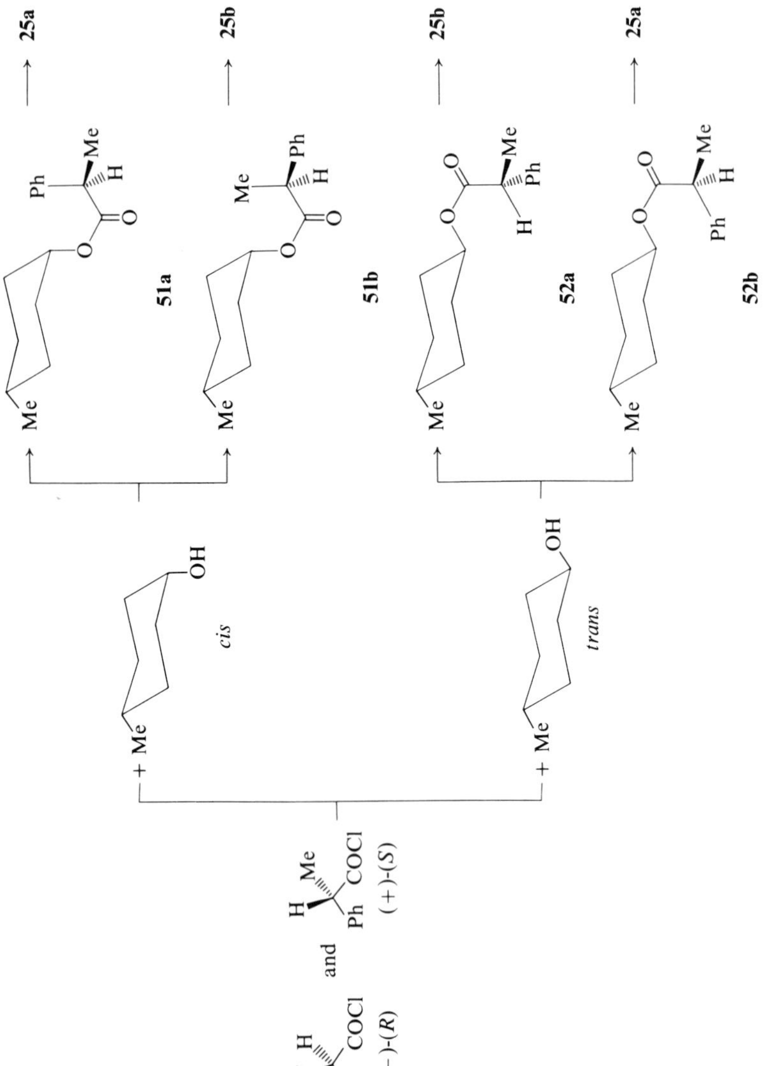

Figure 14. Pyrolysis of each stereoisomeric 4-methylcyclohexyl hydratropate took place with asymmetric selection, giving a small enantiomeric enrichment in the olefinic product (adapted from ref. 30).

5. *A = Hydratropoxy*

Historically, the first experimental demonstration of asymmetric selectivity in an elimination process was carried out with the isomeric esters of the 4-methylcyclohexyl hydratropates, corresponding to **26**, $n = 1$, A = hydratropoxy (29). Each of the four stereoisomeric esters of known absolute configurational features, (−)-(*R*)-*cis*-**51a**, (+)-(*S*)-*cis*-**51b**, (−)-(*R*)-*trans*-**52a**, and (+)-(*S*)-*trans*-**52b**, were prepared and pyrolyzed to enantiomerically enriched 4-methylcyclohexene as indicated in Figure 14.

Because optical yields were very low (less than 1%), and because of the unknown conformational factors in esters, it was not possible to place a transition-state analysis of these results on anything like a firm basis. Low levels of asymmetric selectivity, however, were to be expected, for the chiral alpha–carbon atom is not directly involved in the transition state of the elimination process as is the chiral sulfur and nitrogen atoms in the sulfoxide and amine oxide eliminations. If it is assumed that the level of asymmetric selectivity observed in the pyrolyses of these esters was in fact an accurate reflection of the degree of influence of the chiral alpha-carbon atom, and if the assumptions regarding preferred ester conformations were correct, then the authors' (30) heuristic transition state analysis may prove to be valuable. Given the correctness of these assumptions, the results may serve to distinguish between two different, six-centered, transition-state models, the "parallel" model (**53**) or the "perpendicular" model (**54**), for pyrolytic *syn*-elimination in esters. Within the framework of their assumptions, Goldberg and Lam argued in favor of the "parallel" model (**53**) (Fig. 15).

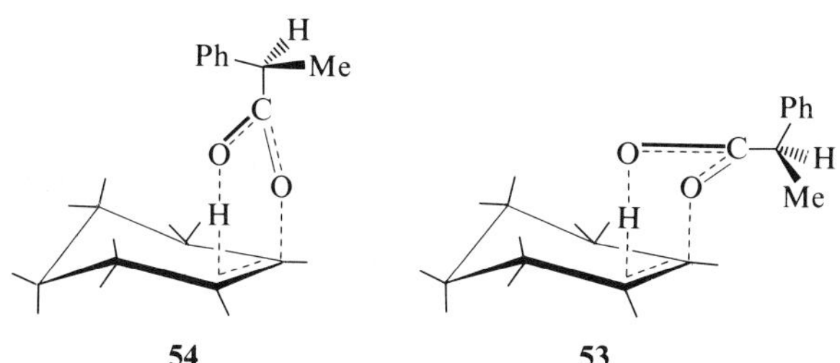

Figure 15. "Parallel" (**53**) and "perpendicular" (**54**) transition state models for pyrolytic *syn* elimination in esters (adapted from ref. 30).

B. Odd-Membered Rings

Using the same principle as in the cases already discussed, it is relatively straightforward to conceive of simple odd-membered ring systems in which elimination of the chiral molecule AH would be expected to exhibit asymmetric selectivity. Some possibilities are illustrated in Figure 16. One of these has been recently realized. See discussion under Section IV.

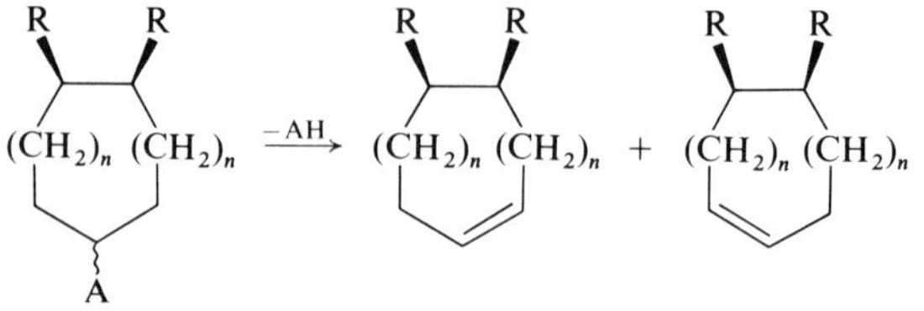

One general system is represented by **55** where the identical R groups and the chiral group A are disposed about the odd-membered ring in manner that does not render any of the ring atoms holding these groups chiral. Thus the two R groups must be *cis*, although A may be either *cis* or *trans* to the R groups. The total number of configurational isomers, therefore, in a given system would be four; and elimination of AH from each stereomer would be expected to provide an enrichment of one or the other enantiomeric (Z)-olefins, **56a** or **56b**. In the systems represented by **55** in which $n > 1$ a complicating feature will be introduced by the formation of (E)-olefin along with or exclusive of (Z)-olefin, for the cyclic (E)-olefinic linkage is itself chiral. This topic is discussed separately under Section II-D.

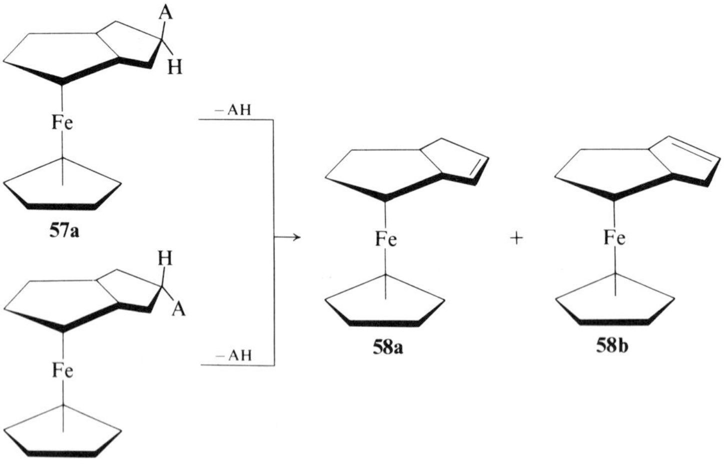

Figure 16. Projected asymmetric selection via elimination of the chiral molecule AH from either *exo*- or *endo*-substituted cyclopentaferrocenes, **57a** or **57b**, to the enantiomeric olefins, **58a** and **58b**.

Another system with no chiral elements other than the group A, but which possesses an odd-membered ring from which AH may be eliminated to give a predominant enantiomeric olefin, **58a** or **58b**, is the cyclopentaferrocene, **57**. Here too four individual configurational isomers, *exo* enantiomers (**57a**) and *endo* enantiomers (**57b**), would be available for separate experiments. A great deal of work on correlation and assignment of configurations in dissymmetric or chiral ferrocenes has been carried out in Professor Schögel's laboratory (31), and this information should be very useful in a study of asymmetric selectivity in elimination reactions of systems represented by or similar to **57**.

C. Bridged Rings

Any number of nondissymmetric bridged ring systems could be devised which would give dissymmetric olefins upon elimination of a chiral substituent. A number of the tropane alkaloids (32) would appear to be convenient sources of such bridged ring systems as seen, for example, in the nondissymmetric, stereomeric forms of the tropinyl (**59**), hyoscinyl (**60**), and granatolinyl (**61**) groups. Each of these would provide a corresponding dissymmetric or chiral, bridged, cyclic olefin through elimination, and with a dissymmetric or chiral substituent as the eliminating moiety there should be, in principle (asymmetric selection), an enantiomeric enrichment during elimination. The alkaloid (−)-hyoscyamine (**62**) meets these requirements and is conveniently provided by nature. Its nondissymmetric tropinyl group is fixed in only one of the two possible configurations, and its chiral tropoxy substituent occurs in the (*R*) configuration.

In independent investigations of Bothner-By, Schultz, Dawson, and Solt (33) and Leete (34) pyrolysis of hyoscyamine to tropidine (**63**) was carried out, but examination of the olefinic product for optical activity was not reported from either laboratory probably because these studies were concerned with other questions. Goldberg and Lam (30), however, carried out several careful pyrolyses of hyoscyamine but were not able to detect the presence of optical activity in the tropidine (**63**) obtained. These experiments were judged, however, to be inconclusive, for it was found that hyoscyamine racemizes rapidly at temperatures considerably lower than those required for a reasonable elimination rate. Subsequent experience (30) indicated that the use of a chiral carbalkoxy group in such work was probably a poor choice as judged from the very low degrees of assymmetric selection observed during pyrolyses of the individual 4-methylcyclohexyl hydratropates, **51a**, **51b**, **52a**, and **52b**. The bridged ring systems, **59**, **60**, and **61**, therefore should be investigated with a chiral substituent such as an arylsulfinyl group.

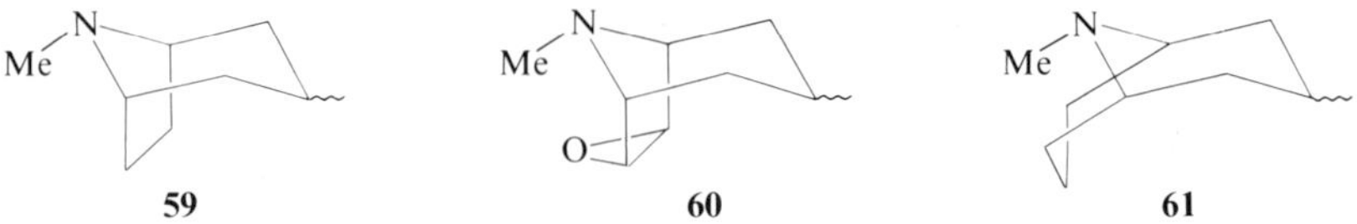

Figure 17. Bridged ring systems related to naturally occurring substances as potential candidates for elimination reactions displaying asymmetric selection.

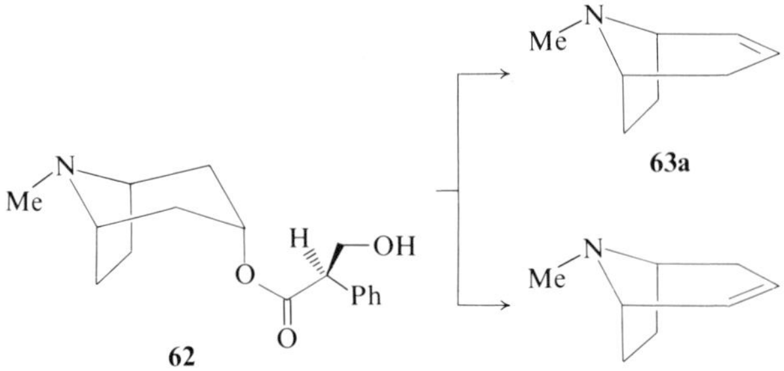

D. Medium-Sized Rings

The ability of rings that are larger than seven-membered to accommodate a *trans* or (*E*) double bond holds special significance in the elimination of a chiral molecule, AH, from such a ring, for the (*E*)-cycloalkene that may be formed is itself chiral or dissymmetric. The chirality of (*E*)-cycloalkenes is, however, a conformational and not a configurational condition; that is to say, the enantiomers of a given (*E*)-cycloalkene, **64a** and **64b**, are interconvertable merely through bond rotations. Breaking and remaking of bonds are not necessary. Interconversion of enantiomeric (*E*)-cycloalkenes (racemization) may be viewed simply in terms of the ability of the chain of methylene groups to loop over the double bond.

(*E*)-Cyclooctene (**64**, $n = 1$) was resolved through its diastereomeric platinum complexes (35) and found to possess a reasonable optical stability. Subsequent work established the correlation of molecular shape with rotatory direction (36). Thus (+)- and (−)-(*E*)-cyclooctene were determined to be (*S*)-**64a** and (*R*)-**64b**, $n = 1$, respectively. The energy barrier to interconversion of these conformational enantiomers was determined from kinetic data to be 35.6 ± 0.9 kcal/mole at 155.3° (37). As would be expected from the racemization mechanism described above, the barrier for interconversion of the enantiomeric (*E*)-cyclononenes (**64**, $n + n = 3$) is less, but perhaps surprisingly a great deal less since the loop of methylenes is increased by only one unit. This barrier was determined to be 19.1 ± 0.2 kcal/mole at −10° (38). The rather dramatic difference in conformational mobility of the two systems is made clearer by the fact that the half-life of (*E*)-cyclooctene enantiomer is near 120 hr at 133°, while that for enantiomeric (*E*)-cyclononene is only about 4 min at 0° (37,38). In view of these data is it not surprising that it was not possible to detect any optical activity in samples of (*E*)-cyclodecene (**64**, $n = 2$) obtained by displacement from diastereomeric platinum complexes even at −80° (38).

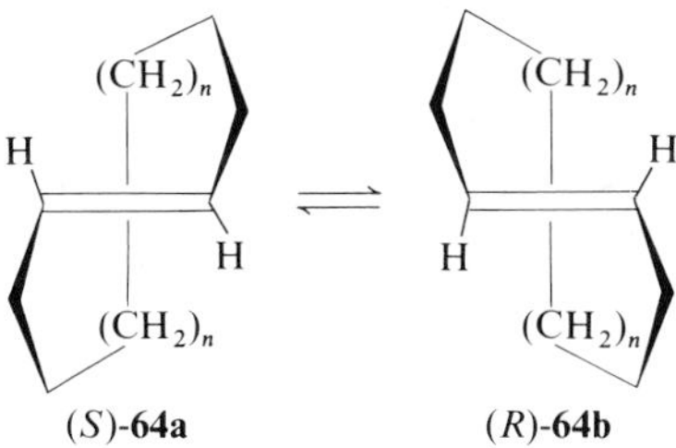

Figure 18. Interconversion or racemization of enantiomeric (*E*)-cycloalkenes involves only conformational changes.

Against the background of this information on the conformational or optical stability of (*E*)-cycloalkenes we may now view the experiments in which medium-sized ring systems were used to observe asymmetric selectivity in elimination reactions. Cope and co-workers (39) have successfully observed asymmetric selectivity during Hofmann elimination of an optically active cyclooctyl quaternary ammonium hydroxide. A great deal of preliminary work (40) was required in order to find that combination of groups on nitrogen which, in addition to rendering the nitrogen chiral, would provide favorable properties for resolution and also give rise to formation of a substantial amount of (*E*)-cyclooctene upon pyrolysis. These requirements were met by *N*-*n*-butyl-*N*-isobutyl-*N*-methylcyclooctylammonium hydroxide (**65**) (39). The compound was resolved to 99.6% enantiomeric purity (isotope dilution technique) and pyrolyzed under reduced pressure to a 53% yield of a mixture of (*Z*)- and (*E*)-cyclooctene (**67** and **66**) in which the latter predominated (33:67). The (*E*)-olefin was separated and found to be optically active. Pyrolysis of (−)-**65** gave (*E*)-olefin in which the levorotatory (*R*)-enantiomer (**66b**) predominated, and the dextrorotatory (*S*)-enantiomer was the predominant dissymmetric olefin produced when (+)-**65** was pyrolyzed. A transition-state analysis of these results, which could conceivably have led to assignment of configuration of nitrogen, was not reported (39). Indeed, there are at least two major factors that would have mitigated against the tenability of such an analysis. First, decisions regarding favored conformations of the *n*-butyl and isobutyl groups attached to nitrogen would have been very difficult and open to serious question, especially since the optical yields were low (0.7–1.4%). The second factor may lay in the fundamental question of how the transition state of the Hofmann elimination should be viewed. This question was emphasized by the fact that stereochemical results opposite to those obtained from the pyrolytic decompositions were produced when the enantiomeric ammonium hydroxides were fragmented in liquid ammonia containing potassium amide (39). Although the E2 mechanism has been generally accepted as the one with which most data on the Hofmann elimination can be most satisfactorily explained (41), recently Sicher and Závada (42) have presented evidence which indicates that in some reactions of cycloalkyltrimethylammonium compounds the (*Z*)-cycloalkenes are formed by *anti*-elimination and the (*E*)-cycloalkenes are produced by *syn*-elimination.

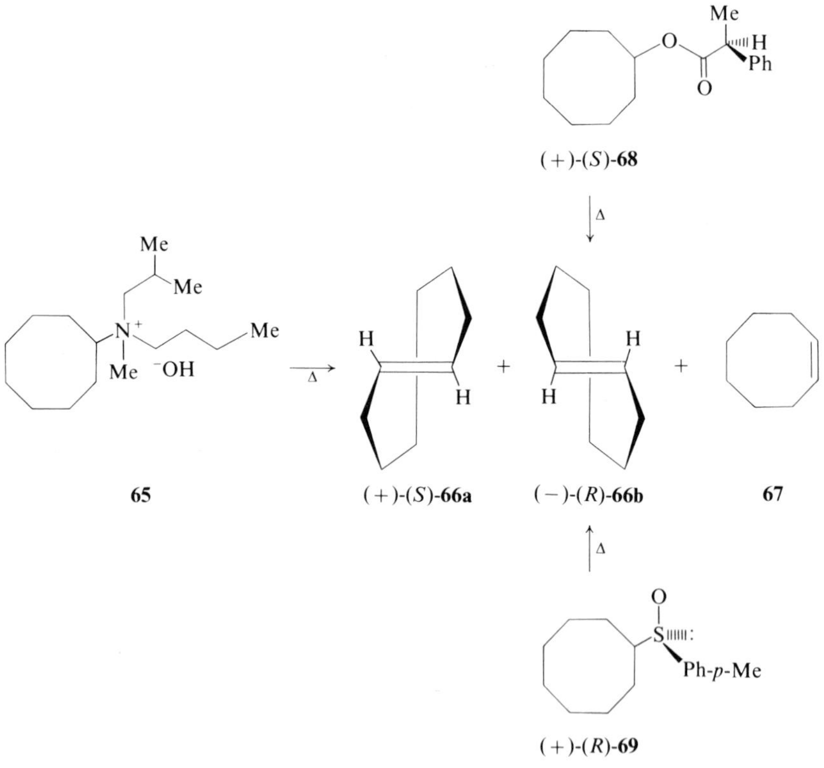

(+)-(*S*)-**68**

65 (+)-(*S*)-**66a** (−)-(*R*)-**66b** **67**

(+)-(*R*)-**69**

Figure 19. Of the cyclooctyl systems tried, only the quaternary ammonium hydroxide (**65**) (39) exhibited asymmetric selectivity in its pyrolytic elimination.

In another experiment designed to effect asymmetric selectivity during elimination, Červinka, Budilová, and Daněček (43) prepared (+)-(*S*)-cyclooctyl hydratropate (**68**) and submitted the optically active ester to pyrolytic elimination conditions (360°) to obtain a mixture of (*Z*)- and (*E*)-cyclooctenes (**67** and **66**) in the ratio of 12:1. Examination of the separated dissymmetric (*E*)-olefin did not reveal the presence of an enantiomeric enrichment since no optical activity could be detected. These authors also carried preparation and pyrolysis of (+)-(*S*)-cyclodecyl hydratopate and, as was expected from the optical stability work of Cope and co-workers (38), no optical activity could be detected in the (*E*)-cyclodecene separated from the pyrolyzate. While it was reasonable to expect the (*E*)-cyclooctene (**66**) obtained from pyrolysis of (+)-(*S*)-(**68**) to be optically stable (38,39), use of the hydratropoxy group as the leaving chiral moiety was probably a poor choice as indicated by the low optical yields obtained in earlier studies (29,30) on the pyrolytic eliminations in 4-methylcyclohexyl hydratropates. Based on this judgement, Goldberg and Davis (44) attempted to use the *p*-tolylsulfinyl group as the chiral moiety attached to cyclooctyl (**69**) since remarkably high degrees of asymmetric selection were obtained with this group in previous work (15,16). This strategy was thwarted, however, when it was learned (44) that only trace amounts (0.2%) of (*E*)-cyclooctene were produced in pyrolyses of cyclooctyl aryl sulfoxides.

III. CONCLUSION

It has been the purpose of this chapter to review the work on those elimination reactions that fall within the boundaries set for asymmetric selectivity. It may be argued that these boundaries are too narrow and that they exclude reactions that are closely related to, and bear importantly on, the very reactions under examination. For example, the Hofmann degradation of optically active *N*,*N*-dimethyl-(*Z*)-cyclooctylamine (**70**) to optical active (*Z*,*E*)-1,5-cyclooctadiene (**71**) (45,46) is clearly within the context of the discussion on medium-sized rings. Similarly, the pyrolytic eliminations of the optically active steriodal sulfoxides, **72** and **74**, to the optically active steriodal olefins, **73** and **75** (47) should not be overlooked in any discussion of the stereochemistry of sulfoxide elimination. Neither should one avoid taking into consideration the steric course of the eliminations followed in the pyrolyses of *N*,*N*-dimethylneomenthylamine oxide (**76**) and *N*,*N*-dimethylmenthylamine oxide (**78**) to 2-menthene (**77**) and a mixture (64.8:35.2) of 2- and 3-menthene (**77** and **79**), respectively (48).

The relevance of these and of other reactions is readily admitted, and it is not intended that they should be ignored on the basis of a technicality.

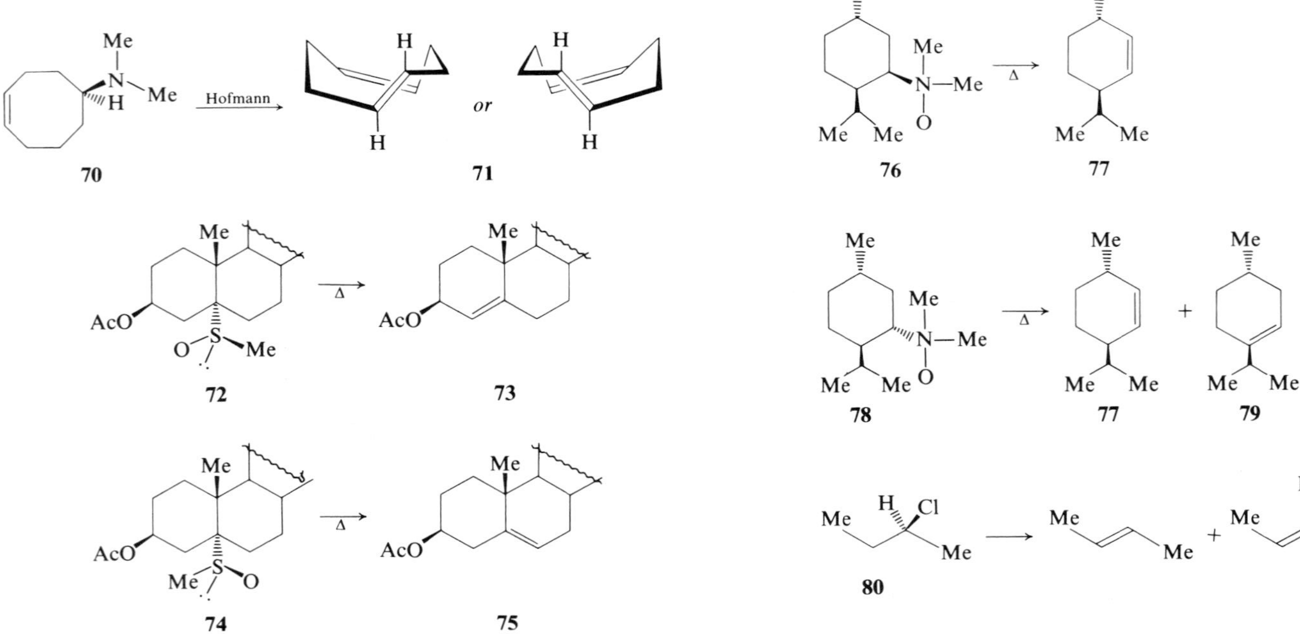

Figure 20. Examples of stereoselective reactions involving optically active components which are inherently incapable of asymmetric selection.

Nevertheless, they do fall outside of the definition of asymmetric selectivity in that they are inherently incapable of enantiomeric enrichment, and it is important to be able to draw distinctions among related reactions on a rational basis. For example, few people are likely to argue that the dehydrochlorination of optically active 2-chlorobutane (**80**) to give a preponderance of (*E*)-2-butene is a process within the context of the present discussion. Yet without the boundaries or criteria suggested here there is no general basis upon which to distinguish this simple stereospecific reaction from any of the special types of stereoselective processes discussed in this review. Decisions based merely on the presence of optical active components are nondiscriminatory, as are choices based on the formation of a new chiral or dissymmetric unit. Only the criterion of enantiomeric enrichment seems to provide the best basis for separate classifications.

Although the number of elimination reactions that experimentally display asymmetric selectivity are not many, it is reasonable to expect that they will grow not only in number and type but also in ingenuity of design. This ought to be the case particularly in the area of eliminations in medium-sized rings to dissymmetric (*E*)-cycloalkenes. Eliminations from cyclononanes and cyclodecanes give the respective (*E*)-cycloalkene as the major olefinic product, but these are, of course, optically (conformationally) unstable. With the proper type and degree of substitution, however, it ought to be possible to increase the energy barrier to interconversion of the conformational enantiomers to such an extent that the rate of interconversion would be reasonably slow. Cope and Fordice (49) have already provided some information along these lines. They found that instead of raising the energy barrier to conformational mobility in (*E*)-cyclononene, a benzene ring fused opposite the double bond actually appeared to introduce a greater degree of conformational mobility. Investigations of other substituents in this context do not appear to have been undertaken.

It may also be expected that some of the elimination reactions embodying asymmetric selectivity will be developed as sensitive probes for gathering information on the detailed structure of transition states; but, perhaps the most important aspect of these novel reactions are the questions, hence the future research, to which they give rise.

IV. Addenda

Since the initial writing of this article some additional work relevant to the topic under discussion has appeared (50). Asymmetric selectivity has been observed during elimination of the elements of water (dehydration) from achiral (nondissymmetric) alcohols corresponding to **26** (Figure 5), R = methyl, $n = 1$, and A = hydroxyl, and corresponding to **55** (Figure

16), R = methyl, n = 0, and A = hydroxyl. This work is of particular interest because, unlike the systems already discussed in this article where elimination of a chiral grouping is the basis of the asymmetric selectivity, these molecules are completely achiral, and the asymmetric selectivity has its origin in the presence of an optically active catalyst which causes formation of the chiral olefinic products in each case with an inequality of enantiomers (optical activity).

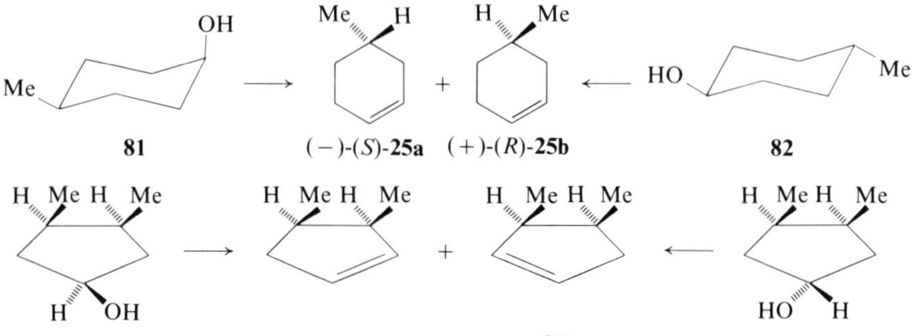

Figure 21. Dehydration of achiral alcohols, catalyzed by the presence of (+)-camphorsulfonic acid, gave optically active olefins (adapted from ref. 50).

Individual heating of each achiral alcohol, **81**, **82**, **83**, and **84**, in the presence of (+)-camphorsulfonic acid gave the corresponding olefin in an optically active state as outlined in Figure 21. It is of further interest to note that while **81** gave a product enriched in (−)-(*S*)-**25a** and its epimer, **82**, provided enrichment of the enantiomeric olefin, (+)-(*R*)-**25b**, the alcohols **83** and **84** each gave the olefin *cis*-3,4-dimethylcyclopentene (**85**) possessing an inequality in favor of the levorotatory enantioer, **85a** ro **85b** since absolute stereochemistry is as yet unknown. Another point is that the overall level of asymmetric selectivity indicated in these experiments by the rotatory magnitude of the olefinic products should, in principle, be the algebraic sum of two separate events. The first is the asymmetric selectivity during the actual olefin-forming process, and the second is the possibility of additional asymmetric selectivity in the form of kinetic resolution brought about by diastereomeric interactions between the individual olefin enantiomers and the chiral acid. These two separate events could be either opposing or reinforcing. They were in fact evaluated by treatment of samples of the racemic olefins **25** and **85** with (+)-camphorsulfonic acid. In these separate experiments only recovered **85** displayed optical activity. The level, however, was much lower than that present in the sample of **85** produced in the dehydration experiments, indicating the presence of a very small kinetic resolution increment in the dehydrations of **83** and **84**, but not in the dehydrations of **81** and **82**.

The preliminary results pose intriguing questions as regards the role of the chiral acid in the acid catalyzed dehydration reaction since this process is usually regarded as one of specific acid catalysis.

Acknowledgment

The author wishes to thank the donors to the Petroleum Research Fund of the American Chemical Society for a grant which provided helpful financial support in the course of the preparation of this chapter.

References

1. Chemical Abstracts Staff, *J. Am. Chem. Soc.*, **90**, 509 (1968).
2. K. Mislow, *Introduction to Stereochemistry*, W. A. Benjamin, New York, 1965, pp. 122ff.
2a. S. I. Goldberg and W. D. Bailey, *J. Am. Chem. Soc.*, **91**, 5685 (1969).
3. S. I. Goldberg and I. Ragade, *J. Org. Chem.*, **32**, 1046 (1967).
4. A. McKenzie and H. B. Thompson, *J. Chem. Soc.*, **1905**, 1004; and many subsequent papers.
5. V. Prelog, *Helv. Chim. Acta*, **36**, 308 (1953).
6. V. Prelog and H. Meier, *Helv. Chim. Acta*, **36**, 320 (1953).
7. W. G. Dauben, D. F. Dickel, O. Jeger, and V. Prelog, *Helv. Chim. Acta*, **36**, 325 (1953).
8. V. Prelog, E. Philbin, E. Watanabe, and M. Wilhelm, *Helv. Chim. Acta*, **39**, 1086 (1956).
9. V. Prelog and H. Scherrer, *Helv. Chim. Acta*, **42**, 2227 (1959).
10. H. C. Brown, N. R. Ayyanger, and G. Zweifel, *J. Am. Chem. Soc.*, **86**, 397 (1964).
11. S. I. Goldberg and F-L. Lam, *J. Org. Chem.*, **31**, 240 (1966).
12. J. D. Morrison, "Asymmetric Reduction," in *Survey of Progress in Chemistry*, Vol. 3, A. F. Scott, Ed., Academic Press, New York, 1966, p. 147.
13. J. Mathieu and J. Weill-Raynal, *Bull. Soc. Chim. France*, **1968**, 1211.
14. D. R. Boyd and M. A. McKervey, *Quart. Rev. (London)*, **22**, 95 (1968).
15. S. I. Goldberg and M. S. Sahli, *Tetrahedron Letters*, **1965**, 4441.
16. S. I. Goldberg and M. S. Sahli, *J. Org. Chem.*, **32**, 2059 (1967).
17. H. F. Herbrandson, R. T. Dickerson, Jr., and J. Weinstein, *J. Am. Chem. Soc.*, **78**, 2576 (1956).
18. K. Mislow, M. M. Green, P. Laur, J. T. Melillo, T. Simmons, and A. L. Ternay, Jr., *J. Am. Chem. Soc.*, **87**, 1958 (1965).
19. P. Bickart, M. Axelrod, J. Jacobus, and K. Mislow, *J. Am. Chem. Soc.*, **89**, 697 (1967).
20. C. A. Kingsbury and D. J. Cram, *J. Am. Chem. Soc.*, **82**, 1810 (1960).
21. G. Berti and G. Bellucci, *Tetrahedron Letters*, **1964**, 3853.
22. A. C. Cope and E. R. Trumbull, *Org. Reactions*, **11**, 362 (1960).
23. S. I. Goldberg and F.-L. Lam, *J. Am. Chem. Soc.*, **91**, 5113 (1969).
24. D. R. Smith, M. Maienthal, and J. Tipton, *J. Org. Chem.*, **17**, 294 (1952).
25. S. I. Goldberg, F.-L. Lam, and M. S. Sahli, *J. Chem. Eng. Data*, **14**, 406 (1969).
26. R. S. Cahn, C. K. Ingold, and V. Prelog, *Angew. Chem., Intern. Ed.*, **5**, 385, 511 (1966); *Experientia*, **12**, 81 (1956); *J. Chem. Ed.*, **41**, 116, 508 (1964).
27. Unpublished preliminary results of a collaborative effort between laboratories at the University of South Carolina (S. I. Goldberg) and Princeton University (K. Mislow).
28. O. Korpium, R. A. Lewis, J. Chickos, and K. Mislow, *J. Am. Chem. Soc.*, **90**, 4842 (1968).
29. S. I. Goldberg and F.-L. Lam, *Tetrahedron Letters*, **1964**, 1893.
30. S. I. Goldberg and F.-L. Lam, *J. Org. Chem.*, **31**, 2336 (1966).

31. K. Schlögl, "Stereochemistry of Metallocenes," in *Topics in Stereochemistry*, Vol. 1, N. L. Allinger and E. L. Eliel, Eds., Interscience, New York, 1967, p. 39.
32. H. L. Holmes, "The Chemistry of the Tropane Alkaloids," in *The Alkaloids, Chemistry and Physiology*, Vol. 1, R. H. F. Manske and H. L. Holmes, Eds., Academic Press, New York, 1950, p. 271.
33. A. A. Bothner-By, R. S. Schultz, R. F. Dawson, and M. L. Solt, *J. Am. Chem. Soc.*, **84**, 52 (1962).
34. E. Leete, *J. Am. Chem. Soc.*, **84**, 55 (1962).
35. A. C. Cope, C. R. Ganellin, H. W. Johnson, Jr., T. V. Van Auken, and J. S. Winkler, *J. Am. Chem. Soc.*, **85**, 3276 (1963).
36. A. C. Cope and S. Mehta, *J. Am. Chem. Soc.*, **86**, 5626 (1964).
37. A. C. Cope and B. A. Pawson, *J. Am. Chem. Soc.*, **87**, 3648 (1965).
38. A. C. Cope, K. Banholzer, H. Keller, B. A. Pawson, J. J. Whang, and H. J. S. Winkler, *J. Am. Chem. Soc.*, **87**, 3644 (1965).
39. A. C. Cope, W. R. Funke, and F. N. Jones, *J. Am. Chem. Soc.*, **88**, 4693 (1966).
40. A. Cope, K. Banholzer, F. N. Jones, and H. Keller, *J. Am. Chem. Soc.*, **88**, 4700 (1966).
41. A. C. Cope and E. R. Trumbull, "Olefins from Amines: The Hofmann Elimination and Amine Oxide Pyrolysis," in *Organic Reactions*, Vol. 11, Wiley, New York, 1960, p. 317.
42. J. Sicher and J. Závada, *Collection Czech. Chem. Commun.*, **33**, 1278 (1968); and references therein.
43. O. Červinka, J. Budilová, and M. Daněček, *Collection Czech. Chem. Commun.*, **32**, 2381 (1967).
44. S. I. Goldberg and J. E. Davis, unpublished work.
45. A. C. Cope, C. F. Howell, and A. Knowles, *J. Am. Chem. Soc.*, **84**, 3190 (1962).
46. A. C. Cope, C. F. Howell, J. Bowers, R. C. Lord, and G. M. Whitesides, *J. Am. Chem. Soc.*, **89**, 4024 (1967).
47. D. N. Jones and M. J. Green, *J. Chem. Soc., (C)*, **1967**, 532; D. N. Jones, M. J. Green, M. A. Saeed, and R. D. Whitehouse, *J. Chem. Soc., (C)*, **1968**, 1362.
48. A. C. Cope and E. M. Acton, *J. Am. Chem. Soc.*, **80**, 355 (1958).
49. A. C. Cope and M. W. Fordice, *J. Am. Chem. Soc.*, **89**, 6187 (1967).
50. S. I. Goldberg and N. C. Miller, *J. Chem. Soc. (D)*, **1969**, 1409.

Subjet Index